Die Grundlehren der mathematischen Wissenschaften

in Einzeldarstellungen
mit besonderer Berücksichtigung
der Anwendungsgebiete

Band 180

N. S. Landkof

Foundations of Modern Potential Theory

Translated from the Russian by

A. P. Doohovskoy

Springer-Verlag New York Heidelberg Berlin 1972

N. S. Landkof

Department of Mathematics, Harkov University
Harkov, U.S.S.R.

Translator

A. P. Doohovskoy

Concord, Massachusetts, U.S.A.

Title of the Russian Original Edition:
Osnovy sovremennoi teorii potenciala. Publisher: "Nauka", Moscow 1966

AMS Subject Classifications (1970)

Primary 31-02, 31AXX, 31BXX, 31C05, 31C15, 31C25, 30A44, 46F10

Secondary 28A10, 41A30, 42A88, 42A96, 46F99, 35J25

ISBN 0-387-05394-8 Springer-Verlag New York-Heidelberg-Berlin
ISBN 3-540-05394-8 Springer-Verlag Berlin-Heidelberg-New York

Library of Congress Catalog Card Number 77-186 131

Preface

For a long time potential theory was necessarily viewed as only another chapter of mathematical physics. Developing in close connection with the theory of boundary-value problems for the Laplace operator, it led to the creation of the mathematical apparatus of potentials of single and double layers; this was adequate for treating problems involving smooth boundaries.

A. M. Lyapunov is to be credited with the rigorous analysis of the properties of potentials and the possibilities for applying them to the solution of boundary-value problems.[1] The results he obtained at the end of the 19th century later received a more detailed and sharpened exposition in the book by N. M. Günter, published in Paris in 1934 and in New York 1967 with additions and revisions.[2]

Of fundamental significance to potential theory also was the work of H. Poincaré, especially his method of sweeping out mass (balayage).

At the beginning of the 20th century the work of S. Zaremba and especially of H. Lebesgue attracted the attention of mathematicians to the unsolvable cases of the classical Dirichlet problem. Through the efforts of O. Kellogg, G. Bouligand, and primarily N. Wiener, by the middle of the 20th century the problem of characterizing the so-called irregular points of the boundary of a region (i.e. the points at which the continuity of the solution of the Dirichlet problem may be violated) was completely solved and a procedure to obtain a generalized solution to the Dirichlet problem was described. Extremely important was the systematic introduction into potential theory of capacities as set functions. These results were presented in the survey by G. Bouligand [2] and in the monograph by O. Kellogg [1].[3]

An important role was played by the 1923 paper of O. Perron.[4]

[1] A. M. Lyapunov: Papers on Potential Theory, Moscow-Leningrad: Gostekhizdat 1949 (in Russian).

[2] N. M. Günther: Potential Theory and its Applications to Basic Problems of Mathematical Physics, New York: Ungar Publ. Co. 1967 (English translation).

[3] Numbers in square brackets refer to the bibliography at the end of the book.

[4] O. Perron: Eine neue Behandlung der ersten Randwertaufgabe für $\Delta u = 0$. Math. Zeitschrift **18**, 42–54 (1923).

Although the method of O. Perron, based on the consideration of so-called upper and lower functions and their envelopes, does not have a potential-theoretic character, nevertheless, as was later shown by Brelot, it can be applied to the balayage problem with great success.

After this the next substantial impulse toward the development of new methods in potential theory was the paper of F. Riesz [1] in 1926, where the close relation between potentials of measures and the concept of sub- and superharmonicity was established. From this moment function-theoretic methods steadily begin to enter into potential theory so that by the early forties this theory should more readily be viewed as a part of the theory of functions and not of mathematical physics. This comment, however, relates to the method of investigations and not to its subject, which as before remains closely tied with physical concepts (capacity, equilibrium distribution of charges, energy, Green functions).

In the thirties Vallée-Poussin [1] modernizes and substantially develops the method of balayage of Poincaré. O. Frostman [1] successfully applies the direct methods of the calculus of variations and proves the existence of a solution to the problems of equilibrium and balayage for arbitrary compact subsets of Euclidean space. In this same paper of O. Frostman there are important generalizations of the theory to the case of certain "non-classical" kernels, depending only on the distance between two points. These generalizations, which led M. Riesz in 1939 to a series of remarkable results presented in his memoir [1], indicate for potential theory a certain breaking away from the problems of mathematical physics. This is related to the fact that the new kernels arise not from the Laplace operator, but from certain non-local integro-differential operators.

At this time (1936—1941) appear the papers of M. V. Keldysh and M. A. Lavrentiev [1—4] and of M. V. Keldysh [1, 2] in which was discovered a new phenomenon of instability of the Dirichlet problem under external variations of the boundary of the region; conditions for stability were also investigated.

Soon after the first papers of M. V. Keldysh and M. A. Lavrentiev, M. Brelot [3, 7], using the method of Perron, established the connection between the stability problem, the problem of balayage of a measure onto an open set, and the harmonic majorants of subharmonic functions.

The papers of H. Cartan [1—6] (1941—1946) begin an important stage in the development of potential theory; he indicates a completely new approach to the basic questions of the theory. This approach is connected with the introduction of a Hilbert-space structure into the set of measures with finite energy; with this begins the systematic incursion of the methods of functional analysis into potential theory.

Later (1950) due to the work of J. Deny [4] potential theory was enriched with the ideas and methods of distributions. On the other hand there begins an equally intense development toward a more abstract theory connected primarily with the names of M. Brelot, G. Choquet, and also the Japanese mathematicians N. Ninomiya and M. Ohtsuka.

G. Choquet in 1955 sets up a general theory of capacities [1] and solves the problem of capacitability for Borel sets, M. Brelot (with partial collaboration with R.-M. Hervé) examines the axiomatic structure of the theory of harmonic and superharmonic functions and potential theory, G. Choquet and J. Deny, and also mathematicians of the Japanese school, investigate the connections between the various "principles" of potential theory, J. Deny searches for the most general types of kernels for which balayage is possible.

In recent years, the papers of J. L. Doob and G. A. Hunt[5] have revealed important connections between potential theory and the theory of Markov processes, attracting to potential theory the attention of many mathematicians who are studying probability theory.

Even from this highly incomplete summary of papers and directions of research, it is apparent that potential theory is at the present an intensively developing area of mathematical analysis, using the ideas and methods of modern function theory, functional analysis, and general topology.

In two courses of lectures M. Brelot [31, 32] gives a compact and quite substantial exposition of that part of potential theory which we characterized above as a "more abstract" tendency of the theory.

However, at the present time there is no monograph on potential theory which at a sufficiently modern level presents the "analytic" part of the theory relating to concrete kernels. This book attempts to eradicate that deficiency in the literature.

The entire exposition (with the exception of the next-to-the last survey Ch. VI, § 2) is devoted to M. Riesz kernels and Green kernels. To justify such a selection of kernels one can say the following. First, M. Riesz kernels include, as special (or limiting) cases, the classical Newtonian and logarithmic kernel. Second, changing the character of the singularity of a kernel leads to, from the point of view of analysis, very deep alterations of the theory; this is because the Laplace differential operator has been replaced by a non-local integro-differential operator.

---

[5] Cf. G. A. Hunt: Markoff Processes and Potentials. Illinois J. Math. 1, 2 (1957, 1958).

With regard to Green kernels, they are essentially a model with which a potential theory for more general elliptic differential operators can be constructed.

The substance of the book is clear from the table of contents. We point out that the Introduction can be used to gain acquaintance with basic facts of measure theory and the theory of distributions. The Ch. I is a collection of numerous facts whose exposition does not require the notion of capacity. The Ch. V is more specialized than the rest. The § 2 of Ch. VI is a survey of the literature and contains no proofs.

Almost all references to the literature are at the end of the book under "Comments and Bibliographic References".

Outside the framework of the book are such important questions as positive harmonic functions and the Martin boundary, axiomatic construction of potential theory and relations with probability theory.

Nevertheless, the author hopes that the material brought together in this book is on the whole truly representative of the ideas and methods of potential theory and will help the reader to pursue the special questions in the current journals.

Knowledge of classical potential theory is not required for the reading of this book.

Although the book is directed to mathematicians being introduced to potential theory for the first time, the author hopes that it will have some interest also for specialists.

In the final draft of the manuscript the author was benefited with the inestimable help of I. V. Ostrovskij whose criticisms eliminated a number of errors and awkward portions of the original text.

The author is sincerely grateful to A. A. Gončar, V. A. Marčenko, and S. Ya. Havinson who read the manuscript and offered several valuable suggestions.

Finally, the author is greatly indebted to E. Ya. Khruslov, G. V. Suzikov, and E. N. Podol'skij for pointing out certain inaccuracies in the manuscript and hence enabling its improvement.

Harkov, 1965

N. S. Landkof

Contents

Introduction

§ 1. Spaces of Measures and Signed Measures. Operations on Measures and Signed Measures

1. Consider in the Euclidean space R^p of dimension $p (p \geq 2)$ a region G, which in particular may coincide with R^p.

Let $B(G)$ denote the σ-algebra of Borel subsets of G, i.e. the smallest σ-algebra, containing all compact subsets of G. In the future we shall as a rule denote compact sets by the letter K.

A *measure* μ in G is a non-negative σ-additive set function defined on $B(G)$ and finite on all compact sets $K \subset G$.

If $\mu(G) < \infty$, then the measure μ is called *finite*.

The set of all measures in G we will denote by $\mathfrak{M}_G^+$; if $G \equiv R^p$ the index G will be omitted.

Let $G_0 = G_0(\mu)$ denote the largest open set for which $\mu(G_0) = 0$ (it may, of course, be empty). Then the closed (relative to G) set

$$S(\mu) = G \setminus G_0(\mu)$$

is called the support of the measure μ.

We will say that the measure μ is concentrated on a set $E \in B(G)$, if

$$\mu(G \setminus E) = 0.$$

Then the support $S(\mu)$ may also be defined as the smallest closed set on which the measure μ is concentrated. It is clear that the measure can be concentrated on a proper subset of the support $S(\mu)$.

The expression "*μ-almost everywhere*" means "everywhere with the exception of a set $e \in B(G)$ for which $\mu(e) = 0$".

The *restriction* or *part* of the measure μ on a set $E \in B(G)$ is the measure μ_E, defined by

$$\mu_E(e) = \mu(e \cap E).$$

It is clear that μ_E is concentrated on E and therefore $S(\mu_E) \subset \overline{E}$.

The set of all measures concentrated on E we will denote by $\mathfrak{M}_E^+$.

By a *signed measure* ν in the region G we mean a real σ-additive set function, defined on $B(G)$ and finite on compact sets $K \subset G$. In order for the signed measure ν to be defined on all sets in $B(G)$, it is necessary to assume that ν takes on only one of the infinite values: $+\infty$ or $-\infty$. In general, without this assumption ν will be defined on all sets of $B(G)$ with a compact closure in G, which is the most essential fact. The set of all signed measures in G we will denote by $\mathfrak{M}_G$.

The difference between two measures is clearly a signed measure. The fact that in this way we can obtain all signed measures comes from the following theorem, which we assume to be known.

Theorem 0.1. *For any signed measure $\nu \in \mathfrak{M}_G$ there exists a decomposition of G into two disjoint sets, G^+ and G^-,*[1] *where*

$$\nu(e) \geq 0 \quad \textit{if} \quad e \subset G^+,$$

$$\nu(e) \leq 0 \quad \textit{if} \quad e \subset G^-.$$

Although the decomposition $G = G^+ \cup G^-$ is not unique, the measures ν^+, ν^-, defined by the formulas

$$\nu^+(e) = \nu(e \cap G^+), \qquad \nu^-(e) = -\nu(e \cap G^-) \tag{0.1.1}$$

do not depend upon the choice, of G^+, G^-. They are called the *positive* and *negative variations* of the signed measure ν. Clearly,

$$\nu = \nu^+ - \nu^- \tag{0.1.2}$$

thus every signed measure is the difference of two measures. Formula (0.1.2) is called the *Jordan decomposition* of the signed measure ν.

The *support* of the signed measure ν is the closed set

$$S(\nu) = S(\nu^+) \cup S(\nu^-).$$

With the help of (0.1.2) the concepts of support, restriction, etc. naturally carry over to signed measures. The measure

$$|\nu| = \nu^+ + \nu^-$$

is called the *complete variation* of the signed measure ν and will sometimes be denoted var ν. If $|\nu|\,(E) = 0$, then $\nu_E \equiv 0$ and we will say that $\nu \equiv 0$ on E.

We note that

$$S(|\nu|) = S(\nu).$$

[1] All the sets mentioned in this book are Borel sets. This will not be explicitely stated in what follows.

Every locally (Lebesgue) integrable function $\psi(x)$ generates a signed measure ν according to the formula

$$\nu(e) = \int_e \psi(x)\, dx.$$

Such a signed measure is called *absolutely continuous*; the function so defined (within equivalence) is called its *density*. When no confusion will result, we will denote an absolutely continuous signed measure and its density by the same letter; that is, we will write

$$\psi(e) = \int_e \psi(x)\, dx.$$

It can be shown that

$$|\psi|\,(e) = \int_e |\psi(x)|\, dx.$$

The support of the function $\psi(x)$ is by definition the support $S(\psi)$ of the corresponding signed measure. This is the smallest closed set outside of which $\psi(x) = 0$ almost everywhere.

We note further that $\mathfrak{M}_G(\mathfrak{M}_E)$ is a linear manifold, while $\mathfrak{M}_G^+(\mathfrak{M}_E^+)$ is a convex cone generating $\mathfrak{M}_G(\mathfrak{M}_E)$.

2. Our aim is to impose on the space $\mathfrak{M}_G$ a *vague topology* using an adequate set of linear functionals.

Consider the set Φ of real, continuous, functions $f(x)$, $x \in G$ with compact support.

Every function $f(x) \in \Phi$ generates in $\mathfrak{M}_G$ a linear functional

$$f(\nu) = \int_{S(f)} f(x)\, d\nu(x), \nu \in \mathfrak{M}_G, \tag{0.1.3}$$

which will be zero only if $f(x) = 0$.

The set Φ, clearly is a linear manifold; if in (0.1.3) we fix the signed measure $\nu \in \mathfrak{M}_G$, then

$$\nu(f) = \int_{S(f)} f(x)\, d\nu(x), \quad f \in \Phi, \tag{0.1.3'}$$

may be viewed as a linear functional on Φ.

Moreover, the functional $\nu(f)$ is *continuous* if in Φ we introduce the notion of convergence as follows.

The sequence $f_n \in \Phi$ converges to $f \in \Phi$, if the supports of the $f_n(x)$ are contained in a fixed compact set K and if $f_n(x)$ uniformly converge to $f(x)$.

In fact,

$$|\nu(f_n) - \nu(f)| \le \int_K |f_n(x) - f(x)|\, d|\nu| \le |\nu|(K) \max_K |f_n(x) - f(x)|.$$

It turns out that the correspondence between the signed measures $\nu \in \mathfrak{M}_G$ and the continuous linear functionals on Φ established by formula (0.1.3′) is one-to-one. In other words, we have

Theorem 0.2. *To every continuous linear functional* $\nu(f)$, $f \in \Phi$ *there corresponds a unique signed measure* $\nu \in \mathfrak{M}_G$ *connected with it by formula* (0.1.3′).

Proof. We use the following classical result from functional analysis.

If $K \subset R^p$ is compact, and C_K is the Banach space of all real continuous functions on K with norm $\|f\| = \max\limits_K |f(x)|$, then the general form of a continuous linear functional on C_K is given by the formula

$$\nu(f) = \int_K f(x)\, d\nu(x) \qquad \nu \in \mathfrak{M}_K, \tag{0.1.4}$$

where $\|\nu\| = |\nu|\,(K)$, and the representation (0.1.4) is unique[2].

Consider now an increasing sequence of compact sets K_n, which exhaust G:

$$K_n \subset \overset{\circ}{K}_{n+1};^{3} \qquad \bigcup_{n=1}^{\infty} K_n = G.$$

Let $\Phi_n \subset \Phi$ be the linear manifold consisting of all functions $f(x)$, for which $S(f) \subset K_n$. Φ_n may be regarded as a subspace of C_{K_n}, where strong convergence in C_{K_n} of elements in Φ_n is equivalent to their convergence in Φ.

Let ν_n be the restriction of a given functional ν on Φ_n. Extending it to a continuous linear functional on C_{K_n}, we obtain according to (0.1.4)

$$\nu_n(f) = \int_{K_n} f(x)\, d\nu_n(x), \qquad f \in \Phi_n,$$

where ν_n is a finite signed measure in $\mathfrak{M}_{K_n}$.

We will show that if $E \subset K_n$, then for any integer $r > 1$ $\nu_{n+r}(E) = \nu_{n+1}(E)$. Clearly, it is sufficient to verify that $\nu_{n+2}(E) = \nu_{n+1}(E)$. Let $f(x) \in \Phi_{n+1} \subset \Phi_{n+2}$; since $\nu_{n+1}(f) = \nu_{n+2}(f) = \nu(f)$, then

$$\int_{S(f)} f(x)\, d\nu_{n+1}(x) = \int_{S(f)} f(x)\, d\nu_{n+2}(x), \qquad f \in \Phi_{n+1},$$

or

$$\int_{S(f)} f(x)\, d\varkappa_n(x) = 0, \quad f \in \Phi_{n+1} \tag{0.1.5}$$

where $\varkappa_n = \nu_{n+1} - \nu_{n+2}$.

---

[2] For a proof of this cf., for example Dunford and Schwartz [1], Ch. 1, General theory, IV.6.3.

[3] $\overset{\circ}{E}$ denotes the interior of the set E, i.e. the largest open subset of E.

Let $E \subset K_n$; we will show that $\varkappa_n(E) = 0$. If $\varkappa_n(E) > 0$, then we take a compact set $F \subset E$ such that $\varkappa_n(F) = \alpha > 0$, and then an open set Ω, $F \subset \Omega \subset \mathring{K}_{n+1}$ such that $|\varkappa_n| (\Omega \setminus F) < \frac{\alpha}{2}$. Consider a continuous function $f(x)$, $0 \leq f(x) \leq 1$, equal to 1 on F and 0 outside of Ω. Since $f \in \Phi_{n+1}$, then according to (0.1.5),

$$\int_f f(x)\, d\varkappa_n(x) + \int_{\Omega \setminus F} f(x)\, d\varkappa_n(x) = 0 .$$

But this is impossible, since the left-hand side is

$$\alpha + \int_{\Omega \setminus F} f(x)\, d\varkappa_n(x) > 0$$

In the same way, we prove that the inequality $\varkappa_n(E) < 0$ is impossible.

We now can build the signed measure $\nu \in \mathfrak{M}_G$ as follows. If for a set $E \subset G$ there is a number n such that $E \subset K_n$, then we put

$$\nu(E) = \nu_{n+1}(E) ,$$

and this definition does not depend on the choice of n. Otherwise we put

$$\nu(E) = \lim_{n\to\infty} \nu^+_{n+1}(E \cap K_n) - \lim_{n\to\infty} \nu^-_{n+1}(E \cap K_n) ,$$

assuming that one of the limits on the right is finite.

Clearly, for any $f(x) \in \Phi$ there is a number n, for which $f \in \Phi_{n+1}$, and consequently,

$$\nu(f) = \nu_{n+1}(f) = \int_{S(f)} f(x)\, d\nu_{n+1}(x) = \int_{S(f)} f(x)\, d\nu(x) .$$

The uniqueness of ν follows from the fact that the signed measure ν_{n+1} is uniquely defined on K_n. ▯[4]

Corollary 1. *The linear functionals* (0.1.3) *in* $\mathfrak{M}_G$ *form a total set of functionals, i.e. if* $\nu \neq 0$, *then there is a function* $f(x) \in \Phi$, *such that*

$$f(\nu) \neq 0 .$$

Otherwise, the functional $\nu(f)$ in Φ, defined by the equation (0.1.3′) would be zero from which it would follow that $\nu \equiv 0$.

Corollary 2. *If* $f \equiv 0$ *on* $S(\nu)$, *then* $\nu(f) = 0$.

Indeed, the integration in (0.1.3′) is actually over the set $S(f) \cap S(\nu)$.

From this is follows that changing $f(x)$ outside of $S(\nu)$ does not affect $\nu(f)$.

We agree to denote by Φ^+ the set of all non-negative functions in Φ.

A functional $\mu(f)$ is called *positive* if $\mu(f) \geq 0$ for $f \in \Phi^+$.

[4] The sign ▯ denotes the end of a proof.

Theorem 0.2'. *A continuous linear functional $\mu(f)$ is positive if, and only if, in the representation* (0.1.3') *the signed measure $\nu \in \mathfrak{M}_G$ is a measure.*

Proof. Sufficiency is obvious. Necessity follows from the fact that if $\nu(E) < 0$ on some set E, then reasoning as we did in the proof of Theorem 0.2, we could construct a function $f(x) \in \Phi^+$, $0 \leq f(x) \leq 1$ for which $\nu(f) < 0$. ☐

Theorem 0.3. *If the functional $\mu(f)$, $f \in \Phi$, is linear and positive, then it is continuous; consequently, it admits a unique representation*

$$\mu(f) = \int_{S(f)} f(x)\, d\mu(x)\,, \quad \mu \in \mathfrak{M}_G^+\,. \tag{0.1.6}$$

Proof. Since $\mu(f)$ is linear, it is sufficient to check its continuity at zero. Suppose $f_n \to 0$ in Φ and $K \subset G$ is a compact set containing all the $S(f_n)$. Denote by $\psi(x)$ some function from Φ, satisfying the conditions: $0 \leq \psi(x) \leq 1$, $\psi(x) \equiv 1$ on K. Moreover, let

$$\varepsilon_n = \max_K |f_n(x)|\,.$$

Then on all of G

$$-\varepsilon_n \psi(x) \leq f_n(x) \leq \varepsilon_n \psi(x)\,,$$

and therefore

$$-\varepsilon_n \mu(\psi) \leq \mu(f_n) \leq \varepsilon_n \mu(\psi)\,,$$

that is

$$|\mu(f_n)| \leq \varepsilon_n \mu(\psi)\,.$$

Since $\varepsilon_n \to 0$, the theorem is proved. ☐

In view of the connection between measures and functionals, we will often denote the quantity $\mu(G) \leq +\infty$ by the symbol $\mu(1)$. The analogous notation $\nu(1)$ we will use for signed measures when it makes sense (i.e. when either $\nu^+(1) < \infty$, or $\nu^-(1) < \infty$).

3. We now introduce a vague topology in $\mathfrak{M}_G$. Since $\mathfrak{M}_G$ is linear, it is sufficient to define a fundamental system of neighborhoods of the null element.

The neighborhood $V = V(f_1, f_2, \ldots, f_n; \varepsilon)$, where $\varepsilon > 0$, $f_k \in \Phi$, of zero in $\mathfrak{M}_G$ is the set of signed measures $\nu \in \mathfrak{M}_G$, for which

$$\left|\int f_k(x)\, d\nu(x)\right| < \varepsilon\,, \quad k = 1, 2, ,,\ldots, n\,.$$

It is not difficult to see that with such a definition the usual neighborhood axioms for a space will be satisfied, and consequently $\mathfrak{M}_G$ becomes a linear topological space; from the corollary to theorem 0.2 it will be separable.

The convergence of a sequence of signed measures ν_n to the signed measure ν in the vague topology or, as we will say below, the *vague convergence* of ν_n to ν means that for any function $f(x) \in \Phi$

$$\lim_{n\to\infty} \int f(x)\, d\nu_n(x) = \int f(x)\, d\nu(x) \tag{0.1.7}$$

Vague convergence will be denoted by

$$\nu_n \dashrightarrow \nu.$$

Let us examine in greater detail vague convergence in $\mathfrak{M}_G^+$, i.e. vague convergence of measures.

First, the vague limit of measures μ_n can only be a measure, not a signed measure. Indeed, for any $f \in \Phi^+$, from $\mu_n(f) \geq 0$ and (0.1.7) it follows that $\mu(f) > 0$, and therefore by Theorem 0.2′ μ is a measure. Furthermore, as the next theorem shows, vague convergence of measures need not be checked on the entire space Φ.

Theorem 0.4. *Suppose that the set $M \subset \Phi^+$ is dense in Φ^+.*[5] *Then, if for any function $\varphi(x) \in M$*

$$\mu_n(\varphi) \to \mu(\varphi), \quad \mu_n, \mu \in \mathfrak{M}_G^+,$$

then

$$\mu_n \dashrightarrow \mu.$$

Proof. For any compact set K we have $\sup\limits_n \mu_n(K) < \infty$.

Indeed, if $\chi_K(x)$ is the characteristic function for K, then in M there is a function $\varphi(x) \geq \chi_K(x)$ and we will be able to write

$$\mu_n(K) = \int \chi_K(x)\, d\mu_n(x) \leq \int \varphi(x)\, d\mu_n(x) = \mu_n(\varphi).$$

But $\sup\limits_n \mu_n(\varphi) < \infty$, since $\mu_n(\varphi) \to \mu(\varphi)$.

Now consider $f(x) \in \Phi^+$ and a sequence $\varphi_n(x) \in M$, converging in Φ to $f(x)$. Let K be a compact set containing $S(f)$ and $S(\varphi_n)$, $n = 1, 2, \ldots$ Then

$$|\mu_n(f) - \mu(f)| \leq |\mu_n(f) - \mu_n(\varphi_m)| + |\mu_n(\varphi_m) - \mu(\varphi_m)| + |\mu(\varphi_m) - \mu(f)|.$$

But

$$|\mu_n(f) - \mu_n(\varphi_m)| \leq \int_K |f - \varphi_m|\, d\mu_n \leq \sup_n \mu_n(K) \max_K |f - \varphi_m|,$$

[5] This means that for any $f(x) \in \Phi^+$ there exists a sequence $\varphi_n(x) \in M$, such that $\varphi_n(x) \to f(x)$ in Φ (it is sometimes said that M is "sequentially" dense in Φ^+). An important example of a set dense in Φ is the set $D \subset \Phi$ of infinitely differentiable functions.

and therefore, selecting first a suitable m, and then n, we obtain for any $\varepsilon > 0$ the inequality

$$|\mu_n(f) - \mu(f)| < \varepsilon. \quad \square$$

In a completely analogous way we prove

Theorem 0.4′. *Suppose the set M is dense in Φ^+ and for any function $\varphi \in M$ the sequence $\{\mu_n(\varphi)\}$ is fundamental. Then the measures μ_n vaguely converge to a certain measure μ.*

The following lemma is useful.

Lemma 0.1. *Suppose the function $f(x) \geq 0$ is lower semicontinuous.*[6] *Then if $\mu_n \dashrightarrow \mu$, $\mu_n \in \mathfrak{M}_G^+$, then*

$$\varliminf_{n\to\infty} \int_G f(x)\, d\mu_n \geq \int_G f(x)\, d\mu \tag{0.1.8}$$

Proof. It is known[7] that the function $f(x)$ is the limit of a non-decreasing sequence of functions $f_m(x) \in \Phi$. From the inequality

$$\int_G f(x)\, d\mu_n \geq \int_G f_m(x)\, d\mu_n$$

we obtain

$$\varliminf_{n\to\infty} \int_G f(x)\, d\mu_n \geq \int_G f_m(x)\, d\mu,$$

which gives (0.1.8) as $m \to \infty$. $\square$

Corollary 1. If $\mu_n \dashrightarrow \mu$, $\mu_n \in \mathfrak{M}_G^+$ and $\Omega \subset G$ is an open set, then

$$\varliminf_{n\to\infty} \mu_n(\Omega) \geq \mu(\Omega).$$

To be convinced if this, it is sufficient to put $f(x) = \chi_\Omega(x)$ in (0.1.8)

Corollary 2. *If $\mu_n \dashrightarrow \mu$, $\mu_n \in \mathfrak{M}_G^+$ and $K \subset G$ is a compact set, then*

$$\varlimsup_{n\to\infty} \mu_n(K) \leq \mu(K).$$

Indeed, put $f(x) = -\chi_K(x)$. Since this is a finite and lower semi-continuous function, (0.1.8) yields the result.

---

[6] The condition $f(x) \geq 0$ may be replaced by the condition that $f(x)$ has compact support; it can be removed if the supports of the measures μ_n and μ are contained in a fixed compact set.

[7] Cf. Hausdorff [1], Ch. IX, § 38, No. 1.

Corollary 3. *Let μ denote a measure. We will call a relatively compact set E μ-almost clopen if the measure of its boundary ∂E is zero:*

$$\mu(\partial E) = \mu(\overline{E} \setminus \mathring{E}) = 0.$$

Then for any μ-almost clopen set E we have

$$\lim_{n\to\infty} \mu_n(E) = \mu(E), \tag{0.1.9}$$

if $\mu_n \dashrightarrow \mu$.

Indeed, if E μ-almost clopen, then

$$\mu(E) = \mu(\overline{E}) = \mu(\mathring{E}).$$

By Corollary 1

$$\underline{\lim}_{n\to\infty} \mu_n(E) \geq \underline{\lim}_{n\to\infty} \mu_n(\mathring{E}) \geq \mu(\mathring{E}) = \mu(E)$$

and by Corollary 2

$$\overline{\lim}_{n\to\infty} \mu_n(E) \leq \overline{\lim}_{n\to\infty} \mu_n(\overline{E}) \leq \mu(\overline{E}) = \mu(E),$$

from which follows (0.1.9).

Corollary 4. *If $\mu_n \dashrightarrow \mu$, $\mu_n \in \mathfrak{M}_G^+$, then*

$$S(\mu) \subset \bigcap_{N=1}^{\infty} \overline{\bigcup_{n=N}^{\infty} S(\mu_n)}. \tag{0.1.10}$$

Indeed, if $x \in S(\mu)$ and V_x is any open neighborhood of x, then $\mu(V_x) > 0$ and by Corollary 1 $\mu_n(V_x) > 0$ for all $n > n_0(V_x)$. Consequently, for any N, $V_x \cap \left(\bigcup_N^{\infty} S(\mu_n)\right) \neq \emptyset$ and therefore x is a cluster point of the set $\bigcup_N^{\infty} S(\mu_n)$.

Theorem 0.5. *If for any μ-almost clopen set E*

$$\lim_{n\to\infty} \mu_n(E) = \mu(E).$$

then $\mu_n \dashrightarrow \mu$.

Proof. Suppose $G \equiv R^p$. There is no more than a countable set of hyperplanes $x_k = \text{const}$ $(k = 1, 2, \ldots, p)$ having positive μ-measure. Consequently, for any $\varepsilon > 0$ there is a decomposition of $R^p = \bigcup_{i=1}^{\infty} I_i$ into a countable set of p-dimensional intervals $I_i = \{x\colon a_k^i \leq x_k \leq b_k^i;\ k = 1, 2, \ldots, p\}$, $b_k^i - a_k^i < \varepsilon$ which are μ-almost clopen sets. Such a decomposition of R^p will be called a normal net.

Now suppose $f(x) \in \Phi$ and I_0 is a μ-almost clopen interval containing $S(f)$. We decompose I_0 into a normal net $I_0 = \bigcup_{i=1}^{N} I_i$ such that the oscillation of $f(x)$ on each I_i does not exceed ε. Then for any points $\xi_i \in I_i$, we will have

$$\left| \int_{I_0} f(x)\, d\mu(x) - \sum_{i=1}^{N} f(\xi_i)\, \mu(I_i) \right| < \varepsilon\mu(I_0),$$

$$\left| \int_{I_0} f(x)\, d\mu_n(x) - \sum_{i=1}^{N} f(\xi_i)\, \mu_n(I_i) \right| < \varepsilon\mu_n(I_0).$$

Since $\mu_n(I_0) \to \mu(I)$, then $\sup \mu_n(I_0) = M < \infty$. Moreover, for any i $\mu_n(I_i) \to \mu(I_i)$ and therefore n may be taken large enough so that

$$\left| \sum_{i=1}^{N} f(\xi_i)\, \mu(I_i) - \sum_{i=1}^{N} f(\xi_i)\, \mu_n(I_i) \right| < \varepsilon.$$

Consequently,

$$\left| \int_{I_0} f(x)\, d\mu(x) - \int_{I_0} f(x)\, d\mu_n(x) \right| < 2\varepsilon M + \varepsilon,$$

that is

$$\lim_{n\to\infty} \int_{S(f)} f(x)\, d\mu_n(x) = \int_{S(f)} f(x)\, d\mu(x).$$

The case of an arbitrary G is completely analogous, except that I_0 is not an interval, but a sum of a finite number of intervals. ▯

Theorem 0.5′. *Suppose that $\mu_n \dashrightarrow \mu$ and E is a μ-almost clopen set. If μ_n', μ' denote the restrictions of the measures μ_n, μ to E, then*

$$\mu_n' \dashrightarrow \mu'$$

Proof. Set

$$b = E \setminus \mathring{E} = \partial E \cap E$$

and denote by μ_n^b, μ^b, correspondingly μ_n^0, μ^0, the restrictions of μ_n, μ to b and correspondingly to $\mathring{E}$. Then

$$\mu_n' = \mu_n^0 + \mu_n^b, \qquad \mu' = \mu^0 + \mu^b.$$

Since

$$\bar{b} \subset \overline{\partial E} \cap \bar{E} = \partial E,$$

then by assumption $\mu(\bar{b}) = 0$ and from this it follows that any subset of b is μ-almost clopen. By Corollary 3 for any e

$$\lim_{n\to\infty} \mu_n(b \cap e) = \mu(b \cap e)$$

or

$$\lim_{n\to\infty} \mu_n^b(e) = \mu^b(e).$$

Consequently, by Theorem 0.5

$$\mu_n^b \dot{\to} \mu^b;$$

it remains to be shown that

$$\mu_n^0 \dot{\to} \mu^0.$$

We will again use Theorem 0.5. Let e be any μ^0-almost clopen set. This means that

$$\mu^0(\partial e) = \mu(\partial e \cap \mathring{E}) = 0.$$

Since $\mathring{E}$ is an open set, then

$$\partial(e \cap \mathring{E}) \subset (\partial e \cap \mathring{E}) \cup \partial\mathring{E} \subset (\partial e \cap \mathring{E}) \cup \partial E.$$

Consequently,

$$\mu[\partial(e \cap \mathring{E})] = 0,$$

i.e. the set $e \cap \mathring{E}$ is μ-almost clopen. Thus we can write the equality

$$\lim_{n\to\infty} \mu_n(e \cap \mathring{E}) = \mu(e \cap \mathring{E})$$

or

$$\lim_{n\to\infty} \mu_n^0(e) = \mu^0(e).$$

According to Theorem 0.5

$$\mu_n^0 \dot{\to} \mu^0,$$

and the theorem is proved. □

Finally, we point out when a set of measures is vaguely compact.

We will call a set of measures $\mathfrak{A} \subset \mathfrak{M}_G^+$ *weakly bounded*, if for any function $f(x) \in \Phi$ there is a constant C_f such that for all $\mu \in \mathfrak{A}$

$$|\mu(f)| < C_f.$$

It is not difficult to see that $\mathfrak{A}$ will be weakly bounded if and only if for any compact set K there is a constant M_K such that

$$\mu(K) < M_K, \quad \mu \in \mathfrak{A}$$

Theorem 0.6. *A weakly bounded set of measures $\mathfrak{A}$ is vaguely compact, i.e. contains a vaguely convergent sequence of measures.*

Proof. Suppose that $G = \bigcup_{i=1}^{\infty} K_i$, where $K_i \subset K_{i+1}$ is a monotone increasing sequence of compact sets. By assumption,

$$\mu(K_i) < M_i \qquad \text{for all } \mu \in \mathfrak{A}.$$

The set of all finite signed measures with supports belonging to K_i, may be regarded as a Banach space dual to the separable Banach space C_{K_i}. By the well-known theorem on the weak compactness of a sphere in the dual space (cf., for example, Dunford and Schwartz [1], Ch. 1, General theory, V.4.2) there is a sequence of measures $\mu_n^{(i)} \in \mathfrak{A}$, vaguely convergent to a certain measure $\mu^{(i)}$. By Corollary 3 to Lemma 0.1 for any $\mu^{(i)}$-almost clopen set $E \subset K_i$,

$$\lim_{n\to\infty} \mu_n^{(i)}(E) = \mu^{(i)}(E).$$

Since each sequence $\{\mu_n^{(i)}\}$ is obtained from the preceding $\{\mu_n^{(i-1)}\}$, we can consider the diagonal sequence of measures $\{\mu_i^{(i)}\}$ and the measure μ uniquely defined by the conditions $\mu \equiv \mu^{(i)}$ on K_i.

Clearly, for any μ-almost clopen set E (consequently, for any $\mu^{(i)}$-almost clopen set, for sufficiently large i), we obtain

$$\lim_{n\to\infty} \mu_i^{(i)}(E) = \mu(E)$$

and by Theorem 0.5 $\mu_i^{(i)} \dashrightarrow \mu$. ☐

We also define an operation of *integration of a measure.* Suppose that we have a family of measures $\mu_x = \mu(E; x) \in \mathfrak{M}_G^+$, depending on a parametric point $x \in G$. We shall assume that it is a Borel family, i.e., that for any function $f \in \Phi$ the function $g(x) = \mu_x(f)$ is Borel measurable. It is easy to see then that the function

$$\mu_x(E), \quad E \in B(G)$$

will also be Borel measurable.

For any $f \in \Phi^+$, consider the nonnegative function

$$g(x) = \int f(y)\, d\mu_x(y) = \mu_x(f).$$

Suppose that the measure $\lambda \in \mathfrak{M}_G^+$ is such that the expression

$$\lambda(g) = \int_G g(x)\, d\lambda(x) = \int_G \mu_x(f)\, d\lambda(x)$$

is finite. Then it is a linear and positive functional an Φ and by Theorem 0.3 there is a unique corresponding measure $\varkappa \in \mathfrak{M}_G^+$. This measure will be called the integral of μ_x with respect to λ; it will be denoted as follows:

$$\varkappa = \int \mu_x \, d\lambda(x); \tag{0.1.11}$$

thus

$$\varkappa(f) = \int \mu_x(f) \, d\lambda(x), \quad f \in \Phi, \tag{0.1.11'}$$

and

$$\varkappa(E) = \int \mu_x(E) \, d\lambda(x), \quad E \in B(G) \tag{0.1.11''}$$

A sufficient condition for the existence of $\varkappa$ is, for example, the boundedness of all the measures μ_x and λ.

In concluding this section, we note that the theory of measures and signed measures can be generalized to any locally compact Hausdorff space. In that case, however, the representation (0.1.6) turns out not to be unique; to obtain the one-to-one correspondence between positive linear functionals and measures, it becomes necessary to restrict ourselves to the class of so-called regular Borel measures. It is not necessary to go into these set-theoretic details if from the beginning a measure is defined as a positive linear functional on Φ, not as a set function and then one need not explore the connection between these concepts. Within such an approach, a measure is usually called a Radon measure. One can construct a theory of integration for Radon measures (cf., for example, Bourbaki [1]).

4. Suppose $G = R^p$. Then in addition to the linear operations, we can also introduce in $\mathfrak{M}$ a convolution operation.

First consider $\mathfrak{M}^+$. Let μ_1, μ_2 be two measures and $f(x) \in \Phi^+$; the function

$$g(x) = \int f(x+y) \, d\mu_1(y) = \mu_1(f_x), \quad \text{where} \quad f_x(y) = f(x+y),$$

is continuous and nonnegative. Set

$$\mu_2(g) = \int g(x) \, d\mu_2(x) \leq +\infty.$$

By the Fubini theorem

$$\mu_2(g) = \int d\mu_1(y) \int f(x+y) \, d\mu_2(x) = \mu_1(h), \tag{0.1.12}$$

where

$$h(y) = \int f(x+y) \, d\mu_2(x).$$

Assume now that for any function $f \in \Phi^+$ the quantity $\mu_2(g) = \mu_1(h)$ is finite. Then the functional (0.1.12) is defined for $f \in \Phi$ and is linear and positive. By Theorem 0.3, it corresponds uniquely to a certain measure which we will call the *convolution of measures* μ_1 and μ_2; it will be denoted by $\mu_1 * \mu_2$. From the definition and (0.1.12), it is clear that $\mu_1 * \mu_2 = \mu_2 * \mu_1$ and if one side of the equation is defined then the other is, too.

A sufficient condition for the existence of $\mu_1 * \mu_2$ is the compactness of the support of one of the measures μ_1, μ_2. Indeed, if, for example, μ_1 has compact support, then

$$g(x) = \int f(x + y)\, d\mu_1(y) \in \Phi^+,$$

and

$$\mu_2(g) < +\infty \quad \text{for any } f \in \Phi^+.$$

Another sufficient condition for the existence of the convolution $\mu_1 * \mu_2$ is the boundedness of both measures μ_1, μ_2, i.e.

$$\mu_1(1) < \infty, \qquad \mu_2(1) < \infty.$$

Convolution reduces to the integration of a measure in the sense of formula (0.1.11). In fact, let us define a family of measures μ_x by the formula

$$\mu_x(E) = \mu(E - x), \quad \mu \in \mathfrak{M}^+, \quad x \in R^p \tag{0.1.13}$$

where $E - x = \{y : x + y \in E\}$.[8] Then,

$$\mu_1 * \mu_2 = \int \mu_{1x}\, d\mu_2(x) = \int \mu_{2x}\, d\mu_1(x). \tag{0.1.13'}$$

From this formula it follows that if both measures μ_1 and μ_2 have compact support, then the measure $\mu_1 * \mu_2$ will also have compact support.

We now prove a few simple lemmas about convolutions.

Lemma 0.2. *Suppose that* $\mu_n \dashrightarrow \mu$ *and that one of the following two statements holds:*

(*i*) *the measure* $\varkappa$ *has compact support,*

(*ii*) *the supports of the measures* μ_n *are contained in a fixed compact set.*

Then

$$\varkappa * \mu_n \dashrightarrow \varkappa * \mu. \tag{0.1.14}$$

Proof. Suppose that (i) holds. For any $f \in \Phi$ the function

$$g(x) = \int f(x + y)\, d\varkappa(y)$$

belongs to Φ, and therefore

$$(\varkappa * \mu_n)(f) = \mu_n(g) \to \mu(g) = (\varkappa * \mu)(f)$$

[8] It is obviously a Borel family since $\mu_x(f)$, $f \in \Phi$ is a continuous function of x.

If we assume that statement (ii) holds, then the sequence of functions

$$g_n(x) = \int f(x+y)\, d\mu_n(y), \quad f \in \Phi,$$

will converge in Φ to

$$g(x) = \int f(x+y)\, d\mu(y)$$

and therefore

$$(\varkappa * \mu_n)(f) = \varkappa(g_n) \to \varkappa(g) = (\varkappa * \mu)(f). \quad \square$$

For measures with noncompact supports the passage to the limit in the convolution is not always possible. The situation is simplified, however, if we consider *monotone* sequences of measures.

A sequence of measures $\{\mu_n\}$ is called *monotone increasing* (*decreasing*), if $\mu_{n+1} - \mu_n$ (correspondingly, $\mu_n - \mu_{n+1}$) is a measure for any $n > 1$. For any function $f(x) \in \Phi^+$ the limit

$$\mu(f) = \lim_{n\to\infty} \mu_n(f) \leq +\infty.$$

exists. If this limit is always finite, then it defines a positive linear functional on Φ and, by Theorem 0.3, a measure μ which is the vague limit of the measures μ_n. This situation is equivalent to the existence of a measure μ_0, which majorizes all the measures μ_n, i.e., for which

$$\mu_0 - \mu_n \in \mathfrak{M}^+; \quad n = 1, 2, \ldots,$$

Lemma 0.3. *Suppose that the sequences* $\{\mu_n\}$, $\{\varkappa_n\}$ *are monotone increasing (or decreasing) and vaguely converge to the measures* μ_0 *and* $\varkappa_0$, *respectively. If also* $\mu_n * \varkappa_n$ *is defined for all n and is majorized by some measure, then*

$$\mu_n * \varkappa_n \xrightarrow{\cdot} \mu_0 * \varkappa_0,$$

Proof. Consider the monotone increasing case. Suppose $\mu_n * \varkappa_n \xrightarrow{\cdot} \alpha$; we will prove that $\mu_0 * \varkappa_0 = \alpha$.

For any $f(x) \in \Phi^+$ we put

$$g_0(x) = \int f(x+y)\, d\mu_0(y)$$

and verify that

$$\alpha(f) = \int g_0(x)\, d\varkappa_0(x) = (\mu_0 * \varkappa_0)(f).$$

We note that the continuous functions

$$g_n(x) = \int f(x+y)\, d\mu_n(y),$$

are monotone increasing and converge to $g_0(x)$. Since $g_0(x)$ is also continuous, then by the well-known Dini theorem, convergence will be uniform on any compact set. Consequently, for any $\varphi(x) \subset \Phi^+$, $\varphi(x) \leq 1$,

$$\int g_0(x)\, \varphi(x)\, d\varkappa_0(x) = \lim_{n\to\infty} \int g_n(x)\, \varphi(x)\, d\varkappa_n(x)$$

$$\leq \lim_{n\to\infty} \int g_n(x)\, d\varkappa_n(x) = \lim_{n\to\infty} (\mu_n * \varkappa_n)\,(f) = \alpha(f).$$

Taking into account that

$$\int g_0(x)\, d\varkappa_0(x) = \sup_{\varphi} \int g_0(x)\, \varphi(x)\, d\varkappa_0(x),$$

wo obtain

$$\int g_0(x)\, d\varkappa_0(x) \leq \alpha(f).$$

This proves the existence of $\mu_0 * \varkappa_0$ and the inequality $\mu_0 * \varkappa_0 \leq \alpha$. The inverse inequality follows from the fact that for any n

$$\mu_n * \varkappa_n \leq \mu_n * \varkappa_0 \leq \mu_0 * \varkappa_0.$$

The monotone decreasing case reduces to the above by the substitutions $\mu_n' = \mu_1 - \mu_n$ and $\varkappa_n' = \varkappa_1 - \varkappa_n$. ▯

We now prove the associativity of the convolution of measures.

Lemma 0.4. *For any $\varkappa, \lambda, \mu \in \mathfrak{M}^+$*

$$(\mu * \varkappa) * \lambda = \mu * (\varkappa * \lambda) \tag{0.1.15}$$

under the condition that one side of (0.1.15) *is defined.*

Proof. For any function $f(x) \in \Phi^+$

$$(\mu * \varkappa) * \lambda(f) = \int d\lambda(y) \int f(x + y)\, d(\mu * \varkappa)\,(x)$$

$$= \int d\lambda(y) \int d\varkappa(x) \int f(x + y + z)\, d\mu(z),$$

and by the Fubini theorem the order of inetgration may be arbitrary. ▯

We may define the operation of convolution in the space of signed measures $\mathfrak{M}$, putting for $\nu_1, \nu_2 \in \mathfrak{M}$

$$\nu_1 * \nu_2 = \nu_1^+ * \nu_2^+ + \nu_1^- * \nu_2^- - \nu_1^+ * \nu_2^- - \nu_1^- * \nu_2^+. \tag{0.1.16}$$

Here it is assumed that all convolutions of measures on the right side of (0.1.16) are defined. For this, it is sufficient that one of the signed measures ν_1, ν_2 has compact support.

We mention the following obvious generalization of lemma 0.4.

Lemma 0.4′. *If $\nu, \tau, \pi \in \mathfrak{M}$, then*

$$(\nu * \tau) * \pi = \nu * (\tau * \pi) \tag{0.1.15′}$$

under the condition that each of the eight convolutions of measures

$$\nu^{\pm} * \tau^{\pm} * \pi^{\pm}$$

is defined for some order of the operations[9].

We will especially consider a convolution with an absolutely continuous signed measure ψ. Assuming that $\nu * \psi$ is defined, we may write, for $f \in \Phi$, the equation

$$(\nu * \psi)(f) = \int d\nu(y) \int f(x + y)\, \psi(x)\, dx$$
$$= \int f(x)\, dx \int \psi(x - y)\, d\nu(y).$$

It shows that $\nu * \psi$ is an absolutely continuous signed measure with density

$$(\nu * \psi)(x) = \int \psi(x - y)\, d\nu(y) = \int \psi(y)\, d\nu_x(y). \tag{0.1.17}$$

If the signed measure ν is also absolutely continuous, then the density of the signed measure $\nu * \psi$ is given by the formula

$$(\nu * \psi)(x) = \int \psi(x - y)\, \nu(y)\, dy = \int \psi(y)\, \nu(x - y)\, dy. \tag{0.1.17'}$$

Recall that the density $(\nu * \psi)(x)$ is defined within equivalence. It may happen that the integral in (0.1.17) or (0.1.17′) may be defined at all points of R^p (or almost all, with the exception of some special set of measure zero). Without further comment we will always identify the density $\nu * \psi$ with the integral (0.1.17) at those points where the latter is defined.

The assumption made at the beginning of this section that G coincides with the entire space R^p is related in an obvious way to the fact that R^p is a group under addition. We cannot define convolution in $\mathfrak{M}_G^+$, without introducing a group structure in G.

5. We consider certain examples and particular types of measures and signed measures.

(a) The *Dirac measure* ε is the unit measure $\varepsilon(1) = 1$ concentrated at the point 0. According to (0.1.13) ε_x will denote the unit measure concentrated at the point x. Thus,

$$\varepsilon_x(f) = \int f(y)\, d\varepsilon_x(y) = f(x)$$

for any function f continuous in a neighborhood of the point x.[10]

---

[9] This condition is equivalent to the existence of the double convolution $|\nu| * |\tau| * |\pi|$.

[10] Cf. Corollary 2 to Theorem 0.2.

The measure ε plays the role of the identity under convolution:

$$\nu * \varepsilon = \nu \tag{0.1.18}$$

for any signed measure ν. Indeed, for any $f \in \Phi$

$$(\nu * \varepsilon)(f) = \nu(\varepsilon_x(f)) = \nu(f).$$

It is also easy to verify the more general formula:

$$\nu * \varepsilon_x = \nu_x \tag{0.1.18'}$$

(b) Let $\{x_k\}$ be an arbitrary sequence of points in R^p and $\{\alpha_k\}$ a sequence of real numbers satisfying only one condition: for any compact set $K \subset R^p$

$$\sum_{x_k \in K} |\alpha_k| < \infty.$$

Then the formula

$$\nu = \sum_{k=1}^{\infty} \alpha_k \varepsilon_{x_k} \tag{0.1.19}$$

defines a signed measure. Moreover

$$\nu(f) = \sum \alpha_k f(x_k), \quad f \in \Phi,$$

and the series converges, since the summation extends only over those k for which $x_k \in S(f)$.

Signed measures of the form (0.1.19) are called *atomic* (or *discrete*). They are characterized by the fact that they are concentrated on a countable set. We note that the support of such a signed measure may coincide with R^p.

(c) Consider the measure $\varepsilon^{(r)}$, obtained by the uniform distribution of a unit mass over the surface of a sphere $\sigma(r)$ of radius r centered at the point 0. This measure may by characterized by the equation

$$\varepsilon^{(r)}(f) = \frac{1}{\omega_p r^{p-1}} \int_{\sigma(r)} f(x)\, dS, \quad f \in \Phi,$$

where dS the element of surface on the sphere $\sigma(r)$, and

$$\omega_p = \frac{2\pi^{p/2}}{\Gamma(p/2)}$$

is the "area" of the unit sphere $|x| = 1$.

In this way, the measure $\varepsilon^{(r)}$ defines the operation of averaging a function over the sphere $|x| = r$.

It is easy to see that as $r \to 0$,

$$\varepsilon^{(r)} \dot{\to} \varepsilon.$$

(d) Denote by $m^{(r)}$ the measure obtained by uniformly distributing a unit mass over the ball $|x| < r$. This measure is absolutely continuous and for its density $m^{(r)}(x)$ we have

$$m^{(r)}(x) = \begin{cases} \dfrac{\Gamma(p/2+1)}{r^p \pi^{p/2}} & \text{for } |x| < r. \\ 0 & \text{for } |x| > r. \end{cases}$$

Thus,

$$m^{(r)}(f) = \frac{\Gamma(p/2+1)}{r^p \pi^{p/2}} \int\limits_{|x|<r} f(x)\, dx,$$

and this is the operation of averaging a function $f(x)$ over the ball $|x| < r$. As $r \to 0$ we have

$$m^{(r)} \xrightarrow{\cdot} \varepsilon.$$

For any signed measure $\nu \in \mathfrak{M}$ the convolution $\nu * m^{(r)}$ exists and, by No. 4, is an absolutely continuous signed measure with density

$$(\nu * m^{(r)})\,(x) = \int m^{(r)}(x-y)\, d\nu(y)$$
$$= \frac{\Gamma(p/2+1)}{r^p \pi^{p/2}} \int\limits_{|y-x|<r} d\nu(y).$$

If the signed measure ν is absolutely continuous, then

$$(\nu * m^{(r)})\,(x) = \frac{\Gamma(p/2+1)}{r^p \pi^{p/2}} \int\limits_{|y-x|<r} \nu(y)\, dy$$

is an almost everywhere differentiable function. Consequently, for any $\nu \in \mathfrak{M}$ the signed measure $\nu * m^{(r)} * m^{(r)}$ will have a differentiable density. Analogously $\nu * m^{(r)} * m^{(r)} * m^{(r)}$ will have a twice differentiable density.

By Lemma 0.2 as $r \to 0$

$$\nu * m^{(r)} \xrightarrow{\cdot} \nu, \qquad \nu * m^{(r)} * m^{(r)} \xrightarrow{\cdot} \nu, \qquad \text{etc.}$$

§ 2. Space of Distributions. Operations on Distributions

6. In the previous paragraph we observed that there is no essential distinction between signed measures and functionals on Φ.

Here we will show that by narrowing the class Φ and strengthening the topology[11], not surprisingly we enlarge the set of continuous linear functionals. As a result we arrive at a generalization of the notion of signed measure which is adequate for potential theory.

---

[11] The necessary definitions relating to the topological notions may be found in Ch. V, § 3.

Denote by D the linear space of all infinitely differentiable functions which have compact support in G. Convergence in D will be defined as follows: a sequence of functions $f_n(x) \in D$ converges to zero in D, if the supports of the $f_n(x)$ are contained in a fixed compact set and for any k all the derivatives $\partial^k f_n(x)$[12] of order $|k|$ tend uniformly to zero.

Let K be a compact set containing all the $S(f_n)$. Denote by $V(m, \varepsilon, K)$ where $\varepsilon > 0$, $m \geq 0$ is an integer, the set of all functions $f(x) \in D$ satisfying the conditions

$$S(f) \subset K, \ |\partial^k f(x)| < \varepsilon \qquad (|k| = 0, 1, \ldots, m).$$

Then in order that $f_n(x)$ converge to zero in D it is necessary and sufficient that for any ε, m

$$f_n(x) \in V(m, \varepsilon, K) \quad \text{for} \quad n \geq n_0(\varepsilon, m).$$

D is a dense subset of Φ. Indeed, if $f \in \Phi$, then extending it by setting it equal to zero on the rest of R^p and forming the convolution $f * \varrho^{(r)}(x)$, where

$$\varrho^{(r)}(x) = \begin{cases} a \exp\,(|x|^2 - r^2)^{-1}, & |x| < r \\ 0, & |x| \geq r \end{cases} \tag{0.2.0}$$

and the constant a is chosen such that $\int \varrho^{(r)}(x)\, dx = 1$, we will obtain for sufficiently small r, a function in D. The supports of these functions are contained in a fixed compact set, and it is easy to see that for $r \to 0$

$$f * \varrho^{(r)}(x) \to f(x)$$

uniformly.

Any continuous linear functional T on D will be called a distribution. The linear space of all distributions we will denote by D_G^*, dropping the index G if $G \equiv R^p$.

Lemma 0.5. *Suppose $T \in D_G^*$ and $K \subset G$ is a compact set. Then there exists an integer $m \geq 0$ such that if $f_n(x) \in D$, $S(f_n) \subset K$, and $\partial^k f_n(x) \to 0$ uniformly for $|k| \leq m$, then*

$$T(f_n) \to 0.$$

Proof. We will show that there exist $\varepsilon > 0$ and m such that

$$\sup |T(f)| < \infty, \quad f \in V(m, \varepsilon, K).$$

[12] Here and below we use the following abbreviations: $k = (k_1, k_2, \ldots, k_p)$ is an integer vector where $k_i \geq 0$; $|k| = k_1 + k_2 + \cdots + k_p$

$$\partial^k = \frac{\partial^{|k|}}{\partial x_1^{k_1} \partial x_2^{k_2} \cdots \partial x_p^{k_p}}.$$

∂ denotes the first derivative with respect to any coordinate x_i.

Otherwise we would have three sequences $\{m_n\}$, $\{\varepsilon_n\}$, $\{f_n\}$ such that $m_n \to \infty$, $\varepsilon_n \to 0$

$$|T(f_n)| \geq n, \quad f_n \in V(m_n, \varepsilon_n, K).$$

Since for any fixed m, ε, $f_n \in V(m, \varepsilon, K)$ for $n \geq n_0(m, \varepsilon)$ we have $f_n \to 0$ in D and consequently, $T(f_n) \to 0$. The resulting contradiction shows that for some m, ε, and $f \in V(m, \varepsilon, K)$

$$\sup |T(f)| = M < \infty.$$

But then, for $f \in V(m, \varepsilon/n, K)$

$$\sup |T(f)| = M/n,$$

and from this easily follows the assertion of the lemma. ▯

A signed measure $\nu \in \mathfrak{M}_G$, if it is considered only on D, generates a distribution. Moreover, since D is dense in Φ, it follows that distinct signed measures yield distinct distributions. Therefore, in order not to complicate notation, the corresponding distribution will also be denoted by the same symbol ν and we will write

$$\mathfrak{M}_G \subset D_G^*.$$

It is easy to give distributions which are not signed measures. For example, for $|k| > 0$ and $x_0 \in G$

$$T(f) = \partial^k f(x_0) \in D_G^* \setminus \mathfrak{M}_G.$$

The concept of support may be extended to any distribution.

We will say that $T = 0$ on the open set $\Omega \subset G$, if $T(f) = 0$ under the condition that $S(f) \subset \Omega$.

The *support* of a distribution T is the complement of the largest open set on which $T = 0$.[13]

From the definition of support, it follows that if $S(f) \cap S(T) = \emptyset$ then $T(f) = 0$.

If a distribution T has compact support, and $\varphi(x)$ is any infinitely differentiable function, then we shall define $T(\varphi)$ by the equation

$$T(\varphi) = T(\alpha\varphi), \tag{0.2.1}$$

where $\alpha(x) \in D$ and $\alpha(x) = 1$ in a neighborhood of the compact set $S(T)$. It is easy to see that this definition does not depend on the choice of the function $\alpha(x)$.

---

[13] The existence of such a largest open set is nontrivial. It follows from the proposition: if $T = 0$ on every open set Ω_α, then $T = 0$ on $\bigcup_\alpha \Omega_\alpha$ (α may range over any index set). For a proof of this fact cf. I. M. Gelfand and G. E. Shilov [1].

Analogously, one may define $T(\varphi)$ under the condition that $S(T) \cap S(\varphi)$ is compact.

Note that although we cannot speak of the value of a distribution T at a given point x_0, it is reasonable to define "the values of T on an open set $\Omega \subset G$". By this we mean the restriction of T to functions $f(x) \in D$ such that $S(f) \subset \Omega$.

In the space D_G^* we define weak convergence $T_n \rightharpoonup T$ by the condition

$$T_n(f) \to T(f) \quad \text{for every} \quad f \in D. \tag{0.2.2}$$

With this definition of convergence, the subset of $\mathfrak{M}_G$ of signed measures is dense in D_G^*. We shall prove a stronger claim. Any function $f(x) \in D$ may be regarded as an absolutely continuous signed measure (cf. end of No. 1) and consequently, $D \subset \mathfrak{M}_G \subset D_G^*$

Theorem 0.7. *The set D is dense in D_G^* (in the sense of weak convergence).*

Proof. Consider an arbitrary distribution $T \in D_G^*$; we shall use it to define a family of functions

$$\varphi^{(r)}(y) = T(\varrho_y^{(r)}),$$

where $\varrho_y^{(r)}(x) = \varrho^{(r)}(x - y)$ (cf. 0.2.0). Let $G^{(r)}$ denote the set of points in G which are at a distance greater than r from the boundary ∂G; then $\varphi^{(r)}(y)$ will be defined and infinitely differentiable in $G^{(r)}$. The differentiability of $\varphi^{(r)}(y)$ follows from the fact that for $y, y + \Delta y \in G^{(r)}$ one can write[14]

$$\frac{1}{\Delta y}\left[\varphi^{(r)}(y + \Delta y) - \varphi^{(r)}(y)\right] = T\left(\frac{\varrho_{y+\Delta y}^{(r)} - \varrho_y^{(r)}}{\Delta y}\right).$$

But the convergence of

$$\frac{\varrho_{y+\Delta y}^{(r)} - \varrho_y^{(r)}}{\Delta y} \to \partial_y \varrho_y^{(r)} \quad (\Delta y \to 0)$$

is easily seen to be convergence in D; therefore

$$\partial \varphi^{(r)}(y) = T(\partial_y \varrho_y^{(r)}).$$

Repeating this argument, we obtain for any $|k| > 0$

$$\partial^k \varphi^{(r)}(y) = T(\partial_y^k \varrho_y^{(r)}).$$

We extend $\varphi^{(r)}(y)$ to G so that it remains infinitely differentiable and is zero outside of $G^{(r/2)}$. Calling the new function again $\varphi^{(r)}(y)$ we have

[14] Δy here denotes an increment in one of the coordinates of the point y; $y + \Delta y$ denotes the point obtained by a translation of Δy along that coordinate axis.

$\varphi^{(r)}(y) \in D$ and for $r \to 0$

$$\varphi^{(r)} \xrightarrow{\cdot} T.$$

Indeed, suppose $f \in D$ and r is so small that $S(f) \subset G^{(r)}$. Then

$$\varphi^{(r)}(f) = \int_{G^{(r)}} f(y)\, \varphi^{(r)}(y)\, dy = T\Big(\int_{G^{(r)}} (f(y)\, \varrho^{(r)}(x-y)\, dy\Big)$$

since convergence of the integral sums to the integral

$$\int_{G^{(r)}} f(y)\, \varrho^{(r)}(x-y)\, dy \tag{0.2.3}$$

is also convergence in D. Further, from the equality

$$\partial^k \int_{G^{(r)}} f(y)\, \varrho^{(r)}(x-y)\, dy = \int_{G^{(r)}} \partial^k f(y)\, \varrho^{(r)}(x-y)\, dy,$$

which is easily verified by integration by parts, it follows that for $r \to 0$ the integral (0.2.3) tends to $f(x)$ in D; consequently,

$$\varphi^{(r)}(f) \to T(f),$$

which is the desired result. ▯

We shall say that a distribution T is positive ($T \geq 0$), if for any function $f(x) \in D$, $f(x) \geq 0$, we have $T(f) \geq 0$.

Theorem 0.8. *A positive distribution T is a measure.*

Proof. By Theorem 0.3 we need only show that T can be extended to a positive linear functional on Φ. Since D is dense in Φ, then for $f(x) \in \Phi$ there is a sequence $f_n(x) \in D$ with $f_n(x) \to f(x)$ in Φ. We will show that the sequence $\{T(f_n)\}$ converges. For this we modify appropriately the proof of Theorem 0.3. Let $S(f_n) \subset K$ and $\psi(x) \in D$ such that $0 \leq \psi(x) \leq 1$ and $\psi(x) \equiv 1$ on K. Then

$$|f_n(x) - f_m(x)| \leq \varepsilon_{n,m}\psi(x),$$

where

$$\varepsilon_{n,m} = \max |f_n(x) - f_m(x)| \to 0 \quad \text{as} \quad n, m \to \infty.$$

From this it easily follows that

$$|T(f_n) - T(f_m)| \leq \varepsilon_{n,m} T(\psi)$$

and, consequently, $\lim\limits_{n\to\infty} T(f_n)$ exists. It obviously does not depend on the choice of the approximating sequence $\{f_n(x)\}$; we take that limit as $T(f)$. To check that this extended functional $T(f)$ is linear and positive presents no difficulties. ▯

7. In many cases, and in particular, when the distribution T is given by a concrete analytic expression, it becomes possible to extend it, preserving linearity and continuity, to a space of functions $E \supset D$, possessing a weaker topology than D.[15] This extended functional will be denoted by the same symbol T. On the other hand, if T is any linear, continuous functional on E, then its restriction to D is obviously a distribution. This situation may be described by writing

$$E^* \subset D_G^*,$$

where by E^* we mean the set of all linear continuous functionals in E. In this way we may distinguish in D_G^* *various classes of distributions* characterized by the imbedding of D into a certain function space E.

As an example consider the space $D^m \supset D$, consisting of the m-times differentiable functions with compact support. Convergence in D^m is defined as follows: $f_n(x) \to 0$ in D^m if their supports contained in a fixed compact set and if

$$\partial^k f_n(x) \to 0 \quad (|k| = 0, 1, \ldots, m)$$

uniformly.

Then the distribution

$$T(f) = \partial^m f(x_0), \quad x_0 \in G,$$

is, of course, a continuous, linear functional on D^m.

In general a distribution from $(D^m)^*$ will be called a distribution of *order not greater than m*. Consequently, signed measures are distributions of order zero.

From Lemma 0.5 it follows that any distribution with compact support has finite order.

In the space D_G^* of distributions we can define a differential operator ∂^n of any order $|n|$. Its definition is the following

$$\partial^n T(f) = (-1)^{|n|}\, T(\partial^n f). \tag{0.2.4}$$

Note that the derivative of an absolutely continuous signed measure with an almost everywhere differentiable density $\psi(x)$ does not always coincide with the absolutely continuous signed measure with density $\partial\psi(x)$. In order for this to hold, the following equality must be true,

$$-\int \psi(x)\, \partial f(x)\, dx = \int \partial\psi(x)\, f(x)\, dx, \qquad f(x) \in D.$$

This will indubitably hold if the function $\psi(x)$ is absolutely continuous on almost all lines parallel to the direction of differentiation.

---

[15] Thus convergence in D implies convergence in E. The reader unfamiliar with general topology may take this as the definition of "weaker topology". Cf. also Ch. V, § 3.

The differential operator is linear and continuous, i.e. from $T_n \dashrightarrow T$, it follows that $\partial T_n \dashrightarrow \partial T$.

If T is a distribution of order $\leq m$, then $\partial^n T$ will have order $\leq m + |n|$. In particular, if ν is a signed measure, then $\partial^n \nu$ will be a distribution of order $\leq n$.

Theorem 0.9. *Every distribution T with compact support may be represented as a finite sum of derivatives of signed measures. Moreover, we may assume that the supports of these signed measures are contained in a preassigned compact neighborhood $\overline{U}$ of the set $S(T)$.*

Proof. As mentioned above, T has finite order $\leq m$. Consequently, for any $\varepsilon > 0$ we can find a $\delta > 0$ such that

$$|T(f)| < \varepsilon \quad \text{if} \quad f \in V(m, \delta, \overline{U}). \tag{0.2.5}$$

We associate with each function $f(x) \in D$, $S(f) \subset \overline{U}$ a finite-dimensional vector-function $\{\partial^k f\}$, $|k| = 0, 1, \ldots, m$. Thus we map the class of functions under consideration into a linear manifold Δ in the Banach space $C^{(N_m)}(\overline{U})$ of continuous vector-functions

$$\varphi(x) = \{\varphi_1(x), \varphi_2(x), \ldots, \varphi_{N_m}(x)\}, \quad x \in \overline{U},$$

with norm

$$\|\varphi(x)\| = \sum_{k=1}^{N_m} \max_{\overline{U}} |\varphi_k(x)|.$$

(Here N_m denotes the number of derivatives ∂^k with $|k| \leq m$.) By (0.2.5) $T(f)$ may be viewed as a continuous (and, hence, bounded) linear functional on Δ. Extending it to all of $C^{(N_m)}(\overline{U})$ and using the Riesz theorem on the representation of a linear functional in $C^{(N_m)}(\overline{U})$, we obtain the formula

$$T(f) = \sum_{|k|=0}^{m} \int \partial^k f(x) \, \partial\nu_k(x),$$

where the supports of all the signed measures ν_k are contained in $\overline{U}$. But, by (0.2.4),

$$\int \partial^k f(x) \, \partial\nu_k(x) = (-1)^{|k|} \, \partial^k \nu_k(f),$$

and therefore

$$T(f) = \sum_{|k|=0}^{m} (-1)^{|k|} \, \partial^k \nu_k(f), \quad S(\nu_k) \subset \overline{U}, \quad S(f) \subset \overline{U}.$$

Now let $f(x)$ be an arbitrary infinitely differentiable function. Select a function $\alpha(x) \in D$ such that $S(\alpha) \subset \bar{U}$ and $\alpha(x) \equiv 1$ in a certain neighborhood S(T). By (0.2.1)

$$T(f) = T(\alpha f) = \sum_{|k|=0}^{m} (-1)^{|k|} \partial^k \nu_k(\alpha f)$$

$$= \sum_{|k|=0}^{m} \int \partial^k(\alpha f)\, d\nu_k = \sum_{|k|=0}^{m} (-1)^{|k|} \int \partial^k f(x)\, \partial \lambda_k(x)$$

$$= \sum_{|k|=0}^{m} \partial^k \lambda_k(f), \quad \lambda_k \in \mathfrak{M}$$

and the theorem is proved. ▯

8. With the idea of defining the convolution of distributions in R^p, we define first the so-called direct product $T_1 \times T_2$ of two distributions. This will be a distribution in $R^p \times R^p$, which we will try to define in such a way that for any $f_1(x) \in D_x$, $f_2(y) \in D_y$

$$(T_1 \times T_2)\,(f_1(x)\, f_2(y)) = T_1 f_1(x)\, T_2 f_2(y). \tag{0.2.6}$$

Note that the set of functions in D_{xy} of the form $F_1(x)\, f_2(y)$ is complete in D_{xy}.[16] Indeed, we take a sequence $\varepsilon_n \to 0$ and construct for a given function $f(x, y) \in D_{xy}$ a sequence of polynomials $P_n(x, y)$ such that on $S(f)$

$$|\partial^k f(x, y) - \partial^k P_n(x, y)| < \varepsilon_n \quad (|k| = 0, 1, 2, \ldots, n).$$

Furthermore, we take functions $\alpha(x)$, $\beta(x) \in D$ such that $\alpha(x)\, \beta(y) = 1$ on $S(f)$. Then the functions

$$\alpha(x)\, \beta(y)\, P_n(x, y)$$

form a sequence of functions of the form $\sum_i f_1^{(i)}(x)\, f_2^{(i)}(y)$, which in D_{xy} converges to $f(x, y)$.

From this it follows that if the distribution $T_1 \times T_2$ exists at all, then it must be uniquely defined. We now show that it does exist and can be computed in one of the following two ways:

$$(T_1 \times T_2)\, f(x, y) = T_1[T_2^{(y)} f(x, y)] = T_2[T_1^{(x)} f(x, y)]. \tag{0.2.7}$$

The upper index of T_i refers to that variable which is considered basic when T_i operates on a function in $R^p \times R^p$.

---

[16] D_{xy} denotes the space of infinitely differentiable functions in $R^p \times R^p$ with compact support, in which convergence has the meaning given at the beginning of No.6.

Consider the expression

$$T_1[T_2^{(y)}f(x, y)], \quad f(x, y) \in D_{xy}. \tag{0.2.8}$$

For any fixed x $f(x, y) \in D_y$, and therefore $T_2^{(y)} f(x, y)$ is defined. Moreover, if x occurs outside a certain compact set $K \subset R^p$, then $f(x, y) \equiv 0$, and therefore $T_2^{(y)}f(x, y)$ is a function of the variable x with compact support. It belongs to D since

$$\partial_x^k T_2^{(y)} f(x, y) = T_2^{(y)} \partial_x^k f(x, y), \tag{0.2.9}$$

which is verified by induction on $|k|$.[17] Consequently, (0.2.8) is defined and is a linear functional in D_{xy}. We will show that it is continuous. For this it is sufficient to show that $f_n(x, y) \to 0$ in D_{xy} implies

$$T_2^{(y)} f_n(x, y) \to 0 \quad \text{in } D.$$

It is clear that the supports of the functions $T_2^{(y)}f_n(x, y)$ are contained in a fixed compact set. If the convergence $T_2^{(y)}f_n(x, y)$ to zero were not uniform in x, then for some $\varepsilon > 0$, we would have a sequence of points x_n such that

$$|T_2^{(y)} f_n(x_n, y)| > \varepsilon. \tag{0.2.10}$$

But from the condition $f_n(x, y) \to 0$ in D_{xy} it follows that for any choice of x_n

$$f_n(x_n, y) \to 0 \quad \text{in } D;$$

this contradicts (0.2.10). Analogously using (0.2.9) it can be checked that $\partial^k T_2^{(y)} f_n(x, y)$ uniformly tend to zero for any k.

Thus, (0.2.8) actually defines a certain distribution in $R^p \times R^p$. The same holds of course for the expression

$$T_2[T_1^{(x)} f(x, y)].$$

It is immediate that they both satisfy condition (0.2.6). From the uniqueness established above, equation (0.2.7) follows.

Analogously one can define a direct product $T_1 \times T_2 \times T_3$ of three (or more) distributions. Moreover, from (0.2.6) and the analogous condition

$$(T_1 \times T_2 \times T_3)\,(f_1(x)\,f_2(y) f(z)) = T_1 f_1(x)\; T_2 f_2(y)\; T_3 f_3(z)$$

associativity easily follows:

$$T_1 \times T_2 \times T_3 = (T_1 \times T_2) \times T_3 = T_1 \times (T_2 \times T_3). \tag{0.2.11}$$

---

[17] We should point out that the difference $\frac{1}{\Delta x}[f(x + \Delta x, y) - f(x, y)] = \partial_x f(x + \theta \Delta x, y)$ tends in D to $\partial_x f(x, y)$.

We assume that the distributions T_1 and T_2 belong to classes E_1^* and E_2^* defined by the specification of two function spaces $E_1 \supset D$ and correspondingly $E_2 \supset D$ (cf. the beginning of No. 7). Then $T_1 \times T_2$ will be a continuous functional on a certain space $E_{12} \supset D_{xy}$, if the following conditions are satisfied:

(a) if $f(x, y) \in E_{12}$ then for fixed y, $f(x, y) \in E_1$, and for fixed x, $f(x, y) \in E_2$.

(b) $T_1^{(x)} f(x, y) \in E_2$, $T_2^{(y)} f(x, y) \in E_1$; moreover, both these operators from E_{12} into E_2 and from E_{12} into E_1 are continuous.

If in addition,

(c) D_{xy} is dense in E_{12},

then $T_1 \times T_2$ is uniquely defined by condition (0.2.6), and the "Fubini formula" (0.2.7) holds.

A similar remark can be made for the double product $T_1 \times T_2 \times T_3$ and formula (0.2.11).

9. We will not be able to define the convolution of two arbitrary distributions T_1, T_2. The reason for this is easily seen from what follows.

Suppose that the class E_{12}, on which the direct product $T_1 \times T_2$ is defined, has the following property:

$$\text{(d)} \quad f(x + y) \in E_2^1 \quad \text{for any } f(x) \in D \tag{0.2.12}$$

and that the mapping $f(x) \to f(x + y)$ is continuous. Supposing that condition (c) also holds, we see that the formula

$$(T_1 \times T_2) f(x + y) = T_1\big(T_2^{(y)} f(x + y)\big) = T_2\big(T_1^{(x)} f(x + y)\big), \quad f \in D \tag{0.2.13}$$

defines a new distribution, called the convolution of T_1 and T_2. We will adopt the following notation:

$$(T_1 * T_2) f = T_1\big(T_2^{(y)} f(x + y)\big). \tag{0.2.13'}$$

Formula (0.2.13) may then be rewritten in the form

$$T_1 * T_2 = T_2 * T_1, \tag{0.2.14}$$

i.e. as the statement of commutativity of convolution.

In order for the formula

$$(T_1 * T_2) * T_3 = T_1 * (T_2 * T_3) \tag{0.2.15}$$

expressing *associativity* of convolution to hold, it is necessary to put certain restrictions on T_1, T_2, T_3. Denoting by E_{ik} the space on which the direct product $T_i \times T_k$ is defined and by E_{123} the space on which $T_1 \times T_2 \times T_3$ is defined, we require that E_{ik} satisfy conditions (a), (b), (c) of No. 8, and (0.2.12); E_{123} must satisfy the following conditions:

(a') if $f(x, y, z) \in E_{123}$, then for fixed x (correspondingly, y, z), $f(x, y, z) \in E_{23}$ (correspondingly E_{13}, E_{12}).

(b') $T_1^{(x)} f(x, y, z) \in E_{23}$, $T_2^{(y)} f(x, y, z) \in E_{13}$, $T_3^{(z)} f(x, y, z) \in E_{12}$; moreover, these operators are continuous.

(c') D_{xyz} is dense in E_{123},

and

(d') $f(x + y + z) \in E_{123}$ for any $f \in D$; (0.2.12')

in addition, the map $f(x) \to f(x + y + z)$ must be continuous. Then (0.2.15) will hold.

Note that if condition (0.2.12) is satisfied not in D but in a certain space $\hat{E} \supset D$, then if the map

$$f(x) \to f(x + y)$$

of $\hat{E}$ into E_{12} is continuous, the convolution $T_1 * T_2$ will belong to the class $\hat{E}^*$. The analogous remark may be made for the multiple convolution $T_1 * T_2 * T_3$.

It is possible to give a more direct definition of the convolution $T_1 * T_2$, $T_1 \in T_1^*$, $T_2 \in E_2^*$ for which, however, commutativity is not guaranteed. Namely, we propose that the following condition be satisfied:

(α) for any $f(x) \in D$

$$T_2^{(y)} f(x + y) \in E_1$$

and this operator from D into E_1 is continuous.

Then formula (0.2.13')

$$(T_1 * T_2) f = T_1(T_2^{(y)} f(x + y))$$

defines a distribution $T_1 * T_2$ which we also call a convolution.

Clearly, the first definition of convolution using formula (0.2.13) is more restricted.

Note that if we replace (α) with the stronger condition,

(α') for any $f(x) \in E_2$, $f(x + y) \in E_2$ and $T_2^{(y)} f(x + y) \in E_1$ and this operator from E_2 into E_1 is continuous, then we have $T_1 * T_2 \in E_2^*$.

Consider now concrete cases of defining convolutions.

First of all, it is clear that $f(x + y) \notin D_{xy}$, if $f(x) \not\equiv 0$, and therefore definition (0.2.13) is not applicable to arbitrary distributions.

Suppose now that one of the distributions, say T_1, has compact support. Then for E_1 we may take the space of all infinitely differentiable functions[18], where by the convergence of $f_n(x)$ to zero ($f_n(x) \to 0$) we mean that for any compact set $K \subset R^p$ and for any k, $\partial^k f_n \to 0$ uniformly on K. Putting $E_2 = D$, we may take for E_{12} the space of all functions $f(x, y)$ infinitely differentiable in $R^p \times R^p$ for which $S(f)$ has a compact intersection with any "strip" $|x| \leq C$. We define convergence in E_{12} in the following way: $f_n(x, y) \to 0$ in E_{12} if, first, for any $K \subset R^p \times R^p$ and for any k $\partial^k f \to 0$ uniformly on K and, second, the intersections of the $S(f_n)$ with the strip $|x| \leq C$ are contained in a fixed compact set.

Then (a), No. 8 is clearly satisfied. We shall verify (b). If $\alpha(x) \in D$ is a function equal to 1 in a neighborhood of $S(T_1)$, then

$$T_1^{(x)} f(x, y) = T_1^{(x)} \alpha(x)\, f(x, y),$$

and since $\alpha(x)\, f(x, y) \in D_{xy}$, then the continuity of this operator from E_{12} into D has been essentially established in No. 8. We also, as in No. 8, prove that formula (0.2.9) holds for the operator $T_2^{(y)} f(x, y)$ and that its range of values is included in the space E_1. Analogously, we check the continuity of this operator; we need only point out that from the conditions that $f_n(x, y) \to 0$ in E_{12} and that the points x_n belong to the compact set $K \subset R^p$, it follows that $f_n(x, y) \to 0$ in D. Finally, (c) holds. Indeed, if $f(x, y) \in E_{12}$ and $\alpha_n(x, y) \in D_{xy}$ satisfies the condition that $\alpha_n(x, y) = 1$ in a sphere of radius n in the space $R^p \times R^p$, then $\alpha_n f \in D_{xy}$ and $\alpha_n f \to f$ in E_{12}.

Thus, if one of the distributions T_1 or T_2 has compact support, then the convolution $T_1 * T_2$ defined according to (0.2.13) is defined and is commutative.

In order to define the multiple convolution $T_1 * T_2 * T_3$ to be equal to the common value in (0.2.15), it is sufficient to assume that two of the distributions T_1, T_2, T_3 have compact support.

Suppose, for example that T_1 and T_2 have compact support. Then E_{12} is the space of all infinitely differentiable functions in $R^p \times R^p$, and E_{13} and E_{23} are defined just as E_{12} was defined above. Then E_{123} is the set of infinitely differentiable functions $f(x, y, z)$ in $R^p \times R^p \times R^p$ with the property that the intersection of $S(f)$ with any "strip" $|x| \leq \text{const}$ is compact. It is not difficult to prove conditions (a), (b), (c) for E and the conditions (a′), (b′), (c′); thus we have the following result.

---

[18] Since T_1 has finite order m, only m-times differentiable functions may be considered.

Lemma 0.6. *For the existence and commutativity of the convolution* $T_1 * T_2$ *it is sufficient that one of the factors has compact support; for the existence of the multiple convolution* $T_1 * T_2 * T_3$ *and its associativity, it is sufficient that two of its factors have compact support.*

The following simple example shows that these conditions cannot be completely removed. Consider in R^1 Heaviside's function

$$h(x) = \begin{cases} 1 & \text{for } x \geq 0 \\ 0 & \text{for } x < 0 \end{cases}$$

and the derivative $\partial\varepsilon$ of the Dirac measure ε. Since the latter has compact support, the convolutions $1 * \partial\varepsilon$ and $\partial\varepsilon * h$ are defined. Moreover, $1 * \partial\varepsilon = 0$,[19] while $\partial\varepsilon * h = \varepsilon$.[20] Consequently,

$$(1 * \partial\varepsilon) * h \neq 1 * (\partial\varepsilon * h).$$

Note that in this example the multiple convolution $1 * \partial\varepsilon * h$ in general has no meaning.

We mention now the following generalization of Lemma 0.2.

Lemma 0.7. *Suppose that* $T_n \dot{\to} T$ *and that one of the following statements is satisfied:*

(*i*) *the distribution* U *has compact support,*

(*ii*) *the supports of the distributions* T_n *are contained in a fixed compact set.*

Then

$$U * T_n \dot{\to} U * T. \qquad (0.2.16)$$

The proof of this lemma proceeds exactly as the proof of Lemma 0.2.

10. In conclusion we consider certain examples of convolutions.

(1) *Convolution with the Dirac measure* ε. For any distribution T

$$T * \varepsilon = T, \qquad (0.2.17)$$

i.e. ε plays the role of the identity. Indeed, according to (0.2.13′)

$$(T * \varepsilon) f = T(\varepsilon^{(y)} f(x + y)) = Tf(x), \quad f \in D.$$

Set

$$T_{x_0} = T * \varepsilon_{x_0}$$

---

[19] Since for $f \in D$, $(1 * \partial\varepsilon) f = -1(f'(x)) = -\int_{-\infty}^{\infty} f'(x)\,dx = 0$.

[20] Since $(h * \partial\varepsilon) f = -\int_0^{\infty} f'(x)\,dx = f(0)$.

Then it is easy to see that

$$T_{x_0}f(x) = T^{(x)}f(x + x_0).$$

(2) *Convolution with the derivative $\partial^k\varepsilon$ of the Dirac measure.* For any distribution T

$$T * \partial^k\varepsilon = \partial^k T. \tag{0.2.18}$$

Indeed,

$$\begin{aligned}(T * \partial^k\varepsilon)f &= T(\partial^k_{(y)}\varepsilon f(x + y)) \\ &= (-1)^{|k|}T(\varepsilon^{(y)}\partial^k f(x + y)) = (-1)^{|k|}T(\partial^k f(x)) = (\partial^k T)f.\end{aligned}$$

From Lemma 0.6 it follows that

$$\partial^k(T * U) = \partial^k T * U = T * \partial^k U,$$

if at least one of the distributions T, U has compact support.

In particular,

$$\partial^k(T * m^{(r)}) = \partial^k T * m^{(r)}, \tag{0.2.19}$$

where the measure $m^{(r)}$ is defined by (d), No. 5.

(3) *Convolution with function from D.* Suppose that $f(x) \in D$. As always, using the same letter f for an absolutely continuous signed measure with density $f(x)$, we obtain

$$T * f = T^{(y)}f(x - y). \tag{0.2.20}$$

This formula shows that $T * f$ is an absolutely continuous signed measure with density defined by the right side of (0.2.20). Although the density of an absolutely continuous signed measure is defined to within equivalence, we have on the right side of (0.2.20) a function defined everywhere in R^p (it is even infinitely differentiable; cf. footnote to (0.2.9)). Just as in § 1, No. 4, we will identify the density of $T * f$ with the right side of (0.2.20).

To verify (0.2.20) it is sufficient to note that for any function $\varphi(x) \in D$, we have

$$\begin{aligned}(T * f)\varphi &= T \int f(y)\, \varphi(x + y)dy = T \int \varphi(y)\, f(y - x)dy \\ &= \int \varphi(y)\, T^{(x)}f(y - x)dy = (T^{(x)}f(y - x))\, \varphi(y).\end{aligned}$$

The admissibility of interchanging the integral and T holds because the function of two variables $f(y - x)$ belongs to a space where the direct product $T \times \varphi$ is defined (cf. No. 8).

The infinitely differentiable function $T * f$ ($f \in D$) is called the *regularization* of T with the help of f. If we choose for f the "delta-shaped" function $\varrho^{(r)}(x)$ (cf. (0.2.20)) and put $r = r_n \to 0$, then we shall obtain a sequence of infinitely differentiable functions weakly converging to T.

Note that (0.2.20) implies the formula

$$T(f) = (T * \check{f})\,(0)\,, \tag{0.2.21}$$

where $\check{f}(x) = f(-x)$, $f \in D$.

If T has compact support, then formula (0.2.20) holds for any infinitely differentiable function $f(x)$.[21] In particular,

$$T * 1 = T(1)$$

More generally, formula (0.2.20) will hold if the function $f(y - x) \in E_{12}$, where E_{12} is a space in which $T \times \varphi$ is defined (cf. No. 8; φ is any element of D).

§ 3. The Fourier Transform of Distributions

11. We will study a special class S^* of distributions which, as mentioned in No. 7, will be defined by specifying a certain space $S \supset D$ with a weaker topology than in D.

Namely, S will denote the linear space of complex infinitely differentiable functions on R^p with the property that for any natural numbers n, k

$$|x|^n \partial^k f \to 0 \quad \text{as} \quad |x| \to \infty\,. \tag{0.3.1}$$

We will call such functions *rapidly decreaing* (at infinity).

In S, the *convergence* of the sequence $f_\nu(x)$ to $f(x)$ is defined in the following way: for any natural numbers n, k

$$(1 + |x|^n)\, \partial^k f_\nu \to (1 + |x|^n) \partial^k f \tag{0.3.2}$$

uniformly in R^p. In particular, if $f_\nu(x) \to 0$ in S, then for any natural numbers n, k

$$(1 + |x|^n) \partial^k f_\nu(x) \to 0 \tag{0.3.2'}$$

uniformly in R^p.

It is easy to establish the following facts:

(i) $D \subset S$, and the topology of S weaker than the topology of D;

(ii) D is dense in S;[22]

(iii) if we denote by $V(n, m, \varepsilon)$ the set of all functions in S which satisfy the conditions

$$(1 + |x|^n)\, |\partial^k f| < \varepsilon\,, \quad |k| \le m\,,$$

---

[21] Or only m times differentiable function $f(x)$, where m is the order of T.

[22] Here we mean the complex space D.

then the convergence of $f_\nu(x)$ to 0 in S is equivalent to the requirement that for any n, m, ε,

$$f_\nu(x) \in V(n, m, \varepsilon), \quad \text{if} \quad \nu \geq \nu_0(n, m, \varepsilon).$$

A set $M \subset S$ is called *bounded, if,* for any $V(n, m, \varepsilon)$ there is a constant $\lambda_0 = \lambda_0(n, m, \varepsilon)$, such that $|\lambda| \leq \lambda_0$ and $f(x) \in M$ imply that $\lambda f(x) \in V(n, m, \varepsilon)$. This condition may be written briefly in the form

$$\lambda M \subset V(n, m, \varepsilon).$$

It is easy to see that M is bounded if, and only if, there exists a collection of positive constants $\{C_{nm}\}$ such that $f(x) \in M$ implies

$$(1 + |x|^n)\, |\partial^k f| < C_{nm}, \quad |k| \leq m.$$

Thus it is easily verified that any sequence convergent in S is a bounded set.

The class S^* mentioned in the beginning is defined as the *set of all continuous linear functionals* on S. It is apparent that every distribution with compact support belongs to S^*. Another interesting example of elements from S^* is given in the following theorem.

Theorem 0.10. *A measure μ belongs to S^* if, and only if, it is slowly increasing, i.e., there exists an integer n such that*

$$\int_{R^p} \frac{d\mu(x)}{(1 + |x|^2)^n} < \infty. \tag{0.3.3}$$

Proof. The sufficiency follows from the fact that with condition (0.3.3) and $f_\nu(x) \in S$, $f_\nu(x) \to 0$, the quantity

$$\mu(f_\nu) = \int f_\nu(x)\, d\mu(x)$$

is defined by (0.3.1) and tends to 0 by (0.3.2′).

Suppose now that $\mu \in S^*$. Reasoning just as in the proof of Lemma 0.5, we can assert the existence of n, m, ε such that for $f \in V(n, m, \varepsilon)$

$$\sup |\mu(f)| = K < \infty.$$

Let $\chi_N(x)$ denote the characteristic function of the ball $|x| \leq N$ and let $\alpha_N(x) = \chi_N * \varrho^{(\varepsilon)}(x) \in D$. It is not difficult to see that there exists a number $\eta = \eta(n, m, \varepsilon) > 0$ such that for all N

$$\frac{\eta \alpha_N(x)}{(1 + |x|z^2)^n} \in V(n, m, \varepsilon).$$

In that case

$$\mu\left(\frac{\eta \alpha_N}{(1 + |x|^2)^n}\right) = \eta \int_{R^p} \frac{\alpha_N(x)}{(1 + |x|^2)^n}\, d\mu(x) < K,$$

that is
$$\int_{R^p} \frac{\varkappa_N(x)}{(1+|x|^2)^n}\, d\mu(x) < \frac{K}{\eta}$$
and as $N \to \infty$ we obtain the required result. ▯

The distributions of S^* are called *slowly increasing*. This name is based on the following result whose proof we omit.

In order that a distribution T belong to the class S^*, it is necessary and sufficient that for any $f \in D$ the regularization $T * f$ and any of its derivatives have polynomial growth as $|x| \to \infty$, i.e.,
$$|\partial^k T * f(x)| < M_k(1 + |x|^{N_k}),$$
where the N_k do not depend on f.

12. In S we introduce the Fourier transform (or operator) $\mathfrak{F}[f(x)]$ by the formula
$$\tilde{f}(y) = \mathfrak{F}[f(x)] = \int_{R_p} f(x)\, e^{-2\pi i xy} dx$$
$$\left(xy = \sum_{k=1}^{p} x_k y_k\right). \tag{0.3.4}$$

Theorem 0.11. *$\mathfrak{F}$ is a one-to-one continuous transformation of the space S onto itself.*

Proof. We will show that $\tilde{f} \in S$. It is infinitely differentiable since the integrals
$$\int_{R^p} f(x)\, (-2\pi i)^{|k|} x^k\, e^{-2\pi i xy}\, dx = \partial^k \tilde{f}(y)\ ^{23}$$
are absolutely convergent. Further, the infinite differentiability of $f(x)x^k$ implies that as $|y| \to \infty$, $\partial_k \tilde{f}(y)$ tends to zero faster than any negative power of $|y|$. Consequently, $\tilde{f}(y) \in S$. The classical inversion formula for Fourier integrals shows that $\mathfrak{F}$ maps S isomorphically onto S.

We shall verify that $\mathfrak{F}$ is continuous. Suppose that $f_\nu(x) \to 0$ in S. For any polynomial $P(y)$, we may integrate by parts and obtain
$$\begin{aligned} P(y)\, \partial^k \tilde{f}_\nu(y) &= P(y) \int_{R^p} f_\nu(x)\, x^k\, (-2\pi i)^{|k|}\, e^{-2\pi i xy}\, dx \\ &= \int_{R^p} f_\nu(x)\, x^k\, [-2\pi i)^{|k|}\, P\left(-\frac{\partial}{2\pi i}\right) e^{-2\pi i xy}\, dx \\ &= \int_{R^p} [P_1(\partial)\, f_\nu(x)\, x^k]\, e^{-2\pi i xy}\, dx, \end{aligned}$$
where $P_1(y)$ is a polynomial of the same degree as $P(y)$.

---

[23] Here we are using the abbreviated notation
$$x^k = x_1^{k_1} x_2^{k_2} \cdots x_p^{k_p}, \quad \text{where} \quad k = (k_1, k_2, \ldots, k_p).$$

Consequently,

$$|P(y)\, \partial^k \tilde{f}_\nu(y)| \le \int_{R^p} |P_1(\partial)\, f_\nu(x)\, x^k|\, dx;$$

since, by (0.3.2′) as $\nu \to \infty$ the right side tends to zero, we have

$$|P(y)\, \partial^k \tilde{f}_\nu(y)| \to 0$$

uniformly in R^p. In other words, $\tilde{f}_\nu \to 0$ in S, and the theorem is proved. ▯

Let us turn to the definition of the Fourier operator on the class S^* of distributions.

Suppose $T \in S^*$. We shall define a new linear functional on S by the formula

$$U\tilde{f} = U\mathfrak{F}[f] = Tf. \tag{0.3.5}$$

By Theorem 0.11 this functional is defined on all of S and is continuous. In other words, $U \in S^*$, and in this way we have defined an operator

$$U = \mathfrak{F}[T],$$

taking S^* into S^* and obviously linear. In order to justify using the same letter $\mathfrak{F}$ as for the Fourier operator on S, we will show that it is an extension of the latter. Consider an absolutely continuous signed measure ψ with density $\psi(x) \in S$. We shall think of it as an element of S^* acting on $f \in S$ according to the rule:

$$\psi(f) = \int_{R^p} f(x)\, \overline{\psi(x)}\, dx.$$

Then, by (0.3.5),

$$\mathfrak{F}[\psi]\, \tilde{f} = \psi(f) = \int_{R^p} f(x)\, \overline{\psi(x)}\, dx.$$

On the other hand, for $\tilde{\psi}(y)$, defined by formula (0.3.4), we have

$$\tilde{\psi}(\tilde{f}) = \int_{R^p} \tilde{f}(y)\, \overline{\tilde{\psi}(y)}\, dy = \int_{R^p} f(x)\, \overline{\psi(x)}\, dx.$$

using the Parseval equality. Thus, we have $\mathfrak{F}[\psi] = \tilde{\psi}$.

Theorem 0.11′. *The operator $\mathfrak{F}$ maps the space S^* isomorphically onto itself. Moreover, $\mathfrak{F}$ and its inverse operator $\mathfrak{F}^{-1}$ are weakly continuous.*

Proof. On S^* define the operator $\mathfrak{F}^{-1}$ by the relation

$$\mathfrak{F}^{-1}[T]\, f = T\tilde{f}, \quad f \in S.$$

Then for any $T \in S^*$ we have

$$\mathfrak{F}[\mathfrak{F}^{-1}[T]] = \mathfrak{F}^{-1}[\mathfrak{F}[T]] = T,$$

and this proves the first assertion of the theorem.

Suppose now that $T_n \dashrightarrow 0$, i.e., for any function $f \in S$, $T_n f \to 0$. Then, according to (0.3.5),

$$\mathfrak{F}[T_n]\tilde{f} = T_n f \to 0,$$

and this holds for any function $\tilde{f} \in S$. Consequently, $\mathfrak{F}[T_n] \dashrightarrow 0$; in the same way it is verified that $\mathfrak{F}^{-1}$ is weakly continuous. ▯

We present several simple examples of Fourier transforms on S^*:

$$\begin{aligned} &\mathfrak{F}[\varepsilon] = 1; && \mathfrak{F}[1] = \varepsilon; \\ &\mathfrak{F}[\varepsilon_{x_0}] = e^{-2\pi i x_0 y}; && \mathfrak{F}[e^{2\pi i x_0 y}] = \varepsilon_{x_0} \\ &\mathfrak{F}[\partial^k \varepsilon] = (2\pi i)^{|k|} y^k; && \mathfrak{F}[\Delta\varepsilon] = -4\pi^2 |y|^2 \text{ [24]}. \end{aligned} \tag{0.3.6}$$

We also mention the formulas

$$\mathfrak{F}[\partial^k T] = (2\pi i)^{|k|} y^k \, \mathfrak{F}[T] \tag{0.3.7$_1$}$$

$$\mathfrak{F}[T_{x_0}] = e^{-2\pi i x_0 y} \, \mathfrak{F}[T] \tag{0.3.7$_2$}$$

$$\mathfrak{F}[\check{T}] = \mathfrak{F}^{-1}[T] \tag{0.3.7$_3$}$$

where $T \in S^*$, and T_{x_0}, $\check{T}$ are defined by the equations

$$T_{x_0} f(x) = Tf(x + x_0), \quad \check{T} f(x) = Tf(-x).$$

We should make a few remarks about these formulas. First, $T \in S^*$ implies that $T_{x_0} \in S^*$, and for any k $\partial^k T \in S^*$; this easily follows from the invariance of the class S under differentiation and translation. Next, we must discover the meaning of the right side of formula (0.3.7$_1$). If $T \in S^*$, then $x^k T$ denotes the element of S^* defined by the formula

$$x^k T(f) = T(x^k f), \quad f \in S.$$

Moreover, it is easy to see that $x^k f \in S$. $e^{-2\pi i x_0 y}\, T$ has the analogous meaning. After this explanation the reader can easily verify formulas (0.3.7$_1$)–(0.3.7$_3$).

13. If $f_1(x), f_2(x) \in S$, then it is elementary to check the formula

$$\mathfrak{F}[f_1 * f_2] = \mathfrak{F}[f_1]\, \mathfrak{F}[f_2];$$

both sides are defined and represent the same function of S.

---

[24] Δ is the Laplace operator, i.e. $\sum_i \partial_i^2$.

The attempt to prove the analogous relation

$$\mathfrak{F}[T_1 * T_2] = \mathfrak{F}[T_1]\,\mathfrak{F}[T_2] \tag{0.3.7}$$

for distributions of S^* naturally runs into difficulties. They arise because the operation of convolution does not always have a meaning in S^* and the operation of multiplication is in general not defined. Therefore, we must require that at least one of the factors T_1, T_2 belongs to a more restricted class $E^* \subset S^*$.

In this connection it is convenient to introduce the following definition.

A function $g(x)$ is called a *multiplier* of type (E_1, E_2) if the multiplication operation $f(x) \to \overline{g(x)}\, f(x)$ defines a continuous transformation of E_1 into E_2.

Any infinitely differentiable function all of whose derivatives grow no faster than a power as $|x| \to \infty$ is a multiplier of type (S, S).

If g is a multiplier of type (E_1, E_2), then for any distribution $T \in E_2^*$ we may define $gT \in E_1^*$ by the formula:

$$gT(f) = T(\bar{g}f), \quad f \in E_1. \tag{0.3.8}$$

In the future we shall assume that $S \subset E \subset S^*$.

Let us agree to denote by $\tilde{E}^*$ the linear space of Fourier transforms of distribution from E^*. Since, by Theorem 0.11′

$$\tilde{S}^* \equiv S^*,$$

then $\tilde{E}^* \subset S^*$, and convergence in $\tilde{E}^*$, induced by S^*, coincides with convergence induced by E^*. We shall also assume that $\tilde{E}$, just as E, is a function space and that the classical Fourier transform realizes an isomorphism $\tilde{E} \leftrightarrow E$.

We will show that

$$\tilde{E}^* \equiv (\tilde{E})^*. \tag{0.3.9}$$

Indeed, suppose that $\mathfrak{F}[T] \in \tilde{E}^*$, i.e. $T \in E^*$. Then by (0.3.5), for any $f \in S$

$$\mathfrak{F}[T]\,\tilde{f} = Tf.$$

But Tf is a continuous linear functional on E, and therefore the left side is also a continuous linear functional on $\tilde{E}$. Consequently, $\mathfrak{F}[T] \in (\tilde{E})^*$ i.e.

$$\tilde{E}^* \subset (\tilde{E})^*.$$

Conversely, suppose $U \in (\tilde{E})^*$. First of all, since $\tilde{E} \supset \tilde{S} = S$ and the topology in $\tilde{E}$ is weaker than in S, it follows that $(\tilde{E})^* \subset S^*$. Setting

$T = \mathfrak{F}^{-1}[U]$, we may write for any $f \in S$ the equation

$$\mathfrak{F}^{-1}[U]\, f = U\tilde{f}.$$

Since $U\tilde{f}$ is a continuous linear functional on $\tilde{E}$, then $\mathfrak{F}^{-1}[U]f$ is also a continuous linear functional on E, i.e. $\mathfrak{F}^{-1}[U] \in E^*$ and $U \in \tilde{E}^*$. Consequently,

$$(\tilde{E})^* \subset \tilde{E}^*,$$

and (0.3.9) is proved.

Theorem 0.12. *Suppose $T_1 \in E_1^*$, $T_2 \in E_2^*$. Let us assume that*

(i) E_2 contains together with each function $f(x)$ any of its translations $f(x + y)$, $y \in R^p$;

(ii) $\mathfrak{F}[T_2]$ is a multiplier of type $(\tilde{E}_2, \tilde{E}_1)$.[25] *Then the convolution $T_1 * T_2$ defined in the sense of (0.2.13′) belongs to E_1^* and*

$$\mathfrak{F}[T_1 * T_2] = \mathfrak{F}[T_1]\, \mathfrak{F}[T_2]. \tag{0.3.7}$$

Proof. We will show that for any $f \in E_2$, $T_2^{(y)} f(x + y) \in E_1$. Using definition (0.3.5) we may write

$$T_2^{(y)} f(x + y) = \mathfrak{F}[T_2]\, \mathfrak{F}^{(y)}[f(x + y)]. \tag{0.3.10}$$

Formula (0.3.7$_2$) shows that

$$\mathfrak{F}^{(y)}[f(x + y)] = e^{2\pi i x z}\tilde{f}(z).$$

If we set $\mathfrak{F}[T_2] = g(z)$, then (0.3.10) may be rewritten in the form

$$T_2^{(y)} f(x + y) = \int_{R^p} e^{2\pi i x z}\, \tilde{f}(z)\, \overline{g(z)}\, dz = \mathfrak{F}^{-1}[\tilde{f}\bar{g}].$$

But by (ii) $\tilde{f}(z)\, \overline{g(z)} \in \tilde{E}_1$, and therefore $T_2^{(y)} f(x + y) \in E_1$. If $f_n(x) \to 0$ in E_1, then by the definition of convergence in $\tilde{E}_1$, $\tilde{f}_n \to 0$ in $\tilde{E}_1$; again by (ii) $]\tilde{f}_n g \to 0$ in $\tilde{E}_2$ and therefore $T_2^{(y)} f_n(x + y) \to 0$ in E_1. Thus, condition (α'), No. 9 is fulfilled and $T_1 * T_2 \in E_1^*$.

Finally, using (0.3.5) once more, we have

$$\begin{aligned} \mathfrak{F}[T_1 * T_2]\, \tilde{f} &= (T_1 * T_2)\, f = T_1(\mathfrak{F}^{-1}[\tilde{f}\bar{g}]) = \mathfrak{F}[T_1]\, (g\tilde{f}) \\ &= g\mathfrak{F}[T_1]\, \tilde{f} = \mathfrak{F}[T_2]\, \mathfrak{F}[T_1]\, \tilde{f} \end{aligned}$$

for any $\tilde{f} \in \tilde{E}_1$; this proves (0.3.7). □

---

[25] This means that $\mathfrak{F}[T_2]$ is an absolutely continuous signed measure and its density is a multiplier of type $(\tilde{E}_2, \tilde{E}_1)$

We remark that the verification of condition (ii) of Theorem 0.12 can rarely be carried out effectively in the first place because it is difficult to formulate necessary and sufficient conditions for membership in the class $\tilde{E}$. We point out, however, a case for which this may be done.

Corollary. *Suppose $T_1 \in S^*$ and T_2 has compact support. Then* (0.3.7) *holds.*

For E_1 and E_2 we take the space S; since by Theorem 0.11 $\tilde{S} \equiv S$, it is sufficient to prove that $\mathfrak{F}[T_2]$ is a multiplier of type (S, S). If T_2 is a signed measure ν, then

$$\mathfrak{F}[T_2] = \int_{S(\nu)} e^{-2\pi i x y}\, d\nu(x)$$

is an infinitely differentiable function, bounded along with all its derivatives.[26] In general, according to Theorem 0.9 and formula $(0.3.7_1)$,

$$\mathfrak{F}[T_2] = \sum_{|k|=0}^{m} y^k g_k(y),$$

where $g_k(y)$ are functions of the specified type. Consequently, $\mathfrak{F}[T_2]$ is an infinitely differentiable function all of whose derivatives grow no faster than some power. As pointed out earlier, such a function is a multiplier of type (S, S).

Theorem 0.12 is far from exhausting all those cases in which equation (0.3.7) can be used. We give one particulier result not covered by that theorem.

Theorem 0.13. *Suppose μ_1, μ_2 are two measures which, together with their convolution $\mu_1 * \mu_2$, belong to S^*. Suppose that*

$$\psi_1 = \mathfrak{F}\mu_1, \quad \psi_2 = \mathfrak{F}\mu_2$$

are absolutely continuous signed measures, that $\psi_1(x), \psi_2(x) \in L^2(K)$, where K is any compact set in R^p, and that $\psi_1(x) \geq 0$. Then

$$\mathfrak{F}[\mu_1 * \mu_2] = \psi_1(x)\, \psi_2(x). \tag{0.3.11}$$

Proof. Since $\psi_1 \in S^*$, then by Theorem 0.10 there exists an even number $2n$ such that

$$\int_{R^p} \frac{\psi_1(x)}{1 + |x|^{2n}}\, dx < \infty.$$

Let us form the function

$$\psi_1(x; \lambda) = \psi_1(x) * \left(\frac{1}{\pi^2}\right)^{pn} \prod_{i=1}^{p} \left(\frac{\sin^2 \pi \lambda x_i}{\lambda x_i^2}\right)^n.$$

---

[26] In fact this is an entire function of exponential type.

It is the Fourier transform of the measure

$$\mu_{1\lambda} = \prod_{i=1}^{p} \varphi^{*n}\left(\frac{x_i}{\lambda}\right) \mu_1$$

having compact support, where

$$\varphi(z) = \begin{cases} 1 - |z|, & \text{if } |z| \leq 1 \\ 0, & \text{if } |z| > 1, \end{cases}$$

and φ^{*n} denotes an n-fold convolution.

By the corollary to Theorem 0.12

$$\mathfrak{F}[\mu_{1\lambda} * \mu_2] = \psi_1(x, \lambda)\, \psi_2(x).$$

Furthermore, as $\lambda \to \infty$, $\mu_{1\lambda} \dot{\to} \mu_1$, monotone inceasing; by Lemma 0.3

$$\mu_{1\lambda} * \mu_2 \dot{\to} \mu_1 * \mu_2.$$

By Theorem 0.11′

$$\mathfrak{F}[\mu_{1\lambda} * \mu_2] \dot{\to} \mathfrak{F}[\mu_1 * \mu_2];$$

it remains for us to verify that

$$\psi_1(x; \lambda)\, \psi_2(x) \dot{\to} \psi_1(x)\, \psi_2(x),$$

i.e. that for any function $f(x) \in D$

$$\lim_{\lambda \to \infty} \int_{S(f)} \psi_1(x, \lambda)\, \psi_2(x)\, f(x)\, dx = \int_{S(f)} \psi_1(x)\, \psi_2(x)\, f(x)\, dx.$$

To be brief, we set

$$S(f) = K, \quad \psi_2(x)\, f(x) = \psi_3(x) \in L^2(K)$$

$$\left(\frac{1}{\pi^2}\right)^{pn} \prod_{i=1}^{p} \left(\frac{\sin^2 \pi\lambda x_i}{\lambda x_i^2}\right)^n = F_\lambda(x).$$

Then

$$\int_{S(f)} \psi_1(x, \lambda)\, \psi_2(x)\, f(x)\, dx = \int_K \psi_3(x)\, dx \int_{R_p} \psi_1(x - y)\, F_\lambda(y)\, dy.$$

By the Fubini theorem

$$\int_K \psi_3(x)\, dx \int_{R^p} \psi_1(x - y)\, F_\lambda(y)\, dy$$

$$= \int_{R^p} F_\lambda(y)\, dy \int_K \psi_1(x - y)\, \psi_3(x)\, dx.$$

But

$$\int_K \psi_1(x-y)\,\psi_3(x)\,dx$$

under the present assumptions is a continuous function of y and $F_\lambda(y) \dot{\to} \varepsilon$, as $\lambda \to \infty$.[27] Consequently,

$$\lim_{\lambda\to\infty} \int \psi_3(x)\,dx \int_{R^p} \psi_1(x-y)\,F_\lambda(y)\,dy = \int_K \psi_1(x)\,\psi_3(x)\,dx$$

which is the required result. ▯

---

[27] Moreover this convergence holds not only on continuous functions, with compact support, but also on continuous functions $f(x)$ for which $\dfrac{f(x)}{1+|x|^{2n}} \in L(R^p)$.

Chapter I

Potentials and Their Basic Properties

In this chapter, potentials will be defined either on a Euclidean space R^p ($p \geq 2$) or on some open region $G \subseteq R^p$.

We shall see that a potential is some integral operator, acting on a space of signed measures. Such an operator is defined by a kernel $K(x, y)$ which in our exposition will usually depend only on the distance $|x - y|$:

$$K(x, y) = k(|x - y|).$$

The nature of the theory will therefore be determined first of all by the choice of kernel. We will consider M. Riesz kernels which include as special and limiting cases the kernels of the classical theory (Newtonian and logarithmic kernels), and also Green kernels $G(x, y)$ associated with a region $G \subset R^p$.

§ 1. M. Riesz Kernels

1. Consider in R^p ($p \geq 2$) a function of the point x

$$k_\alpha(x) = A(p, \alpha)\, |x|^{\alpha - p},$$

where the positive constant $A(p, \alpha)$ is simply a normalization factor and will be specified later. This function, which we shall call an *M. Riesz kernel,* is positive, continuous for $x \neq 0$, and in general is lower semicontinuous. The latter (which must be checked for $\alpha < p$) follows from the fact that $k_\alpha(x)$ is the limit as $\eta \to 0$ of a monotone increasing sequence of continuous "truncated" kernels

$$k_\alpha^{(\eta)}(x) = \begin{cases} A(p, \alpha)\, |x|^{\alpha - p} & \text{for} \quad |x| > \eta, \\ A(p, \alpha)\, \eta^{\alpha - p} & \text{for} \quad |x| \leq \eta. \end{cases}$$

Let us first restrict ourselves to values of the parameter which satisfy the inequality $0 < \alpha < p/2$. Then $k_\alpha(x)$ will be a locally integrable function in R^p and will consequently define an absolutely continuous measure k_α.

We shall show that its Fourier transform has the form

$$\mathfrak{F}[k_\alpha(x)] = A(p, \alpha)\, \pi^{\frac{p}{2}-\alpha} \frac{\Gamma\left(\frac{\alpha}{2}\right)}{\Gamma\left(\frac{p-\alpha}{2}\right)} |y|^{-\alpha}. \tag{1.1.1}$$

First we compute the integral

$$g(y) = \int_{R^p} e^{2\pi i x y} |x|^{\alpha-p}\, dx.$$

Changing to spherical coordinates and using well-known formulas (cf., for example, S. Bochner [1])[1] we obtain

$$g(y) = \frac{2\pi}{|y|^{\frac{p}{2}-1}} \int_0^\infty \varrho^{\alpha-\frac{p}{2}} J_{\frac{p}{2}-1}(2\pi |y|\, \varrho)\, d\varrho$$

$$= \frac{(2\pi)^{\frac{p}{2}-\alpha}}{|y|^\alpha} \int_0^\infty u^{\alpha-\frac{p}{2}} J_{\frac{p}{2}-1}(u)\, du = \frac{\pi^{\frac{p}{2}-\alpha}\Gamma\left(\frac{\alpha}{2}\right)}{\Gamma\left(\frac{p-\alpha}{2}\right)} \frac{1}{|y|^\alpha}.$$

Let us convince ourselves that we have indeed found the Fourier transform in the sense of distribution theory. According to (0.3.5), we must establish that for any function $f(x) \in S$

$$\int_{R^p} g(y)\, \tilde{f}(y)\, dy = \int_{R^p} |x|^{\alpha-p} f(x)\, dx.$$

But

$$\int_{R^p} |x|^{\alpha-p} f(x)\, dx = \int_{R^p} |x|^{\alpha-p}\, dx \int_{R^p} \tilde{f}(y)\, e^{2\pi i x y}\, dy,$$

and after interchanging the order of integration we obtain the required result. This interchange is easily justified if we take into account the bounded convergence of the integral

$$\int_{|x|<N} e^{2\pi i x y} |x|^{\alpha-p}\, dx$$

to $g(y)$. Thus, formula (1.1.1) is established for $0 < \alpha < p/2$.

Now we fix the constant $A(p, \alpha)$, putting

$$A(p,\alpha) = \pi^{\alpha-\frac{p}{2}} \frac{\Gamma\left(\frac{p-\alpha}{2}\right)}{\Gamma\left(\frac{\alpha}{2}\right)}. \tag{1.1.2}$$

Then (1.1.1) gives

$$\mathfrak{F}[k_\alpha(x)] = |y|^{-\alpha}$$

[1] The result also holds for $0 < \alpha < \frac{p+1}{2}$.

Let us now suppose that $0 < \alpha < p/2$, $0 < \beta < p/2$. Since the convolution $k_\alpha * k_\beta$ of the measures k_α, k_β exists, then Theorem 0.13 leads us to the following important rule of composition for Riesz kernels:

$$k_\alpha(x) * k_\beta(x) = k_{\alpha+\beta}(x). \tag{1.1.3}$$

2. For future applications it is important to free formula (1.1.3) of unnecessary restrictions imposed on the parameters α, β. This may be achieved by using the method of analytic continuation on α and β. Then of course we will regard α, β as complex numbers, and the kernels $k_\alpha(x)$ become, in general, distributions k_α.

Consider an arbitrary function $\varphi \in D$ and form the expression

$$\psi(x,\alpha) = k_\alpha^{(y)} \varphi(x+y) = \pi^{\alpha - \frac{p}{2}} \frac{\Gamma\left(\frac{p-\alpha}{2}\right)}{\Gamma\left(\frac{\alpha}{2}\right)} \int_{R^p} \frac{\varphi(x+y)}{|y|^{p-\alpha}}\, dy.$$

For any $x \in R^p$ this function is an analytic function of α, regular in the half plane $\operatorname{Re}\alpha > 0$ with singular points $\alpha = p + 2n$ $(n = 0, 1, \ldots)$.

We will show that it can be continued analytically to the entire half-plane $\operatorname{Re}\alpha \leq 0$.

Denoting by T the unit ball $|y| < 1$, for $\operatorname{Re}\alpha > 0$, $\alpha \neq p + 2n$ we are able to write

$$\begin{aligned}\psi(x,\alpha) &= A(p,\alpha) \int_{R^p \setminus T} \frac{\varphi(x+y)}{|y|^{p-\alpha}}\, dy \\ &\quad + A(p,\alpha) \int_T \frac{\varphi(x+y) - \varphi(x)}{|y|^{p-\alpha}}\, dy + A(p,\alpha)\, \varphi(x) \int_T \frac{dy}{|y|^{p-\alpha}} \\ &= A(p,\alpha) \int_{R^p \setminus T} \frac{\varphi(x+y)}{|y|^{p-\alpha}}\, dy \\ &\quad + A(p,\alpha) \int_T \frac{\varphi(x+y) - \varphi(x)}{|y|^{p-\alpha}}\, dy + \frac{A(p,\alpha)}{\alpha}\, \varphi(x)\, \omega_p,\end{aligned} \tag{1.1.4}$$

where

$$\omega_p = \frac{2\pi^{p/2}}{\Gamma(p/2)}$$

is the "surface area" of the unit hypersphere in R^p. It is easy to see that the first and third terms in (1.1.4) are regular in the half-plane $\operatorname{Re}\alpha \leq 0$. To investigate the second term we introduce the average value

$$\tilde{\varphi}(x,r) = \frac{1}{\omega_p} \int_{|y|=r} \varphi(x+y)\, d\omega_y.$$

of the function φ on a sphere of radius r with center at x.

Then it is easy to see that

$$\int_T \frac{\varphi(x+y)-\varphi(x)}{|y|^{p-\alpha}}\,dy = \omega_p \int_0^1 \{\tilde{\varphi}(x,r)-\varphi(x)\}\, r^{\alpha-1}\,dr. \qquad (1.1.5)$$

Let us now apply the well-known Pizzetti formula (cf., for example, Courant [1]):

$$\tilde{\varphi}(x,r) = \varphi(x) + \frac{\Delta\varphi(x)}{2\cdot 1!p} r^2 + \cdots$$
$$\cdots + \frac{\Delta^m \varphi(x)}{2^m m! p(p+2)\cdots(p+2m-2)} r^{2m} + O(r^{2m+2})\ (r\to 0).$$

From this it follows that $\tilde{\varphi}(x,r)-\varphi(x) = O(r^2)$ and therefore, the integral (1.1.5) is regular in the strip $-2 < \operatorname{Re}\alpha \le 0$. Thus, the function $\psi(x,\alpha) = k_\alpha^{(y)}\varphi(x+y)$ is analytically continued to the strip $-2 < \operatorname{Re}\alpha \le 0$.

Note that within this strip

$$\frac{A(p,\alpha)}{\alpha}\varphi(x)\,\omega_p = -A(p,\alpha)\,\varphi(x)\int_{R^p\setminus T}\frac{dy}{|y|^{p-\alpha}},$$

and therefore formula (1.1.4) becomes

$$\begin{aligned}\psi(x,\alpha) &= A(p,\alpha)\int_{R^p}\frac{\varphi(x+y)-\varphi(x)}{|y|^{p-\alpha}}\,dy \\ &= A(p,\alpha)\,\psi_p\int_0^\infty \{\tilde{\varphi}(x,r)-\varphi(x)\}\,r^{\alpha-1}\,dr.\end{aligned} \qquad (1.1.6)$$

Moreover, if we take into account that

$$\lim_{\alpha\to 0}\frac{A(p,\alpha)}{\alpha} = \frac{1}{\omega_p}$$

then formula (1.1.4) gives

$$\psi(x,0) = \varphi(x)$$

Consequently,

$$k_0 = \lim_{\alpha\to 0} k_\alpha = \varepsilon. \qquad (1.1.7)$$

In order to continue $\psi(x,\alpha)$ into the next strip $-4 < \operatorname{Re}\alpha \le -2$, we will represent (1.1.6) in the form

$$\begin{aligned}\psi(x,\alpha) &= A(p,\alpha)\,\omega_p\int_0^1\left\{\tilde{\varphi}(x,r)-\varphi(x)-\frac{\Delta\varphi(x)}{2p}r^2\right\}r^{\alpha-1}\,dr \\ &+\frac{A(p,\alpha)}{2p(\alpha+2)}\,\omega_p\,\Delta\varphi(x) + A(p,\alpha)\,\omega_p\int_1^\infty\{\tilde{\varphi}(x,r)-\varphi(x)\}\,r^{\alpha-1}\,dr.\end{aligned} \qquad (1.1.8)$$

By the Pizzetti formula

$$\tilde{\varphi}(x, r) - \varphi(x) - \frac{\Delta\varphi(x)}{2p} r^2 = O(r^4),$$

and therefore all the terms on the right-hand side of (1.1.8) are regular in the strip $-4 < \operatorname{Re}\alpha \leq -2$. It is not difficult to verify also that within this strip

$$\psi(x, \alpha) = A(p, \alpha)\, \omega_p \int_0^\infty \left\{\tilde{\varphi}(x, r) - \varphi(x) - \frac{\Delta\varphi(x)}{2p} r^2\right\} r^{\alpha-1}\, dr$$

and that

$$\psi(x, -2) = -\frac{1}{4\pi^2} \Delta\varphi(x),$$

that is

$$k_{-2} = -\frac{1}{4\pi^2} \Delta\varepsilon. \tag{1.1.9}$$

Continuing this reasoning, we see that $\psi(x, \alpha)$ can be analytically continued into any strip

$$-2m - 2 < \operatorname{Re}\alpha \leq -2m$$

and admits there a representation

$$\begin{aligned}\psi(x, \alpha) &= A(p, \alpha)\, \omega_p \int_0^\infty \left\{\tilde{\varphi}(r, x) - \varphi(x) - \cdots \right.\\ &\qquad \left. \cdots - \frac{\Delta^m \varphi(x) r^{2m}}{2^m m!\, p(p+2) \cdots (p+2m-2)}\right\} r^{\alpha-1}\, dr \\ &= A(p, \alpha) \lim_{\eta\to 0} \left\{ \int\limits_{|y|>\eta} \frac{\varphi(x+y)}{|y|^{p-\alpha}}\, dy + \sum_{k=0}^{m} H_k \Delta^k \varphi(x) \frac{\eta^{\alpha+2k}}{\alpha+2k} \right\} \\ &= A(p, \alpha) \int\limits_{R^p} \left\{\varphi(x+y) - \sum_{k=0}^{m} H_k\, \Delta^k \varphi(x)\, |y|^{2k}\right\} \frac{dy}{|y|^{p-\alpha}}\end{aligned} \tag{1.1.10}$$

where

$$H_k = \frac{\omega_p}{2^k k!\, p(p+2) \cdots (p+2k-2)}.$$

Moreover,

$$\psi(x, -2m) = \left(-\frac{1}{4\pi^2}\right)^m \Delta^m \varphi(x),$$

that is

$$k_{-2m} = \left(-\frac{1}{4\pi^2}\right)^m \Delta^m \varepsilon.$$

Note also that the analytic continuation $\psi(x, \alpha)$ is possible even when $\varphi \notin D$. Namely, from the preceding analysis, it is clear that if $\varphi(x)$ has continuous derivatives of order $2m$ and if $\varphi(x) = O(|x|^{-N})$ as $|x| \to \infty$, then $\psi(x, \alpha)$ will be regular in the strip

$$-2m < \operatorname{Re} \alpha < N$$

with the exception of the points $\alpha = p + 2n$ $(n = 0, 1, 2, \ldots)$.

Consider now for $\varphi \in D$ the expression

$$(k_\alpha * k_\beta)\,(\varphi) = k_\beta^{(x)}\big(k_\alpha^{(y)}\varphi(x + y)\big) = k_\beta \psi(x, \alpha)\,.$$

Since for any $\alpha \neq p + 2n$ $\psi(x, \alpha)$ is infinitely differentiable in x, and since for $|x| \to \infty$ it follows from (1.1.10) that

$$\psi(x, \alpha) = O\left(\frac{1}{|x|^{p - \operatorname{Re}\alpha}}\right),$$

then $k_\beta \psi(x, \alpha)$ will be an analytic function of β (and from symmetry, also of α) under the conditions

$$\operatorname{Re}\,(\alpha + \beta) < p; \quad \alpha, \beta \neq p + 2n \quad (n = 0, 1, \ldots,). \tag{1.1.11}$$

This shows that the relation

$$k_\alpha * k_\beta = k_{\alpha+\beta} \tag{1.1.12}$$

established earlier only for $0 < \alpha < p/2$, $0 < \beta < p/2$, remains true under the conditions (1.1.11).

In particular, for all $\alpha \neq \pm(p + 2n)$ $(n = 0, 1, \ldots)$

$$k_\alpha * k_{-\alpha} = \varepsilon\,. \tag{1.1.12'}$$

Let us consider in greater detail the meaning and possible applications of equation (1.1.12) and its corollary (1.1.12′).

For any α, denote by $2m_\alpha \geq 0$ the smallest even number satisfying the inequality

$$2m > |\operatorname{Re} \alpha|,$$

if $\operatorname{Re} \alpha \leq 0$, and zero, if $\operatorname{Re} \alpha > 0$.

Furthermore, let E_α denote the linear space of continuous functions $f(x)$ which, in a neighborhood of the point 0, have continuous derivatives of order $2m_\alpha$, integrable with weight $(1 + |x|^{p - \operatorname{Re}\alpha})^{-1}$. The convergence $f_n(x) \to 0$ in E_α signifies that

$$\int_{R^p} \frac{|f_n(x)|}{1 + |x|^{p - \operatorname{Re}\alpha}}\, dx \to 0$$

and in some neighborhood of zero

$$\partial^k f_n(x) \to 0; \quad |k| \leq 2m_\alpha$$

uniformly. Then the distribution k_α will, as is easily verified, belong to the class E^*_α.

Let $E_{\alpha\beta}$ denote the linear space of functions $f(x, y)$ in $R^p \times R^p$ which have $2(m_\alpha + m_\beta)$ continuous derivatives and together with the above derivatives integrable with weight $(1 + |x|^{p-\mathrm{Re}\alpha})^{-1}$ $(1 + |y|^{p-\mathrm{Re}\beta})^{-1}$. The convergence $f_n(x, y) \to 0$ in $E_{\alpha\beta}$ signifies that

$$\int\limits_{R^p \times R^p} \frac{|\partial^k f_n(x, y)|}{(1 + |x|^{p-\mathrm{Re}\alpha})\,(1 + |y|^{p-\mathrm{Re}\beta})}\, dx\, dy \to 0,$$

$$|k| \leq 2(m_\alpha + m_\beta),$$

and in some neighborhood of zero $\partial^k f_n(x, y) \to 0$ uniformly, $|k| \leq 2(m_\alpha + m_\beta)$. Using formula (1.1.10), it is not difficult to verify that the conditions (a), (b), (c), listed at the end of No. 8 of the Introduction, are satisfied.

Finally, if condition (1.1.11) is satisfied and $f(x) \in D^{2(m_\alpha + m_\beta)}$, then

$$f(x) \to f(x + y)$$

is a continuous mapping of $D^{2(m_\alpha + m_\beta)}$ into $E_{\alpha\beta}$. Let us check that $f(x + y) \in E_{\alpha\beta}$. Let K be the support of $f(x)$ and set $M = \max |f(x)|$; then we may write

$$\int\limits_{R^p} \frac{|f(x + y)|}{1 + |y|^{p-\mathrm{Re}\beta}}\, dy \leq M \int\limits_{K+x} \frac{dy}{1 + |y|^{p-\mathrm{Re}\beta}} = O\left(\frac{1}{1 + |x|^{p-\mathrm{Re}\beta}}\right)$$

whence

$$\int\limits_{R^p \times R^p} \frac{|f(x + y)|\, dx\, dy}{(1 + |x|^{p-\mathrm{Re}\alpha})\,(1 + |y|^{p-\mathrm{Re}\beta})} \leq M_1 \int\limits_{R^p} \frac{dx}{1 + |x|^{2p-\mathrm{Re}\alpha-\mathrm{Re}\beta}}$$

and for $\mathrm{Re}\,(\alpha + \beta) < p$ the integral converges. Similarly one can verify the convergence of the integrals of the derivatives $\partial^k f(x + y)$, and the continuity of the map $f(x) \to f(x + y)$ is easily verified.

Consequently, $k_\alpha * k_\beta$ belongs to the class $[D^{2(m_\alpha + m_\beta)}]^*$ and for any function $f(x) \in D^{2(m_\alpha + m_\beta)}$, (0.2.13) holds.

In particular, using (1.1.12′), we may assert (since $\mathrm{Re}\,\alpha > 0$), that for any function $\varphi(x) \in D^{2m_{-\alpha}}$ one may form the function

$$\nu(x) = k_{-\alpha} * \varphi(x) = k^{(y)}_{-\alpha}\, \varphi(x - y)$$

and that then the convolution $k_\alpha * \nu$ is defined and moreover

$$k_\alpha * \nu(x) = \varphi(x) \quad (\mathrm{Re}\,\alpha > 0). \tag{1.1.13}$$

It is also useful to note that for $|x| \to \infty$,

$$\nu(x) = O(|x|^{-p-\mathrm{Re}\,\alpha})$$

3. Returning to real values of the parameter α, let us examine the behavior of the M. Riesz kernel $k_\alpha(x)$ as $\alpha \to p$.[2] Since the factor $A(p, \alpha)$ has a pole at $\alpha = p$, then $k_\alpha(x)$ as $\alpha \to p$ does not have a limit even in the vague sense.

However, on the subspace Φ, orthogonal to the unity, vague convergence of $k_\alpha(x)$ will hold, and (in this sense) the vague limit of $k_\alpha(x)$ as $\alpha \to p$ is the *logarithmic kernel*

$$k_p(x) = \omega_p \ln \frac{1}{|x|}. \tag{1.1.14}$$

This means the following. If a function $f(x) \in \Phi$ satisfies the relation

$$\int_{R^p} f(x)\, dx = 0,$$

then

$$\lim_{\alpha \to p} \int_{R^p} k_\alpha(x) f(x)\, dx = \omega_p \int_{R^p} \ln \frac{1}{|x|} f(x)\, dx. \tag{1.1.15}$$

Indeed, setting $p - \alpha = \delta$, we will have

$$\int_{R^p} k_\alpha(x) f(x)\, dx = A(p, p-\delta) \int \frac{f(x)}{|x|^\delta}\, dx$$

$$= \pi^{\frac{p}{2}-\delta} \frac{\Gamma\left(\frac{\delta}{2}\right)}{\Gamma\left(\frac{p-\delta}{2}\right)} \int_{R^p} \{|x|^{-\delta} - 1\} f(x)\, dx$$

$$= \pi^{\frac{p}{2}-\delta} \frac{\Gamma\left(\frac{\delta}{2}\right)}{\Gamma\left(\frac{p-\delta}{2}\right)} \int_{R^p} \left\{\delta \ln \frac{1}{|x|} + O\left(\delta^2 \ln^2 \frac{1}{|x|}\right)\right\} f(x)\, dx,$$

and since $f(x)$ has a compact support, then this yields (1.1.15).

Since a shift in the argument of the function leaves it in the subspace orthogonal to the unit, (1.1.15) may be rewritten in the form

$$\lim_{\alpha \to p} (k_\alpha * f) = k_p * f.$$

It is also clear that in the preceding argument the function $f(x)$ with compact support may be replaced by any signed measure ν with compact

---

[2] A similar investigation can be made in case $\alpha \to p + 2n$.

support, satisfying the condition $\nu(1) = 0$, and hence

$$\lim_{\alpha \to p} (k_\alpha * \nu) = k_p * \nu, \quad \nu(1) = 0^3. \tag{1.1.16}$$

From this and the formula $k_\alpha = k_{\alpha/2} * k_{\alpha/2}$ which holds for $\alpha < p$, an important consequence is obtained.

If ν is a signed measure with compact support, satisfying the condition $\nu(1) = 0$, then

$$k_p * \nu = (k_{p/2} * k_{p/2}) * \nu .$$

In other words, the formula

$$k_p = k_{p/2} * k_{p/2} \tag{1.1.17}$$

holds if we apply it only to signed measures ν of the above type.

Note that, unlike M. Riesz kernels, the logarithmic kernel $k_p(x)$ is first, not positive, and second, has a singularity at infinity. These circumstances give certain definite characteristics of the theory of the logarithmic potential which we will repeatedly discover in what follows.

The logarithmic kernel plays an especially important role in the case of a plane ($p = 2$). This is connected with the fact that for $p = 2$, it is the *fundamental solution* for the Laplace operator:

$$-\Delta k_2(x) = -\Delta * k_2 = 4\pi^2 \varepsilon .$$

The *Newtonian kernel* plays a similar role for $p > 2$

$$k_2(x) = A(p,2)\, |x|^{2-p} = \frac{4\pi^2}{(p-2)\,\omega_p}\, |x|^{2-p}, \tag{1.1.18}$$

and also satisfies the equation (cf. (1.1.9) and (1.1.12))

$$-\Delta * k_2 = 4\pi^2 \varepsilon .$$

We will call these kernels classical.

In the theory of M. Riesz kernels, the role of the Laplace operator, which has a local character, is taken (in view of (1.1.12′)) by a non-local integral operator $k_{-\alpha}$. This circumstance often substantially complicates the theory (cf., for example, § 6).

---

[3] It should be noted that (1.1.16) holds almost everywhere, namely at those points x for which

$$l(x) = \int_{R^p} \ln^2 \frac{1}{|x-y|}\, d\,|\nu|\,(y) < \infty .$$

Since $|\nu|$ has compact support, $l(x)$ is locally integrable and hence is bounded almost everywhere. Moreover (1.1.16) may be understood in the sense of vague convergence of $k_\alpha * \nu$ to $k_p * \nu$.

Of interest to mathematical physics is the replacement of the classical kernels by fundamental solutions (or Green functions) for other elliptic differential operators. We shall mention this possibility in Ch. VI, § 2.

§ 2. Superharmonic Functions

4. A function $f(x)$ is called *superharmonic in* R^p if

(a) $-\infty < f(x) \leq +\infty$ and $f(x) \not\equiv +\infty$,

(b) $f(x)$ is lower semicontinuous,

(c) for any $x \in R^p$ and $r > 0$

$$f(x) \geq \frac{1}{\omega_p r^{p-1}} \int_{S_r} f(y)\, dS, \tag{1.2.1}$$

where S_r is the sphere of radius r, center at x, and $\omega_p r^{p-1}$ is the surface area of S_r.

The inequality (1.2.1) may be rewritten in the form

$$\int_{R^p} f(y)\, d\sigma_x^{(r)}(y) \geq 0 \tag{1.2.2}$$

or

$$f * \sigma^{(r)} \geq 0,\text{[4]}$$

where $\sigma_x^{(r)} = \varepsilon_x - \varepsilon_x^{(r)}$, $\sigma^{(r)} \equiv \sigma_0^{(r)}$ (for definitions cf. Introduction, § 1, No. 5).

Integrating (1.2.1) along r we easily obtain the inequality

$$\int_{R^p} f(y)\, d\tau_x^{(r)}(y) \geq 0 \tag{1.2.3}$$

or

$$f * \tau^{(r)} \geq 0$$

where $\tau_x^{(r)} = \delta_x - m_x^{(r)}$, and $m_x^{(r)}$ is a unit mass uniformly distributet throughout the interior of S_r. The inequality (1.2.3) or its equivalent

$$f(x) \geq \frac{p}{\omega_p r^p} \int_{T_r} f(y)\, dy \tag{1.2.3'}$$

where T_r is the ball bounded by S_r may also serve as the defining characteristic of superharmonic functions. From this, incidentally, one easily obtains the fact that a superharmonic function is locally integrable.

---

[4] In this section the convolution of the function $f(x)$ with some measure should be viewed not as a measure, but as a function of x, defined everywhere in R^p. Therefore, we may not use assertions of the type of Lemma 0.4.

Let us now show that if $f(x)$ is superharmonic, then at each point $x \in R^p$

$$f(x) = \lim_{r \to 0} f * \varepsilon^{(r)} = \lim_{r \to 0} f * m^{(r)}. \tag{1.2.4}$$

Indeed, from (1.2.2) and (1.2.3) it follows that

$$\overline{\lim_{r \to 0}} f * \varepsilon^{(r)} \leq f(x), \quad \overline{\lim_{r \to 0}} f * m^{(r)} \leq f(x).$$

On the other hand, the lower semicontinuity of $f(x)$ yields

$$f(x) \leq \underline{\lim_{r \to 0}} f * \varepsilon^{(r)}, \quad f(x) \leq \underline{\lim_{r \to 0}} f * m^{(r)},$$

which implies (1.2.4).

Suppose that $\overline{G} \subset R^p$ is some closed bounded region and that $f(x)$ is a superharmonic function. Since it is lower semicontinuous, the function $f(x)$ attains its minimum in $\overline{G}$, say α. If the equation $f(x_0) = \alpha$ holds at an interior point of G, then $f(x) \equiv \alpha$ in $\overline{G}$.

Indeed, the set

$$E = \{x : f(x) = \alpha,\ x \in G\}$$

on the one hand is closed relative to G (since f is semicontinuous), and on the other hand is open relative to G. The latter is easily established with the help of (1.2.3) and (1.2.4). Since G is connected, $E = G$. It follows from this that $f(x) = \alpha$ on ∂G, and hence that $f(x) \equiv \alpha$ in $\overline{G}$.

In a completely analogous way we prove the following

Principle of the Harmonic Minorant. *Suppose that $f(x)$ is superharmonic, and that $h(x)$ is harmonic in the bounded region G, continuous in $\overline{G}$. If the inequality*

$$f(x) \geq h(x)$$

holds on ∂G, then it holds in G; equality at some point in G is possible only if $f(x) \equiv h(x)$ in $\overline{G}$.

If in the region G, the Dirichlet problem is solvable for any continuous boundary functions (such regions are called regular, cf. Ch. IV), then it is not difficult to construct in $\overline{G}$ the *best harmonic minorant* $H(x)$ for $f(x)$ which will majorize any function $h(x)$ with the properties mentioned in the formulation of the principle of the harmonic minorant. For this it is sufficient to construct a nondecreasing sequence of functions $\{\varphi_k(x)\}$ continuous on ∂G which converge on ∂G to $f(x)$[5]; then ex-

---

[5] Since $f(x)$ is lower semicontinuous, such a sequence exists.

amine the corresponding solutions to the Dirichlet problem $h_k(x)$ and set

$$H(x) = \lim_{k\to\infty} h_k(x), \quad x \in \overline{G}.$$

In particular, if for G we take the ball

$$|x - x_0| < R,$$

then the indicated construction will bring us to the Poisson integral

$$H(x) = \frac{1}{\omega_p R} \int_{|y-x_0|=R} f(y) \frac{R^2 - r^2}{\varrho^p} dS_y,$$

where $r = |x - x_0|$, $\varrho = |y - x|$.

Now it is easy to prove that for fixed x, the mean

$$f^{(r)}(x) = f * \varepsilon^{(r)}$$

is a *nonincreasing* function of r. In fact, if $H(y)$ is the best harmonic minorant of $f(x)$ in the sphere $|y - x| < r$, then for $r_1 < r$ we have

$$f^{(r_1)}(x) = f * \varepsilon^{(r_1)} \geq H * \varepsilon^{(r_1)} = H * \varepsilon^{(r)} = f * \varepsilon^{(r)} = f^{(r)}(x).$$

In exactly the same way, the mean $f * m^{(r)}$ is a nonincreasing function of r.

From the finiteness of the quantity

$$f * m^{(R)} = \frac{1}{R^p} \int_0^R r^{p-1} f^{(r)}(x)\, dr$$

and the monotonicity of $f^{(r)}(x)$ follows the finiteness of $f^{(r)}(x)$ for any $r > 0$.

5. The above definition of superharmonicity does not have a local character since in (1.2.1) the radius $r > 0$ is arbitrary. Let us replace condition (c) by the formally weaker condition:

(c_1) for any $x \in R^p$ and $0 < r < r_0(x)$

$$f(x) \geq \frac{1}{\omega_p r^{p-1}} \int_{S_r} f(y)\, dS.$$

It turns out that we still obtain the same class of functions in R^p. In fact, under condition (c_1) it is not difficult to see that the harmonic minorant principle is preserved, from which (c) follows easily.

The local character of condition (c_1) allows us to define in a natural way the class of *superharmonic functions in the region* $G \subset R^p$. This term refers to functions defined in G and satisfying conditions (a), (b), (c_1). Moreover $r_0(x)$ in (c_1) may be any number less than the distance from x to ∂G.

It is easy to see that the results of No. 4 are preserved for superharmonic functions in a region. We shall present in a more general form the harmonic minorant principle.

Theorem 1.1. *Suppose that $f(x)$ is superharmonic in a bounded region G, and that $h(x)$ is harmonic in G and continuous in $\overline{G}$, if at each point s of the boundary ∂G, the condition*

$$\varliminf_{x \in G; x \to s} f(x) \geq h(s),$$

is fulfilled, then $f(x) \geq h(x)$ throughout G, where equality at any point of G implies the identity $f(x) \equiv h(x)$.

Proof. The difference

$$f_1(x) = f(x) - h(x)$$

is obviously superharmonic in G and

$$\varliminf_{x \in G, x \to s} f_1(x) \geq 0, \quad s \in \partial G.$$

Set $m = \inf\limits_G f_1(x)$. It is easily seen that m is finite. If $f_1(x) \equiv m$ in G, then clearly, $m \geq 0$, i.e.

$$f_1(x) = f(x) - h(x) \geq 0, \quad x \in G$$

and the theorem is proved. Otherwise at every point in G $f_1(x) > m$, and therefore there is a sequence $x_n \in G$, which may be taken to converge at some point $s \in \partial G$, such that

$$\lim_{n \to \infty} f_1(x_n) = m.$$

This again shows that $m \geq 0$, which implies that the theorem holds in this case as well. ▯

If in Theorem 1.1 we take for $h(x)$ a constant, then the corresponding assertion is called the *minimum principle* for superharmonic functions.

Denote by $\mathfrak{S}(G)$ the class of functions superharmonic in G. The class $\mathfrak{S}(G)$ is closed with respect to addition and multiplication by a positive constant, i.e. it is a convex cone in the linear space of all functions on G. Furthermore, $\mathfrak{S}(G)$ is closed under the following operations:

(i) uniform passage to the limit (on any compact subset of G);

(ii) passage to the limit in a monotone increasing sequence (if the limit is $\not\equiv +\infty$);

(iii) the operation inf, applied to a finite number of elements.

Let us prove (ii), for example. Suppose that $f(x) = \lim_{n\to\infty} f_n(x)$, $f_n(x) \leq f_{n+1}(x)$; $f_n(x) \in \mathfrak{S}(G)$. First of all, we know that $f(x)$ is lower semicontinuous.

Furthermore, $f_n(x) \geq f_n * \varepsilon^{(r)}$ which implies $f(x) \geq f_n * \varepsilon^{(r)}$; by reason of the monotonicity we may pass to the limit as $n \to \infty$ and obtain $f(x) \geq f * \varepsilon^{(r)}$.

Finally, let us prove another theorem, which essentially is a consequence of the convexity of the cone $\mathfrak{S}(R^p)$.

Theorem 1.2. *Suppose that $f(x) \geq 0$ and is superharmonic in R^p, and that $\mu \in \mathfrak{M}^+$. Then, if*

$$f * \mu = \int_{R^p} f(x - y)\, d\mu(y) \not\equiv +\infty,$$

then this function is also superharmonic in R^p.

Proof. Suppose that μ_n is the restriction of μ to the ball $|x| < n$. Then $f * \mu_n \leq f * \mu_{n+1}$ and $f * \mu = \lim_{n\to\infty} f * \mu_n$ and, according to (ii), it is sufficient to prove the theorem for the case of a measure μ with compact support.

The semicontinuity of $f * \mu$ can be obtained by approximating $f(x)$ by a monotone nondecreasing sequence $\{\varphi_n(x)\}$ of continuous functions. Then $\{\varphi_n * \mu\}$ is a sequence of the same type which converges to $f * \mu$. Furthermore, at any point $x \in R^p$, according to the Fubini theorem,

$$(f * \mu)(x) \geq \left((f * \varepsilon^{(r)}) * \mu\right)(x) = \left((f * \mu) * \varepsilon^{(r)}\right)(x),$$

as was required. ▯

Remark. The constraint $f(x) \geq 0$ may be removed if the measure μ has compact support. In general, the constraint may be weakened (for example, replaced by the condition that $f(x)$ have a harmonic minorant).

Corollary. *Any function $f(x)$ superharmonic in R^p is the limit of a monotone increasing sequence $\{f_n(x)\}$ of continuous superharmonic functions.*

Indeed, it is sufficient to put

$$f_n(x) = f * m^{(1/n)}$$

and to use Theorem 1.2, the remark at the end of No. 4 and (1.2.4). The continuity of $f_n(x)$ is established without difficulty.

An important example of a superharmonic function is a function $f(x)$ having continuous second-order derivatives and satisfying the inequality

$$\Delta f(x) \leq 0, \quad x \in G.$$

To prove that $f(x)$ is superharmonic it is sufficient to verify the minimum principle in the following weaker form: if $G' \subset G$ and $f(x) \geq \alpha$ on $\partial G'$, then $f(x) \geq \alpha$ in G'. In fact, if we apply this principle to the difference $f(x) - H(x)$, where $H(x)$ is the Poisson integral for the sphere S_r, we will obtain (c).

Thus, suppose that

$$f(x_0) < \alpha, \quad x_0 \in G'.$$

Take $\varepsilon > 0$ sufficiently small such that

$$\varepsilon |x|^2 < \alpha - f(x_0), \quad x \in \partial G'$$

and form the function

$$f_1(x) = f(x) - \varepsilon |x|^2.$$

Since on $\partial G'$

$$f_1(x) \geq \alpha - \varepsilon |x|^2 > f(x_0)$$

and

$$f_1(x_0) = f(x_0) - \varepsilon |x_0|^2 \leq f(x_0),$$

the function $f_1(x)$ attains its minimum at a certain point $x_1 \in G'$. This, however, is impossible since

$$\Delta f_1(x_1) = \Delta f(x_1) - 2\varepsilon p < 0.$$

The above argument still holds even if $f(x)$ has isolated discontinuities at which $f(x) = +\infty$.

Along with the superharmonic functions that we are investigating in this section, it is also useful to introduce the notion of a *subharmonic function.*

A function $f(x)$ is called *subharmonic,* if $-f(x)$ is superharmonic. In other words, to say that $f(x)$ is subharmonic means

(a_1) $-\infty \leq f(x) < \infty$, $f(x) \not\equiv -\infty$;

(b_1) $f(x)$ is upper semicontinuous;

(c_1) for any $x \in G$ and $0 < r < r_0(x)$

$$f(x) \leq \frac{1}{\omega_p r^{p-1}} \int\limits_{S_r} f(y)\, dS.$$

It is clear that a function simultaneously super- and subharmonic in G, must be harmonic in G.

We note that for every assertion about superharmonic functions there is an equivalent assertion about subharmonic functions.

§ 3. Definition of Potentials and Their Simplest Properties

6. Suppose that G is a region in R^p, possibly coincident with R^p. If $K(x, y)$ is a real function, defined on $G \times G$, and ν is a signed measure with support in G, then we may formally consider the expression

$$U_K^\nu(x) = \int_{S(\nu)} K(x, y)\, d\nu(y) \tag{1.3.1}$$

which we will call the *K-potential of the signed measure* ν. Of course, in order to insure that $U_K^\nu(x)$ exists and to construct a theory which could be considered as a substantial extention of the classical potential theory, we must impose certain restrictions on the kernel $K(x, y)$.

In this chapter we shall consider only certain concrete kernels; a survey of the results in the case of more general kernels will be given in Ch. VI, § 2.

If G coincides with R^p ($p \geq 3$), then we will put

$$K(x, y) = k_\alpha(x - y) \quad (0 < \alpha < p)$$

and the corresponding potential

$$U_\alpha^\nu(x) = \int_{S(\nu)} k_\alpha(x - y)\, d\nu(y) = A(p, \alpha) \int_{S(\nu)} \frac{d\nu(y)}{|x - y|^{p-\alpha}} \tag{1.3.2}$$

will be called the *Riesz potential* of order α. Also we will often omit the normalization factor $A(p, \alpha)$ and will simply write

$$U_\alpha^\nu(x) = \int \frac{d\nu(y)}{|x - y|^{p-\alpha}}. \tag{1.3.2'}$$

In rare cases when confusion might arise, we will use the corresponding stipulations.

Note that the expression (1.3.2) is a convolution:

$$U_\alpha^\nu(x) = k_\alpha * \nu. \tag{1.3.3}$$

One should keep in mind, however, that $k_\alpha * \nu$ as the density of a certain signed measure (cf. Introduction, No. 1) represents essentially only a class of equivalent functions in R^p, while definition (1.3.2) gives $U_\alpha^\nu(x)$ as a well-defined function of the point x in its domain of definition. This distinction causes us to name a potential theory *coarse* if the potentials are considered as signed measures (or densities of signed measures), and *thin* if the potentials are considered as functions of a point, defined by equation (1.3.2).

We shall study the thin theory, although we will sometimes use the coarse definition (1.3.3), remembering the distinction pointed out above.

In the case of R^2, in addition to the Riesz potentials, we shall also examine the *logarithmic potential*

$$U_2^\nu(x) = \int_{S(\nu)} k_2(x - y)\, d\nu(y) = 2\pi \int_{S(\nu)} \ln \frac{1}{|x-y|}\, d\nu(y). \tag{1.3.4}$$

According to (1.1.16) with the condition $\nu(1) = 0$, the logarithmic potential is the vague limit of Riesz potentials $U_\alpha^\nu(x)$ as $\alpha \to 2$.

In the future, in speaking of the classical case $\alpha = 2$, for $p = 2$ we will always mean the case of the logarithmic potential.

Finally, if $G \subset R^p$ $(p \geq 2)$ is a region with a sufficiently smooth boundary ∂G, then for a kernel we will take the *Green function* of the region G.

Recall that by this we mean a function

$$G(x, y) = k_2(x - y) + h(x, y) \tag{1.3.5}$$

(where $h(x, y)$ is harmonic in G), which for any fixed $y \in G$ is continuous for $x \in \overline{G} \setminus \{y\}$, harmonic in $G \setminus \{y\}$ and is equal to zero on ∂G. It is known (cf., for example, Courant [1]) that under certain conditions of smoothness of ∂G, the Green function exists. In what follows we will free ourselves of the smoothness requirement by introducing a generalized Green function; we shall also introduce an analogous function in the non-classical case $\alpha \neq 2$. However, in this chapter we will consider only "nice" regions G, for which $G(x, y)$ is known to exist.[6]

The corresponding potential

$$U_g^\nu(x) = \int_G G(x,y)\, d\nu(y), \quad x \in \overline{G}, \tag{1.3.6}$$

will be called a *Green potential*. Clearly, on ∂G, this potential is equal to zero.

In those cases when we wish to speak of a potential of any of the above-mentioned types, then we shall sometimes simply write $U^\nu(x)$.

7. Let us examine the simplest properties of a potential of a measure μ. First suppose thst μ has compact support.

Then the potential U^μ is lower semicontinuous and for any point $x \in R^p$

$$U^\mu(x) = \varliminf_{y \to x} U^\mu(y). \tag{1.3.7}$$

---

[6] We note that the exposition in Ch. IV does not presume the knowledge of any classical results in potential theory. However, not wishing to dissect the presentation and introduce repetitions, in Ch.I we will work with the "classical" Green function.

We convince ourselves of this by introducing the continuous "truncated" kernel

$$k_\alpha^{(\eta)}(x) = \begin{cases} k_\alpha(x) & \text{for} \quad |x| > \eta \\ \text{const} & \text{for} \quad |x| \leq \eta. \end{cases}$$

Then

$$U_\alpha^\mu(x) = \lim_{\eta \to 0} \int_{S(\mu)} k_\alpha^{(\eta)}(x - y)\, d\mu(y)$$

is a representation of $U_\alpha^\mu(x)$ as the limit of a monotone increasing sequence of continuous functions. The result (1.3.7) is also true for Green potentials, since

$$U_g^\mu(x) = U_2^\mu(x) + \int_{S(\mu)} h(x, y)\, d\mu(y), \tag{1.3.8}$$

and the second term is obviously continuous.

We give two examples of discontinuous potentials.

Example 1. Consider in R^3 a sequence of points x_n, $|x_n| \to 0$ and a sequence of positive numbers α_n such that

$$\sum_{n=1}^{\infty} \frac{\alpha_n}{|x_n|} < \infty.$$

The atomic measure $\mu = \sum_1^\infty \alpha_n \varepsilon_{x_n}$ generates a potential $U_2^\mu(x)$, finite at the point 0 and infinite at all the points x_n.

Thus,

$$U_2^\mu(0) < \overline{\lim_{\alpha \to 0}}\, U_2^\mu(x) = +\infty.$$

This example may be slightly modified, for having got a finite jump in the potential.

Example 2. Preserving the notation of Example 1, let us suppose that for $n \neq k$ $\dfrac{|x_n|}{|x_n - x_n|} < M$. Enclose each point x_n in asphere S_n of radius r_n, selecting r_n such that:

a) the spheres S_n are pairwise disjoint;

b) $\sum_1^\infty \dfrac{r_n}{|x_n|} < 1$;

c) $\sum_{n \neq k} \dfrac{r_n}{|x_n - x_k|} < 1$.

It is not difficult to see that such a choice is possible. Form the measure

$$\mu = \sum_{n=1}^{\infty} r_n \varepsilon_{x_n}^{(r_n)}.$$

Then $U_2^\mu(0) < 1$, while within S_k we have the inequality

$$1 \leq U_2^\mu(x) < 2.$$

Consequently,

$$U_2^\mu(0) < \overline{\lim_{x \to 0}}\, U_2^\mu(x),$$

where the right-hand side is finite.

Let us return to the general properties of potentials of measures. Obviously,

$$-\infty < U^\mu(x) \leq +\infty$$

and in all cases except the logarithmic one

$$0 \leq U^\mu(x) \leq +\infty.$$

More precisely,

$$0 < U_\alpha^\mu(x) \leq +\infty, \quad \text{if} \quad \mu \not\equiv 0, \tag{1.3.9}$$

while

$$0 \leq U_g^\mu(x) \leq +\infty, \tag{1.3.9'}$$

where for $\mu \not\equiv 0$, $U_g^\mu(x) = 0$ exactly at the points of ∂G.

The set of points where $U^\mu(x) = +\infty$ will be studied in detail below (cf. Ch. II and III, § 1). For the moment we note that it has, in any case, *Lebesgue measure zero*, since the potential $U^\mu(x)$ is locally integrable. To verify this last assertion, by (1.3.8) we may limit ourselves to potentials $U_\alpha^\mu(x)$. In that case for any $r > 0$ we will have, by the Fubini theorem,

$$\int\limits_{|x|<r} dx \int\limits_{S(\mu)} k_\alpha(x-y)\, d\mu(y) = \int\limits_{S(\mu)} d\mu(y) \int\limits_{|x|<r} k_\alpha(x-y)\, dx < \infty.$$

In the future, for Riesz kernels we will also consider measures without compact support. The only constraint which is essential here is the requirement that the potential $U_\alpha^\mu(x)$ not be identically equal to infinity.

We will show that a *necessary and sufficient condition for finiteness almost everywhere of the potential $U_\alpha^\mu(x)$ is the inequality*

$$\int\limits_{|y|>1} \frac{d\mu(y)}{|y|^{p-\alpha}} < \infty. \tag{1.3.10}$$

Indeed, if (1.3.10) is not satisfied, then for any $r > 0$

$$\int\limits_{|y|>r} \frac{d\mu(y)}{|y|^{p-\alpha}} = \infty,$$

and

$$\int_{R^p} \frac{d\mu(y)}{|x-y|^{p-\alpha}} \geq \int_{|y|>|x|+1} \frac{d\mu(y)}{|x-y|^{p-\alpha}} > \left(\frac{|x|+1}{2|x|+1}\right)^{p-\alpha} \int_{|y|>|x|+1} \frac{d\mu(y)}{|y|^{p-\alpha}} = \infty.$$

On the other hand, since for any $r > 0$

$$\int_{|x|<r} \frac{dx}{|x-y|^{p-\alpha}} = O\left(\frac{1}{|y|^{p-\alpha}}\right), \quad |y| \to \infty,$$

then if (1.3.10) is satisfied

$$\int_{R^p} d\mu(y) \int_{|x|<r} k_\alpha(x-y)\, dx < \infty,$$

which implies, as above, the integrability of $U_\alpha^\mu(x)$.

In all cases, when we consider measures or signed measures ν, without compact support, we shall assume that condition (1.3.10) is satisfied for ν^+ and ν^-, without explicit mention on each occasion. Therefore, the potential of a signed measure without compact support

$$U_\alpha^\nu(x) = U_\alpha^{\nu^+}(x) - U_\alpha^{\nu^-}(x)$$

will be defined in all cases almost everywhere.

Note that the potential $U_\alpha^\mu(x)$ of a measure is lower semicontinuous not only relative to the point x, but also relative to the measure $\mu \in \mathfrak{M}^+$. This means that we have the so-called

Principle of Descent. *If $\mu_n \dot{\to} \mu$, then*

$$U_\alpha^\mu(x) \leq \varliminf_{n\to\infty} U_\alpha^{\mu_n}(x).$$

We shall prove a more general statement, expressing the semicontinuity of $U_\alpha^\mu(x)$ in $R^p \times \mathfrak{M}^+$.

Theorem 1.3. *If $\mu_n \dot{\to} \mu$ $(\mu_n \in \mathfrak{M}^+)$ and $x_n \to x_0$, then*

$$U_\alpha^\mu(x_0) \leq \varliminf_{n\to\alpha} U_\alpha^{\mu_n}(x).$$

Proof. We shall first assume that the measures μ_n, μ have supports contained in some fixed compact set. We introduce, as in No. 7, the truncated kernel $k_\alpha^{(\varepsilon)}(x)$ and set

$$[U_\alpha^\mu(x)]_\varepsilon = \int_{R^p} k_\alpha^{(\varepsilon)}(x-y)\, d\mu(y).$$

The family of functions $[U_\alpha^{\mu_n}(x)]_\varepsilon$ for fixed ε is uniformly continuous, and

$$\lim_{n\to\infty} [U_\alpha^{\mu_n}(x_0)]_\varepsilon = [U_n^\mu(x_0)]_\varepsilon.$$

Consequently,

$$[U_\alpha^\mu(x_0)]_\varepsilon = \lim_{n\to\infty} [U_\alpha^{\mu_n}(x_n)]_\varepsilon \leq \varliminf_{n\to\infty} U_\alpha^{\mu_n}(x_n),$$

and as $\varepsilon \to 0$ we obtain the assertion of the theorem.

In the general case we select a sequence of positive numbers N_k such that $N_k \to \infty$ and the balls $|x| < N_k$ are μ-clopen. If $\mu_n^{(k)}$, $\mu^{(k)}$ are restrictions of μ_n, μ to the ball $|x| < N_k$, then, according to Theorem 0.5′, as $n \to \infty$

$$\mu_n^{(k)} \dot{\to} \mu^{(k)}.$$

From what has been proved

$$U_\alpha^{\mu^{(k)}}(x_0) \leq \varliminf_{n\to\infty} U_\alpha^{\mu_n^{(k)}}(x_n).$$

Passing to the limit as $k \to \infty$ first on the right-hand side and then on the left of this inequality, we conclude the proof of the theorem. ▯

Corollary. *If $\mu_n \dot{\to} \mu$ and are monotonically increasing, then*

$$\lim_{n\to\infty} U_\alpha^{\mu_n}(x) = U_\alpha^\mu(x).$$

From the viewpoint of coarse potential theory the equation

$$\lim_{n\to\infty} U_\alpha^{\mu_n}(x) = U_\alpha^\mu(x)$$

holds unconditionally, if the supports of the measures μ_n are contained in a fixed compact set and $\mu_n \dot{\to} \mu$. This follows directly from Lemma 0.2 (ii). However, the existence of the limit and its equality with $U_\alpha^\mu(x)$ is not at all guaranteed for all points of R^p. Below (cf. Ch. IV, § 2) we will be able to easily specify examples where at some points of R^p

$$U_\alpha^\mu(x) < \varliminf_{n\to\infty} U_\alpha^{\mu_n}(x),$$

although $\mu_n \dot{\to} \mu$ and the supports of the measures μ_n are contained in a fixed compact set.

Staying for the moment within the framework of the coarse theory, we prove the following lemma:

Lemma 1.0. *Suppose that $\{U_\alpha^{\mu_n}(x)\}$ is a nondecreasing sequence of potentials of measures. Then either*

$$\lim_{n\to\infty} U_\alpha^{\mu_n}(x) \equiv +\infty$$

or

$$\lim_{n\to\infty} U_\alpha^{\mu_n}(x) = U_\alpha^{\mu}(x) + A,$$

almost everywhere in R^p, *where* $\mu \in \mathfrak{M}^+$ *and* $A \geq 0$.

Proof. Suppose that

$$\lim_{n\to\infty} U_\alpha^{\mu_n}(x) \not\equiv +\infty.$$

In that case:

(1) the sequence $\{\mu_n\}$ is weakly bounded, i.e. for any compact set K,

$$\mu_n(K) < \text{const},$$

(2) for any $r > 0$

$$\int_{|x|>r} \frac{d\mu_n(x)}{|x|^{p-\alpha}} < \text{const}.$$

Indeed, if (1) were not true, then for some K and subsequence $\{n_i\}$ $\mu_{n_i}(K) \to \infty$. But then

$$\lim_{k\to\infty} U_\alpha^{\mu_{n_k}}(x) \geq \overline{\lim_{k\to\infty}} \int_K \frac{d\mu_{n_k}(y)}{|x-y|^{p-\alpha}} \equiv +\infty,$$

which contradicts the hypothesis. Analogously, (2) may be proved, using the inequality

$$\int_{R^p} \frac{d\mu_n(y)}{|x-y|^{p-\alpha}} \geq \left(\frac{|x|+1}{2|x|+1}\right)^{p-\alpha} \int_{|y|>|x|+1} \frac{d\mu_n(y)}{|y|^{p-\alpha}},$$

obtained in No. 7 in the proof of the necessity of condition (1.3.10).

By Theorem 0.6 the sequence of measures $\{\mu_n\}$ is vaguely compact, and therefore we may assume $\mu_n \dashrightarrow \mu$. Put

$$F(x) = \lim_{n\to\infty} U_\alpha^{\mu_n}(x)$$

and consider an arbitrary function $f(x) \in \Phi$, satisfying the condition

$$\int_{R^p} f(x)\, dx = 0.$$

From (1) and (2) it follows that $F(x)$ is locally integrable, since for any n

$$\int_{|x|<r} U_\alpha^{\mu_n}(x)\, dx = \int d\mu_n(y) \int_{|x|<r} \frac{dx}{|x-y|^{p-\alpha}} < \text{const}.$$

Therefore

$$\lim_{n\to\infty} \int_{R^p} U_\alpha^{\mu_n}(x) f(x)\, dx = \int_{R^p} F(x) f(x)\, dx.$$

On the other hand,

$$\int_{R^p} U_\alpha^{\mu_n}(x)\, f(x)\, dx = \int_{R^p} U_\alpha^f(x)\, d\mu_n(x).$$

But it is not difficult to see, expanding $|x - y|^{\alpha - p}$ in a series of negative powers of $|x|$, that

$$U_\alpha^f(x) = \int_{S(f)} \frac{f(y)\, dy}{|x - y|^{p-\alpha}} = O\left(\frac{1}{|x|^{p+1-\alpha}}\right).$$

Consequently,

$$\int_{|x|>r} U_\alpha^f(x)\, d\mu_n(x) < Q_1 \int_{|x|>r} \frac{d\mu_n(x)}{|x|^{p+1-\alpha}} < \frac{Q_1}{r} \int_{|x|>r} \frac{d\mu_n(x)}{|x|^{p-\alpha}},$$

and this expression, by (2), uniformly tends to zero as $r \to \infty$. If we also take into account the continuity of $U_\alpha^f(x)$,[7] then we obtain

$$\lim_{n\to\infty} \int_{R^p} U_\alpha^f(x)\, d\mu_n(y) = \int_{R^p} U_\alpha^f(x)\, d\mu(x) = \int_{R^p} U_\alpha^\mu(x)\, f(x)\, dx.$$

Thus, for any $f(x) \in \Phi$, orthogonal to 1,

$$\int_{R^p} F(x)\, f(x)\, dx = \int_{R^p} U_\alpha^\mu(x)\, f(x)\, dx.$$

From this it easily follows that

$$F(x) - U_\alpha^\mu(x) = A,$$

almost everywhere in R^p, where A is a constant.[8] That A is nonnegative follows from the principle of descent. The lemma is proved. □

Below (cf. §§ 5 and 6) we will sharpen the conclusion of this lemma.

8. We will now study the connection between the potential of a measure and sub- and superharmonicity and the maximum principle which follows from this.

---

[7] Cf., for example, Courant [1].

[8] Let $f_0(x)$ be a fixed function of Φ, for which

$$\int f_0(x)\, dx = 1.$$

Then for any $f(x) \in \Phi$ we have the unique representation $f(x) = \lambda f_0(x) + \varphi(x)$ where $\varphi(x)$ is orthogonal to 1. Furthermore,

$$\int_{R^p} (F - U_\alpha^\mu) f\, dx = \lambda \int_{R^p} (F - U_\alpha^\mu) f_0\, dx = \lambda A = \int_{R^p} A f(x)\, dx$$

where

$$A = \int_{R^p} (F - U_\alpha^\mu) f_0\, dx.$$

Theorem 1.4. *For $2 < \alpha < p$, the potential $U^\mu_\alpha(x)$ of a measure is a superharmonic function in R^p.*

For $\alpha = 2$ the potential $U^\mu_2(x)$ is superharmonic in R^p and harmonic outside of $S(\mu)$.[9]

For $\alpha < 2$ the potential $U^\mu_\alpha(x)$ is subharmonic outside of $S(\mu)$.

Proof. First of all

$$\Delta k_\alpha(x) = A(p, \alpha) \frac{(p-\alpha)(2-\alpha)}{|x|^{p-\alpha+2}}.$$

Therefore for $\alpha \geq 2$ $k_\alpha(x)$ will be superharmonic in R^p, since $\Delta k_\alpha(x) \leq 0$ (cf. No. 5). The superharmonicity of $U^\mu_\alpha(x)$ in all cases except the logarithmic, follows from Theorem 1.2. In the logarithmic case we may put

$$U^\mu_2(x) = U^{\mu_1}_2(x) + U^{\mu_2}_2(x),$$

where μ_1 is the restriction of μ to a certain disc $|x| < r$. Then $U^{\mu_1}_2(x)$ is superharmonic by the remark to Theorem 1.2, and $U^{\mu_2}_2(x)$ will be harmonic in the disc $|x| < r$.

The result for $\alpha < 2$ follows from the fact for $|x| > 0$ $\Delta k_\alpha(x) \geq 0$, and therefore, outside of $S(\mu)$

$$\Delta U^\mu_\alpha(x) \geq 0. \quad \square$$

Corollary. *For $\alpha \geq 2$ the minimum principle holds: if*

$$U^\mu_\alpha(x) \geq m,$$

on the boundary ∂G of a bounded region G, then this inequality holds also within G.

Theorem 1.5. *If*

$$U^\mu_\alpha(x) \leq M \quad (0 < \alpha < p),$$

holds on $S(\mu)$, then

$$U^\mu_\alpha(x) \leq 2^{p-\alpha} M$$

in the whole space R^p.

Proof. Suppose that $x \notin S(\mu)$ and that x' is a point of $S(\mu)$ closest to x. Then for any point $y \in S(\mu)$ we have

$$|y - x'| \leq |y - x| + |x' - x| \leq 2\,|y - x|,$$

from which

$$|y - x'|^{\alpha-p} \geq 2^{\alpha-p}\,|y - x|^{\alpha-p}$$

[9] This also holds for Green potentials.

and

$$U_\alpha^\mu(x) \le 2^{p-\alpha}\, U_\alpha^\mu(x') \le 2^{p-\alpha} M. \quad \square$$

Under the condition $\alpha \le 2$, a much more precise result holds, namely:

If the inequality

$$U_\alpha^\mu(x) \le M \quad (0 < \alpha \le 2) \tag{1.3.11}$$

holds on $S(\mu)$ or is true μ-almost everywhere[10], *then it is satisfied everywhere (first maximum principle).*

We remark that although outside of $S(\mu)$, $U_\alpha^\mu(x)$ is a subharmonic function (Theorem 1.4), the above theorem does not follow directly from the maximum principle for subharmonic functions. This is so because the inequality (1.3.11) does not directly imply the inequality

$$\overline{\lim_{y \notin S(\mu);\, y \to x}}\; U_\alpha^\mu(y) \le M,$$

necessary for the above-mentioned maximum principle.

Consider first the case of the logarithmic potential ($\alpha = p = 2$), where the proof can be given by elementary methods.

Theorem 1.6. (*First maximum principle for logarithmic potentials.*) *Suppose that μ is a measure with compact support,*

$$U_2^\mu(x) = \int_{S(\mu)} \ln \frac{1}{|x-y|}\, d\mu(y)$$

and at all points of $S(\mu)$ the inequality

$$U_2^\mu(x) \le M$$

holds. Then this inequality holds everywhere.

Proof. Suppose that G is a connected component of the complement of $S(\mu)$. Since $U_2^\mu(x)$ is harmonic in G, then it is sufficient to show that

$$\overline{\lim_{x \to y_0}}\, U_2^\mu(x) \le M, \quad x \notin S(\mu), \quad y_0 \in S(\mu).$$

Indeed if G is a bounded region, then the inequality

$$U_2^\mu(x) \le M, \quad x \in G,$$

[10] The second condition is distinct from the first only in form. In fact, using the semicontinuity of $U_\alpha^\mu(x)$, it is easy to see that if (1.3.11) holds μ-almost everywhere, then it holds at any point of $S(\mu)$.

will follow from the maximum principle for harmonic functions; if G includes infinity, then the maximum principle needs to be applied within a disc $|x| < R$, R sufficiently large such that

$$U_2^\mu(x) \leq M \quad \text{for} \quad |x| = R.$$[11]

Since the potential $U_2^\mu(x)$ is bounded on $S(\mu)$, the μ-measure of any point is zero. Therefore, given $\eta > 0$, one may find a disc B with center at the point y_0, such that $\mu(B) < \eta$. Set $B_1 = B \cap S(\mu)$; $B_2 = B \setminus B_1$. Let $x \in B_2$ (consequently, $x \notin S(\mu)$, and $y_1 \in B_1$ be a point of B_1 closest to x:

$$|x - y_1| \leq |x - y| \quad \text{for} \quad y \in B_1.$$

Then for $y \in B_1$

$$|y_1 - y| \leq |x - y_1| + |x - y| \leq 2\,|x - y|$$

$$\frac{1}{|x - y|} \leq \frac{2}{|y_1 - y|}$$

and consequently,

$$\begin{aligned} U_2^\mu(x) &= \int_{B_1} \ln \frac{1}{|x - y|}\, d\mu(y) + \int_{S(\mu)\setminus B_1} \ln \frac{1}{|x - y|}\, d\mu(y) \\ &\leq \mu(B_1) \ln 2 + \int_{B_1} \ln \frac{1}{|y_1 - y|}\, d\mu(y) + \int_{S(\mu)\setminus B_1} \ln \frac{1}{|x - y|}\, d\mu(y) \\ &\leq \eta \ln 2 + U_2^\mu(y_1) + \int_{S(\mu)\setminus B_1} \left[\ln \frac{1}{|x - y|} - \ln \frac{1}{|y_1 - y|}\right] d\mu(y) \\ &\leq \eta \ln 2 + M + \int_{S(\mu)\setminus B_1} \ln \frac{|y_1 - y|}{|x - y|}\, d\mu(y). \end{aligned}$$

Since as $x \to y_0$ we have $y_1 \to y_0$, then $\ln \frac{|y_1 - y|}{|x - y|} \to 0$ uniformly with respect to $y \in S(\mu) \setminus B_1$. Thus

$$\overline{\lim_{x \to y_0}}\, U_2^\mu(x) \leq \eta \ln 2 + M,$$

and since η was arbitrary

$$\overline{\lim_{x \to y_0}}\, U_2^\mu(x) \leq M. \quad \square$$

A similar theorem holds also for Green potentials.

---

[11] This can be done, of course, since as $|x| \to \infty$, $U_2^\mu(x) \to -\infty$ uniformly.

Theorem 1.6′. *Suppose that measure μ has compact support in a plane region G,*

$$U_g^\mu(x) = \int_{S(\mu)} G(x, y)\, d\mu(y)$$

is its Green potential, and $U_g^\mu(x) \leq M$ for $x \in S(\mu)$. Then

$$U_g^\mu(x) \leq M \quad \text{in } G.$$

Proof. From (1.3.5) it follows that

$$U_g^\mu(x) = U_2^\mu(x) + H(x),$$

where $H(x)$ is continuous in $\overline{G}$ and harmonic in $G \setminus S(\mu)$. Let Ω be any connected component of $G \setminus S(\mu)$. If $\partial\Omega$ consists only of points of $S(\mu)$, then from the inequality

$$U_2^\mu(x) \leq M - H(x), \quad x \in \partial\Omega,$$

as in the preceding proof, we obtain

$$U_2^\mu(x) \leq M - H(x), \quad x \in \Omega.$$

If $\partial\Omega$ includes ∂G, then since for $x \in \partial G$, $U_g^\mu(x) = 0$ and since $M \geq 0$, we obtain

$$U_2^\mu(x) = -H(x) \leq M - H(x), \quad x \in \partial G,$$

from which, as above

$$U_2^\mu(x) \leq M - H(x), \quad x \in \Omega. \quad \square$$

Before proving the first maximum principle for any $\alpha \leq 2$, we establish certain continuity properties of potentials.

Theorem 1.7. *If the potential $U_\alpha^\mu(x)$ of the measure μ is continuous on the support $S(\mu)$, then it is continuous in R^p.*

Proof. Clearly one need only prove that $U_\alpha^\mu(x)$ is continuous at the points of $S(\mu)$. Suppose that $x_0 \in S(\mu)$, that μ_ε is the restriction of the measure μ to the ball $|y - x_0| < \varepsilon$ $(\varepsilon > 0)$, and that $\mu_1 = \mu - \mu_\varepsilon$. Then

$$U_\alpha^\mu(x) = U_\alpha^{\mu_\varepsilon}(x) + U_\alpha^{\mu_1}(x)$$

and

$$|U_\alpha^\mu(x_0) - U_\alpha^\mu(x)| \leq U_\alpha^{\mu_\varepsilon}(x_0) + U_\alpha^{\mu_\varepsilon}(x) + |U_\alpha^{\mu_1}(x_0) - U_\alpha^{\mu_1}(x)|.$$

Since the potential $U_\alpha^{\mu_1}(x)$ is continuous at the point x_0, then it is sufficient to establish that the potential $U_\alpha^{\mu_\varepsilon}(x)$ uniformly tends to zero as $\varepsilon \to 0$. Since $U_\alpha^\mu(x_0)$ is finite, for any given $\eta > 0$ one can find $\varepsilon_1 > 0$ such that $U_\alpha^{\mu_{\varepsilon_1}}(x_0) < \eta/2$. But the potential $U_\alpha^{\mu_{\varepsilon_1}}(x)$ on $S(\mu)$ is continuous at the point x_0; hence at all points of $S(\mu)$ for which $|x - x_0| < \varepsilon(\eta) < \varepsilon_1$ the inequality

$$U_\alpha^{\mu_{\varepsilon_1}}(x) < \eta$$

holds; moreover,

$$U_\alpha^{\mu_\varepsilon}(x) < \eta.$$

Since this is true at all points of $S(\mu)$, then by Theorem 1.5

$$U_\alpha^{\mu_\varepsilon}(x) < 2^{p-\alpha}\eta$$

throughout R^p and the theorem is proved. □

This theorem together with the classical theorem of Lusin (cf., for example, Halmos [1], p. 243) enables us to obtain the following results.

Theorem 1.8. *Suppose that μ has compact support and a μ-almost everywhere finite potential $U_\alpha^\mu(x)$. Then for any $\varepsilon > 0$ a compact set $K \subset S(\mu)$ can be found such that the restriction μ' of the measure μ to K will have the following properties:*

(i) $U_\alpha^{\mu'}(x)$ is continuous in R^p;

(ii) $\mu(1) - \mu'(1) < \varepsilon$.

Proof. By Lusin's theorem one can find $K \subset S(\mu)$ such that (ii) holds and $U_\alpha^\mu(x)$ is continuous on K. Then the potential $U_\alpha^\mu(x)$ will also be continuous on K. To be convinced of this it is sufficient to write down the equation

$$U_\alpha^{\mu'}(x) = U_\alpha^\mu(x) - U_\alpha^{\mu-\mu'}(x)$$

and note that its left-hand side is lower semicontinuous while its right-hand side is upper semicontinuous on K. By Theorem 1.7, $U_\alpha^{\mu'}(x)$ is continuous in R^p. □

Theorem 1.9. *Let μ be a measure with a μ-almost everywhere finite potential. Then there exists an increasing sequence of measures $\{\mu_n\}$ with compact supports such that:*

(i) $U_\alpha^{\mu_n}(x)$ is continuous in R^p;

(ii) $\lim\limits_{n\to\infty} U_\alpha^{\mu_n}(x) = U_\alpha^\mu(x)$.

Proof. First suppose that the measure μ has compact support. Take a sequence of numbers $\{\varepsilon_n\}$ which is monotone decreasing and tends to zero. Applying Theorem 1.8 to μ and ε_1, we find a measure λ_1 (the restriction of μ to a certain compact set $K \subset S(\mu)$)with a continuous potential and for which $\mu(1) - \lambda_1(1) < \varepsilon_1$.

Now we apply Theorem 1.8 to the measure $\mu - \lambda_1$ and the number ε_2. We obtain a measure λ_2 with a continuous potential, which is a restriction of $\mu - \lambda_1$, and $\mu(1) - \lambda_1(1) - \lambda_2(1) < \varepsilon_2$. Repeating this argument and setting $\mu_n = \sum_{k=1}^{n} \lambda_k$, we obtain a monotone increasing sequence of measures $\{\mu_n\}$ with continuous potentials, such that $\mu_n(1) \to \mu(1)$. From this it of course follows that $\mu_n \dot{\to} \mu$, and therefore (ii) holds (cf. Corollary to Theorem 1.3).

In the case of a measure μ with noncompact support we represent it in the form

$$\mu = \sum_{k=1}^{\infty} \mu_k,$$

where μ_k is the restriction of μ to the layer $k - 1 \le |x| < k$; then for each measure μ_k we construct a sequence $\{\mu_{kn}\}$ for which $S(\mu_{kn}) \subset S(\mu_k)$, $U_\alpha^{\mu_{kn}}(x)$ is continuous, and $\lim_{n\to\infty} U_\alpha^{\mu_{kn}}(x) = U_\alpha^{\mu_{kn}}(x)$. Set

$$\lambda_n = \sum_{k=1}^{n} \mu_{kn}.$$

The potentials $U_\alpha^{\lambda_n}(x)$ are continuous in R^p, the measures λ_n with compact support form a monotone increasing sequence, and it is easy to see that $\lambda_n \dot{\to} \mu$. Consequently,

$$U_\alpha^{\mu}(x) = \lim_{n\to\infty} U_\alpha^{\lambda_n}(x),$$

and the theorem is proved. □

Remark. The requirement that $U_\alpha^\mu(x)$ be finite μ-almost everywhere cannot be removed. This is shown by the example of the measure ε, whose potential cannot, of course, be represented in the way required by Theorem 1.9.

Theorem 1.9 provides a simple way of proving the maximum principle.

Theorem 1.10. (*First Maximum Principle.*) *Suppose that* μ *is an arbitrary measure and that*

$$U_\alpha^\mu(x) \le M \qquad (0 < \alpha \le 2). \tag{1.3.11}$$

μ*-almost everywhere. Then this inequality holds throughout all of* R^p.

Proof. Suppose that $\{\mu_n\}$ is a sequence of measures with compact support satisfying the conditions of Theorem 1.9. Then

$$U_\alpha^{\mu_n}(x) \le M.$$

μ-almost everywhere and in fact μ_n-almost everywhere. By continuity, this inequality holds at all points of $S(\mu_n)$. In each connected component of the complement of $S(\mu_n)$ the potential $U_\alpha^{\mu_n}(x)$ is subharmonic (cf. Theorem 1.4) and continuous on the boundary where it is $\leq M$. Applying the maximum principle for subharmonic functions (if Ω is an infinite component of the complement of $S(\mu_n)$, then it can be applied to $\Omega \cap \{x;\, |x| < N\}$ for sufficiently large N), we obtain $U_\alpha^{\mu_n}(x) \leq M$ everywhere. But then, by (ii) of Theorem 1.9,

$$U_\alpha^\mu(x) \leq M$$

throughout R^p, and the theorem is proved. □

9. Let us now consider certain general properties of potentials of signed measures.

Theorem 1.11. *The spatial mean of a potential*

$$U_{\alpha,r}^\nu(x) = U_\alpha^\nu * m^{(r)}$$

is a continuous function of x; moreover

$$U_\alpha^\nu(x) = \lim_{r\to 0} U_{\alpha,r}^\nu(x) \tag{1.3.12}$$

at any point x where $U_\alpha^\nu(x)$ is defined.

Proof. Using the representation $\nu = \nu^+ - \nu^-$, we need only prove the theorem for measures μ.

The continuity of $U_{\alpha,r}^\mu(x)$ is a consequence of the local integrability of $U_\alpha^\mu(x)$ (cf. No. 7).

We will show that at any point $x \in R^p$

$$U_\alpha^\mu(x) = \lim_{r\to 0} U_{\alpha,r}^\mu(x)\,.^{12} \tag{1.3.12'}$$

For this we establish the inequality

$$U_{\alpha,r}^\mu(x) \leq A\, U_\alpha^\mu(x)\,, \tag{1.3.13}$$

where A depends only on α and p.

Denoting by v_r the volume of a ball of radius r, we have

$$U_{\alpha,r}^\mu(x) = \frac{1}{v_r} \int\limits_{|z-x|<r} dz \int\limits_{S(\mu)} \frac{d\mu(y)}{|z-y|^{p-\alpha}}$$

$$= \frac{1}{v_r} \int\limits_{S(\mu)} \frac{d\mu(y)}{|x-y|^{p-\alpha}} \int\limits_{|z-x|<r} \frac{|x-y|^{p-\alpha}}{|z-y|^{p-\alpha}}\, dz\,.$$

[12] For $\alpha \geq 2$ this follows immediately from Theorem 1.4 and (1.2.4).

The integral

$$\frac{1}{v_r} \int\limits_{|z-x|<r} \frac{|x-y|^{p-\alpha}}{|z-y|^{p-\alpha}}\, dz$$

is invariant under similarity transformations; therefore we may assume that $r = 1$. Furthermore,

$$\frac{1}{v_1} \int\limits_{|z-x|<1} \frac{|x-y|^{p-\alpha}}{|z-y|^{p-\alpha}}\, dz$$

is a continuous function of $|x-y|$ equal to zero when $|x-y| = 0$ and tends to 1 as $|x-y| \to \infty$. Consequently, it has a finite maximum A, and

$$U^{\mu}_{\alpha,r}(x) < A \int\limits_{S(\mu)} \frac{d\mu(y)}{|x-y|^{p-\alpha}} = A U^{\mu}_{\alpha}(x).$$

Now the relation (1.3.12) is easily verified. If $U^{\mu}_{\alpha}(x) = \infty$, then the result follows from the lower semicontinuity of a measure potential. If $U^{\mu}_{\alpha}(x) < \infty$, then, given $\varepsilon > 0$, we find a ball T_δ with center x and radius δ such that

$$U^{\mu'}_{\alpha}(x) < \varepsilon$$

for the restriction μ' of μ to T_δ. By (1.3.13)

$$U^{\mu'}_{\alpha,r}(x) < A\varepsilon.$$

Then, by the continuity of the potential $U^{\mu-\mu'}_{\alpha}$ at the point x,

$$\lim_{r\to 0} U^{\mu-\mu'}_{\alpha,r}(x) = U^{\mu-\mu'}_{\alpha}(x).$$

In view of this, we may write

$$U^{\mu}_{\alpha}(x) - (A+1)\,\varepsilon \le \varliminf_{r\to 0} U^{\mu}_{\alpha,r}(x) \le \varlimsup_{r\to 0} U^{\mu}_{\alpha,r}(x)$$

$$\le U^{\mu}_{\alpha}(x) + (A+1)\,\varepsilon,$$

and since $\varepsilon > 0$ is arbitrary, then this implies (1.3.12′) and (1.3.12) for those points where $U^{\nu}_{\alpha}(x)$ is defined. ▯

Remark. If μ is a measure, then, applying the Fubini theorem, we see that

$$U^{\mu}_{\alpha,r}(x) = U^{\mu_r}_{\alpha}(x),$$

where

$$\mu_r = \mu_* \, m^{(r)}.$$

Further denoting by $\varkappa_r$ the restriction of μ_r to the ball $|x| < 1/r$, we have,

$$U_\alpha^\mu(x) = \lim_{r\to 0} U_\alpha^{\varkappa_r}(x). \tag{1.3.14}$$

In fact, since $r < 1$, we write

$$U_\alpha^{\mu_r}(x) - U_\alpha^{\varkappa_r}(x) = U_\alpha^{\mu_r - \varkappa_r}(x)$$

$$= \int\limits_{|y|>\frac{1}{r}} \frac{d\mu_r(y)}{|x-y|^{p-\alpha}} \leq \frac{1}{\omega_p r^p} \int\limits_{|z|<r} dz \int\limits_{|y|>\frac{1}{r}-1} \frac{d\mu(y)}{|x-z-y|^{p-\alpha}}.$$

But by (1.3.10)

$$\lim_{r\to 0} \int\limits_{|y|>\frac{1}{r}-1} \frac{d\mu(y)}{|x-z-y|^{p-\alpha}} = 0$$

uniformly with respect to z in any finite region; this implies (1.3.14).

Let us now prove the important *uniqueness theorem* which we will strengthen below.

Theorem 1.12. *If the potential of a signed measure ν is equal to zero almost everywhere, then $\nu \equiv 0$, and consequently, $U_\alpha^\nu(x) \equiv 0$.*

In the proof of this theorem we shall utilize a fact arising from No. 2 (cf. (1.1.13)), which for future reference we formulate in the form of a separate lemma.

Lemma 1.1. *If a function $f(x)$ has compact support and has continuous derivatives of order $p + 2$, then it is representable in the form of a potential*

$$f(x) = U_\alpha^\psi(x) = k_\alpha * \psi \quad (0 < \alpha < p). \tag{1.3.15}$$

Moreover, the signed measure ψ is absolutely continuous and its density $\psi(x)$, defined by the formula

$$\psi(x) = k_{-\alpha} * f(x),$$

satisfies the condition [13]

$$\psi(x) = O\left(\frac{1}{|x|^{p+\alpha}}\right) \quad \text{as} \quad |x| \to \infty. \tag{1.3.16}$$

---

[13] It should be noted that in this lemma we may utilize not the coarse but the thin definition of the potential $U_\alpha^\psi(x)$. Indeed, the convolution $k_\alpha * \psi(x)$ is continuous, and therefore equation (1.3.15) holds everywhere and not only almost everywhere.

Remark. In the classical case ($\alpha = 2$) it is sufficient to require that $f(x)$ have three continuous derivatives. Then, as is known[14], the function

$$\psi(x) = -\frac{1}{4\pi^2}\Delta f(x)$$

(also with compact support) satisfies the relation

$$f(x) = U_2^{\psi}(x).$$

Proof of Theorem 1.12. It is sufficient to verify the equation

$$\nu(\varphi) = 0$$

for a set of functions $\varphi \in \Phi$ dense in Φ. For such a subset of Φ we take the set of functions $f(x) \in \Phi$, having continuous derivatives of order $p + 2$. Applying lemma 1.1 to the function $\check{f}(x) = f(-x)$, we obtain

$$\check{f} = k_\alpha * \psi$$

and

$$\nu(f) = \nu * \check{f}(0) = \nu * (k_\alpha * \psi)\,(0).$$

We will show that

$$\nu * (k_\alpha * \psi) = (\nu * k_\alpha) * \psi. \tag{1.3.17}$$

For this, as we saw in No. 4 of the Introduction, it is sufficient to establish that the double convolutions $\nu^\pm * k_\alpha * \psi^\pm$ of measures are defined for some order of operations. But by (1.3.16)

$$|\psi^\pm(x)| < \frac{A}{1 + |x|^{p+\alpha}}$$

and therefore

$$\begin{aligned}
|k_\alpha * \psi^\pm(x)| &< A \int\limits_{R^p} \frac{dy}{|x-y|^{p-\alpha}\,(1 + |y|^{p+\alpha})} \\
&= A \int\limits_{|y-x|>\frac{1}{2}|x|} \frac{dy}{|x-y|^{p-\alpha}\,(1 + |y|^{p+\alpha})} \\
&\quad + A \int\limits_{|y-x|<\frac{1}{2}|x|} \frac{dy}{|x-y|^{p-\alpha}\,(1 + |y|^{p+\alpha})} \\
&< A \left(\frac{2}{|x|}\right)^{p-\alpha} \int\limits_{R^p} \frac{dy}{1 + |y|^{p+\alpha}}
\end{aligned}$$

---

[14] Cf. Courant [1].

$$+ \frac{A}{1 + \left|\frac{x}{2}\right|^{p-\alpha}} \int\limits_{|y-x|<\frac{1}{2}|x|} \frac{dy}{|x-y|^{p-\alpha}} = O\left(\frac{1}{|x|^{p-\alpha}}\right)$$

$$+ \frac{A}{1 + \left|\frac{x}{2}\right|^{p+\alpha}} O\left(|x|^2\right) = O\left(\frac{1}{|x|^{p-\alpha}}\right).$$

In view of condition (1.3.10), which is satisfied by the measures $\nu^{\pm}$, we see that the expressions $\nu^{\pm} * (k_\alpha * \psi^{\pm})$ are defined.

Thus, equation (1.3.17) is proved, and

$$\nu(f) = (k_\alpha * \nu) * \psi(0) = U^\nu_\alpha * \psi(0) = 0,$$

since $U^\nu_\alpha(x) = 0$ almost everywhere. This proves the theorem. ☐

The corresponding theorem for logarithmic potentials is simpler to prove.

Theorem 1.12'. *If the signed measure ν has compact support and $U^\nu_2 = 0$ almost everywhere in R^2, then $\nu \equiv 0$.*

Proof. As in the proof of Theorem 1.12, we shall verify the equation $\nu(\varphi) = 0$ for a dense subset of the class of functions Φ. Consider the functions $f(x) \in \Phi$, having three continuous derivatives. Then

$$\check{f}(x) = k_2 * \psi$$

and $\psi = -\frac{1}{4\pi_2}\Delta\check{f}$ will also have compact support. Equation (1.3.17) quickly follows from this (cf. Lemma 0.6 of the Introduction); the proof is completed as above. ☐

In the classical case (i.e. for logarithmic and Newtonian potentials), the uniqueness theorem admits the following generalization.

Theorem 1.13. *Suppose that in the region G the potential of a signed measure $U^\nu_2(x)$ is equal to a harmonic function $h(x)$ almost everywhere. Then $\nu \equiv 0$ in G.*

Proof. As in the proofs of the preceding theorems, it is sufficient to verify that

$$\nu(f) = 0$$

for all functions f with three continuous derivatives and which have compact support in G. Setting $\psi = -\frac{1}{4\pi^2}\Delta\check{f}$, we have, using Green's formula,

$$\nu(f) = \nu * \check{f}(0) = \nu * (k_2 * \psi)(0) = (k_2 * \nu) * \psi(0)$$
$$= (k_2 * \nu)(\check{\psi}) = \int_G U_2^\nu(x)\, \check{\psi}(x)\, dx$$
$$= -\frac{1}{4\pi^2} \int_G U_2^\nu(x)\, \Delta f\, dx = -\frac{1}{4\pi^2} \int_G h(x)\, \Delta f(x)\, dx$$
$$= -\frac{1}{4\pi^2} \int \Delta h(x)\, f(x)\, dx = 0,$$

as required. □

Now it is easy to prove the uniqueness theorem for Green potentials.

Theorem 1.14. *Suppose that the Green potential*

$$U_g^\nu(x) = \int_G G(x, y)\, d\nu(y)$$

is equal to zero almost everywhere in a bounded region G.[15] *Then* $\nu \equiv 0$.

Proof. By (1.3.5)

$$U_g^\nu(x) = U_2^\nu + \int_G h(x, y)\, d\nu(y).$$

The function

$$H(x) = \int_G h(x, y)\, d\nu(y)$$

is harmonic in G and by the condition of the theorem

$$U_2^\nu(x) = -H(x)$$

almost everywhere in G. Then according to Theorem 1.13, $\nu \equiv 0$. □

§ 4. Energy. Potentials with Finite Energy

10. Every kernel $K(x, y)$ generates in the space $\mathfrak{M}$ of signed measures a quadratic functional

$$I_K[\nu] = \int_{R^p \times R^p} K(x, y)\, d\nu(x)\, d\nu(y) \tag{1.4.1}$$

which we shall call the *energy* of the signed measure ν.

If $K(x, y) \geq 0$, then the Fubini theorem shows that $I_K[\nu]$ will be defined (although it may equal $+\infty$) under the condition

$$\int_{R^p} U_K^{\nu+}(x)\, d\nu^-(x) < \infty, \quad \int_{R^p} U_K^{\nu-}(x)\, d\nu^+(x) < \infty, \tag{1.4.2}$$

[15] Or is equal to a harmonic function.

which is understood to be always satisfied. Then

$$I_K[\nu] = \int_{R^p} U_K^\nu(x)\, d\nu(x).$$

For a logarithmic kernel this remark still holds if we limit ourselves to signed measures with compact support. In this connection and in view of certain other considerations, *in the case of a logarithmic kernel, we shall allow only signed measures with compact support.*

The *mutual energy* of signed measures ν_1, ν_2 is defined as the bilinear functional

$$I_K[\nu_1, \nu_2] = \int_{R^p \times R^p} K(x, y)\, d\nu_1(x)\, d\nu_2(y).$$

Under the conditions

$$\int_{R^p} U_K^{\nu_2^\pm}(x)\, d\nu_1^\mp(x) < \infty \tag{1.4.2'}$$

the expression $I_K[\nu_1, \nu_2]$ is well-defined.

If the kernel $K(x, y)$ is *symmetric*, i.e. $K(y, x) = K(x, y)$, then

$$I_K[\nu_1, \nu_2] = I_K[\nu_2, \nu_1]. \tag{1.4.3}$$

We note that all the kernels we have introduced (cf. No. 6) are symmetric.

Equation (1.4.3), which may also be written in the form

$$\int_{R^p} U_K^{\nu_1}(x)\, d\nu_2(x) = \int_{R^p} U_K^{\nu_2}(x)\, d\nu_1(x),$$

is called the *reciprocity law.*

If the kernel $K(x, y)$ is nonnegative and lower semicontinuous in $R^p \times R^p$, then $I_K[\nu_1, \nu_2]$ is a *lower semicontinuous functional* on $\mathfrak{M}^+ \times \mathfrak{M}^+$. This means that if

$$\mu_n \xrightarrow{\cdot} \mu, \quad \lambda_n \xrightarrow{\cdot} \lambda \quad (\mu_n, \lambda_n \in \mathfrak{M}^+),$$

then

$$I_K[\mu, \lambda] \leq \varliminf_{n\to\infty} I_K[\mu_n, \lambda_n]. \tag{1.4.4}$$

Indeed, the measures $\pi_n = \mu_n \times \lambda_n$ in the space $R^{2p} = R^p \times R^p$ vaguely converge to $\pi = \mu \times \lambda$ and therefore (1.4.4) follows directly from Lemma 0.1 of the Introduction.

In particular, from (1.4.4) it follows that if $\mu_n \xrightarrow{\cdot} \mu$, $\mu_n \in \mathfrak{M}^+$ then

$$I_K[\mu] \leq \varliminf_{n\to\infty} I[\mu_n]. \tag{1.4.5}$$

Inequalities (1.4.4) and (1.4.5) still hold for a logarithmic kernel, if we assume that the supports of all the measures λ_n and μ_n are contained in a fixed compact set. Indeed, it is sufficient to apply first (1.4.4) to the kernel $K(x, y) = \ln \frac{d}{|x-y|}$, where d is chosen sufficiently large such that $K(x, y) > 0$ on all the $S(\mu_n) \times S(\lambda_n)$.

We note one case in which the limits on the right-hand sides of (1.4.4) and (1.4.5) exist and equality holds.

Suppose that for all n $U_K^{\mu_n}(x) \leq U_K^{\mu}(x)$, $U_K^{\lambda_n}(x) \leq U_K^{\lambda}(x)$. Then

$$I_K[\mu, \lambda] = \lim_{n\to\infty} I_K[\mu_n, \lambda_n]. \qquad (1.4.6)$$

In fact,

$$\begin{aligned} I_K[\mu_n, \lambda_n] &= \int U_K^{\mu_n}(x)\, d\lambda_n(x) \leq \int U_K^{\mu}(x)\, d\lambda_n(x) \\ &= \int U_K^{\lambda_n}(x)\, d\mu(x) \leq \int U_K^{\lambda}(x)\, d\mu(x) = I_K[\mu, \lambda] \end{aligned}$$

and therefore

$$\overline{\lim_{n\to\infty}}\, I_K[\mu_n, \lambda_n] \leq I_K[\mu, \lambda].$$

This together with (1.4.4) gives (1.4.6).

In particular (1.4.6) holds if both sequences $\{\mu_n\}$ and $\{\lambda_n\}$ are *monotone increasing*.

11. All the kernels which we are considering have a positive definite character in a sense made precise by the theorems following below.

Theorem 1.15. *For Riesz kernels $k_\alpha(x)$, $0 < \alpha < p$, the inequality*

$$I_\alpha[\nu] = \int_{R^p} U_\alpha^\nu(x)\, d\nu(x) \geq 0$$

holds for any signed measure $\nu \in \mathfrak{M}$.[16]

Moreover $I_\alpha[\nu] = 0$ only if $\nu \equiv 0$.

Proof. We start with the formula

$$k_\alpha = k_{\alpha/2} * k_{\alpha/2}$$

(cf. (1.1.3)) and suppose that $I_\alpha[\nu] < \infty$. Then by condition (1.4.2),

$$I_\alpha[|\nu|] = \int_{R^p \times R^p} k_\alpha\, (x-y)\, d|\nu|(x) d|\nu|(y) < \infty,$$

---

[16] Condition (1.4.2) is assumed to be satisfied. In the future we shall make no special mention of this.

and with the help of the Fubini theorem it is easily verified that

$$\begin{aligned}
I_\alpha[\nu] &= \int_{R^p \times R^p} d\nu(x)\, d\nu(y) \int_{R^p} k_{\alpha/2}(x-y-z)\, k_{\alpha/2}(z)\, dz \\
&= \int_{R^p \times R^p} d\nu(x)\, d\nu(y) \int_{R^p} k_{\alpha/2}(x-z)\, k_{\alpha/2}(z-y)\, dz \\
&= \int_{R^p} dz \int_{R^p} k_{\alpha/2}(x-z)\, d\nu(x) \int_{R^p} k_{\alpha/2}(z-y)\, d\nu(y) \\
&= \int_{R^p} [U^\nu_{\alpha/2}(z)]^2\, dz.
\end{aligned}$$

Consequently, $I_\alpha[\nu] \geq 0$. If $I_\alpha[\nu] = 0$, then $U^\nu_{\alpha/2}[z] = 0$ ν-almost everywhere, and by Theorem 1.12, $\nu = 0$. □

Corollary. *If* $U^\nu_\alpha(x) = 0$ *ν-almost everywhere (in particular, everywhere on* $S(\nu)$*), then* $\nu = 0$.[17]

Consider now the logarithmic kernel in R^2:

$$k_2(x) = 2\pi \ln \frac{1}{|x|}.$$

Under certain restrictions on the class of admissible signed measures ν it will also be positive. Recall that in considering a logarithmic kernel all signed measures are assumed to have compact support.

Theorem 1.16. *Suppose that one of the following conditions is satisfied:*

(*a*) $\nu(1) = 0$,

(*b*) $S(\nu)$ *is contained in the disc* $|x| < 1$;

then we have the inequality

$$I_2[\nu] = \int_{R^2 \times R^2} 2\pi \ln \frac{1}{|x-y|}\, d\nu(x)\, d\nu(y) \geq 0,$$

and $I_2[\nu] = 0$ *implies* $\nu = 0$.

Proof. Suppose that (a) is satisfied. Then according to (1.1.17) we may use the formula

$$k_2 = k_1 * k_1.$$

Assuming that $I_2[\nu] < \infty$, we obtain, as in the proof of Theorem 1.15, the formula

$$I_2[\nu] = \int_{R^2} [U^\nu_1(x)]^2\, dx,$$

from which follow the assertions of the theorem.

---

[17] It should be remembered that we have assumed the finiteness of $I_\alpha[\nu^+, \nu^-]$. In Ch.III and IV we shall obtain results independent of this restriction

Consider case (b). Denote by $\varepsilon^{(1)}$ the measure obtained by uniformly distributing a unit mass on the circumference $|x| = 1$. It is not difficult to see that for $|x| < 1$, $U_2^{\varepsilon^{(1)}}(x) = 0$.[18] We now put

$$\nu_1 = \nu - \nu(1)\, \varepsilon^{(1)}.$$

The signed measure ν_1 with compact support obviously satisfies condition (a), and is $\not\equiv 0$, if $\nu \not\equiv 0$. By what has been proved

$$I_2[\nu_1] > 0.$$

But

$$I_2[\nu_1] = I_2[\nu] - 2\nu(1)\, I_2[\nu, \varepsilon_1] + \nu^2(1)\, I_2[\varepsilon_1] = I_2[\nu], \qquad (1.4.7)$$

since $I_2[\nu_1, \varepsilon^{(1)}] = I_2[\varepsilon^{(1)}] = 0$ by the indicated property of the potential $U_2^{\varepsilon^{(1)}}(x)$. Thus, if $\nu \not\equiv 0$, then $I_2[\nu] > 0$ and the theorem is proved. ▯

Remark. If the signed measure ν is contained in the disc $|x| < R$, then analogous to the above we obtain the bound

$$I_2[\nu] > 2\pi\nu^2(1) \ln \frac{1}{R},$$

where the quantity $I_2[\nu]$ for $R > 1$ may be negative. Thus, for $\varepsilon^{(R)}$ we have

$$I_2[\varepsilon^{(R)}] = 2\pi \ln \frac{1}{R} < 0.$$

Regarding the positive definiteness of Green kernels, we defer those results to Ch. IV, § 1, No. 2.

12. Denote by $\mathscr{E}_K$ the vector space of all signed measures having finite energy, i.e. those for which

$$I_K[\nu] < \infty.$$

In the case of M. Riesz kernels we shall write $\mathscr{E}_\alpha$ instead of $\mathscr{E}_{k_\alpha}$, and for Green kernels, we shall write $\mathscr{E}_g$.

The simplest examples of signed measures in $\mathscr{E}_K$ are:

(a) signed measures μ, with compact supports, whose potential is bounded on $S(\mu)$;

(b) signed measures heaving a bounded potential with compact support.[19]

---

[18] This reduces to the Poisson integral

$$\int_0^\pi \ln (1 - 2\varrho \cos \varphi + \varrho^2)\, d\varphi = 0 \quad \text{for} \quad \varrho < 1.$$

[19] In the classical case, every signed measure of type (b) is also of type (a). This follows from Theorem 1.13.

Because of the positive definiteness theorem of the preceding section, in $\mathscr{E}_K$ we may introduce a Hilbert space structure, setting

$$(\nu, \lambda) = I_K[\nu, \lambda].$$

The inner product axioms are verified in a straightforward manner. One should only remember that for logarithmic kernels signed measures (with compact support) in $\mathscr{E}_2$ must satisfy one of the conditions (a) or (b) of Theorem 1.16. The set of signed measures in $\mathscr{E}_2$ satisfying condition (a), correspondingly (b), will be denoted by $\mathscr{E}_2^{(0)}$ and $\mathscr{E}_2^{(1)}$ respectively.

The positiveness of (ν, ν) allows us to establish the *Schwarz-Bunyakovsky inequality*

$$|(\lambda, \nu)| \leq (\lambda, \lambda) \cdot (\nu, \nu),$$

from which follows the finiteness of (λ, ν). The quantity

$$||\nu|| = \sqrt{(\nu, \nu)}$$

will be called the *norm of the signed measure* ν.

Convergence of signed measures in the norm will be called *strong convergence*. The Hilbert structure generates another form of convergence for signed measures which we shall call *weak convergence*. Namely, a sequence ν_n *weakly converges* to ν if $(\nu_n, \lambda) \to (\nu, \lambda)$ for any $\lambda \in \mathscr{E}_K$.

Strong convergence implies weak convergence. This is shown in the usual way using the Schwarz-Bunyakovsky inequality.

We now prove, two lemmas from which it will follow that the set of measures with compact support, which have continuous (or even smooth) potentials is dense in $\mathscr{E}_\alpha$.

Lemma 1.2. *Suppose that* $\nu \in \mathscr{E}_\alpha$ $(0 < \alpha < p)$. *Then the sequences of signed measures*

$$\nu_n = \nu * m^{(1/n)}$$

and $\hat{\nu}_n$, *where* $\hat{\nu}_n$ *is the restriction of* ν_n *to the ball* $|x| < n$, *strongly converge to* ν.

Proof. In view of the decompositions $\nu = \nu^+ - \nu^-$, $\nu_n = \nu^+ * m^{(1/n)} - \nu^- * m^{(1/n)}$ it is sufficient to prove the lemma for measures $\mu \in \mathscr{E}_\alpha^+$. We begin with the formula

$$||\mu - \mu_n||^2 = \int_{R^p} [U^\mu_{\alpha/2}(x) - U^{\mu_n}_{\alpha/2}(x)]^2 \, dx, \tag{1.4.8}$$

obtained in the proof of Theorem 1.15.

By (1.3.12) for all x

$$\lim_{n\to\infty} U^{\mu_n}_{\alpha/2}(x) = U^{\mu}_{\alpha/2}(x).$$

Furthermore, in view of inequality (1.3.13), we may write

$$[U^{\mu}_{\alpha/2}(x) - U^{\mu}_{\alpha/2}(x)]^2 \leq 2[U^{\mu}_{\alpha/2}(x)]^2 + 2[U^{\mu_n}_{\alpha/2}(x)]^2$$
$$< 2(1 + A^2)\,[U^{\mu}_{\alpha/2}(x)]^2,$$

where $A > 0$ depends only on α, p. This permits us to pass to the limit on the right side of (1.4.8) as $n \to \infty$, and we obtain

$$\lim_{n\to\infty} ||\mu - \mu_n|| = 0.$$

In complete analogy, using the remark to Theorem 1.11 and the inequality

$$\left[U^{\hat{\mu}_n}_{\alpha/2}(x)\right]^2 \leq \left[U^{\mu_n}_{\alpha/2}(x)\right]^2 < A^2\left[U^{\mu}_{\alpha/2}(x)\right]^2,$$

we prove that

$$\lim_{n\to\infty} ||\mu - \hat{\mu}_n|| = 0. \quad \square$$

Lemma 1.2′. *Suppose that* $\nu \in \mathscr{E}_2^{(1)}$ $(p = 2)$. *Then the signed measures*

$$\nu_n = \nu * m^{(1/n)}$$

strongly converge to ν.

Proof. As in Lemma 1.2, we need only consider measures.

Knowing that the signed measures $\mu - \mu_n$ satisfies condition (a) of Theorem 1.16 we may write

$$||\mu - \mu_n||^2 = \int_{R^2} [U^{\mu}_1(x) - U^{\mu_n}_1(x)]^2\,dx.$$

The proof is completed just as in Lemma 1.2. $\square$

The same fact holds also for the space $\mathscr{E}_G$ of signed measures with finite Green energy

$$||\nu||_g^2 = \int_{G\times G} G(x, y)\,d\nu(x)\,d\nu(y).$$

This easily follows from formula (1.3.5).

Corollary 1. *The set of signed measures with compact supports, whose potentials have continuous derivatives up to a fixed order* N, *is dense in* $\mathscr{E}_\alpha$ *(or* $\mathscr{E}_g$*).*

Indeed, it is sufficient in Lemmas 1.2 and 1.2′ to replace $\hat{\nu}_n$ (correspondingly ν_n) by the signed measure $\hat{\nu}_n^{(N)}$, obtained by N-fold convolution of $\hat{\nu}_n$ with $m^{(1/n)}$; this provides a sequence of signed measures strongly converging to ν, whose potentials have the required smoothness. From this we also obtain

Corollary 2. *The set of absolutely continuous signed measures with compact support, which have a smooth density is dense in* $\mathscr{E}_\alpha$ $(\mathscr{E}_g)$.

It turns out that the property of *signed measures* to have *compact supports* in the preceding assertion may be replaced by the condition that the *potentials* have *compact support.*

Theorem 1.17. *The set of signed measures whose potentials have compact supports and continuous derivatives of order* N *is dense in* $\mathscr{E}_\alpha$.

Proof. We shall assume from the beginning that $N \geq p + 2$. It is sufficient to prove that every signed measure ν with compact support having continuous derivatives of order N is the strong limit of signed measures satisfying the condition of the theorem.

Take $R > 0$ such that $S(\nu)$ is contained in the ball $|x| < R/2$; define the function $f(x)$ with compact support in the following way:

$$f(x) = U_\alpha^\nu(x)\, m_R(x), \tag{1.4.9}$$

where

$$m_R(x) = \begin{cases} 1, & |x| \leq R, \\ \left[1 - \left(\frac{|x| - R}{R}\right)^{N+1}\right]^{N+1}, & R \leq |x| < 2R, \\ 0, & |x| \geq 2R. \end{cases} \tag{1.4.9'}$$

It is clear that $m_R(x)$ has continuous derivatives of order N in R^p; moreover for any order k, we have

$$|\partial^k m_R(x)| < \frac{Q}{R^k} \quad (k = 0, 1, \ldots, N),$$

where the constant Q depends only on N.

Hence, the function $f(x)$ also has continuous derivatives up through order N, and

$$\partial^k f(x) = \sum_{j=0}^{k} A_{jk}\, \partial^j U_\alpha^\nu(x)\, \partial^{k-j} m_R(x).$$

If we take into account that for large $|x|$

$$U_\alpha^\nu(x) \sim \frac{\nu(1)}{|x|^{p-\alpha}}, \qquad \partial^j U_\alpha^\nu(x) \sim \frac{A_j}{|x|^{p-\alpha+j}}, \tag{1.4.10}$$

where the A_j are easily defined by the signed measure ν, then the preceding formulas allow us to conclude that in the spherical shell $R/2 \leq |x|$

$\leq 2R$, we have the following bounds:

$$|\partial^k f(x)| < \frac{Q}{R^{p-\alpha+k}} \quad (k = 0, 1, \ldots, N), \tag{1.4.11}$$

where the constant Q[20] depends only on ν and N.

The function $f(x)$, by Lemma 1.1, is a potential

$$f(x) = U_\alpha^\psi(x),$$

where the absolutely continuous signed measure ψ is in $\mathscr{E}_\alpha$, since

$$||\psi||^2 = \int_{|x|<2R} f(x)\, \psi(x)\, dx < \infty.$$

Consequently,

$$F(x) = f(x) - U_\alpha^\nu(x) = U_\alpha^{\psi-\nu}(x),$$

and we must evaluate the quantity $||\psi - \nu||$.

For this we note that

$$||\psi - \nu||^2 = \int_{R^p} F(x)\, d(\psi - \nu) = \int_{|x|>R} F(x)\, d(\psi - \nu)$$

$$= \int_{|x|>R} F(x)\, \psi(x)\, dx.\,^{21} \tag{1.4.12}$$

The density $\psi(x)$ for $\alpha = 2m$ is defined by the formula

$$\psi(x) = k_{-2m} * f = \left(-\frac{1}{4\pi^2}\right)^m \Delta^m f(x), \tag{1.4.13}$$

and if α is not equal to an even number, then by the formula

$$\psi(x) = A(p, -\alpha) \int_{R^p} \left\{ f(x+y) - \sum_{k=0}^{m} H_k\, \Delta^k f(x)\, |y|^{2k} \right\} \frac{dy}{|y|^{p+\alpha}},$$

where $2m < \alpha < 2m + 2$ (cf. § 1, No. 2). We remark that from the inequalities $\alpha < p$, $N \geq p + 2$ it follows that $N > 2m + 2$, therefore $\Delta^k f(x)$ is defined for all $k \leq m + 1$.

We shall now evaluate $\psi(x)$. Consider first the more complex situation when α is not an even number. Suppose that $|x| > 3R$. Then by (1.4.9) and (1.4.9′)

$$|\psi(x)| = \left| A(p, -\alpha) \int_{|z|<2R} \frac{f(z)\, dz}{|z - x|^{p+\alpha}} \right|$$

$$< \frac{Q_1}{|x|^{p+\alpha}} \int_{|z|<2R} |f(z)|\, dz$$

---

[20] Just as all the numbered constants Q appearing below.

[21] The density of an absolutely continuous signed measure ψ we have denoted with the same letter ψ.

since in the region of integration $|x| < \text{const}\ |z - x|$. Furthermore from the definition of $f(x)$ and (1.4.10), there exists a constant Q_2, such that

$$|f(z)| < \frac{Q_2}{|z|^{p-\alpha}}.\ ^{22} \tag{1.4.14}$$

Consequently,

$$\int_{|z|<2R} |f(z)|\, dz < Q_2\omega_p \int_0^{2R} \frac{r^{p-1}dr}{r^{p-\alpha}} < Q_3R^\alpha.$$

Thus, for $|x| > 3R$,

$$|\psi(x)| < \frac{Q_4R^\alpha}{|x|^{p-\alpha}}. \tag{1.4.15}$$

Suppose now that $|R| < |x| \leq 3R$. We first evaluate

$$\psi_1(x) = A(p, -\alpha) \int_{|y|<R/2} \left\{f(x+y) - \sum_{k=0}^{m} H_k\, \Delta^k f(x)\, |y|^{2k}\right\} \frac{dy}{|y|^{p+\alpha}}.$$

Since $f(x)$ has continuous derivatives of order $2m + 2$, then, expanding $f(x + y)$ by Taylor's formula with remainder of order $2m + 2$ with respect to $|y|$, we obtain

$$|\psi_1(x)| < Q_5 \int_{|y|<R/2} \left|\sum_j c_j\, \partial_j^{2m+2} f(\xi)\right| \frac{dy}{|y|^{p-\alpha-2m-2}},$$

where under the modulus sign is a certain linear combination of derivatives of order $2m + 2$. Since the point ξ is in the ball of radius $R/2$ with center x, then $|\xi| > R/2$, and, by (1.4.11)

$$\left|\sum_j c_j\partial_j^{2m+2} f(\xi)\right| < \frac{Q_6}{R^{p-\alpha+2m+2}}.$$

Consequently,

$$|\psi_1(x)| < \frac{Q_7}{R^{p-\alpha+2m+2}} \int_0^{R/2} \frac{r^{p-1}dr}{r^{p+\alpha-2m-2}} = \frac{Q_8}{R^p}.$$

Let us evaluate

$$\psi_2(x) = A(p, -\alpha) \times \int_{|y|>R/2} \left\{f(x+y) - \sum_{k=0}^{m} H_k\Delta^k f(x)\, |y|^{2k}\right\} \frac{dy}{|y|^{p+\alpha}}.$$

First of all, by (1.4.11)

$$\int_{|y|>R/2} \Delta^k f(x) \frac{dy}{|y|^{p+\alpha-2k}} < \frac{Q_9}{R^{p-\alpha+2k}} \int_{R/2}^{\infty} \frac{r^{p-1}dr}{r^{p+\alpha-2k}} = \frac{Q_{10}}{R^p},$$

---

[22] It is essential that Q_2 does not depend on R.

and therefore

$$\left| \int_{|y|>R/2} \sum_{k=0}^{m} H_k \Delta^k f(x) \, |y|^{2k} \frac{dy}{|y|^{p+\alpha}} \right| < \frac{Q_{11}}{R^p}.$$

Furthermore, using (1.4.14), we obtain

$$\left| \int_{|y|>R/2} f(x+y) \frac{dy}{|y|^{p+\alpha}} \right| = \left| \int_{\substack{|z-x|>R/2 \\ |z|<2R}} \frac{f(z)\, dz}{|z-x|^{p+\alpha}} \right|$$

$$< Q_2 \int_{\substack{|z-x|>R/2 \\ |z|<2R}} \frac{dz}{|z|^{p-\alpha} |z-x|^{p+\alpha}}$$

$$= \frac{Q_2}{R^p} \int_{\substack{|z_1-x_1|>1/2 \\ |z_1|<2}} \frac{dz_1}{|z_1|^{p-\alpha} |z_1-x_1|^{p+\alpha}},$$

where $x_1 = \frac{1}{R} x$, $1 < |x_1| < 3$. Consequently,

$$\left| \int_{|y|>R/2} f(x+y) \frac{dy}{|y|^{p+\alpha}} \right| < \frac{Q_{12}}{R^p}$$

and therefore

$$|\psi_2(x)| < \frac{Q_{13}}{R^p}.$$

Finally, for $R < |x| < 3R$ we have the bound

$$|\psi(x)| < \frac{Q_{14}}{R^p}. \tag{1.4.16}$$

Let us return to the formula (1.4.12), writing it in the form

$$||\psi - \nu||^2 = \int_{R<|x|<3R} F(x) \, \psi(x) \, dx + \int_{|x|>3\mathrm{R}} F(x) \, \psi(x) \, dx.$$

In view of (1.4.16), (1.4.15), and also the bound

$$|F(x)| < \frac{Q_{15}}{|x|^{p-\alpha}}, \tag{1.4.17}$$

which follows from the definition of $F(x)$, (1.4.10), and (1.4.11), we have

$$\left| \int_{R<|x|<3R} F(x) \, \psi(x) \, dx \right| \leq \frac{Q_{16}}{R^p} \int_{R<|x|<3R} \frac{dx}{|x|^{p-\alpha}}$$

$$= \frac{Q_{16}\omega^p}{R^p} \int_R^{3R} \frac{r^{p-1} dr}{r^{p-\alpha}} = Q_{16}\omega_p \frac{(3R)^\alpha - R^\alpha}{\alpha R^p} = O\left(\frac{1}{R^{p-\alpha}}\right)$$

and

$$\left| \int_{|x|>3R} F(x)\,\psi(x)\,dx \right| \le Q_{17} R^{\alpha} \int_{|x|>3R} \frac{dx}{|x|^{2p}}$$

$$= Q_{17}\omega_p R^{\alpha} \int_{3R}^{\infty} \frac{r^{p-1}dr}{r^{2p}} = O\left(\frac{1}{R^{p-\alpha}}\right).$$

This shows that as $R \to \infty$

$$\|\psi - \nu\| \to 0$$

and the theorem is established in the case that α is not even.

Now suppose that $\alpha = 2m$. Then from (1.4.13) and (1.4.11) it follows that for $R < |x| < 2R$

$$|\psi(x)| < \frac{Q}{R^{p-\alpha+2m}} = \frac{Q}{R^p},$$

and for $|x| > 2R$

$$\psi(x) \equiv 0.$$

Therefore (1.4.12) and (1.4.17) give

$$\|\psi - \nu\|^2 = \int_{R<|x|<2R} F(x)\,\psi(x)\,dx$$

$$\le \frac{Q}{R^p} \int_{R}^{2R} \frac{r^{p-1}dr}{r^{p-2m}} = O\left(\frac{1}{R^{p-2m}}\right),$$

and since $2m < p$, then as above,

$$\|\psi - \nu\| \to 0 \quad \text{as} \quad R \to \infty.$$

Thus, the proof of the theorem is complete[23]. □

13. Consider now the convex cone $\mathscr{E}_\alpha^+ \subset \mathscr{E}_\alpha$ of measures with finite energy.[24]

The following assertions hold for $\mathscr{E}_\alpha^+$.

Lemma 1.3. *Weak convergence of the measures μ_n to μ implies their vague convergence to μ*

Proof. Weak convergence of μ_n to μ means that for any signed measure $\nu \in \mathscr{E}_\alpha$

$$\lim_{n\to\infty} \int_{R^p} U_\alpha^\nu(x)\,d\mu_n(x) = \int_{R^p} U_\alpha^\nu(x)\,d\mu(x).$$

---

[23] The case of the logarithmic kernel in R^2 is treated just as the case for α even.

[24] In the case of a logarithmic kernel $\mathscr{E}_2^+ \subset \mathscr{E}_2^{(1)}$. Analogous results hold for $\mathscr{E}_g^+$.

From Lemma 1.1 it follows that for any function $f(x)$ with compact support having $p + 2$ continuous derivatives

$$\lim_{n\to\infty} \int_{R^p} f(x)\, d\mu_n(x) = \int_{R^p} f(x)\, d\mu.$$

Since the set of these functions $f(x)$ is dense in Φ, then by Theorem 0.4 of the Introduction, $\mu_n \dot{\to} \mu$. □

Lemma 1.4. *Suppose that the measures $\mu_n \in \mathscr{E}_\alpha^+$ vaguely converge to μ. If also $||\mu_n|| < M$, then $\mu \in \mathscr{E}_\alpha^+$ and the μ_n weakly converge to μ.*

Proof. That μ belongs to $\mathscr{E}_\alpha^+$ follows from (1.4.5) which may be written in the form

$$||\mu|| \le \varliminf_{n\to\infty} ||\mu_n||.$$

If ψ is a signed measure with continuous potential $U_\alpha^\psi(x)$ having compact support, then

$$(\mu_n, \psi) = \int U_\alpha^\psi(x)\, d\mu_n(x) \to \int U_\alpha^\psi(x)\, d\mu(x) = (\mu, \psi).$$

According to Theorem 1.17 such signed measures form a dense set in $\mathscr{E}_\alpha$. Therefore, given $\nu \in \mathscr{E}_\alpha$ and $\varepsilon > 0$ we choose the signed measure ψ such that $||\nu - \psi|| < \varepsilon$; then we may write for $n > n_0(\varepsilon)$

$$|(\mu_n, \nu) - (\mu, \nu)| \le |(\mu_n, \psi) - (\mu, \psi)| + |(\mu_n, \nu - \psi)|$$

$$+ |(\mu, \nu - \psi)| \le |(\mu_n, \psi) - (\mu, \psi)| + 2M\varepsilon < (2M + 1)\,\varepsilon.$$

In this way

$$\lim_{n\to\infty} (\mu_n, \nu) = (\mu, \nu), \quad \nu \in \mathscr{E}_\alpha,$$

and the lemma is proved. □

Our next goal is to prove the important theorem asserting the completeness of $\mathscr{E}_\alpha^+$. Basic to that proof is

Lemma 1.5. *If $\{\mu_n\}$ is a fundamental sequence in $\mathscr{E}_\alpha^+$ and $\mu_n \dot{\to} \mu$, then $\mu \in \mathscr{E}_\alpha^+$ and μ_n strongly converges to μ.*

Proof. Since $\{\mu_n\}$ is fundamental, there exists a number M satisfying the inequalities

$$||\mu_n|| < M \quad \text{for all } n.$$

Thus, by Lemma 1.4, $\mu \in \mathscr{E}_\alpha^+$ and the μ_n weakly converge to μ. Consequently,

$$||\mu - \mu_n||^2 = (\mu - \mu_n, \mu - \mu_n)$$

$$= \lim_{p\to\infty} (\mu - \mu_n, \mu_p - \mu_n) \le \varliminf_{p\to\infty} ||\mu - \mu_n|| \cdot ||\mu_p - \mu_n||,$$

from which

$$||\mu - \mu_n|| \leq \varliminf_{p\to\infty} ||\mu_p - \mu_n|| < \varepsilon$$

for $n > n_0(\varepsilon)$, and the lemma is proved. □

Theorem 1.18. *The metric space* $\mathscr{E}_\alpha^+$ *is complete.*

Proof. By Lemma 1.5 it is sufficient to establish that for every fundamental sequence $\{\mu_n\}$, $\mu_n \in \mathscr{E}_\alpha^+$, a vague limit μ exists.

First of all, for any $\nu \in \mathscr{E}_\alpha$ the number sequence $\{(\mu_n, \nu)\}$ converges. This follows from the inequality

$$|(\mu_n - \mu_m, \nu)| \leq ||\mu_n - \mu_m||\, ||\nu||.$$

Consequently, by Lemma 1.1 for any function $f(x)$ with compact support having $p + 2$ continuous derivatives

$$\lim_{n\to\infty} \int f(x)\, d\mu_n(x)$$

exists. Since the set of these functions is dense in Φ^+, then by Theorem 0.4′ of the Introduction we conclude that $\mu_n \dot{\to} \mu \in \mathfrak{M}^+$. □

On the other hand, in contrast to Theorem 1.18, we have the following negative result.

Theorem 1.19. *The metric space* $\mathscr{E}_\alpha (\alpha > 1)$ *is not complete.*

Proof. We construct a fundamental sequence of signed measures $\{\nu_n\}$ which do not have a limit in $\mathscr{E}_\alpha$.

Let $\{r_k\}$ be a monotone increasing sequence of positive numbers tending to 1. We shall soon make the choice of the r_k more precise. Consider the sequence of signed measures

$$\nu_n = n\varepsilon^{(1)} - \sum_{k=1}^{n} \varepsilon^{(r_k)}.$$

It is easy to see that the potential $U_\alpha^{\varepsilon^{(r)}}(x)$ as a function of $r > 0$ is uniformly continuous with respect to $x \in R^p$.[25] Since $\nu_n - \nu_{n-1} = \varepsilon^{(1)} - \varepsilon^{(r_n)}$, then we may choose the numbers r_n, such that

$$|U_\alpha^{\nu_n - \nu_{n-1}}(x)| < \frac{1}{2^n}.$$

Then

$$||\nu_n - \nu_{n-1}||^2 = \int U_\alpha^{\nu_n - \nu_{n-1}}(x)\, d(\varepsilon^{(1)} - \varepsilon^{(r_n)}) < \frac{1}{2^{n-1}},$$

---

[25] In the case of a logarithmic potential in R^2 one would consider $x \in K$, where K is any compact set.

and it is easy to see that the sequence of signed measures $\{\nu_n\}$ is fundamental.

Nevertheless, the strong limit of this sequence does not exist. Indeed, if $||\nu_n - \nu|| \to 0$, then by Lemma 1.1 it would follow that for any function $f(x)$ with compact support having $p + 2$ continuous derivatives

$$\lim_{n\to\infty} \int f(x)\, d\nu_n = \int f(x)\, d\nu.$$

Take for $f(x)$ the function satisfying the inequality $0 \le f(x) \le 1$, equal to 1 in the ball B_ϱ: $|x| \le \varrho < 1$, and equal to 0 for $|x| \ge \frac{1+\varrho}{2}$. Then

$$\nu(B_\varrho) + \int\limits_{\varrho<|x|<\frac{1+\varrho}{2}} f(x)\, d\nu < -\sum_{r_k\le\varrho} 1,$$

from which

$$-\nu^-(B_\varrho) - \int\limits_{\varrho<|x|<\frac{1+\varrho}{2}} f(x)\, d\nu^- < -\sum_{r_k\le\varrho} 1,$$

and

$$\nu^-\left(B_{\frac{1+\varrho}{2}}\right) \ge \nu^-(B_\varrho) + \int\limits_{\varrho<|x|<\frac{1+\varrho}{2}} f(x)\, d\nu^- > \sum_{r_k\le\varrho} 1.$$

If we let ϱ tend to 1, then we obtain

$$\nu^-(T) = +\infty,$$

where T is the open ball $|x| < 1$. But this is impossible and the theorem is proved. ▯

14. In the classical case the energy integral $I_2[\nu]$, $\nu \in \mathscr{E}_2$ (or $\nu \in \mathscr{E}_2^{(0)}$ if $p = 2$), admits one other interesting representation. Suppose that the signed measure ν has compact support and a smooth density

$$\nu(x) = \frac{-1}{4\pi^2} \Delta U_2^\nu.$$

Then, since $U_2^\nu(x)$ decreases at infinity, we apply Green's formula to obtain[26]

$$I_2[\nu] = \int\limits_{S(\nu)} U_2^\nu(x)\, \nu(x)\, dx = -\frac{1}{4\pi^2} \int\limits_{S(\nu)} U_2^\nu(x)\, \Delta U_2^\nu(x)\, dx$$

$$= \frac{1}{4\pi^2} \int\limits_{R^p} |\operatorname{grad} U_2^\nu|^2\, dx. \qquad (1.4.18)$$

[26] Under analogous conditions on ν_1 and ν_2

$$I_2[\nu_1, \nu_2] = \frac{1}{4\pi^2} \int\limits_{R^p} (\operatorname{grad} U_2^{\nu_1}, \operatorname{grad} U_2^{\nu_2})\, dx. \qquad (1.4.18')$$

Analogously, for a bounded region G with a smooth boundary ∂G and signed measure ν, $S(\nu) \subset G$ with smooth density, we may write

$$\nu(x) = \frac{-1}{4\pi^2} \Delta U_g^\nu(x),$$

where

$$U_g^\nu(x) = \int_{S(\nu)} G(x, y)\, d\nu(y)$$

is the Green potential; noticing that $U_g^\nu = 0$ on ∂G, we obtain

$$I_g[\nu] = \frac{1}{4\pi^2} \int_G |\text{grad } U_g^\nu|^2\, dx. \tag{1.4.19}$$

Denote by Λ the linear manifold of signed measures with compact support and smooth density. By Corollary 2 to Lemma 1.2, Λ is dense in $\mathscr{E}_2$ (or $\mathscr{E}_G$). On the other hand, let D denote the linear space of continuously differentiable functions $f(x)$ in R^p (correspondingly, in the region G), equal to zero at infinity (correspondingly, on ∂G), for which the *Dirichlet norm*

$$||f||^2 = \frac{1}{4\pi^2} |\text{grad } f|^2\, dx$$

is finite. Then formula (1.4.18) (or (1.4.19)) defines an isomorphism from Λ to D. The natural questions arise: does this isomorphism extend to all of $\mathscr{E}_2$? To what extension of the space D belong the potentials $U_2^\nu(x)$, $\nu \in \mathscr{E}_2$?

We begin with an answer to the second question.

Lemma 1.6. *Any potential* $U_2^\nu(x)$, $\nu \in \mathfrak{M}$, *satisfying the finiteness condition* (1.3.10), *is absolutely continuous on almost all lines parallel to the axes* Ox_i $(i = 1, 2, \ldots, p)$. *Moreover almost everywhere*

$$\frac{\partial U_2^\nu}{\partial x_i} = \int_{R^p} \frac{\partial}{\partial x_i} k_2(x - y)\, d\nu(y) = A(p, 2) \int_{R^p} \frac{\partial}{\partial x_i} \frac{1}{|x - y|^{p-2}}\, d\nu(y).^{27} \tag{1.4.20}$$

Proof. Clearly, it is sufficient to consider the case when ν is a measure. We will show that the integral

$$F_i(x) = \int_{R^p} \frac{\partial}{\partial x_i} \frac{1}{|x - y|^{p-2}}\, d\nu(y)$$

---

[27] The analogous formulas hold, if $p = 2$, for a logarithmic potential and signed measure ν with compact support.

is locally integrable and, consequently, is finite almost everywhere. The inequality

$$\left|\frac{\partial}{\partial x_i}\frac{1}{|x-y|^{p-2}}\right| \leq \frac{p-2}{|x-y|^{p-1}}$$

shows that $\frac{\partial}{\partial x_i}\frac{1}{|x-y|^{p-2}}$ is locally integrable. By the Fubini theorem, we may write, for any bounded region $\omega \subset R^p$, the equation

$$\int_\omega dx \int_{R^p} \left|\frac{\partial}{\partial x_i}\frac{1}{|x-y|^{p-2}}\right| d\nu(y) = \int_{R^p} d\nu(y) \int_\omega \left|\frac{\partial}{\partial x_i}\frac{1}{|x-y|^{p-2}}\right| dx. \tag{1.4.21}$$

But

$$\int_\omega \left|\frac{\partial}{\partial x_i}\frac{1}{|x-y|^{p-2}}\right| dx = O\left(\frac{1}{|y|^{p-1}}\right),$$

and by (1.3.10) the above multiple integrals are finite. From this follows the local integrability of $F_i(x)$, and, again using the Fubini theorem, we conclude that for almost all lines parallel to the axis $0x_i$, and any finite a, b

$$\int_a^b |F_i(x)|\, dx_i < \infty.$$

In addition, on these lines

$$\begin{aligned}\int_a^b F_i(x)\, dx_i &= \int_{R^p} d\nu(y) \int_a^b \frac{\partial}{\partial x_i}\frac{1}{|x-y|^{p-2}}\, dx_i \\ &= \frac{1}{A(p,2)}\left[U_2^\nu(b) - U_2^\nu(a)\right],\end{aligned} \tag{1.4.22}$$

and consequently, $U_2^\nu(x)$ is absolutely continuous there. Since (1.4.20) results from (1.4.22) by differentiation along b, the lemma is proved. □

Remark 1. The lemma remains true for Green potentials of any signed measure ν with compact support, $S(\nu) \subset G$. Also, $U_g^\nu(x)$ will be absolutely continuous on the intervals belonging to G of almost all lines parallel to a coordinate axis. To see this, in the previous proof take G for ω and note that the quantity $\left|\frac{\partial}{\partial x_i} h(x,y)\right|$ is bounded on $S(\nu)$ (cf. 1.3.5).[28]

[28] One could also consider any signed measure ν in G and obtain the absolute continuity of $U_g^\nu(x)$ on any internal lying entirely within G, of almost all lines parallel to $0x_i$.

Remark 2. Utilizing spherical coordinates in R^p, one can prove that on almost all rays from the point 0, the potential $U_2^\nu(x)$ is an absolutely continuous function of $r = |x|$.

An analogous result can be obtained for Green potentials. However, for this it is necessary to introduce certain auxiliary concepts which are of independent interest.

Within the region G, we fix the point 0 and consider the Green function

$$G_0(x) = G(x, 0)$$

with a pole at this point. The *level surfaces*

$$G_0(x) = r \quad (0 \leq r < \infty)$$

for $r > 0$ are smooth (even analytic) surfaces if only they do not contain points of the set

$$N = \{x : \text{grad } G_0(x) = 0\}.^{29}$$

Let us now examine the so-called *Green lines*. These are orthogonal trajectories of the family of level surfaces, or, alternatively, integral curves of the field grad $G_0(x)$. Through every point of G except 0, there is a unique Green line. Since in a sufficiently small neighborhood of the pole 0

$$\text{grad } G_0(x) = \text{grad}\left(\frac{1}{|x|^{p-2}} + h(x)\right) \neq 0,$$

through every point of this neighborhood (ecxept 0) there is a unique Green line.

In connection with these lines, the following can be shown:

(a) at 0 each of these lines has a one-sided tangent;

(b) to each ray issuing from 0 there corresponds a Green line touching it at the point 0;

(c) the correspondence between the points of the surface $G_0(x) = r$ (r sufficiently large) and the rays issuing from 0, by which to each ray is associated the point of intersection of the corresponding Green line with the surface $G_0(x) = r$, is a homeomorphism.

Let L denote the set of all Green lines issuing from 0, and let L_1 be the set of tangent rays (or unit vectors). Using the correspondence between L and L_1, one can introduce into L a measure by taking as the measure of $E \subset L$ $\frac{1}{4\pi}$ the measure of the intersection of the corresponding set $E_1 \subset L_1$ with the unit sphere.

[29] It can be shown that N consists of no more than a countable sum of analytic manifolds of dimension $\leq p - 2$.

A Green line issuing from 0 is called *regular*, if it terminates on ∂G, i.e. if the value G changes along the line (monotonically) from $+\infty$ to 0. The following assertions hold:

(1) almost all lines in L are regular,

(2) except for a set $N_1 \supset N$ of Lebesgue measure zero, through each point of the region G, there is a unique regular Green line.

This allows us to introduce coordinates into $G \setminus N_1$ by associating with each point $x \in G \setminus N_1$ the number r, equal to the value $G_0(x)$ at that point, and the unit vector $l \in L_1$, corresponding to that regular Green line which passes through the point x. The coordinates (r, l) will be called the *Green coordinates* of the point x.

Note that if s denotes arc length along a Green line from the point 0, then at any point of $G \setminus N_1$

$$\frac{\partial}{\partial s} = \frac{\partial G}{\partial n} \frac{\partial}{\partial r},$$

where $\partial G/\partial n$ is the derivative along the external normal to the level surface passing through the point x.

Modifying somewhat the proof of Lemma 1.6, one can show that for almost all Green lines, the potential $U_g^\nu(x)$ is an absolutely continuous function of the parameter r (or s).

Now let us consider potentials $U_2^\nu(x)$ (correspondingly $U_g^\nu(x)$) with finite energy; first we prove the following lemma.

Lemma 1.7. *If* $\mu \in \mathscr{E}_2^+$ $(\mathscr{E}_g^+)$, *then*

$$\frac{1}{4\pi^2} \int_{R^p} |\operatorname{grad} U_2^\mu|^2\, dx \le ||\mu||^2 < \infty,$$

correspondingly,

$$\frac{1}{4\pi^2} \int_G |\operatorname{grad} U_g^\mu|^2\, dx \le ||\mu||_g^2 < \infty.$$

Proof. We prove the case $\mu \in \mathscr{E}_2^+$, leaving to the reader the case $\mu \in \mathscr{E}_g^+$.

First suppose that the measure μ has compact support. Set

$$\mu_n = \mu * m^{(1/n)} * m^{(1/n)} * m^{(1/n)}.$$

Then the potential

$$U_2^{\mu_n}(x) = U_2^\mu(x) * m^{(1/n)} * m^{(1/n)} * m^{(1/n)}$$

(cf. remark to Theorem 1.11) will have three continuous derivatives, and, according to (1.4.18)

$$\frac{1}{4\pi^2} \int_{R^p} |\operatorname{grad} U_2^{\mu_n}|^2\, dx = ||\mu_n||^2.$$

Since $U_2^\mu(x)$ is superharmonic, it follows that

$$U_2^{\mu_n}(x) \le U_2^\mu(x),$$

and therefore $||\mu_n|| \le ||\mu||$.[30] In fact

$$(\mu, \mu) - (\mu_n, \mu_n) = (\mu - \mu_n, \mu) + (\mu_n, \mu - \mu_n)$$
$$= \int U_2^{\mu-\mu_n}\, d\mu + \int U_2^{\mu-\mu_n}\, d\mu_n \ge 0.$$

We will show that almost everywhere in R^p we have

$$\lim_{n\to\infty} |\mathrm{grad}\, U_2^{\mu_n}|^2 = |\mathrm{grad}\, U_2^\mu|^2. \tag{1.4.23}$$

Of course, it is sufficient to establish that

$$\lim_{n\to\infty} \frac{\partial U_2^{\mu_n}}{\partial x_i} = \frac{\partial U_2^\mu}{\partial x_i}. \tag{1.4.23'}$$

almost everywhere in R^p. From formula (0.2.19) of the Introduction it follows that

$$\frac{\partial U_2^{\mu_n}}{\partial x_i} = \frac{\partial U_2^\mu}{\partial x_i} * m^{(1/n)} * m^{(1/n)} * m^{(1/n)},$$

where by the continuity of both sides, this equation holds everywhere in R^p (and not only almost everywhere as an equation between two distributions).[31]

This formula immediately implies (1.4.23') and, consequently, (1.4.23). Now the Fatou lemma yields

$$\frac{1}{4\pi^2} \int_{R^p} \mathrm{grad}\, U_2^\mu|^2\, dx \le \varliminf_{n\to\infty} \frac{1}{4\pi^2} \int_{R^p} |\mathrm{grad}\, U_2^{\mu_n}|^2\, dx \le ||\mu||,$$

and the assertion is proved for measures with compact support.

If μ is an arbitrary measure from $\mathscr{E}_2^+$, then denoting by $\mu^{(n)}$ the restriction of μ to the ball $|x| \le n$, by Lemma 1.6, we may write the equations

$$\frac{\partial U_2^\mu}{\partial x_i} = A(p, 2) \int_{R^p} \frac{\partial}{\partial x_i} \frac{1}{|x-y|^{p-2}}\, d\mu\,(y),$$

$$\frac{\partial U_2^{\mu^{(n)}}}{\partial x_i} = A(p, 2) \int_{R^p} \frac{\partial}{\partial x_i} \frac{1}{|x-y|^{p-2}}\, d\mu^{(n)}(y)$$
$$= A(p, 2) \int_{|x|<n} \frac{\partial}{\partial x_i} \frac{1}{|x-y|^{p-2}}\, d\mu(y),$$

[30] It can be shown that $||\mu_n|| \to ||\mu||$ as $n \to \infty$.

[31] By Lemma 1.6 the generalized derivative of $U_2^\mu(x)$ along x_i coincides (in the sense of distribution theory) with $\partial U_2^\mu/\partial x_i$.

which hold almost everywhere for all n. It then follows from these that

$$\lim_{n\to\infty} |\operatorname{grad} U_2^{\mu^{(n)}}|^2 = |\operatorname{grad} U_2^{\mu}|^2,$$

almost everywhere; again applying the Fatou lemma, we obtain

$$\frac{1}{4\pi^2}\int_{R^p} |\operatorname{grad} U_2^{\mu}|\, dx \le \varliminf_{n\to\infty} \frac{1}{4\pi^2}\int_{R^p} |\operatorname{grad} U_2^{\mu^{(n)}}|^2\, dx$$

$$\le \varliminf_{n\to\infty} \|\mu^{(n)}\|^2 \le \|\mu\|,$$

as required. □

Remark. Using spherical (or Green) coordinates, in a similar manner we could have established the inequality

$$\frac{1}{4\pi^2}\int_{R^p} \left(\frac{\partial U_2^{\mu}}{\partial r}\right)^2 dx \le \|\mu\|^2, \quad \mu\in\mathscr{E}_2^+,$$

or correspondingly

$$\frac{1}{4\pi^2}\int_{G} \left(\frac{\partial U_2^{\mu}}{\partial s}\right)^2 dx \le \|\mu\|_g^2, \quad \mu\in\mathscr{E}_G^+$$

From this and Remark 2 to Lemma 1.6 one can conclude that for almost all rays (or Green lines) issuing from the point 0,

$$\lim_{|x|\to\infty} U_2^{\mu}(x), \quad \text{correspondingly} \quad \lim_{r\to 0} U_g^{\mu}(x).$$

exists.

From the following it will be seen that this limit is equal to zero for almost all rays (Green lines).

We now prove the basic theorem of this section.

Theorem 1.20. *If* $\nu\in\mathscr{E}_2$, *correspondingly* $\mathscr{E}_G$, *then*

$$\|\nu\|^2 = I_2[\nu] = \frac{1}{4\pi^2}\int_{R^p} |\operatorname{grad} U_2^{\nu}(x)|^2\, dx, \tag{1.4.24}$$

correspondingly

$$I_g[\nu] = \frac{1}{4\pi^2}\int_{G} |\operatorname{grad} U_g^{\nu}(x)|^2\, dx.$$

Conversely, if the integral on the right side of (1.4.24) *is finite, then* $\nu\in\mathscr{E}_2$ *and equation* (1.4.24) *holds.*

Proof. If we take into account the equations

$$I_2[\nu] = I_2[\nu^+] - 2I_2[\nu^+,\nu^-] + I_2[\ ^-]$$

and

$$\int_{R^p} |\text{grad } U_2^{\nu}|^2 \, dx = \int_{R^p} |\text{grad } U_2^{\nu^+}|^2 \, dx$$

$$- 2 \int_{R^p} (\text{grad } U_2^{\nu^+}, \text{grad } U_2^{\nu^-}) \, dx + \int_{R^p} |\text{grad } U_2^{\nu^-}|^2 \, dx$$

and keep Lemma 1.7 in mind, then it is sufficient to prove, for any two measures μ_1, μ_2 for which

$$|\text{grad } U_2^{\mu_1}| \in L^2(R^p), \qquad |\text{grad } U_2^{\mu_2}| \in L^2(R^p)$$

that

$$I_2[\mu_1, \mu_2] = \frac{1}{4\pi^2} \int_{R^p} (\text{grad } U_2^{\mu_1}, \text{grad } U_2^{\mu_2}) \, dx. \tag{1.4.25}$$

Preserving the notation $\mu^{(n)}$ for the restriction of μ to the ball $|x| < n$, we set

$$\mu_n = \mu^{(n)} * m^{(1/n)} * m^{(1/n)} * m^{(1/n)}.$$

Then according to (1.4.18′)

$$I_2[\mu_{1n}, \mu_{2m}] = \frac{1}{4\pi^2} \int_{R^p} (\text{grad } U_2^{\mu_{1n}}, \text{grad } U_2^{\mu_{2m}}) \, dx, \tag{1.4.25′}$$

and we only have to pass to the limits as $m \to \infty$ and $n \to \infty$.

First of all, since the potentials $U_2^{\mu_{1n}}(x)$ and $U_2^{\mu_{2m}}(x)$ are monotone increasing and tend to $U_2^{\mu_1}(x)$ and $U_2^{\mu_2}(x)$, respectively, as $n \to \infty$ and $m \to \infty$, then

$$\lim_{n\to\infty} \lim_{m\to\infty} I_2[\mu_{1n}, \mu_{2m}] = \lim_{n\to\infty} \lim_{m\to\infty} \int_{R^p} U_2^{\mu_{2m}}(x) \, d\mu_{1n}(x)$$

$$= \lim_{n\to\infty} \int_{R^p} U_2^{\mu_2}(x) \, d\mu_{1n}(x) = \lim_{n\to\infty} \int_{R^p} U_2^{\mu_{1n}}(x) \, d\mu_2(x)$$

$$= \int_{R^p} U_2^{\mu_1}(x) \, d\mu_2(x) = I_2[\mu_1, \mu_2].$$

We now show that we can pass to the limit on the right side of (1.4.25′), i.e., that

$$\lim_{n\to\infty} \lim_{m\to\infty} \int_{R^p} \frac{\partial U_2^{\mu_{1n}}}{\partial x_i} \frac{\partial U_2^{\mu_{1m}}}{\partial x_i} \, dx = \int_{R^p} \frac{\partial U_2^{\mu_1}}{\partial x_i} \frac{\partial U_2^{\mu_2}}{\partial x_i} \, dx.$$

Since for fixed n, by (1.4.18)

$$\frac{\partial U_2^{\mu_{1n}}}{\partial x_i} \in L^2(R^p),$$

and, as it has been shown that

$$\lim_{m\to\infty} \frac{\partial U_2^{\mu_{2m}}}{\partial x_i} = \frac{\partial U_2^{\mu_2}}{\partial x_i},$$

almost everywhere, then in order to justify passing to the limit as $m \to \infty$, it is sufficient to verify the boundedness of the integrals

$$\int_{R^p} \left(\frac{\partial U_2^{\mu_{2m}}}{\partial x_i} \right)^2 dx$$

(cf. Vallée-Poussin [6]). But by (1.4.18)

$$\int_{R^p} \left(\frac{\partial U_2^{\mu_{2m}}}{\partial x_i} \right)^2 dx \leq \int_{R^p} |\text{grad } U_2^{\mu_{2m}}|^2 \, dx = 4\pi^2 \, ||\mu_{2m}||^2 \leq 4\pi^2 \, ||\mu||^2.$$

Analogously, we can justify the second passage to the limit as $n \to \infty$. Now on both sides of (1.4.25′) we pass to the limit first with m, and then with n; thus we obtain (1.4.25) and, consequently, (1.4.24). □

That part of the proof of Theorem 1.20 relating to signed measures in $\mathscr{E}_G$ is carried through in a completely analogous (with the obvious modifications) manner, and is therefore omitted here.

15. The above theorem estrablishes an important connection between the class of signed measures $\mathscr{E}_2$ (respectively $\mathscr{E}_G$) and functions $f(x)$ in R^p (respectively, in G), satisfying the two conditions:

(a) On almost all lines parallel to the axis $0x_i$ $(i = 1, 2. \ldots, p)$, $f(x)$ is an absolutely continuous function of the variable x_i;[32]

(b) $|\text{grad } f(x)| \in L^2(R^p)$ (respectively $L^2(G)$).

The class of functions satisfying (a) and (b) is usually called the *class BL*. The quantity

$$||f||^2 = (\int |\text{grad } f|^2 \, dx)^{1/2} \tag{1.4.26}$$

is a seminorm in BL.

It is not difficult to see that the equation $||f|| = 0$ implies that $f(x) = \text{const}$ almost everywhere. Indeed, first of all, grad $f = 0$ almost everywhere in R^p (correspondingly in G). Then, from (a) it follows that there exists *a* set A with full measure, containing along with each of its points, lines parallel to the axes $0x_i$, passing through that point; moreover, on each such line $f(x)$ is absolutely continuous.[33] Excluding, possibly, a null-set of these lines, one may obtain a new set A_1 with

[32] In the case of G, one speaks of absolute continuity on any line interval lying entirely within G.

[33] Cf. the previous footnote.

analogous structure, for which $\operatorname{grad} f = 0$ almost everywhere on each of its lines. Consequently, $f(x) = c$ on each line of A_1 and it is easy to see that the value c must be the same for all lines of A_1.

We now distinguish from BL the linear manifolds on which $\|f\|$ is a norm. In the case of the whole space, we impose on $f(x)$ the additional condition

$$\text{(c)} \qquad \lim_{r\to\infty} \frac{1}{r^{p-1}} \int_{|x|=r} f(x)\, dS = 0,$$

in the case of a finite region G, the corresponding condition is:

$$\text{(c}'\text{)} \qquad \lim_{r\to 0} \int_{S_r} f(x) \frac{\partial G_0(x)}{\partial n}\, dS = 0,$$

where S_r is the level surface $G_0(x) = r$ and the integration is over the surface area.

The set BL_0 of functions satisfying conditions (a), (b), and (c) (correspondingly (c′)) is a normed space with norm (1.4.26).

Let us return to the mappings

$$\nu \to U_2^\nu(x) \quad \text{or} \quad \nu \to U_g^\nu(x)$$

of the class $\mathscr{E}_2$, correspondingly $\mathscr{E}_G$ into BL.[34] We will show that they are mappings into BL_0.

We verify condition (c) for $f(x) = U_2^\nu(x)$. Here, we may of course assume that ν is a measure. Within a constant factor, the integral in (c) is equal to

$$\int_{R^p} U_2^\nu(x)\, d\varepsilon^{(r)} = \int_{R^p} U_2^{\varepsilon^{(r)}}(x)\, d\nu.$$

But

$$U_2^{\varepsilon^{(r)}}(x) = \begin{cases} \dfrac{1}{|x|^{p-2}} & \text{for } |x| \geq r, \\[2ex] \dfrac{1}{r^{p-2}} & \text{for } |x| \leq r \end{cases}$$

and, consequently, as $r \to \infty$, it tends, monotonically decreasing, to zero. This allows us to pass to the limit under the integral sign, and therefore,

$$\lim_{r\to\infty} \int_{R^p} U_2^\nu(x)\, d\varepsilon^{(r)} = 0.$$

To verify condition (c′) in the case $f(x) = U_g^\nu(x)$ we introduce the measure $\varepsilon_g^{(r)}$, concentrated on S_r and having a surface density $\frac{\partial G_0}{\partial n}$. Then

---

[34] Here we take the "coarse" point of view, since the functions in the class BL are defined only almost everywhere in R_p (or G).

the integral in (c′) may be rewritten in the form

$$\int_G U_g^\nu(x)\, d\varepsilon_g^{(r)} = \int_G U_g^{\varepsilon_g^{(r)}}(x)\, d\nu .$$

Applying Green's formula, one can easily show that

$$U_g^{\varepsilon_g^{(r)}}(x) = \int_{S_r} G(x, y) \frac{\partial G_0(y)}{\partial n}\, dS_y = \begin{cases} G_0(x) & \text{outside } S_r, \\ r & \text{inside } S_r. \end{cases}$$

Therefore, as $r \to 0$, $U_g^{\varepsilon_g^{(r)}}(x)$ tends decreasing to zero; as above this implies (c′).

If we now take into consideration the remark made after Lemma 1.7, then using the Fatou lemma, we immediately obtain the following proposition.

Theorem 1.21. *The potentials $U_2^\nu(x)$, $U_g^\nu(x)$ where $\nu \in \mathscr{E}_2$ correspondingly $\nu \in \mathscr{E}_G$, tend to zero on almost all rays, correspondingly, Green lines.*

In concluding this section we note that the completion of $\mathscr{E}_2$ (or $\mathscr{E}_G$) with the norm $\sqrt{(\nu, \nu)}$ yields a class of distributions such that the class of corresponding potentials coincides with BL_0. We shall return to this question in Ch. VI.

§ 5. Representation of Superharmonic Functions by Potentials

16. In § 3 we saw that the classical potential of a measure is a superharmonic function in R^p. It turns out that under certain conditions the converse is also true. In this section we shall prove certain local theorems on the representation of superharmonic functions.

Theorem 1.22. *Suppose that $f(x)$ is a superharmonic function in R^p. Then for any bounded region Ω there is a measure μ_Ω concentrated on Ω, and a function $h_\Omega(x)$ harmonic in Ω, such that*

$$f(x) = \int_\Omega k_2(x - y)\, d\mu_\Omega(y) + h_\Omega(x) = U_2^{\mu_\Omega}(x) + h_\Omega(x), \quad x \in \Omega. \quad (1.5.1)$$

Moreover, given Ω, the measure μ_Ω and the function $h_\Omega(x)$ are unique.

Proof. If $f(x)$ were sufficiently smooth, then the desired measure μ, as one can easily see, would have density $\mu(x)$ defined by the formula

$$\mu(x) = -\frac{1}{4\pi^2} \Delta f(x) .$$

We will show that in the general case in analogous formula holds, in which, however, Δf must be interpreted as a distribution.

In this connection, consider the signed measures

$$\frac{1}{r^2}(\varepsilon - \varepsilon^{(r)});$$

we show that as $r \to 0$

$$\frac{1}{r^2}(\varepsilon - \varepsilon^{(r)}) \dot{\to} -\frac{1}{2p}\Delta\varepsilon.$$

Indeed, for any $\varphi \in D$, the Pizzetti formula (cf. p. 46) gives

$$\lim_{r\to 0}\frac{1}{r^2}(\varepsilon - \varepsilon^{(r)})(\varphi) = \lim_{r\to 0}\frac{1}{r^2}\{\varphi(0) - \tilde{\varphi}(0, r)\}$$

$$= \lim_{r\to 0}\frac{1}{r^2}\left\{-\frac{r^2}{2p}\Delta\varphi(0) + O(r^4)\right\} = -\frac{1}{2p}\Delta\varphi(0) = -\frac{1}{2p}(\Delta\varepsilon)(\varphi).$$

Let us now form the distribution

$$\mu = -\frac{1}{4\pi^2}\Delta f = -\frac{1}{4\pi^2}(\Delta\varepsilon) * f.$$

Since the supports of the signed measures $\varepsilon - \varepsilon^{(r)}$ are contained in a fixed compact set, we have the formula (cf. Lemma 0.7)

$$\frac{1}{r^2}(\varepsilon - \varepsilon^{(r)}) * f \dot{\to} -\frac{1}{2p}(\Delta\varepsilon) * f. \tag{1.5.2}$$

But the left side of (1.5.2) is, by (1.2.2), a nonnegative, locally integrable function from which by Theorem 0.8 it follows that the right side of (1.5.2), and consequently μ, is a measure in R^p.

Let μ_Ω be the restriction of μ to Ω, i.e.

$$\mu_\Omega(e) = \mu(e \cap \Omega).$$

Form the potential $U_2^{\mu_\Omega}(x)$ of this measure with compact support and put

$$H_\Omega(x) = f(x) - U_2^{\mu_\Omega}(x). \tag{1.5.3}$$

Regarding $H_\Omega(x)$ as a distribution and using Lemma 0.6 on the associativity of convolution, we may write

$$\Delta\varepsilon * H_\Omega(x) = \Delta\varepsilon * f - \Delta\varepsilon * (k_2 * \mu_\Omega)$$

$$= -4\pi^2\mu - (\Delta\varepsilon * k_2) * \mu_\Omega = 4\pi^2(\mu_\Omega - \mu).$$

Thus, in Ω,

$$\Delta\varepsilon * H_\Omega = \Delta H_\Omega = 0.$$

Then we will show that the distribution $H_\Omega(x)$ coincides in Ω (almost everywhere) with an ordinary harmonic function $h_\Omega(x)$.

Consider the region Ω_ε, consisting of those points of Ω whose distance from $\partial\Omega$ is greater than $\varepsilon > 0$; we form the regularization (cf. Introduction, § 2, No. 10, (3))

$$H_\Omega^{(\varepsilon)}(x) = H_\Omega * \varrho_\varepsilon(x) \in D.$$

According to Lemma 0.6 we have in Ω_ε,

$$\Delta H_\Omega^{(\varepsilon)}(x) = \Delta\varepsilon * H_\Omega^{(\varepsilon)} = (\Delta\varepsilon * H_\Omega) * \varrho_\varepsilon(x) = (\Delta H_\Omega)^{(y)} \varrho_\varepsilon(x-y) = 0,$$

since the support of $\varrho_\varepsilon(x-y)$ for $x \in \Omega_\varepsilon$ is a subset of Ω. Consequently, $H_\Omega^{(\varepsilon)}(x)$ is a harmonic function in Ω_ε. We now show that in an ε-neighborhood of any point of $\Omega_{2\varepsilon}$, $H_\Omega(x)$ coincides (as a distribution) with $H_\Omega^{(\varepsilon)}$. Suppose that $x_0 \in \Omega_{2\varepsilon}$ and $\varphi(x)$ is any function in D whose support $S(\varphi)$ is contained in the ball $|x - x_0| < \varepsilon$. Then (cf. (0.2.21))

$$H_\Omega^{(\varepsilon)}(\varphi) = H_\Omega * \varrho_\varepsilon * \check{\varphi}(0) = H_\Omega * \check{\varphi} * \varrho_\varepsilon(0).$$

But the function $H_\Omega * \check{\varphi}(x) \in D$ is harmonic in the ball $|x| < \varepsilon$, since

$$\Delta(H_\Omega * \check{\varphi}) = (\Delta\varepsilon * H_\Omega)^{(y)} \varphi(y-x) = 0.^{35}$$

By the well-known theorem on the arithmetic mean for harmonic functions

$$H_\Omega^{(\varepsilon)}(\varphi) = H_\Omega * \check{\varphi}(\varrho_\varepsilon) = H_\Omega * \check{\varphi}(0) = H_\Omega(\varphi).$$

Consequently (cf. Introduction, § 2, No. 6, footnote on p. 21), $H_\Omega \equiv H_\Omega^{(\varepsilon)}$ in $\Omega_{2\varepsilon}$; since $\varepsilon > 0$ was arbitrary, this proves that in Ω

$$H_\Omega(x) = h_\Omega(x),$$

where $h_\Omega(x)$ is a function harmonic in Ω.

Returning to (1.5.3), we obtain

$$f(x) = U_2^{\mu_\Omega}(x) + h_\Omega(x)$$

for almost all $x \in \Omega$.

But from this together with (1.2.4) it follows that (1.5.1) holds everywhere in Ω. Thus we have proved the existence of μ_Ω and $h_\Omega(x)$.

In order to establish the uniqueness of μ_Ω and $h_\Omega(x)$, suppose that in Ω we have

$$U_2^{\mu'_\Omega}(x) + h'_\Omega(x) = U_2^{\mu''_\Omega}(x) + h''_\Omega(x).$$

Then the signed measure $\nu = \mu'_\Omega - \mu''_\Omega$, concentrated on Ω, has the potential

$$U_2^\nu(x) = h''_\Omega(x) - h'_\Omega(x),$$

---

[35] The latter conclusion is based on the fact that for $|x| < \varepsilon$, the support of $\varphi(y-x)$ is contained in Ω.

harmonic in Ω. By Theorem 1.13 this means that $\nu \equiv 0$, and consequently $h'_\Omega(x) \equiv h''_\Omega(x)$. $\square$

Now consider the function $f(x)$ superharmonic in a region $G \subset R^p$ which does not coincide with R^p. In this case, for any region Ω, $\overline{\Omega} \subset G$, there is also a representation of $f(x)$ in the form of a classical potential or a Green potential. Namely, we have

Theorem 1.22′. *Suppose $f(x)$ is superharmonic in G. Then for any region Ω, compact in G, i.e. such that $\overline{\Omega} \subset G$, there exist a unique measure μ_Ω and harmonic functions $h_\Omega(x)$ and $h'_\Omega(x)$ such that*

$$f(x) = \int_\Omega k_2(x - y)\, d\mu_\Omega(y) + h_\Omega(x) \quad (x \in \Omega), \tag{1.5.1}$$

$$f(x) = \int_\Omega G(x, y)\, d\mu_\Omega(y) + h'_\Omega(x) \qquad (x \in \Omega). \tag{1.5.1′}$$

Proof. By (1.3.5)

$$\int_\Omega G(x, y)\, d\mu_\Omega(y) = \int_\Omega k_2(x - y)\, d\mu_\Omega(y) + \int_\Omega h(x, y)\, d\mu_\Omega(y)$$

and since the last term is a function harmonic in G, then the formulas (1.5.1) and (1.5.1′) hold or fail simultaneously. We will prove (1.5.1).

Since $f(x)$ is locally integrable in G, it may be regarded as a distribution in G and we may form

$$\mu = -\frac{1}{4\pi^2} \Delta f.$$

This is also a distribution in G and for any $\varphi \in D$, $S(\varphi) \subset G$,

$$\mu(\varphi) = -\frac{1}{4\pi^2} f(\Delta\varphi) = -\frac{1}{4\pi^2} \int_{S(\varphi)} f(x)\, \Delta\varphi(x)\, dx.$$

We will show that this is a measure in G. By Theorem 0.8 it is sufficient to verify that $\mu(\varphi) \geq 0$ for $\varphi \geq 0$. Suppose that V is a compact neighborhood of $S(\varphi)$ with smooth boundary, i.e. $S(\varphi) \subset V \subset \overline{V} \subset G$. If $\delta > 0$ is the distance of V from ∂G, then for $1/n < \delta$ the result $f_n(x)$ of triple averaging $f(x)$ on the ball of radius $1/(3n)$ will be a superharmonic function in $\overline{V}$; moreover, by (1.2.4), $f_n(x) \to f(x)$, monotonically increasing in $\overline{V}$.

Therefore

$$\mu(\varphi) = -\frac{1}{4\pi^2} \lim_{n\to\infty} \int_{S(\varphi)} f_n(x)\, \Delta\varphi(x)\, dx.$$

Applying Green's formula in the region V and noting that on ∂V $\varphi = 0$, $\partial\varphi/\partial n = 0$, we obtain

$$\int_{S(\varphi)} f_n(x)\, \Delta\varphi(x)\, dx = \int_{S(\varphi)} \varphi(x)\, \Delta f_n(x)\, dx.$$

But $\Delta f_n(x) \leq 0$ in V, and $\varphi(x) \geq 0$; consequently,

$$\int\limits_{S(\varphi)} f_n(x)\, \Delta\varphi(x)\, dx \leq 0$$

and $\mu(\varphi) \geq 0$.

As in the proof of Theorem 1.22, by putting

$$H_\Omega(x) = f(x) - U_2^{\mu_\Omega}(x),$$

we obtain in Ω,

$$\Delta H_\Omega = \Delta f - \Delta U_2^{\mu_\Omega}(x) = -4\pi^2\mu - \Delta\varepsilon * (k_2 * \mu_\Omega) = 0,$$

the remainder of the proof is exactly the same as the proof of Theorem 1.22. ☐

17. Let us clarify the conditions under which the superharmonic function $f(x)$ admits a representation of the type (1.5.1) or (1.5.1′) in the *entire region where it is superharmonic.*

Consider first the case when $f(x)$ is superharmonic everywhere in R^p ($p \geq 3$). Suppose that there is a representation

$$f(x) = U_2^\mu(x) + h(x), \quad x \in R^p. \tag{1.5.4}$$

In that case $h(x)$ is a harmonic minorant of $f(x)$ in R^p. Soon we will see that the condition of existence of a harmonic minorant in R^p is also sufficient for the to be a representation (1.5.4) and that the function $h(x)$ in (1.5.4) is the best harmonic minorant of $f(x)$ in R^p.

Let us recall one general property of potentials $U_2^\mu(x)$, established at the end of § 4: the spherical mean tends to zero under the condition that the radius of the sphere tends to infinity. This fact may be written in the form

$$\lim_{r\to\infty} \varepsilon^{(r)}(U_2^\mu) = 0. \tag{1.5.5}$$

It turns out that this together with superharmonicity characterizes potentials of measures. We prove first a preliminary lemma.

Lemma 1.8. *If the function $f(x)$ is nonnegative and superharmonic in R^p, and*

$$\lim_{r\to\infty} \varepsilon^{(r)}(f) = \alpha,^{36}$$

then $f(x) \geq \alpha$ everywhere.

Proof. For the sphere $|y| = r$ we form the Poisson integral

$$H(x) = \frac{1}{\omega_p r} \int\limits_{|y|=r} f(y) \frac{r^2 - |x|^2}{|x-y|^p}\, dS_y \quad (|x| < r).$$

[36] Recall that $\varepsilon^{(r)}(f)$ is a monotone increasing function of r, and hence this limit exists.

As was shown in § 2.4 for $|x| < r$

$$f(x) \geq H(x).$$

For $|x| < \theta r$, where $0 < \theta < 1$, we have the bound

$$\frac{1}{\omega_p r} \frac{r^2 - |x|^2}{|x-y|^p} \geq \frac{1}{\omega_p r^{p-1}} \frac{1-\theta}{(1+\theta)^{p-1}}.$$

Therefore, since $f(y)$ is nonnegative, we obtain

$$f(x) \geq H(x) \geq \frac{1-\theta}{(1+\theta)^{p-1}} \varepsilon^{(r)}(f),$$

and as $r \to \infty$ this yields

$$f(x) \geq \frac{1-\theta}{(1+\theta)^{p-1}} \alpha.$$

Since θ may be taken arbitrary small, we obtain $f(x) \geq \alpha$ as required. ☐

Corollary. *If $f(x) \geq 0$ and is superharmonic, then the quantity*

$$\lim_{r\to\infty} \varepsilon_{x_0}^{(r)}(f) = \alpha$$

does not depend on the position x_0 of the center of the sphere over which the function is averaged.

Theorem 1.23. *A function $f(x)$ superharmonic in R^p ($p \geq 3$) is a Newtonian potential of measure if, and only if, it is nonnegative and*

$$\lim_{r\to\infty} \varepsilon^{(r)}(f) = 0. \tag{1.5.6}$$

Proof. Since the necessity of (1.5.6) has already been established (cf. (1.5.5)), we shall prove the sufficiency of the conditions of the theorem.

We form the measure

$$\mu = -\frac{1}{4\pi^2} (\Delta\varepsilon) * f,$$

which we utilized in the proof of Theorem 1.22, and denote by μ_r its restriction to the ball $|x| \leq r$.

The difference

$$f(x) - U_2^{\mu_r}(x),$$

is, by Theorem 1.22, a harmonic function in the ball $|x| < r$. We will show that it is superharmonic in R^p. For a neighborhood of any point x_1,

$|x_1| \geq r$, we may, by Theorem 1.22, write

$$f(x) = U_2^{\mu_{r_1}}(x) + h_1(x),$$

where $|r_1| > x_1$ and $h_1(x)$ is a harmonic function in the ball $|x| < r_1$. Consequently,

$$f(x) = U_2^{\mu_r}(x) + U_2^{\mu_{r_1}-\mu_r}(x) + h_1(x),$$

$$f(x) - U_2^{\mu_r}(x) = U_2^{\mu_{r_1}-\mu_r}(x) + h_1(x),$$

and we see that the right side of the last equation is superharmonic in the neighborhood of the point x_1.

Now by (1.5.5), (1.5.6), and the Corollary to Lemma 1.8, we note that for any point x_0

$$\lim_{R\to\infty} \varepsilon_{x_0}^{(R)}(f - U_2^{\mu_r}) = 0.$$

Since $\varepsilon_{x_0}^{(R)}(f - U_2^{\mu_r})$ is monotonic, we obtain

$$f(x_0) - U_2^{\mu_r}(x_0) = \lim_{R\to 0} \varepsilon_{x_0}^{(R)}(f - U_2^{\mu_r}) \geq 0,$$

and for $r \to \infty$

$$U_2^{\mu}(x_0) \leq f(x_0).$$

Thus, the function $f(x) - U_2^{\mu}(x)$ is nonnegative and harmonic in R^p. Consequently, it is a constant. But

$$\lim_{r\to\infty} \varepsilon^{(r)}(f - U_2^{\mu}) = 0,$$

and therefore

$$f(x) \equiv U_2^{\mu}(x). \quad \square$$

Corollary 1. *If the function $f(x)$ is nonnegative, superharmonic in R^p $(p \geq 3)$ and is majorized by a potential*

$$f(x) \leq U_2^{\nu}(x)$$

then it is itself a potential of measure

$$f(x) = U_2^{\mu}(x), \quad \mu \in \mathfrak{M}^+.$$

Corollary 2. *If $f(x)$ is nonnegative and superharmonic in R^p $(p \geq 3)$, then*

$$f(x) = U_2^{\mu}(x) + \alpha, \tag{1.5.7}$$

where

$$\alpha = \lim_{r\to\infty} \varepsilon^{(r)}(f) \geq 0.$$

Indeed, by Lemma 1.8 the function

$$f_1(x) = f(x) - \alpha$$

is also nonnegative and superharmonic and

$$\lim_{r\to\infty} \varepsilon^{(r)}(f_1) = 0.$$

By Theorem 1.23 $f_1(x) = U_2^\mu(x)$, $\mu \in \mathfrak{M}^+$, and this proves (1.5.7).

Theorem 1.24. *The function $f(x)$, $x \in R^p$ $(p \geq 3)$ admits a representation*

$$f(x) = U_2^\mu(x) + h(x) \tag{1.5.4}$$

everywhere in R^p, where $\mu \in \mathfrak{M}^+$ and $h(x)$ is harmonic in R^p if, and only if, $f(x)$ is superharmonic in R^p and has a harmonic minorant.

Proof. The necessity of the specified conditions is trivial. Their sufficiency follows so: If $f(x)$ is superharmonic in R^p and $h_1(x)$ is its harmonic minorant, then $f(x) - h_1(x)$ is a nonnegative superharmonic function. By Corollary 2

$$f(x) - h_1(x) = U_2^\mu(x) + \alpha,$$

which gives (1.5.4).

From this it also follows that the function h in the representation (1.5.4) is the best harmonic minorant of the function $f(x)$.

On the other hand, it is not difficult to see that for a function $f(x)$ superharmonic in R^p the representation (1.5.4) is possible if, and only if,

$$U_2^\mu(x) \not\equiv +\infty,$$

where $\mu = -\frac{1}{4\pi^2}\Delta f$. Thus the condition that $U_2^\mu(x)$ be finite is necessary and sufficient for the existence of a harmonic minorant for $f(x)$. ☐

We point out another condition of this type.

Theorem 1.25. *A function $f(x)$ superharmonic in R^p has a harmonic minorant, if, and only if,*

$$\lim_{r\to\infty} \varepsilon^{(r)}(f) > -\infty. \tag{1.5.8}$$

Proof. The necessity of the condition (1.5.8) is clear since for a harmonic function $h(x)$

$$\varepsilon^{(r)}(h) = h(0).$$

To prove that (1.5.8) is sufficient, consider the Poisson integral

$$H_r(x) = \frac{1}{\omega_p r} \int_{|y|=r} f(y) \frac{r^2 - |x|^2}{|x-y|^p} dS_y,$$

which in the ball $|x| \leq r$ is the best harmonic minorant for $f(x)$. Since $H_r(x)$ is also a harmonic minorant in the ball $|x| \leq r_1$ for any $r_1 < r$, then $H_{r_1}(x) \geq H_r(x)$, i.e. as r increases, the harmonic functions $H_r(x)$ decrease. Consequently, for any x,

$$\lim_{r \to \infty} H_r(x) = h(x)$$

exists. By the well-known theorem of Harnack (cf. for example, Brelot [31], Applications, § 7, No. 20) $h(x)$ either $\equiv -\infty$, or is a harmonic function which is obviously a harmonic minorant for $f(x)$ in R^p. But

$$H_r(0) = \varepsilon^{(r)}(f),$$

and therefore the condition (1.5.8) guarantees that $h(x) \not\equiv -\infty$. □

In conclusion, we formulate several theorems relating to functions $f(x)$ which are superharmonic in a region $G \subset R^p$ $(p \geq 2)$.

Theorem 1.24′. *The function $f(x)$, $x \in G$, admits in G a representation*

$$f(x) = \int_G G(x, y)\, d\mu(\eta) + h(x), \tag{1.5.9}$$

where $\mu \in \mathfrak{M}_G^+$, and $h(x)$ is a function harmonic in G, if and only if, $f(x)$ is superharmonic in G and has a harmonic minorant.

Moreover, the function $h(x)$ in (1.5.9) is the best harmonic minorant of $f(x)$ in the region G.

Theorem 1.25′. *A function $f(x)$ superharmonic in G has a harmonic minorant if, and only if,*

$$\lim_{r \to 0} \int_{S_r} f(x) \frac{\partial G_0(x)}{\partial n}\, dS > -\infty,$$

where $G_0(x) = G(x, x_0)$ is the Green function for the region G with a pole at $x_0 \in G$, and S_r is the level surface $G_0(x) = r$.

18. We point out two interesting applications of Theorem 1.23. The first concerns the limit of a monotone increasing sequence of potentials of measures and is contained in the following theorem.

Theorem 1.26. *Suppose that the sequence $\{U_2^{\mu_n}(x)\}$, $\mu_n \in \mathfrak{M}^+$, is monotone increasing and is majorized by a potential of a measure. Then*

$$\lim_{n \to \infty} U_2^{\mu_n}(x) = U_2^{\mu}(x), \quad \mu \in \mathfrak{M}^+, \tag{1.5.10}$$

and $\mu_n \dashrightarrow \mu$.

Proof. From proposition (ii), § 2, No. 5 it follows that $\lim_{n\to\infty} U_2^{\mu_n}(x)$ is a function superharmonic in R^p. Since it is majorized by a potential, it satisfies condition (1.5.6) and (1.5.10) follows from Theorem 1.23. Furthermore, by the well-known Levi theorem on monotonic convergence we have the equation

$$\lim_{n\to\infty} \int U_2^{\mu_n}(x)\, d\lambda = \int U_2^{\mu}(x)\, d\lambda$$

or

$$\lim_{n\to\infty} \int U_2^{\lambda}(x)\, d\mu_n = \int U_2^{\lambda}(x)\, d\mu$$

for any measure λ. Consequently, it also holds for any signed measure λ, under the condition that $\int U_2^{\lambda}\, d\mu$ is defined. Then the remark to Lemma 1.1 shows that for any compact function $f(x)$ with compact support having three continuous derivatives,

$$\lim_{n\to\infty} \mu_n(f) = \mu(f).$$

But then from Theorem 0.4 it follows that $\mu_n \dashrightarrow \mu$ and the assertion is proved. □

The second application of Theorem 1.23 is the so-called *principle of domination* for measures with finite energy.

Theorem 1.27. *Suppose that μ is a measure with finite energy, and that $f(x)$ is a nonnegative function $f(x)$ superharmonic in R^p. Then the inequality*

$$U_2^{\mu}(x) \leq f(x) \tag{1.5.11}$$

holds everywhere, if it holds μ-almost everywhere.

Proof. Set

$$V(x) = \inf\big(U_2^{\mu}(x), f(x)\big).$$

Since it is superharmonic in R^p and is majorized by the potential $U_2^{\mu}(x)$, by Theorem 1.23 it is itself a potential of some measure λ. Clearly, the theorem will be established if we prove $\lambda \equiv \mu$.

Since $U_2^{\lambda}(x) \leq U_2^{\mu}(x)$, then

$$\int U_2^{\lambda}(x)\, d\lambda \leq \int U_2^{\mu}(x)\, d\lambda = \int U_2^{\lambda}(x)\, d\mu = \|\mu\|^2,$$

i.e. λ also has finite energy. Setting

$$e = \{x : U_2^{\mu}(x) > f(x)\} = \{x : U_2^{\mu}(x) > U_2^{\lambda}(x)\},$$

we obtain $\mu(e) = 0$; since on the complement of e $U_2^\mu(x) = U_2^\lambda(x)$, we have

$$\|\mu - \lambda\|^2 = \int_e (U_2^\mu - U_2^\lambda)\,(d\mu - d\lambda) = -\int_e (U_2^\mu - U_2^\lambda)\,d\lambda \leq 0.$$

Consequently, $\|\mu - \lambda\| = 0$ and by Theorem 1.15, $\mu \equiv \lambda$. ▯

From this theorem we obtain the so-called.

Second Maximum Principle. *Suppose that μ is a measure with finite energy, and that λ is any measure. Then if the inequality*

$$U_2^\mu(x) \leq U_2^\lambda(x), \tag{1.5.12}$$

holds μ-almost everywhere, it must hold everywhere in R^p.

This is clear, since in (1.5.11) we may set $f(x) = U_2^\lambda(x)$.

Just as easily from Theorem 1.27 we deduce the first maximum principle (cf. Theorem 1.10) for Newtonian potentials.

§ 6. Superharmonic Functions of Fractional Order

19. In § 3 (Theorem 1.4) we established that a potential of a measure $U_\alpha^\mu(x)$ under the condition $\alpha \geq 2$ is a superharmonic function in R^p. However, this is not true if $\alpha < 2$.

In this section we introduce a generalization of the notion of superharmonicity, depending on the parameter $\alpha < 2$ and coextensive with potentials of order α, $0 < \alpha < 2$ in the sense that namely those potentials $U_\alpha^\mu(x)$, $\mu \in \mathfrak{M}^+$ and essentially only those, are superharmonic.

The usual definition of superharmonicity (cf. § 2) compares the quantity $f(x)$ with its mean $\varepsilon_x^{(r)}(f)$. In order to understand the direction of the needed generalization, let us prove the superharmonicity of $U_2^\mu(x)$, $\mu \in \mathfrak{M}^+$, as follows:

$$\varepsilon_x^{(r)}(U_2^\mu) = \int U_2^\mu(y)\,d\varepsilon_x^{(r)}(y) = U_2^\mu * \varepsilon^{(r)}$$
$$= \int U_2^{\varepsilon_x^{(r)}}(y)\,d\mu(y) = U_2^{\varepsilon^{(r)}} * \mu.$$[37]

But

$$U_2^{\varepsilon^{(r)}}(x) = \begin{cases} \dfrac{1}{|x|^{p-2}} & \text{for} \quad |x| \geq r, \\[2ex] \dfrac{1}{r^{p-2}} & \text{for} \quad |x| < r, \end{cases}$$

[37] Recall, that the convolution here and below is understood not as a signed measure but as a point function, defined throughout R^p.

therefore

$$U_2^{\varepsilon^{(r)}}(x) \leq U_2^{\varepsilon}(x) = k_2(x)$$

and

$$\varepsilon_x^{(r)}(U_2^{\mu}) \leq k_2 * \mu = U_2^{\mu}(x).$$

everywhere in R^p.

Since $U_2^{\varepsilon^{(r)}}(x) = k_2(x)$ outside the ball $|x| < r$, we easily conclude that outside of $S(\mu)$ the potential $U_2^{\mu}(x)$ is harmonic.[38]

In order to use the analogous reasoning in the case $\alpha < 2$, one needs to find a measure $\varepsilon_\alpha^{(r)}$, $\varepsilon_\alpha^{(r)}(1) = 1$, with the potential

$$U_\alpha^{\varepsilon_\alpha^{(r)}}(x) \leq k_\alpha(x) \quad \text{everywhere in } R^p$$

and (1.6.1)

$$U_\alpha^{\varepsilon_\alpha^{(r)}}(x) = k_\alpha(x) \quad \text{outside of } |x| < r.$$

Moreover, we will also require that within the ball $|x| < r$, $\varepsilon_\alpha^{(r)} \equiv 0$. We encounter here a particular case of the so-called *problem of balayage*, which we will study in Ch. IV.

The desired measure is absolutely continuous and has a density defined by the formula

$$\varepsilon_\alpha^{(r)}(x) = \begin{cases} 0, & |x| < r, \\ \dfrac{\Gamma(p/2)}{\pi^{p/2+1}} \sin \dfrac{\pi\alpha}{2} r^\alpha (|x|^2 - r^2)^{-\alpha/2} |x|^{-p}, & |x| > r. \end{cases} \tag{1.6.2}$$

The relations (1.6.1) are verified in the Appendix, (A. 3) and (A. 4). It is not difficult to find that as $\alpha \to 2$, $\varepsilon_\alpha^{(r)} \dot{\to} \varepsilon^{(r)}$, and as $r \to 0$, $\varepsilon_\alpha^{(r)} \dot{\to} \varepsilon$. The measure $\varepsilon_\alpha^{(r)}$ determines an averaging operation

$$\varepsilon_\alpha^{(r)}(f) = \int_{R^p} f(x)\, \varepsilon_\alpha^{(r)}(x)\, dx, \tag{1.6.3}$$

applicable to locally integrable functions $f(x)$, satisfying the condition

$$\int_{|x|>1} \frac{|f(x)|}{|x|^{p+\alpha}}\, dx < \infty.^{39} \tag{1.6.4}$$

---

[38] The analogous reasoning holds for $U_\alpha^{\mu}(x)$, $\alpha > 2$, using the fact that throughout R^p $U_\alpha^{\varepsilon^{(r)}}(x) < k_\alpha(x)$. The latter follows from the superharmonicity of the kernel $k_\alpha(x)$, $\alpha > 2$, as proved in Theorem 1.4. However, in this case, the equation $k_\alpha(x) = U_\alpha^{\varepsilon^{(r)}}(x)$ outside the ball $|x| < r$ does not hold.

[39] Under this condition

$$\int_{r_1}^{r_2} \varepsilon_\alpha^{(r)}(f)\, dr < \infty$$

for any $0 < r_1 < r_2$ such that $\varepsilon_\alpha^{(r)}(f)$ is finite for almost all $r > 0$.

It is not difficult to verify condition (1.6.4) for a potential $U^\mu_\alpha(x)$ of a measure μ which, as always, satisfies (1.3.10). Indeed,

$$\int\limits_{|x|>1} \frac{U^\mu_\alpha(x)}{|x|^{p+\alpha}}\,dx = \int d\mu(y) \int\limits_{|x|>1} \frac{dx}{|x|^{p+\alpha}\,|x-y|^{p-\alpha}}.$$

But

$$\int\limits_{|x|>1} \frac{dx}{|x|^{p+\alpha}\,|x-y|^{p-\alpha}} = O\left(\frac{1}{|y|^{p-\alpha}}\right)$$

(cf. proof of Theorem 1.12, § 3.9), and by (1.3.10)

$$\int\limits_{|x|>1} \frac{U^\mu_\alpha(x)}{|x|^{p+\alpha}}\,dx < \infty.$$

Now using the same approach as in the beginning of this section, we are able to prove that throughout R^p

$$\varepsilon^{(r)}_\alpha * U^\mu_\alpha \leq U^\mu_\alpha; \tag{1.6.5}$$

moreover, at points $x \notin S(\mu)$ for sufficiently small $r > 0$

$$\varepsilon^{(r)}_\alpha * U^\mu_\alpha(x) = U^\mu_\alpha(x).$$

20. We now introduce the following definitions.

A function $f(x)$ is called *α-superharmonic in* R^p $(0 < \alpha < 2)$, if

(a) $f(x) \geq 0$; $f(x) \not\equiv +\infty$;

(b) $f(x)$ is lower semicontinuous;

(c) $f(x)$ satisfies condition (1.6.4) and for any point $x \in R^p$

$$\varepsilon^{(r)}_\alpha * f(x) = \varepsilon^{(r)}_{\alpha x}(f) \leq f(x) \quad (r > 0).$$

Furthermore, a function $f(x)$ satisfying condition (1.6.4), will be called *α-harmonic* at the point x_0 if it is continuous in a neighborhood of this point and for sufficiently small r

$$f(x_0) = \varepsilon^{(r)}_\alpha * f(x_0) = \varepsilon^{(r)}_{\alpha x_0}(f). \tag{1.6.6}$$

If $f(x)$ is α-harmonic at each point of the set Ω, then we will say that $f(x)$ is *α-harmonic* in Ω. Note that these definitions are not of a local nature since the support of $\varepsilon^{(r)}_\alpha$ is the whole exterior of the sphere $|x| \geq r$.

In the previous section we established that a potential $U^\mu_\alpha(x)$ of a measure μ is an α-superharmonic function in R^p and an α-harmonic function in the complement of $S(\mu)$.

We shall point out a few simple properties of α-superharmonic functions.

(i) At each point $x \in R^p$

$$\lim_{r\to 0} f * \varepsilon_\alpha^{(r)} = f(x).$$

This is easy to prove, using the semicontinuity of $f(x)$ and the formula $\varepsilon_\alpha^{(r)} \dot{\to} \varepsilon$.

(ii) If at some point

$$f(x_0) = \inf_{R^p} f(x),$$

then $f(x) \equiv f(x_0)$.

Indeed, if for some point x_1

$$f(x_1) > m = \inf_{R^p} f(x),$$

then

$$f(x) > m$$

in some neighborhood of x_1. But then for sufficiently small $r > 0$,

$$\varepsilon_{\alpha x_0}^{(r)}(f) > m = f(x_0)$$

which contradicts condition (c).

(iii) If $\{f_n(x)\}$ is a monotonically nondecreasing sequence of α-superharmonic functions, then the function

$$f(x) = \lim_{n\to\infty} f_n(x)$$

either is $\equiv +\infty$, or is also α-superharmonic.

This is proved exactly as the analogous property for ordinary superharmonic functions (cf. § 2, No. 5).

(iv) If $\mu \in \mathfrak{M}^+$, and $f(x)$ is an α-superharmonic function, then

$$f * \mu(x) = \int_{R^p} f(x-y)\, d\mu(y)$$

is either $\equiv +\infty$, or is also α-superharmonic.

This is the analogue of Theorem 1.2 (§ 2) and is proved in exactly the same way.

From this, in particular, it follows that for fixed $r > 0$, the function

$$f^{(r)}(x) = f * \varepsilon_\alpha^{(r)}$$

will be α-superharmonic.

We will now establish an analogue of the principle of the harmonic minorant (§ 2, No. 4).

Theorem 1.28. *Suppose that in R^p we have two functions: the α-superharmonic function $f(x)$ and a function $h(x)$, α-harmonic in an open set Ω*

which does not contain the point at infinity[40] *and is continuous in* $\bar{\Omega}$. *Then, if* $f(x) \geq h(x)$ *in the complement of* Ω, *then this inequality holds throughout* R^p. *If at some point* $x_0 \in \Omega$, $f(x_0) = h(x_0)$, *then* $f(x) \equiv h(x)$ *almost everywhere in* R^p.

Proof. Consider the difference

$$d(x) = f(x) - h(x).$$

It is lower semicontinuous on the closed set $\bar{\Omega}$. At any point s on the boundary of Ω,

$$\varliminf_{x \to s} d(x) \geq 0 \quad (x \in \Omega),$$

since

$$\varliminf_{x \to s} d(x) = \varliminf_{x \to s} f(x) - \lim_{x \to s} h(x) \geq f(s) - h(s).$$

Consequently, if $d(x)$ takes on negative values, then there is a point $x_0 \in \Omega$, where $d(x)$ achieves its absolute minimum. In that case

$$d(x_0) < \varepsilon_{\alpha x_0}^{(r)}(d) = \varepsilon_{\alpha x_0}^{(r)}(f) - \varepsilon_{\alpha x_0}^{(r)}(h).$$

But $f(x_0) \geq \varepsilon_{\alpha x_0}^{(r)}(f)$, and subtracting this inequality from the previous one, we obtain

$$-h(x_0) < -\varepsilon_{\alpha x_0}^{(r)}(h).$$

For sufficiently small $r > 0$ this inequality contradicts the α-harmonicity of $h(x)$ at the point x_0.

Thus, at all points of R^p

$$d(x) \geq 0,$$

and the first assertion of the theorem is proved.

If we assume that $d(x_0) = 0$ for $x_0 \in \Omega$, then for sufficiently smal $r > 0$ we have

$$0 = d(x_0) \geq \varepsilon_{\alpha x_0}^{(r)}(d),$$

and therefore $d(x) = 0$ almost everywhere. ☐

The theorem just proved together with the approximation Theorem 1.9 will allow us to prove for Riesz potentials for $\alpha < 2$ the principle of domination which in § 5 was established only for Newtonian potentials with finite energy.

Theorem 1.29. *Suppose that* μ *is a measure whose potential* $U_\alpha^\mu(x)$ *is finite* μ*-almost everywhere,*[41] *and that* $f(x)$ *is an* α*-superharmonic function.*

[40] The boundary of Ω may contain the point at infinity.

[41] It is clear that if μ has finite energy, then this condition will be satisfied.

Then if the inequality

$$U_\alpha^\mu(x) \leq f(x)$$

holds μ-almost everywhere, then it holds everywhere.

Proof. Suppose that $\{U_\alpha^{\mu_n}(x)\}$ is a sequence of continuous potentials constructed as in Theorem 1.9. As is seen from the proofs of Theorems 1.8 and 1.9, the measures μ_n may be chosen such that $S(\mu_n)$ belongs to a set of full μ-measure on which the inequality $U_\alpha^\mu(x) \leq f(x)$ is satisfied. But outside of $S(\mu_n)$ the potential $U_\alpha^{\mu_n}(x)$ is α-harmonic. Therefore, by Theorem 1.28, the inequality

$$U_\alpha^{\mu_n}(x) \leq f(x)$$

holds throughout R^p. The required result follows as we let $n \to \infty$. □

21. We now establish the connection between α-superharmonicity and the distribution $k_{-\alpha}$. This is based on the following lemmas.

Lemma 1.9. *As $r \to 0$*

$$\frac{1}{r^\alpha}(\varepsilon - \varepsilon_\alpha^{(r)}) \dot{\to} Qk_{-\alpha},$$

where $Q > 0$ depends only on α and p.

Proof. For any function $\varphi(x) \in D$, we have

$$(\varepsilon - \varepsilon_\alpha^{(r)})(\varphi) = \int\limits_{|x|>r} [\varphi(0) - \varphi(x)]\, \varepsilon_\alpha^{(r)}(x)\, dx$$

$$= \Gamma\left(\frac{p}{2}\right) \pi^{-\frac{p}{2}-1} \sin\frac{\pi\alpha}{2} r^\alpha \int\limits_r^\infty \frac{d\varrho}{\varrho(\varrho^2 - r^2)^{\alpha/2}} \int\limits_{|x|=1} [\varphi(0) - \varphi(\varrho x)]\, d\omega$$

$$= \Gamma\left(\frac{p}{2}\right) \pi^{-\frac{p}{2}-1} \sin\frac{\pi\alpha}{2} r^\alpha \omega_p \int\limits_r^\infty \frac{\varphi(0) - \tilde{\varphi}(0,\varrho)}{\varrho(\varrho^2 - r^2)^{\alpha/2}}\, d\varrho,$$

where

$$\tilde{\varphi}(0,\varrho) = \frac{1}{\omega_p} \int\limits_{|x|=\varrho} \varphi(x)\, d\omega$$

is the mean value of $\varphi(x)$ on the sphere of radius ϱ with center at 0. Consequently,

$$\frac{1}{r^\alpha}(\varepsilon - \varepsilon_\alpha^{(r)})\varphi = -\Gamma\left(\frac{p}{2}\right) \pi^{-\frac{p}{2}-1} \sin\frac{\pi\alpha}{2} \omega_p \int\limits_r^\infty \frac{\tilde{\varphi}(0,\varrho) - \varphi(0)}{\varrho(\varrho^2 - r^2)^{\alpha/2}}\, d\varrho,$$

and since $\varphi(x)$ is smooth

$$\lim_{r\to 0} \frac{1}{r^\alpha} (\varepsilon - \varepsilon_\alpha^{(r)}) (\varphi)$$

$$= -\Gamma\left(\frac{p}{2}\right) \pi^{-\frac{p}{2}-1} \sin\frac{\pi\alpha}{2}\, \omega_p \int_0^\infty \frac{\tilde{\varphi}(0, \varrho) - \varphi(0)}{\varrho^{\alpha+1}}\, d\varrho .$$

Since, according to (1.1.6)

$$k_{-\alpha}(\varphi) = A(p, -\alpha)\, \omega_p \int_0^\infty \frac{\tilde{\varphi}(0, \varrho) - \varphi(0)}{\varrho^{\alpha+1}}\, d\varrho ,$$

then by (1.1.2)

$$\lim_{r\to 0} \frac{1}{r^\alpha} (\varepsilon - \varepsilon_\alpha^{(r)}) (\varphi) = \frac{2\pi^\alpha}{\alpha} \frac{\Gamma\left(\frac{p}{2}\right)}{\Gamma\left(\frac{\alpha}{2}\right) \Gamma\left(\frac{p+\alpha}{2}\right)} k_{-\alpha}(\varphi) . \ \square$$

Remark. In order for the last formula to hold, it is sufficient to assume that $\varphi(x) \in E_{-\alpha}$ (cf. § 1, No. 2), i.e. using the inequality $0 < \alpha < 2$, that $\varphi(x)$ in a neighborhood of zero has two continuous derivatives and satisfies condition (1.6.4).

Lemma 1.10. *If $f(x)$ is an α-superharmonic function, then*

$$\mu = k_{-\alpha} * f$$

is a measure. The convolution $k_{-\alpha} * f$ is considered as a distribution defined by the equation

$$(k_{-\alpha} * f)\, (\varphi) = k_{-\alpha} \int f(y)\, \varphi(x + y)\, dy .$$

Proof. By Theorem 0.8 it is sufficient to prove that for any function $\varphi(x) \in D$, $\varphi(x) \geq 0$,

$$(k_{-\alpha} * f)\, (\varphi) \geq 0 .$$

Since

$$\psi(x) = \int f(y)\, \varphi(x + y)\, dy \in E_{-\alpha},$$

then by the previous remark

$$(k_{-\alpha} * f)\, (\varphi) = k_{-\alpha}(\psi) = \lim_{r\to 0} \frac{1}{Qr^\alpha} (\varepsilon - \varepsilon_\alpha^{(r)})\, (\psi) ,$$

and therefore it is sufficient to establish that $\psi(x)$ is an α-superharmonic function. This follows directly from property (iv). $\square$

22. The following important theorem holds.

Theorem 1.30. *Any function $f(x)$ which is α-superharmonic in R^p ($p \geq 3$) admits a unique representation in the form*

$$f(x) = U_\alpha^\mu(x) + A \tag{1.6.7}$$

where $\mu \in \mathfrak{M}^+$, *and the constant* $A \geq 0$. *Moreover, if* $f(x)$ *is* α*-harmonic in* Ω, *then* $\mu(\Omega) = 0$.

The basis for the proof of Threoem 1.30 is in the following sequence of lemmas.

Lemma 1.11. *Suppose that* $\alpha > 0$, $\beta > 0$, $\alpha + \beta < p$ *and that* $f(x)$ *is a locally integrable function satisfying the condition*

$$f(x) = O\left(\frac{1}{|x|^{p-\beta}}\right) \quad (x \to \infty). \tag{1.6.8}$$

Then

$$f(x) = k_\alpha * (k_{-\alpha} * f). \tag{1.6.9}$$

almost everywhere.

Proof. We must prove that the equation

$$(k_\alpha * k_{-\alpha}) * f = k_\alpha * (k_{-\alpha} * f)$$

holds as an equation between distributions.

For this, according to § 2, Nos. 8 and 9 of the Introduction, we must examine the spaces E_1, E_2, E_3, E_{12}, E_{23}, E_{123}, on which the corresponding distributions k_α, $k_{-\alpha}$, f, $k_\alpha * k_{-\alpha}$, $k_{-\alpha} * f$, $k_\alpha * k_{-\alpha} * f$ could be defined, and verify that they satisfy conditions (a), (b), (c), (0.2.12) and (a'), (b'), (c'), (0.2.12'). It is not difficult to see that these spaces may be chosen such that:

(i) E_1 consists of continuous functions $\varphi(x)$ for which

$$\int_{R^p} \frac{|\varphi(x)|}{|x|^{p-\alpha}}\, dx < \infty$$

Moreover, $\varphi_n(x) \to 0$ in E_1 if

$$\int_{R^p} \frac{|\varphi_n(x)|}{|x|^{p-\alpha}} \to 0.$$

E_3 is defined in exactly the same way replacing α by β.

(ii) E_2 consists of continuous functions having two continuous derivatives in a neighborhood of the origin and satisfying the condition

$$\int_{R^p} \frac{|\varphi(x)|}{1 + |x|^{p+\alpha}}\, dx < \infty.$$

Moreover, $\varphi_n(x) \to 0$ in E_2, if

$$\int_{R^p} \frac{|\varphi_n(x)|}{1 + |x|^{p+\alpha}}\, dx \to 0, \quad \text{and} \quad \partial^k \varphi_n \to 0, \ |k| \leq 2,$$

uniformly in some neighborhood of the origin.

(iii) E_{12} consists of functions $\varphi(x, y)$ which have two continuous derivatives in $R^p \times R^p$ and satisfy the condition

$$\iint_{R^p \times R^p} \frac{|\partial^k \varphi(x, y)|}{|x|^{p-\alpha}(1 + |y|^{p+\alpha})} dx\, dy < \infty, \quad |k| \leq 2.$$

Moreover, $\varphi_n(x, y) \to 0$ in E_{12}, if

$$\iint_{R^p \times R^p} \frac{|\partial^k \varphi_n(x, y)|}{|x|^{p-\alpha}(1 + |y|^{p+\alpha})} dx\, dy \to 0, \quad |k| \leq 2,$$

and

$$\partial^k \varphi_n \to 0, \quad |k| \leq 2$$

uniformly in a certain neighborhood of the origin in $R^p \times R^p$.

(iv) E_{23} consists of functions $\varphi(y, z)$ which have two continuous derivatives in $R^p \times R^p$ and satisfy the condition

$$\iint_{R^p \times R^p} \frac{|\partial^k \varphi(y, z)|}{(1 + |y|^{p+\alpha})\, |z|^{p-\beta}} dy\, dz < \infty, \quad |k| \leq 2.$$

Convergence in E_{23} is defined in a way analogous to (iii).

(v) E_{123} consists of functions $\varphi(x, y, z)$ which have two continuous derivatives in $R^p \times R^p \times R^p$ and which satisfy the condition

$$\iiint_{R^p \times R^p \times R^p} \frac{|\partial^k \varphi(x, y, z)|}{|x|^{p-\alpha}(1 + |y|^{p+\alpha})\, |z|^{p-\beta}} dx\, dy\, dz < \infty, \quad |k| \leq 2.$$

Convergence in E_{123} is defined just as in (iii).

The verification of conditions (a), (b), (c), (a'), (b'), (c') and (0.2.12) is rather simple and we omit this. We shall verify (0.2.12'). Suppose that $\varphi(x) \in D$, and we need to prove that $\varphi(x + y + z) \in E_{123}$. We have

$$\int_{R^p} \frac{|\varphi(x + y + z)|}{|x|^{p-\alpha}} dx = \int_{S(\varphi)} \frac{|\varphi(\xi)|}{|\xi - y - z|^{p-\alpha}} d\xi = O\left(\frac{1}{1 + |y + z|^{p-\alpha}}\right),$$

$$\iint_{R^p \times R^p} \frac{|\varphi(x + y + z)|\, dx\, dy}{|x|^{p-\alpha}(1 + |y|^{p+\alpha})} = \int_{R^p} O\left(\frac{1}{1 + |y + z|^{p-\alpha}}\right) \frac{dy}{1 + |y|^{p+\alpha}}$$

$$= O\left(\frac{1}{1 + |z|^{p-\alpha}}\right),\text{[42]}$$

$$\iiint_{R^p \times R^p \times R^p} \frac{|\varphi(x + y + z)|\, dx\, dy\, dz}{|x|^{p-\alpha}(1 + |y|^{p+\alpha})\, |z|^{p-\beta}} = \int_{R^p} O\left(\frac{1}{1 + |z|^{2p-\alpha-\beta}}\right) dz.$$

[42] Cf. proof of Theorem 1.12.

Consequently, under the conditions of the lemma, the integral is finite. Similarly one can check the derivatives $\partial^k \varphi(x + y + z)$, $|k| \leq 2$; thus, condition (0.2.12′) is verified.

Thus, equation (1.6.9) holds almost everywhere and the lemma is established. $\square$

Lemma 1.11′. *Suppose that $f(x)$ is an α-superharmonic function which, as $x \to \infty$, satisfies the condition*[43]

$$f(x) = \frac{a}{|x|^{p-\alpha}} + O\left(\frac{1}{|x|^{p-\beta}}\right), \quad \beta < p - \alpha .$$

Then

$$f(x) = k_\alpha * \mu = U_\alpha^\mu(x), \tag{1.6.9′}$$

where $\mu = k_{-\alpha} * f$.

Proof. Set

$$f_1(x) = f(x) - \frac{a}{|x|^{p-\alpha}} .$$

This function satisfies condition (1.6.8) and by Lemma 1.11

$$f_1(x) = k_\alpha * (k_{-\alpha} * f_1) .$$

almost everywhere. But

$$k_{-\alpha} * f_1 = k_{-1} * f - a_1 \varepsilon ,$$

where $a_1 = a \cdot A^{-1}(p, \alpha)$ (cf. (1.1.2)). Consequently,

$$f_1(x) = k_\alpha * (k_{-\alpha} * f) - a_1 k_\alpha$$

or

$$f(x) = k_\alpha * \mu = U_\alpha^\mu(x)$$

almost everywhere. But then, by property (i) No. 20 and the α-superharmonicity of the potential $U_\alpha^\mu(x)$, we obtain the equation (1.6.9) for all points of R^p. $\square$

Lemma 1.12. *Suppose that $\{U_\alpha^{\mu_n}(x)\}$ is a nondecreasing sequence of potentials of measures. Then either*

$$\lim_{n \to \infty} U_\alpha^{\mu_n}(x) \equiv + \infty ,$$

or throughout R^p

$$\lim_{= \infty} U_\alpha^{\mu_n}(x) = U_\alpha^\mu(x) + A , \tag{1.6.10}$$

where $\mu \in \mathfrak{M}^+$ and $A \geq 0$.

---

[43] For $p \geq 4$, we can clearly ignore the first term and then use Lemma 1.11 with $\beta = \alpha$ directly.

Proof. If $\lim U_{\alpha}^{\mu_n}(x) \not\equiv +\infty$, then equation (1.6.10) holds almost everywhere in R^p. This is the content of Lemma 1.0. Furthermore, if $\alpha < 2$, then $\lim U_{\alpha}^{\mu_n}(x)$, by property (iii), No. 20 is an α-superharmonic function. Therefore, it follows directly from property (i), No. 20, that equation (1.6.10) holds at any point of R^p.

If $\alpha \geq 2$, then both sides of (1.6.10) are ordinary superharmonic functions and this also implies the lemma. ▯

23. To complete the proof of Theorem 1.30 we shall need a special approximation technique based on a generalization of the Poisson integral.

The following function $\check{P}_r(u, x)$ may be viewed as an analogue of the Poisson kernel[44] on the exterior of the sphere S_r with radius r and center at the origin:

$$\check{P}_r(y, x) = \Gamma\left(\frac{p}{2}\right)\pi^{-\frac{p}{2}-1} \sin\frac{\pi\alpha}{2} \frac{(|x|^2 - r^2)^{\alpha/2}}{(r^2 - |y|^2)^{\alpha/2}} \frac{1}{|x-y|^p} \tag{1.6.11}$$

$$(|y| < r;\ |x| > r).$$

According to the Appendix, (cf. (A. 6)), for any position of the point y_0 within the sphere S_r, outside of S_r we have the equation

$$\frac{1}{|y_0 - x|^{p-\alpha}} = \int_{|y|<r} \frac{\check{P}_r(y, x)}{|y_0 - y|^{p-\alpha}}\, dy, \quad |x| > r. \tag{1.6.12}$$

Suppose that ν is a signed measure with support $S(\nu)$, lying outside of the sphere S_r. We shall define an absolutely continuous signed measure ν', with support $S(\nu')$, lying within S_r, by the formula

$$\nu'(y) = \int_{S(\nu)} \check{P}_r(y, x)\, d\nu(x), \qquad |y| < r.$$

Then, integrating (1.6.12) with respect to $d\nu(x)$, we obtain

$$U_{\alpha}^{\nu}(y_0) = U_{\alpha}^{\nu'}(y_0), \quad |y_0| < r. \tag{1.6.13}$$

In this way we have solved the so-called problem of balayage on the sphere S_r: any signed measure ν outside of S_r may be replaced by a signed measure ν' within S_r such that the potential within S_r is unchanged.

---

[44] It can be shown that as $\alpha \to 2$, $\check{P}_r(y, x)$ vaguely converges to the classical Poisson kernel.

In a similar way we may introduce the analogue of the Poisson kernel for the interior of the sphere S_r. This is the function

$$\hat{P}_r(y, x) = \Gamma\left(\frac{p}{2}\right) \pi^{-\frac{p}{2}-1} \sin \frac{\pi\alpha}{2} \frac{(r^2 - |x|^2)^{\alpha/2}}{(|y|^2 - r^2)^{\alpha/2}} \frac{1}{|x - y|^p} \quad (1.6.11')$$

$$(|y| > r, |x| < r),$$

which for $x = 0$ coincides with $\varepsilon_\alpha^{(r)}(y)$ (cf. (1.6.2)).

For any point y_0, $|y_0| > r$, within the sphere S_r we have the equation (cf. Appendix, (A. 5))

$$\frac{1}{|y_0 - x|^{p-\alpha}} = \int\limits_{|y|>r} \frac{\hat{P}_r(y, x)}{|y_0 - y|^{p-\alpha}} \, dy, \quad |x| < r. \quad (1.6.12')$$

From this, as above, we are able to obtain a solution of the problem of balayage of the signed measure ν within S_r to the exterior of the sphere:

$$U_\alpha^\nu(y_0) = U_\alpha^{\nu'}(y_0), \quad |y_0| > r \quad (1.6.13')$$

where

$$\nu'(y) = \int\limits_{S(\nu)} \hat{P}_r(y, x) \, d\nu(x), \quad |y| > r,$$

is the signed measure for $|y| \geq r$.

The analogy between the function $\check{P}_r(y, x)$ and the Poisson kernel is most completely seen in the following lemma.

Lemma 1.13. *Let $f(x)$ be any measurable function within S_r for which*

$$\int\limits_{|x|<r} \frac{|f(x)|}{(r^2 - |x|^2)^{\alpha/2}} \, dx < \infty.$$

Set

$$\check{f}(x) = \begin{cases} f(x), & |x| \leq r, \\ \int\limits_{|y|<r} f(y) \, \check{P}_r(y, x) \, dy, & |x| > r. \end{cases} \quad (1.6.14)$$

Then $\check{f}(x)$ will be an α-harmonic function in the region $|x| > r$.

Proof. The function $\check{f}(x)$ is obviously continuous for $|x| > r$ and as $x \to \infty$

$$\check{f}(x) = O\left(\frac{1}{|x|^{p-\alpha}}\right), \quad (1.6.15)$$

so that condition (1.6.4) is satisfied.

We will show that for any δ, $0 < \delta < |x| - r$,

$$\check{f}(x) = \varepsilon_{\alpha x}^{(\delta)}(\check{f}), \tag{1.6.16}$$

from which the lemma will follow. First suppose that

$$f(x) = U_\alpha^\nu(x),$$

where the signed measure ν is concentrated on the ball $|x| < r$. Integrating equation (1.6.12) with respect to $d\nu(y_0)$, we obtain for $|x| > r$ the equation

$$U_\alpha^\nu(x) = \int\limits_{|y|<r} U_\alpha^\nu(y)\, \check{P}_r(y, x)\, dy.$$

In other words,

$$\check{U}_\alpha^\nu(x) \equiv U_\alpha^\nu(x)$$

But, as we have seen in No. 19, $U_\alpha^\nu(x)$ is an α-harmonic function outside of $S(\nu)$ and in particular, for $|x| > r$.

Suppose now that the function $f(x)$, given within the ball $|x| \le r$, has there $p + 2$ continuous derivatives. Preserving the continuity of all these derivatives, we may extend it to a function with compact support defined on all of R^p. By Lemma 1.1

$$f_1(x) = U_\alpha^\psi(x), \quad x \in R^p.$$

Suppose that ψ_1 is the restriction of the signed measure ψ outside the sphere S_r. Sweeping out ψ_1 onto the sphere S_r according to (1.6.13), we will obtain a new signed measure ψ_1'. Setting $\nu = \psi - \psi_1 + \psi_1'$, we obtain a signed measure concentrated on the ball $|x| < r$; moreover,

$$f(x) = U_\alpha^\nu(x), \quad |x| < r.$$

As we have seen, $\check{f}(x)$ will in this case be α-harmonic in the region $|x| > r$.

Finally, suppose that $f(x)$ is any function satisfying the conditions of the lemma. There is a sequence of functions $f_n(x)$ having $p + 2$ continuous derivatives which converges to $f(x)$ in the space $L(S_r)$ with weight $(r^2 - |x|^2)^{-\alpha/2}$, i.e. such that

$$\lim_{n\to\infty} \int\limits_{|x|<r} \frac{|f_n(x) - f(x)|}{(r^2 - |x|^2)^{\alpha/2}}\, dx = 0. \tag{1.6.17}$$

By what has been proved $\check{f}_n(x)$ satisfies the condition (1.6.16) for any n, i.e.

$$\check{f}_n(x) = \varepsilon_{\alpha x}^{(\delta)}(\check{f}_n), \quad |x| > r.$$

But from (1.6.17) it follows, first, that for any fixed x, $|x| > r$,

$$\lim_{n\to\infty} \check{f}_n(x) = \check{f}(x), \tag{1.6.18}$$

and second, that

$$\lim_{n\to\infty} \varepsilon_{\alpha x}^{(\delta)}(\check{f}_n) = \varepsilon_{\alpha x}^{(\delta)}(\check{f}). \tag{1.6.18'}$$

The first result is obvious; we shall prove the second. We have (cf. (1.6.3))

$$\begin{aligned}\varepsilon_{\alpha x}^{(\delta)}(\check{f}_n) &= \int \varepsilon_\alpha^{(\delta)}(x-y)\, \check{f}_n(y)\, dy \\ &= \int\limits_{|y|<r} \varepsilon_\alpha^{(\delta)}(x-y)\, f_n(y)\, dy + \int\limits_{|y|>r} \varepsilon_\alpha^{(\delta)}(x-y)\, dy \\ &\quad \times \int\limits_{|z|<r} f_n(z)\, \check{P}_r(z,y)\, dz = \int\limits_{|y|<r} \varepsilon_\alpha^{(\delta)}(x-y)\, f_n(y)\, dy \\ &\quad + \int\limits_{|z|<r} f_n(z)\, dz \int\limits_{|y|>r} \varepsilon_\alpha^{(\delta)}(x-y)\, \check{P}_r(z,y)\, dy.\end{aligned}$$

Since the function $\varepsilon_\alpha^{(\delta)}(x-y)$ is bounded on the ball $|y| < r$, the first integral, by (1.6.17), tends to $\int\limits_{|y|<r} \varepsilon_\alpha^{(\delta)}(x-y)\, f(y)\, dy$. Furthermore,

$$\begin{aligned}&\int\limits_{|y|>r} \varepsilon_\alpha^{(\delta)}(x-y)\, \check{P}_r(z,y)\, dy \\ &\quad = \Gamma\left(\frac{p}{2}\right) \pi^{-\frac{p}{2}-1} \sin\frac{\pi\alpha}{2} (r^2 - |z|^2)^{-\alpha/2} \\ &\qquad \times \int\limits_{|y|>r} \varepsilon_\alpha^{(\delta)}(x-y)\, (|y|^2 - r^2)^{\alpha/2} \frac{dy}{|z-y|^p}.\end{aligned}$$

For any $\eta > 0$ and $|z| < r$,

$$\int\limits_{|y|>r+\eta} \varepsilon_\alpha^{(\delta)}(x-y)\, (|y|^2 - r^2)^{\alpha/2} \frac{dy}{|z-y|^p} = O(1).$$

Taking η such that the spheres $|y| < r + \eta$ and $|x - y| < \delta$ do not intersect, and using the boundedness of $\varepsilon_\alpha^{(\delta)}(x-y)$ in the annulus $|r| < |y| < r + \eta$, we obtain

$$\begin{aligned}&\int\limits_{r<|y|<r+\eta} \varepsilon_\alpha^{(\delta)}(x-y)\, (|y|^2 - r^2)^{\alpha/2} \frac{dy}{|z-y|^p} \\ &\quad < Q \int\limits_{r<|y|<r+\eta} (|y|^2 - r^2)^{\alpha/2} \frac{dy}{|z-y|^p} = O(1),\end{aligned}$$

since the last integral uniformly converges for $|z| \leq r$.

Thus,

$$\int_{|z|<r} f_n(z)\,dz \int_{|y|>r} \varepsilon_\alpha^{(\delta)}(x-y)\,\check{P}_r(z,y)\,dy$$

$$= \int_{|z|<r} f_n(z)\,O\big((r^2-|z|^2)^{-\alpha/2}\big)\,dz,$$

and, by (1.6.17), this integral tends to

$$\int_{|z|<r} f(z)\,dz \int_{|y|>r} \varepsilon_\alpha^{(\delta)}(x-y)\,\check{P}_r(z,y)\,dy.$$

This proves (1.6.18′). Together with (1.6.18) this yields (1.6.16) and the lemma is proved. □

The analogous result can be proved for the interior of the sphere S_r: the function

$$\hat{f}(x) = \begin{cases} f(x), & |x| \geq r, \\ \int_{|y|>r} f(y)\,\hat{P}_r(y,x)\,dy, & |x| < r, \end{cases} \tag{1.6.19}$$

under natural constraints on the function $f(x)$ given outside S_r, will be α-harmonic *within* S_r.

24. Let us return to α-superharmonic functions and establish several other properties for them.

Lemma 1.14. *Suppose that* $f(x)$ *is an* α*-superharmonic function in* R^p, *and* $\check{f}(x)$ *and* $\hat{f}(x)$ *are defined by equations* (1.6.14) *and* (1.6.19), *respectively. Then*

$$\check{f}(x) \leq f(x), \quad \hat{f}(x) \leq f(x) \tag{1.6.20}$$

Proof. If $f(x)$ is continuous, then (1.6.20) is a consequence of Lemma 2.13 and Theorem 1.28.[45]

In the general case consider the sequence

$$f_n(x) = f * m^{(1/n)}$$

of continuous functions, which according to property (iv), No. 20, will be α-superharmonic. As was just indicated

$$\check{f}_n(x) \leq f_n(x), \quad \hat{f}_n(x) \leq f_n(x) \tag{1.6.20′}$$

As $n \to \infty$, $f_n(x) \to f(x)$ almost everywhere in R^p, and in addition,

$$\int_{|x|<r} \frac{|f_n(x)-f(x)|}{(r^2-|x|^2)^{\alpha/2}}\,dx \to 0, \qquad \int_{|x|>r} \frac{|f_n(x)-f(x)|}{|x|^p\,(|x|^2-r^2)^{\alpha/2}}\,dx \to 0.$$

[45] In the case of $\check{f}(x)$, Theorem 1.28 should be applied to the region $r < |x| < \infty$.

From the latter relations it follows that

$$\hat{f}_n(x) \to \hat{f}(x), \quad |x| < r,$$

$$\check{f}_n(x) \to \check{f}(x), \quad |x| > r.$$

Passing to the limit in (1.6.20′) as $n \to \infty$, we obtain that (1.6.20) holds almost everywhere in R^p. In order to prove (1.6.20) at any point of R^p, it is sufficient to verify the inequality $\check{f}(x) \leq f(x)$ only for points $|x| > r$, and the inequality $\hat{f}(x) \leq f(x)$ only for points $|x| < r$. Since, from what has been proved, it follows that we have the inequalities

$$\varepsilon_{\alpha x}^{(r)}(\check{f}) \leq \varepsilon_{\alpha x}^{(r)}(f), \quad \varepsilon_{\alpha x}^{(r)}(\hat{f}) \leq \varepsilon_{\alpha x}^{(r)}(\hat{f}),$$

and at the specified points, the functions $\check{f}(x)$ and $\hat{f}(x)$, are continuous; then as $r \to 0$, by property (i), No. 20, we obtain

$$\check{f}(x) \leq f(x), \qquad \hat{f}(x) \leq f(x),$$

as required. □

Corollary. *If $f(x)$ is an α-superharmonic function, then for fixed x*

$$\varepsilon_{\alpha}^{(r)} * f(x) = \varepsilon_{\alpha x}^{(r)}(f)$$

is a nonincreasing function of the variable r.

Indeed, suppose that $r_1 < r$. For simplicity of notation we will assume that $x = 0$. Constructing as in (1.6.19) the function $\hat{f}(x)$ α-harmonic in the ball $|x| < r$, and using (1.6.20), we obtain

$$\hat{f}(0) = \varepsilon_{\alpha}^{(r)}(f) = \varepsilon_{\alpha}^{(r)}(\hat{f}) \leq \varepsilon_{\alpha}^{(r_1)}(f).$$

For what follows it is important to replace property (c) in the definition of α-superharmonic function by the apparently weaker requirement:

(c′) $f(x)$ satisfies condition (1.6.4), and for any point $x \in R^p$ one can find $r(x)$ such that for any r, $0 < r < r(x)$,

$$\varepsilon_{\alpha}^{(r)} * f(x) = \varepsilon_{\alpha x}^{(r)}(f) \leq f(x).$$

Denote by $\mathfrak{S}_\alpha$ the class of α-superharmonic functions, and by $\mathfrak{S}'_\alpha$ the class of functions satisfying conditions (a) and (b), No. 20, and the condition (c′) just introduced.

The inclusion $\mathfrak{S}_\alpha \subset \mathfrak{S}'_\alpha$ is obvious. But in fact $\mathfrak{S}_\alpha \equiv \mathfrak{S}'_\alpha$. In order to be convinced of this, it is sufficient to establish that:

(1) the functions of $\mathfrak{S}'_\alpha$ satisfy the principle of the α-harmonic minorant (Theorem 1.28);

(2) a function $f(x)$ which satisfies conditions (a), (b), (1.6.4), No. 20 and the principle of the α-harmonic minorant belongs to $\mathfrak{S}_\alpha$.

Assertion (1) follows from the fact that in the proof of Theorem 1.28 we actually used property (c′) and not (c).

We shall prove (2). Suppose that $f_n(x)$ is a nondecreasing sequence of functions continuous in R^p, which converges to $f(x)$. Then the functions $\hat{f}_n(x)$ will also be continuous in R^p and α-harmonic within the sphere S_r: $|x| = r$. Since for $|x| \geq r$

$$\hat{f}_n(x) = f_n(x) \leq f(x),$$

we obtain by applying the principle of the α-harmonic minorant

$$\hat{f}_n(0) = \varepsilon_\alpha^{(r)}(f_n) \leq f(0),$$

and from this, by the monotonic convergence of $f_n(x)$ to $f(x)$,

$$\varepsilon_\alpha^{(r)}(f) \leq f(0).$$

But $r > 0$ is arbitrary, and the point 0 may, of course, be replaced by any point of R^p. Consequently $f(x)$ is α-superharmonic.

Thus we have proved that

$$\mathfrak{S}'_\alpha \equiv \mathfrak{S}_\alpha.$$

This allows us to give a very simple proof of the following assertion.

Lemma 1.15. *If $f(x)$ is an α-superharmonic function, then the functions $\hat{f}(x)$ and $\check{f}(x)$ are also α-superharmonic.*

Proof. For definiteness we will consider the function $\check{f}(x)$. We will show that it is lower semicontinuous. This need only be checked on points of the sphere $|x| = r$. Consequently, we need to show that if $x_n \to x$, $|x_n| > r$, $|x| = r$, then

$$\check{f}(x) = f(x) \leq \varliminf_{n\to\infty} \check{f}(x_n).$$

But this easily follows from the formula

$$\check{f}(x_n) = \int\limits_{|y|<r} f(y)\, \check{P}_r(y, x_n)\, dy$$

the semicontinuity of $f(y)$ and the fact that for $x_n \to x$

$$\check{P}_r(y, x_n) \dot{\to} \varepsilon_x.$$

We now check condition (c′). At points x, $|x| \leq r$, even the stronger condition (c) is satisfied, since for any $\delta > 0$ by (1.6.20),

$$\check{f}(x) = f(x) \geq \varepsilon_{\alpha x}^{(\delta)}(f) \geq \varepsilon_{\alpha x}^{(\delta)}(\check{f}).$$

At points x, $|x| > r$, by Lemma 1.13, the function $\check{f}(x)$ will be α-harmonic, i.e. for $r < r(x)$

$$\check{f}(x) = \varepsilon_{\alpha x}^{(r)}(\check{f}).$$

Consequently, condition (c') is satisfied everywhere, and $\check{f}(x) \in \mathfrak{S}'_\alpha \equiv \mathfrak{S}_\alpha$. This proves the lemma. ☐

We now present the

Proof of Theorem 1.30. Consider the function $f_n(x)$, coincident with $\check{f}(x)$, constructed for $|x| > n$. By Lemma 1.15 it is α-superharmonic. Since, in addition, as $x \to \infty$

$$\check{P}_r(y, x) = \frac{C_1(y)}{|x|^{p-\alpha}} + O\left(\frac{1}{|x|^{p+1-\alpha}}\right),$$

which is not difficult to obtain from a series expansion of the right side of (1.6.11) we have by (1.6.14)

$$f_n(x) = \frac{Q}{|x|^{p-\alpha}} + O\left(\frac{1}{|x|^{p+1-\alpha}}\right).$$

But since $\alpha < 2$, then $2\alpha - 1 < p$, and by Lemma 1.11',

$$f_n(x) = U_\alpha^{\mu_n}(x),$$

where $\mu_n = k_{-\alpha} * f_n$ is a measure with compact support concentrated on the ball $|x| \leq n$. We will show that

$$f_n(x) \leq f_{n+1}(x). \tag{1.6.21}$$

Then using the obvious relation

$$\lim_{n\to\infty} f_n(x) = f(x)$$

we obtain the result from Lemma 1.12.

Since $f_{n+1}(x)$ is α-superharmonic (Lemma 1.15), and $f_n(x)$ is continuous and α-harmonic on the set $|x| \geq n + 1$, then by Theorem 1.28 the inequality (1.6.21) need be verified for $|x| \leq n + 1$ and at infinity.

But for $|x| \leq n$ (1.6.21) becomes an equation, since $f_n(x) = f_{n+1}(x) = f(x)$; for $n < |x| \leq n + 1$ it follows from (1.6.20), since in this region $f_{n+1}(x) = f(x)$, and at infinity, both functions are zero. Consequently, inequality (1.6.21), and therefore the existence of the representation (1.6.7) in Theorem 1.30 is established.

Furthermore, using No. 19, it is not difficult to show that

$$\lim_{r\to\infty} \varepsilon_\alpha^{(r)}(U_\alpha^\mu) = 0.$$

Therefore the quantity

$$A = \lim_{r\to\infty} \varepsilon_\alpha^{(r)}(f)$$

is precisely that constant which appeared in the representation (1.6.7). This proves the uniqueness of the representation (1.6.7). The last assertion of Theorem 1.30 easily follows from the construction of the measure μ. ▯

Corollary. *In order that an α-superharmonic function $f(x)$ be an α-potential of a measure μ, it is necessary and sufficient, that*

$$\lim_{r\to\infty} \varepsilon_\alpha^{(r)}(f) = 0.$$

This corollary is a generalization of Theorem 1.23 to M. Riesz potentials of order $\alpha < 2$.

From this, just as in No. 18, we obtain the principle of domination (cf. Theorem 1.27), and with that both of the maximum principles.

25. Suppose that $f_1(x)$ and $f_2(x)$ are α-superharmonic functions ($\alpha \leq 2$). Set

$$g(x) = \inf\,[f_1(x), f_2(x)].$$

Then for $g(x)$ it is completely elementary to prove conditions (a), (b), (c) of No. 20, and in this way, $g(x)$ becomes an α-superharmonic function. Using Theorem 1.23 (in the case $\alpha = 2$) and the previous corollary (in the case $\alpha < 2$) we immediately arrive at the following result.

Theorem 1.31. *If $f(x) \geq 0$ is any α-superharmonic function ($\alpha \leq 2$), and μ is a measure, then*

$$\inf\,[f(x), U_\alpha^\mu(x)]$$

is a potential of some measure.

In particular, if μ_1, μ_2 are two measures, then

$$\inf\,[U_\alpha^{\mu_1}(x), U_\alpha^{\mu_2}(x)] = U_\alpha^{\mu_3}(x)$$

where μ_3 is also a measure. It is also clear that if even one of the measures μ_1, μ_2 has finite energy, then μ_3 will also belong to $\mathscr{E}_\alpha^+$.

The analogous result holds for any finite system of measures.

In conclusion we prove a theorem on the behavior of a potential $U_\alpha^\nu(x)$ at infinity, which for $\alpha = 2$ has already been established at the end of § 4.

Theorem 1.32. *For $2 \leq \alpha < p$,*

$$\lim_{r\to\infty} \varepsilon^{(r)}(U_\alpha^\nu) = 0.$$

Proof. Clearly, it is sufficient to consider only a measure μ. First of all,

$$\varepsilon^{(r)}(U_\alpha^\mu) = \int U_\alpha^\mu(x)\, d\varepsilon^{(r)}(x) = \int U_\alpha^{\varepsilon^{(r)}}(x)\, d\mu(x).$$

Let us compute the potential $U_\alpha^{\varepsilon^{(r)}}(x)$ on the sphere $|x| = r$. For any $|x|$ we have

$$\begin{aligned} U_\alpha^{\varepsilon^{(r)}}(x) &= A(p, \alpha) \int \frac{d\varepsilon^{(r)}(y)}{|x - y|^{p-\alpha}} \\ &= A(p, \alpha) \frac{\omega_{p-1}}{\omega_p} \int_0^\pi \frac{\sin^{p-2}\theta\, d\theta}{(r^2 + |x|^2 - 2r\,|x| \cos\theta)^{(p-\alpha)/2}}; \end{aligned}$$

consequently, for $|x| = r$

$$U_\alpha^{\varepsilon^{(r)}}(x) = \frac{2^{\alpha-2}\pi^{\alpha-\frac{p+1}{2}}\,\Gamma\left(\frac{p-\alpha}{2}\right)\Gamma\left(\frac{\alpha-1}{2}\right)\Gamma\left(\frac{p}{2}\right)}{\Gamma\left(\frac{\alpha}{2}\right)\Gamma\left(\frac{\alpha+p}{2} - 1\right)} \frac{1}{r^{p-\alpha}} = w(p, \alpha) \frac{1}{r^{p-\alpha}}.$$

Therefore by Theorem 1.5 throughout the whole space we have

$$U_\alpha^{\varepsilon^{(r)}}(x) \le 2^{p-\alpha} w(p, \alpha) \frac{1}{r^{p-\alpha}}. \tag{1.6.22}$$

But $U_\alpha^{\varepsilon^{(r)}}(x)$ is a symmetric function of r and $|x|$; therefore we also have the inequality

$$U_\alpha^{\varepsilon^{(r)}}(x) \le 2^{p-\alpha} w(p, \alpha) \frac{1}{|x|^{p-\alpha}}. \tag{1.6.23}$$

Since

$$\int_{|x|>1} \frac{d\mu(x)}{|x|^{p-\alpha}} < \infty,$$

then (1.6.23) shows that the family of functions $\{U_\alpha^{\varepsilon^{(r)}}(x)\}$ has a μ-integrable majorant, and from (1.6.22) it follows that

$$\lim_{r\to\infty} U_\alpha^{\varepsilon^{(r)}}(x) = 0.$$

Hence

$$\lim_{r\to\infty} \varepsilon^{(r)}(U_\alpha^\mu) = \lim_{r\to\infty} \int U_\alpha^{\varepsilon^{(r)}}(x)\, d\mu(x) = 0,$$

and the theorem is proved. ▯

Chapter II

Capacity and Equilibrium Measure

In this chapter we shall associate with each kernel $k_\alpha(x)$ $(0 < \alpha \leq p)$ a certain non-negative set function, defined on the σ-field of Borel sets in the space R^p. This function $C_\alpha(E)$, which we shall call the *α-capacity*, for fixed α is determined by the "geometry" of the set E, i.e. it is invariant with respect to the group of motions in R^p. In this sense it is analogous to Lebesgue measure. However, unlike a measure, it is not additive.

The significance of the concept of capacity for potential theory is contained in the fact that in many circumstances sets of *zero capacity* and only these can appear as sets of singularities for certain classes of functions (cf. Ch. III) or as null-sets which can be ignored.

Of course, of greatest interest is the classical case $\alpha = 2$ (Newtonian or logarithmic potential).

First (§§ 1—3) we shall assume that $\alpha < p$. The case $\alpha = p = 2$ will be considered later in § 4.

§ 1. Equilibrium Measure and Capacity of a Compact Set

1. Consider a compact set $K \subset R^p$ and denote by $\mathfrak{M}_K^+$ the set of all measures whose support lies in K. The subset of $\mathfrak{M}_K^+$ consisting of measures μ, normalized by the condition $\mu(1) = 1$, will be denoted by $\mathring{\mathfrak{M}}_K^+$. It is not difficult to see that $\mathring{\mathfrak{M}}_K^+$ is convex and vaguely compact (cf. Theorem 0.6).

The latter circumstance allows us by classical argument to prove the existence of a *minimizing measure* $\lambda_K \in \mathring{\mathfrak{M}}_K^+$ for which

$$||\lambda_K|| \leq ||\mu||, \quad \mu \in \mathring{\mathfrak{M}}_K^+$$

Recall that $||\mu||^2$ is the energy integral

$$I_\alpha(\mu) = \iint_{K \times K} k_\alpha(x - y)\, d\mu(x)\, d\mu(y),$$

which gives to the posed minimum-problem a certain physical meaning.

Denote by $W_\alpha(K)$ the quantity $\inf I_\alpha(\mu)$, where the inf is taken over $\mathring{\mathfrak{M}}_K^+$. Since

$$k_\alpha(x-y) \geq \frac{\text{const}}{d^{p-\alpha}}$$

where d is the diameter of K, then

$$0 < W_\alpha(K) \leq \infty.^1 \tag{2.1.1}$$

If $\{\mu_n\}$, $\mu_n \in \mathring{\mathfrak{M}}_K^+$ is a *minimizing sequence*, i.e.

$$\lim_{n\to\infty} I_\alpha(\mu_n) = W_\alpha(K)$$

then by the vague compactness of $\mathring{\mathfrak{M}}_K^+$ we may assume that $\mu_n \dot{\to} \lambda_K \in \mathring{\mathfrak{M}}_K^+$. By the descent principle (cf. (1.4.5)), we obtain

$$I_\alpha(\lambda_K) \leq \lim_{n\to\infty} I_\alpha(\mu_n) = W_\alpha(K)$$

and consequently,

$$I_\alpha(\lambda_K) = W_\alpha(K),$$

that is,

$$||\lambda_K|| \leq ||\mu||$$

for any $\mu \in \mathring{\mathfrak{M}}_K^+$.

The variational problem under discussion can equally well be put in the class of measures

$$\mathring{\mathscr{E}}^+ = \mathring{\mathfrak{M}}_K^+ \cap \mathring{\mathscr{E}}_\alpha^+$$

with finite energy. Because of this we can obtain an even stronger result. Namely, we will show that λ_K is the *strong limit* of the minimizing sequence and is the *unique* solution of the variational problem.

Indeed, if $\{\mu_n\}$ is a minimizing sequence, then so is the sequence

$$\frac{1}{2}(\mu_n + \mu_m) \in \mathring{\mathscr{E}}_K^+ \quad (n, m \to \infty).$$

This follows from the inequality

$$\sqrt{W_\alpha(K)} \leq \left\| \frac{1}{2}(\mu_n + \mu_m) \right\| \leq \frac{1}{2}(||\mu_n|| + ||\mu_m||)$$

[1] If $W_\alpha(K) = \infty$, there is, of course, no minimum problem. Therefore, in speaking below of minimizing sequences and measure, we assume that $W_\alpha(K)$ is finite.

Hence, the indentity

$$||\mu_n - \mu_m||^2 = 2\,||\mu_n||^2 + 2\,||\mu_m||^2 - 4\left|\left|\frac{1}{2}(\mu_n + \mu_m)\right|\right|^2$$

shows that $\{\mu_n\}$ is a *fundamental sequence*. By the completeness of $\mathscr{E}_\alpha^+$ (Theorem 1.18) there exists a $\lambda \in \mathring{\mathscr{E}}_\alpha^+$ which is the strong limit of μ_n. As noted in Ch. I, § 13 (Lemma 1.3), it follows from this that $\mu_n \dot{\rightarrow} \lambda$, and this in turn implies that $\lambda \in \mathring{\mathfrak{M}}_K^+$. Consequently, $\lambda \in \mathring{\mathscr{E}}_K^+$ i.e. $\mathring{\mathscr{E}}_K^+$ is strongly closed. The fact that λ is a solution of the minimum problem is clear, since

$$||\lambda||^2 = \lim_{n\to\infty} ||\mu_n||^2 = W_\alpha(K)$$

The uniqueness of λ follows since for any other solution λ', we would have

$$\left|\left|\frac{\lambda + \lambda'}{2}\right|\right| = W_\alpha(K)$$

and

$$||\lambda - \lambda'||^2 = 2\,||\lambda||^2 + 2\,||\lambda'||^2 - 4\left|\left|\frac{\lambda + \lambda'}{2}\right|\right|^2 = 0.$$

In the future we shall always denote this solution of the variational problem by λ_K, sometimes omitting the index K.

2. We introduce the following definition.

Definition. *The capacity (the α-capacity) of a compact set K is the number*

$$C_\alpha(K) = W_\alpha^{-1}(K).$$

From (2.1.1) it follows that $C_\alpha(K)$ is *finite and nonnegative.* If $W_\alpha(K) = \infty$, i.e. if $\mathring{\mathscr{E}}_K^+$ is empty, then according to this definition $C_\alpha(K) = 0$.

Compact sets which have zero capacity will play an especially important part in the theory. If some property holds everywhere with the possible exception of a set of points belonging to a compact set of zero capacity, then we will say that the property holds *approximately everywhere.* We point out at once the following simple theorem.

Theorem 2.1. *The equation $C_\alpha(K) = 0$ holds if, and only if,*

$$\mu(K) = 0$$

for any measure $\mu \in \mathscr{E}_\alpha^+$.

Proof. If $\mu(K) = 0$ for all $\mu \in \mathscr{E}_\alpha^+$, then this means that $\mathring{\mathscr{E}}_K^+$ is empty, i.e. $C_\alpha(K) = 0$. Conversely, if $C_\alpha(K) = 0$ and $\mu \in \mathscr{E}_K^+$, then the restriction

μ_K is in $\mathscr{E}_K^+$, since $\|\mu_K\| \leq \|\mu\| < \infty$. But $W_\alpha(K) = \infty$, and therefore, $\mu_K \equiv 0$, i.e. $\mu(K) = 0$. $\square$

Corollary. *If* $C_\alpha(K) = 0$, *then* $\nu(K) = 0$ *for any signed measure* $\nu \in \mathscr{E}_\alpha$.

From Theorem 2.1 it follows that:

(a) If $C_\alpha(K) = 0$, then mes $K = 0$.[2]

Indeed, the measure $\mu(E) = \text{mes}\,(E \cap K)$ has finite energy, since its potential

$$U_\alpha^\mu(x) = C(p, \alpha) \int \frac{dy}{|x - y|^{p-\alpha}}$$

is continuous and bounded on K.

We point out another property of compact sets of zero capacity.

(b) If $C_\alpha(K) = 0$, $\mu \in \mathring{\mathfrak{M}}_K^+$, then

$$\sup_{S(\mu)} U_\alpha^\mu(x) = +\infty.$$

Otherwise we would have

$$\|\mu\|^2 = I_\alpha(\mu) = \int_{S(\mu)} U_\alpha^\mu(x) d\mu(x) < +\infty,$$

i.e. $C_\alpha(K) > 0$.

We note that the condition $\mu \in \mathring{\mathfrak{M}}_K^+$ can be replaced by $\mu \in \mathfrak{M}_K^+$ and $\mu \neq 0$, since the normalization $\mu(1) = 1$ obviously doesn't play a role.

Below (No. 4) we will significantly sharpen property (b) and we shall also establish the converse proposition: if $\sup\limits_{S(\mu)} U_\alpha^\mu(x) = \infty$ for any $\mu \in \mathring{\mathfrak{M}}_K^+$, then $C_\alpha(K) = 0$.

The following definition will also be useful in the future.

Definition. *A signed measure* ν *will be called C-absolutely continuous, if for any compact set* K *of zero capacity*, $\nu(K) = 0$.

We have seen that every signed measure with finite energy is C-absolutely continuous. That the converse is *not true* is clear from the following example.

Consider in R^3 an absolutely continuous measure with integrable density

$$f(x) = \begin{cases} |x|^{-5/2} & \text{for} \quad 0 < |x| < 1, \\ 0 & \text{for} \quad |x| \geq 1. \end{cases}$$

[2] "mes" denotes Lebesque measure.

According to (a) this measure will be C-absolutely continuous. For its Newtonian potential at any point x, $|x| < 1$, we have

$$U_2^f(x) \geq \frac{1}{|x|} \int_0^{|x|} r^{-5/2}\, r^2\, dr = \frac{2}{\sqrt{|x|}}\,.$$

Consequently,

$$I_2(f) = \int_{|x|<1} U_2^f(x)\, f(x)\, dx = 8\pi \int_0^1 r^{-1/2}\, r^{-5/2}\, r^2\, dr = \infty\,,$$

and the measure with density $f(x)$ has infinite energy.

From the definition it follows that if ν is a C-absolutely continuous signed measure, and $f(x)$ is a ν-integrable function, then

$$\int_E f(x)\, d\nu(x) = \int_{E \setminus K} f(x)\, d\nu(x)\,,$$

if $C_\alpha(K) = 0$. In other words:

(c) In integration with respect to a C-absolutely continuous signed measure (and in particular, with respect to a signed measure $\nu \in \mathscr{E}_\alpha$), one may ignore any compact set of capacity zero.

3. We shall investigate the potential U_α^λ of the minimizing measure $\lambda = \lambda_K$. Here we will encounter exceptional sets of points which are not compact. For this reason we introduce a class of sets E, which, similar to the compact sets of capacity zero, can often be viewed as inessential sets.

Definition. *The set E is said to have inner capacity zero, if for any compact set $K \subset E$, $C_\alpha(K) = 0$.*

The term *approximately everywhere* will in the future mean "everywhere, except for a set of inner capacity zero".

It is not difficult to see that Theorem 2.1 admits the following generalization.

Theorem 2.2. *E has inner capacity zero if, and only if, $\mu_*(E) = 0$ for any measure $\mu \in \mathscr{E}_\alpha^+$.*

Indeed, sufficiency follows since for any $K \subset E$ we have $\mu(K) = 0$, $\mu \in \mathscr{E}_\alpha^+$, i.e. $C_\alpha(K) = 0$. Conversely, if $C_\alpha(K) = 0$ for any $K \subset E$, then $\mu(K) = 0$ for $\mu \in \mathscr{E}_\alpha^+$ which implies $\mu_*(E) = 0$.

From this, as in § 2, it follows that if E has inner capacity zero, then $\mathrm{mes}_* E = 0$. In particular, such a set has no interior points.

If ν is a C-absolutely continuous signed measure and E has inner capacity zero, then $\nu_*^+(E) = 0$ and $\nu_*^-(E) = 0$. Indeed, for any $K \subset E$ we have $\nu^+(K) = \nu^-(K) = 0$.

Thus, in integrating over a C-absolutely continuous signed measure ν, we may ignore any ν -measurable set with inner capacity zero.

Corollary. *If the Borel sets E_i, $i = 1, 2, \ldots$, have inner capacity zero, then $E = \bigcup_1^\infty E_i$ also have inner capacity zero.*

Clearly, we may assume that the E_i do not intersect. In that case, for any measure $\mu \in \mathscr{E}_\alpha^+$ we will have

$$\mu(E) = \sum_{i=1}^{\infty} \mu(E_i) = 0,$$

and this establishes the corollary.

Now let us turn to the properties of the potential $U_\alpha^\lambda(x)$.

(a) Approximately everywhere in K

$$U_\alpha^\lambda(x) \geq W_\alpha(K) = \|\lambda\|^2.$$

Otherwise we would find a compact set $K_0 \subset K$, $C_\alpha(K_0) > 0$ at all of whose points

$$U_\alpha^\lambda(x) < \|\lambda\|^2.$$

Suppose that $\nu \in \mathring{\mathscr{E}}_{K_0}^+$ (this set of measures is not empty). Then, on the one hand,

$$(\nu, \lambda) = \int U_\alpha^\lambda(x)\, d\nu(x) < \|\lambda\|^2.$$

On the other hand, since for any a, $0 \leq a \leq 1$,

$$a\nu + (1 - a)\,\lambda \in \mathring{\mathscr{E}}_K^+$$

then

$$\|a\nu + (1 - a)\,\lambda\| \geq \|\lambda\|. \tag{2.1.2}$$

From this it follows that

$$a^2 \|\nu\|^2 + 2a(1 - a)\,(\nu, \lambda) + (1 - a)^2 \|\lambda\|^2 \geq \|\lambda\|^2,$$

that is

$$2a[(\nu, \lambda) - \|\lambda\|^2] + O(a^2) \geq 0,$$

and, consequently,

$$(\nu, \lambda) \geq \|\lambda\|^2.$$

The obtained contradiction proves that, $C_\alpha(K_0) = 0$.

(b) At all points of $S(\lambda) \subset K$

$$U_\alpha^\lambda(x) \leq W_\alpha(K) = \|\lambda\|^2.$$

In fact, if we suppose that

$$U_\alpha^\lambda(x_0) > W_\alpha(K), x_0 \in S(\lambda),$$

then from the semicontinuity of the potential

$$U_\alpha^\lambda(x) > W_\alpha(K)$$

in some neighborhood V of the point x_0. Moreover $\lambda(V) > 0$. Then

$$||\lambda||^2 = \int_V U_\alpha^\lambda(x)\, d\lambda(x) + \int_{S(\lambda)\setminus V} U_\alpha^\lambda(x)\, d\lambda(x)$$

$$> W_\alpha \lambda(V) + W_\alpha[1 - \lambda(V)],$$

since in the last integral $U_\alpha^\lambda(x) \geq W_\alpha$ approximately everywhere according to (a). But this inequality is impossible since $||\lambda||^2 = W_\alpha$, and thus (b) is proved.

From (a) and (b) follows

(c) The equation

$$U_\alpha^\lambda(x) = W_\alpha$$

holds approximately everywhere in $S(\lambda)$; moreover,

$$U_\alpha^\lambda(x) \leq W_\alpha$$

everywhere in $S(\lambda)$.

Property (c) together with the normalization condition $\lambda(1) = 1$ completely determines the minimizing measure λ. Indeed, since (c) implies the C-absolute continuity of λ, we have

$$||\lambda||^2 = \int_{S(\lambda)} U_\alpha^\lambda(x)\, d\lambda(x) = W_\alpha \lambda(1) = W_\alpha,$$

thus λ is actually a minimizing measure.

If $\alpha \leq 2$, then the *maximum principle* holds (cf. Ch. I, § 3, Theorem 1.10), and then it follows from (a) and (b) that

(c′) Approximately everywhere on K

$$U_\alpha^\lambda(x) = W_\alpha$$

and throughout R^p

$$U_\alpha^\lambda(x) \leq W_\alpha.$$

We note that of all the measures belonging to $\mathring{\mathscr{E}}_K^+$, λ is the only measure whose potential is constant approximately everywhere on K.

Indeed, if $\lambda_1 \in \mathscr{E}_K^+$ and $U_\alpha^{\lambda_1} = A$ approximately everywhere on K, then

$$||\lambda_1||^2 = \int_K U_\alpha^{\lambda_1}(x)\, d\lambda_1 = A \geq W_\alpha(K) = ||\lambda||^2$$

and

$$||\lambda_1 - \lambda||^2 = ||\lambda_1||^2 + ||\lambda||^2 - 2(\lambda_1, \lambda)$$
$$= A + W_\alpha - 2 \int_K U_\alpha^{\lambda_1}(x)\, d\lambda(x) = W_\alpha - A \leq 0 .$$

Consequently,

$$\lambda_1 \equiv \lambda \quad \text{and} \quad A = W_\alpha(K) .$$

Often in place of λ we shall consider a measure proportional to it

$$\gamma = \gamma_K = W_\alpha^{-1}(K)\, \lambda = C_\alpha(K)\, \lambda .$$

We shall call this the *equilibrium measure* of the compact set K, and its potential $U_\alpha^\gamma(x)$ the *equilibrium potential*. From (a), (b), (c) we easily obtain the following properties of γ:

(a_1) $U_\alpha^\gamma(x) \geq 1$, approximately everywhere on K.

(b_1) $U_\alpha^\gamma(x) = 1$, approximately everywhere on $S(\gamma)$; moreover, $U_\alpha^\gamma(x) \leq 1$ throughout $S(\gamma)$.

In addition

$$||\gamma||^2 = \gamma(1) = C_\alpha(K) . \tag{2.1.3}$$

For $\alpha \leq 2$ $U_\alpha^\gamma(x) = 1$ approximately everywhere on K and $U_\alpha^\gamma(x) \leq 1$ everywhere. In this case we shall say that the measure γ solves the *equilibrium problem of Robin* and this will be called the *Robin case.*

Theorem 2.3. *Suppose $C_\alpha(K) > 0$. Then the equilibrium measure γ may be viewed as the (unique) solution of the following variational problems:*

(*i*) *In the subset of measures $\mu \in \mathfrak{M}_K^+$ satisfying the condition*

$$\sup_{S(\gamma)} U_\alpha^\mu(x) = 1 ,$$

find the measure γ, for which $\gamma(1) = \max \mu(1)$.

(*ii*) *Among the measures $\mu \in \mathfrak{M}_K^+$ normalized by the condition*

$$\mu(1) = C_\alpha(K)$$

find the measure γ, for which

$$\sup_{S(\gamma)} U_\alpha^\gamma(x) = \min \sup_{S(\mu)} U_\alpha^\mu(x) .$$

(*iii*) *Among all the measures $\mu \in \mathfrak{M}_K^+$, find the measure γ, for which*

$$||\gamma||^2 - 2\gamma(1) = \min \{||\mu||^2 - 2\mu(1)\}.$$

Proof. We will show that the equilibrium measure γ solves problem (ii). By (b_1) we have for the measure γ

$$\sup_{S(\gamma)} U_\alpha^\gamma(x) = 1 .$$

Further, if $\mu(1) = C_\alpha(K)$ and $\sup\limits_{S(\mu)} U_\alpha^\mu(x) \leq 1$, then

$$||\mu||^2 = \int\limits_{S(\mu)} U_\alpha^\mu(x)\, d\mu(x) \leq C_\alpha(K) = ||\gamma||^2,$$

from which

$$||C_\alpha^{-1}(K)\, \mu||^2 \leq ||C_\alpha^{-1}(K)\, \gamma||^2 = ||\lambda||^2.$$

But $C_\alpha^{-1}(K)\, \mu \in \mathring{\mathfrak{M}}_K^+$, and by No. 1

$$C_\alpha^{-1}(K)\, \mu = \lambda, \quad \text{i.e. } \mu \equiv \gamma.$$

Thus, the minimum in problem (ii) is equal to 1, and the unique solution of the problem is the equilibrium measure γ.

Now consider the functional

$$\Phi(\mu) = \frac{\mu(1)}{\sup\limits_{S(\mu)} U_\alpha^\mu(x)}, \qquad \mu \in \mathfrak{M}_K^+.$$

In view of the equation $\Phi(a\mu) = \Phi(\mu)$ $(a > 0)$ we see that the maximal value of $\Phi(\mu)$ in $\mathfrak{M}_K^+$ is attained for $\mu = a\gamma$ $(a > 0)$ and is equal to $C_\alpha(K)$. Therefore, taking the subclass of $\mathfrak{M}_K^+$ determined by the normalization $\sup\limits_{S(\mu)} U_\alpha^\mu(x) = 1$, we see that the measure γ solves problem (i).

To prove the final assertion of the theorem, we note that for any measure $\mu \in \mathfrak{M}_K^+$, by (a_1),

$$||\mu - \gamma||^2 = ||\mu||^2 - 2\int\limits_K U_\alpha^\gamma(x)\, d\mu(x) + ||\gamma||^2$$

$$\leq ||\mu||^2 - 2\mu(K) + ||\gamma||^2,$$

from which

$$||\mu||^2 - 2\mu(1) \geq ||\mu - \gamma||^2 - ||\gamma||^2.$$

Consequently, the quantity $||\mu||^2 - 2\mu(1)$ achieves its smallest value only for $\mu = \gamma$, and this value is equal to $-\,||\gamma||^2 = -C_\alpha(K)$. ▯

Remark. Since the maximum in problem (i) is equal to $C_\alpha(K)$, we arrive at the *definition of the capacity of a compact set due to Vallée-Poussin:*

$$C_\alpha(K) = \max \mu(1),$$

where the maximim is taken over the set of measures $\mu \in \mathfrak{M}_K^+$, satisfying the inequality

$$U_\alpha^\mu(x) \leq 1, \quad x \in S(\mu).$$

In the case $\alpha \leq 2$ the requirement $x \in S(\mu)$ in the above condition is redundant.

4. From Theorem 2.2 it is easy to concludethat if the set E is of inner capacity zero, then for any measure $\mu \not\equiv 0$, $S(\mu) \subset E$,

$$\sup_{S(\mu)} U^{\mu}_{\alpha}(x) = \infty .$$

Using the semicontinuity of the potential, one can strengthen this result.

Theorem 2.4. *If E has inner capacity zero, and $\mu \not\equiv 0$ is a measure concentrated on E, then there is a point x_0 in E such that*

$$U^{\mu}_{\alpha}(x_0) = \infty .$$

Proof. We need consider only the non-zero part of μ concentrated on some compact set and therefore we may assume that $S(\mu) \subset E$. The potential $U^{\mu}_{\alpha}(x)$ cannot be bounded on any portion of $S(\mu)$. Indeed, if $U^{\mu}_{\alpha}(x) < M$ on $V \cap S(\mu)$, where V is an open set, then, denoting by $\mu_1 \neq 0$ the restriction of μ to this portion, we would have $U^{\mu_1}_{\alpha}(x) < M$. But this would mean that $C_{\alpha}(\overline{V} \cap S(\mu)) > 0$, and this is impossible, since E is of inner capacity zero.

We now are able to construct a decreasing sequence of neighborhoods

$$V(x_1) \supset V(x_2) \supset V(x_3) \supset \cdots,$$

such that on the set $S(\mu) \cap \overline{V(x_n)}$ $U^{\mu}_{\alpha}(x) > n$. Indeed, $V(x_1)$ exists since there is a point x_1 at which $U^{\mu}_{\alpha}(x_1) > 1$; by lower semicontinuity there is a closed neighborhood $\overline{V(x_1)}$, where $U^{\mu}_{\alpha}(x) > 1$. We now use induction. If $V(x_i)$, $i = 1, 2, \ldots, n$, have already been constructed, then in $V(x_n)$, by assumption there is a point x_{n+1} where $U^{\mu}_{\alpha}(x_{n+1}) > n + 1$. By lower semicontinuity there is a neighborhood $V(x_{n+1})$, $\overline{V(x_{n+1})} \subset V(x_n)$, in which $U^{\mu}_{\alpha}(x) > n + 1$. There is a point x_0 belonging to all the sets $S(\mu) \cap \overline{V(x_n)}$. Clearly, $U^{\mu}_{\alpha}(x_0) = \infty$.[3] ▯

Remark. From the proof it can be seen that the points at which $U^{\mu}_{\alpha}(x) = \infty$ are everywhere dense in $S(\mu)$.

Corollary. *If $U^{\mu}_{\alpha}(x) < \infty$, then the measure μ is C-absolutely continuous.*

5. Let us return to the properties of the capacity of a compact set. First of all, $C_{\alpha}(K)$ is a *monotone increasing function*, i.e.

$$C_{\alpha}(K_1) \leq C_{\alpha}(K_2)$$

---

[3] The reader can easily establish the connection between this theorem and the well-known theorem of Osgood (cf. Riesz and Nagy [1].)

if $K_1 \subset K_2$. This follows simply from

$$\mathfrak{M}^+_{K_1} \subset \mathfrak{M}^+_{K_2}$$

and means that $W_\alpha(K_1) \geq W_\alpha(K_2)$.

We shall now prove the following properties of $C_\alpha(K)$:

(1) C_α is *subadditive,* i e. if $K = \bigcup_{i=1}^{\infty} K_i$, then

$$C_\alpha(K) \leq \sum_{i=1}^{\infty} C_\alpha(K_i).$$

(2) C_α is *continuous on the right.* This means that for any $\varepsilon > 0$ there is a neighborhood G of the compact set K such that for $K \subset K' \subset G$

$$C_\alpha(K') - C_\alpha(K) < \varepsilon.$$

(3) If $\alpha \leq 2$, then C_α is a *convex function,* i.e. for any K_1, K_2

$$C_\alpha(K_1 \cup K_2) + C_\alpha(K_1 \cap K_2) \leq C_\alpha(K_1) + C_\alpha(K_2).$$

If we put $X = K_1 \cap K_2$, this property can be expressed in the form

$$C_\alpha(X) - C_\alpha(X \cup K_1) - C_\alpha(X \cup K_2) + C_\alpha(X \cup K_1 \cup K_2) \leq 0,$$

which gives rise to the term "convexity".

Proof of Subadditivity. Let γ be the equilibrium measure for K, $\gamma(1) = C_\alpha(K)$. Denote by γ_i the restriction of γ to K_i; it is easy to see that $\gamma_i(1) \leq C_\alpha(K_i) \sup\limits_{S(\gamma_i)} U_\alpha^{\gamma_i}(x)$. Therefore

$$C_\alpha(K) \leq \sum_{i=1}^{\infty} \gamma_i(1) \leq \sum_{i=1}^{\infty} C_\alpha(K_i) \sup_{S(\gamma_i)} U_\alpha^{\gamma_i}(x)$$

$$\leq \sum_{i=1}^{\infty} C_\alpha(K_i) \sup_{S(\gamma)} U_\alpha^{\gamma}(x) = \sum_{i=1}^{\infty} C_\alpha(K_i). \quad \square$$

Proof of Continuity of the Right. If the assertion were false, then there would exist $\varepsilon > 0$ and a monotone decreasing sequence of compact sets $K_i \supset K$ such that $\bigcap K_i = K$ and $C_\alpha(K_i) \geq C_\alpha(K) + \varepsilon$.

Let γ_i denote the equilibrium measure for K_i, $||\gamma_i||^2 = \gamma_i(1) = C_\alpha(K_i)$. The set of measures $\{\gamma_i\}$ is vaguely compact (cf. Theorem 0.6) and therefore contains a vaguely convergent subsequence $\{\gamma_{n_i}\}$ such that $\gamma_{n_i} \to \mu \in \mathfrak{M}^+_K$.

By the principle of descent (cf. Ch. I, formula (1.4.5))

$$||\mu||^2 \leq \lim ||\gamma_{n_i}||^2 = \lim ||\gamma_i||^2.$$

Further, $\mu(1) = \lim \gamma_{n_i}(1) \geq C_\alpha(K) + \varepsilon$. Therefore we must have the inequality

$$||\mu||^2 \geq C_\alpha^{-1}(K)\,[C_\alpha(K) + \varepsilon]^2 > C_\alpha(K) + 2\varepsilon.^4$$

From the previous inequality and the monotone decreasing $||\gamma_i||^2 = C_\alpha(K_i)$ we obtain that $C_\alpha(K_i) > C_\alpha(K) + 2\varepsilon$. Repeating the argument n times we obtain $C_\alpha(K_i) > C_\alpha(K) + 2^n\varepsilon$, which for sufficiently large n is absurd.[5] □

Proof of Convexity. For $\alpha \leq 2$ the equilibrium measure γ, as indicated in No. 3, solves the Robin problem, i.e. $U_\alpha^\gamma(x) \leq 1$ everywhere and $U_\alpha^\gamma(x) = 1$ approximately everywhere on K. This allows us to prove the following lemma.

Lemma 2.1. *If the signed measure $\nu \in \mathscr{E}_\alpha$ with compact support satisfies $U_\alpha^\nu(x) \geq 0$ approximately everywhere on $S(\nu)$, then $\nu(1) \geq 0$.*

Proof. Denote by $\varkappa$ the equilibrium measure on $S(\nu)$. Then

$$0 \leq \int_{S(\nu)} U_\alpha^\nu(x)\, d\varkappa = \int_{S(\nu)} U_\alpha^\varkappa(x)\, d\nu = \nu(1).$$

Denote by $h(X)$ the equilibrium potential for X and consider the "first difference"

$$d_1(X, K_1) = h(X \cup K_1) - h(X)$$

and the "second difference"

$$d_2(X, K_1, K_2) = d_1(X, K_1) - d_1(X \cup K_2, K_1)$$
$$= h(X \cup K_1) + h(X \cup K_2) - h(X) - h(X \cup K_1 \cup K_2).$$

Since $d_1(X, K_1) = 0$ approximately everywhere on K, then the same holds true for $d_2(X, K_1, K_2)$. Furthermore, $d_1(X \cup K_2, K_1) = 0$ approximately everywhere on K_2, and, using the second maximum principle, we obtain that $d_2(X, K_1, K_2) \geq 0$. By the symmetry of d_2 with respect to K_1 and K_2, the same inequality holds approximately everywhere on K_1. Thus, the inequality $d_2(X, K_1, K_2) \geq 0$ holds approximately everywhere on $X \cup K_1 \cup K_2$. Since all the equilibrium measures under consideration belong to $\mathfrak{M}^+_{X \cup K_1 \cup K_2}$, $d_2(X, K_1, K_2)$ is the potential of a signed measure with support in $X \cup K_1 \cup K_2$.

---

[4] We assume that $C_\alpha(K) > 0$. If $C_\alpha(K) = 0$ the contradiction is immediate since the measure $\mu \in \mathfrak{M}_K^+$, $\mu(1) \geq \varepsilon$ has finite energy.

[5] In the case $\alpha \leq 2$ the proof can be simplified somewhat using the maximum principle.

By Lemma 2.1 the total mass of the corresponding signed measure is nonnegative. Taking into account that the total mass of an equilibrium measure is equal to the capacity, we obtain

$$C_\alpha(X) + C_\alpha(X \cup K_1 \cup K_2) \leq C_\alpha(X \cup K_1) + C_\alpha(X \cup K_2),$$

which is the required result. ☐

It is not difficult to prove by induction that the "higher differences"

$$d_n(X, K_1, K_2, \ldots, K_n) = d_{n-1}(X, K_1, K_2, \ldots, K_{n-1}) - d_{n-1}(X \cup K_n, K_1, \ldots, K_{n-1}),$$

constructed for equilibrium potentials or for capacities will also be nonnegative.

§ 2. Inner and Outer Capacities and Equilibrium Measures. Capacitability

6. For an arbitrary set $E \subset R^p$ we define the *inner* capacity $\underline{C}_\alpha(E)$ by the formula

$$\underline{C}_\alpha(E) = \sup_{K \subset E} C_\alpha(K) \quad K \text{ is compact}$$

and the *outer* capacity $\overline{C}_\alpha(E)$ by the formula

$$\overline{C}_\alpha(E) = \inf_{G \supset E} \underline{C}_\alpha(G) \quad \text{G is open.}$$

It is not difficult to see that the definition of $\underline{C}_\alpha(E)$ is equivalent to the following (Vallée-Poussin): $\underline{C}_\alpha(E) = \sup \mu(1)$, where the supremum is taken over all measures μ, for which $S(\mu) \subset E$, $\sup\limits_{S(\mu)} U_\alpha^\mu(x) = 1$.

Both capacities introduced are monotone increasing set functions taking finite values on bounded sets. For open sets

$$\underline{C}_\alpha(G) = \overline{C}_\alpha(G), \tag{2.2.1}$$

and in general

$$\underline{C}_\alpha(E) \leq \overline{C}_\alpha(E). \tag{2.2.2}$$

Indeed for any $\varepsilon > 0$ we can find $K \subset E \subset G$ such that

$$\underline{C}_\alpha(E) - \varepsilon < C_\alpha(K), \quad \underline{C}_\alpha(G) < \overline{C}_\alpha(E) + \varepsilon.$$

Moreover, since $C_\alpha(K) \leq \underline{C}_\alpha(G)$, then

$$\underline{C}_\alpha(E) - \varepsilon < \overline{C}_\alpha(E) + \varepsilon,$$

and because ε was arbitrary we obtain the required result. If

$$\underline{C}_\alpha(E) = \overline{C}_\alpha(E),$$

then the set E will be called *capacitable* and we will write $C_\alpha(E)$ instead of $\underline{C}_\alpha(E)$ or $\overline{C}_\alpha(E)$. This notation is correct in view of the following elementary theorem on capacitability.

Theorem 2.5. *Every compact set and every open set is capacitable.*

Proof. For open sets this has already been mentioned (cf. (2.2.1)); for compact sets this follows from upper semicontinuity (No. 5). ☐

Moreover, it turns out that in general all Borel sets in R^p are capacitable. We shall prove this important theorem shortly.

If something holds for all points of R^p, with the exception of a set of *outer capacity zero,* then we will say that it holds *quasi-everywhere.*

It is clear that a theorem in which some property holds quasi-everywhere is stronger than a corresponding theorem in which that property holds "approximately everywhere". However, by Theorem 2.8, this distinction is revealed quite rarely.

Both capacities introduced have the property of countable subadditivity:

$$\underline{C}_\alpha\left(\bigcup_1^\infty E_n\right) \leq \sum \underline{C}_\alpha(E_n); \quad \overline{C}_\alpha\left(\bigcup_1^\infty E_n\right) \leq \sum_1^\infty \overline{C}_\alpha(E_n), \tag{2.2.3}$$

where in the first of the inequalities in (2.2.3) the sets E_n should be assumed Borel; in the second they may be arbitrary.

The first inequality is proved just as the countable subadditivity of the capacity of compact sets. For γ one need only take the equilibrium measure for a compact set $K \subset \bigcup_1^\infty E_n$, which is chosen such that $C_\alpha(K) > \underline{C}_\alpha\left(\bigcup_1^\infty E_n\right) - \varepsilon$ (or $C_\alpha(K) > N$, if $\underline{C}_\alpha\left(\bigcup_1^\infty E_n\right) = \infty$), and then use the definition of Vallée-Poussin (cf. above). The E_n must be Borel in order to obtain the inequality

$$\gamma(1) \leq \sum_{n=1}^\infty \gamma_n(1).$$

To obtain the second of the inequalities in (2.2.3) it is sufficient to note that for open sets

$$C_\alpha\left(\bigcup_1^\infty G_n\right) \leq \sum_1^\infty C_\alpha(G_n).$$

in view of the first inequality in (2.2.3). Assuming that $G_n \supset E_n$, one needs now to take the lower bounds.

It is not difficult to establish also that $\overline{C}_\alpha(E)$ is continuous on the right and for $\alpha \leq 2$ convex[6].

From the countable subadditvitiy of the outer capacity we obtain two useful consequences:

(1) The countable union of sets with outer capacity zero also has outer capacity zero.

(2) Adding or eliminating a set of outer capacity zero does not change the outer capacity of the original set.

7. In this section we prove the following theorem.

Theorem 2.6. *Suppose that* $\underline{C}_\alpha(E) < \infty$. *Then there is a unique measure* $\gamma_E = \gamma$ *with support in* $\overline{E}$ *which satisfies the following conditions:*

(*a*) $\|\gamma\|^2 = \gamma(1) = \underline{C}_\alpha(E)$,

(*b*) $U_\alpha^\gamma(x) \geq 1$ *approximately everywhere in* E,

(*c*) $U_\alpha^\gamma(x) \leq 1$ *everywhere on the support* $S(\gamma)$ *of the measure* γ.

Since the case $\underline{C}_\alpha(E) = 0$ is trivial, in what follows we shall assume that $\underline{C}_\alpha(E) > 0$. This measure γ will be called the *inner equilibrium measure* for E. We note that from (b) and (c) it follows that

$$U_\alpha^\gamma(x) = 1 \quad \gamma\text{-almost everywhere},$$

and for $\alpha \leq 2$, $U_\alpha^\gamma(x) = 1$ approximately everywhere in E; $U_\alpha^\gamma(x) \leq 1$ throughout R^p.

Before proceeding with the proof of this theorem, we shall establish several auxiliary results.

We denote by Γ_E the subset of $\mathscr{E}_\alpha^+$, consisting of those measures μ for which $U_\alpha^\mu(x) \geq 1$ approximately everywhere in E. Clearly, Γ_E is convex. We will show that it is *strongly closed.* Otherwise, there exists a sequence $\mu_n \in \Gamma_E$ and a measure $\mu \notin \Gamma_E$ such that the convergence $\mu_n \to \mu$ is strong. Consequently, there exists a compact set $K \subset E$, $C_\alpha(K) > 0$, on which $U_\alpha^\mu(x) < 1$. If $\gamma = \gamma_K$ is the equilibrium measure on K, then, on the one hand,

$$\lim_{n\to\infty} (\mu_n, \gamma) = (\mu, \gamma) = \int_K U_\alpha^\mu(x)\, d\gamma < \gamma(1) = C_\alpha(K),$$

and on the other hand

$$(\mu_n, \gamma) = \int_K U_\alpha^{\mu_n}(x)\, d\gamma \geq \gamma(1) = C_\alpha(K).$$

[6] For the proof of convexity one should first prove the inequality

$$C_\alpha(G_1 \cup G_2) + C_\alpha(G_1 \cap G_2) \leq C_\alpha(G_1) + C_\alpha(G_2)$$

for open sets G_1, G_2.

The resulting contradiction proves that Γ_E is strongly closed.

We will show for any $\mu \in \Gamma_E$ that

$$||\mu||^2 \geq \underline{C}_\alpha(E). \tag{2.2.4}$$

It is sufficient to establish for any $K \subset E$

$$||\mu||^2 > C_\alpha(K).$$

Suppose $\gamma = \gamma_K$ is the equilibrium measure on K. Then

$$C_\alpha(K) = \gamma(1) \leq \int_K U_\alpha^\mu(x) d\gamma \leq ||\gamma|| \cdot ||\mu|| = \sqrt{C_\alpha(K)} \cdot ||\mu||,$$

and this gives the required result.

Lemma 2.2. *Suppose $\lambda \in \Gamma_E$ is a minimal measure, i.e.*

$$||\lambda|| = \inf_{\mu \in \Gamma_E} ||\mu||.$$

Then for any measure $\mu \in \Gamma_E$

$$||\mu - \lambda||^2 \leq ||\mu||^2 - ||\lambda||^2.$$

Proof. Since for any t, $0 \leq t \leq 1$,

$$\nu = (1 - t)\,\lambda + t\mu = \lambda + t(\mu - \lambda) \in \Gamma_E,$$

then

$$||\nu||^2 = ||\lambda||^2 + 2t(\mu - \lambda, \lambda) + t^2\,||\mu - \lambda||^2 \geq ||\lambda||^2.$$

Consequently, $(\mu - \lambda, \lambda) \geq 0$, i.e. $(\mu, \lambda) \geq ||\lambda||^2$, whence

$$||\mu - \lambda||^2 = ||\mu||^2 - 2(\mu, \lambda) + ||\lambda||^2 \leq ||\mu||^2 - ||\lambda||^2. \quad \square$$

Corollary. *The minimal measure in Γ_E is unique, and by* (2.2.4) *and* (*a*) *the inner equilibrium measure for E is also unique.*

Lemma 2.3. *Suppose that $\mu_n \in \mathfrak{M}^+$*

$$\sup_{S(\mu_n)} U_\alpha^{\mu_n}(x) \leq 1$$

and $\mu_n \rightharpoonup \mu$. Then

$$\sup_{S(\mu)} U_\alpha^\mu(x) \leq 1.$$

Proof. This follows from the general form of the descent principle (Theorem 1.3). Namely, the formula

$$S(\mu) \subset \bigcap_{N=1}^{\infty} \overline{\left(\bigcup_{n=N}^{\infty} S(\mu_n)\right)}$$

shows that for every point $x \in S(\mu)$, there is a sequence $x_k \to x$, where $x_k \in S(\mu_{n_k})$. Hence

$$U_\alpha^\mu(x) \le \varliminf_{k\to\infty} U_\alpha^{\mu_{n_k}}(x_k) \le 1,$$

and the lemma is proved. ▯

Proof of Theorem 2.6. Consider an increasing sequence of compact sets $K_n \subset K_{n+1} \subset E$ such that

$$\lim_{n\to\infty} C_\alpha(K_n) = \underline{C}_\alpha(E), \tag{2.2.5}$$

and denote by γ_n the equilibrium measure on K_n. By (a_1), No. 3

$$\gamma_n \in \Gamma_{K_p} \quad \text{for} \quad n \ge p. \tag{2.2.6}$$

Juxtaposing (2.2.6) and (2.1.3) (No. 3) we arrive at the conclusion that γ_p is the minimal measure in Γ_{K_p}. But then, by Lemma 2.2, we obtain $(n \ge p)$

$$\|\gamma_n - \gamma_p\|^2 \le \|\gamma_n\|^2 - \|\gamma_p\|^2 = C_\alpha(K_n) - C_\alpha(K_p).$$

Since we assume that $\underline{C}_\alpha(E) < \infty$, the sequence $\{\gamma_n\}$ is fundamental. Using the completeness of $\mathscr{E}_\alpha^+$, we see that $\gamma \in \mathscr{E}_\alpha^+$ exists as the strong limit of the γ_n. But then, $\gamma_n \dashrightarrow \gamma$, and since $S(\gamma_n) \subset K_n \subset \overline{E}$, we have $S(\gamma) \subset \overline{E}$. We will show that γ satisfies all the conditions of the theorem.

First of all, from (b_1), No. 3 and Lemma 2.3 we immediately can obtain (c).

Further,

$$\|\gamma\|^2 = \lim_{n\to\infty} \|\gamma_n\|^2 = \lim_{n\to\infty} C_\alpha(K_n) = \underline{C}_\alpha(E). \tag{2.2.7}$$

In addition,

$$\gamma(1) \le \varlimsup_{n\to\infty} \gamma_n(1) = \varlimsup_{n\to\infty} C_\alpha(K_n) = \underline{C}_\alpha(E),$$

and by (c) and (2.2.7)

$$\underline{C}_\alpha(E) = \|\gamma\|^2 = \int_{S(\gamma)} U_\alpha^\gamma(x)\, d\gamma \le \gamma(1).$$

In this way

$$\|\gamma\|^2 = \gamma(1) = \underline{C}_\alpha(E),$$

and (a) is proved.

It remains to prove (b). Since Γ_{K_p} is strongly closed and $\gamma_n \to \gamma$ strongly, we have by (2.2.6)

$$\gamma \in \Gamma_{K_p} \quad \text{for any } p.$$

This means that

$$U_\alpha^\gamma(x) \geq 1$$

approximately everywhere on each K_p and, consequently, approximately everywhere throughout the set

$$H = \bigcup_{p=1}^{\infty} K_p \subset E.$$

In other words, $\gamma \in \Gamma_H$.

Since

$$\lim_{p\to\infty} C_\alpha(K_p) \leq \underline{C}_\alpha(H) \leq \underline{C}_\alpha(E),$$

then by (2.2.5)

$$\underline{C}_\alpha(H) = \underline{C}_\alpha(E). \tag{2.2.8}$$

We assert that $U_\alpha^\gamma(x) \geq 1$ approximately everywhere in E, i.e. approximately everywhere on any compact set $K \subset E$. Set

$$H' = H \cup K = \bigcup_{p=1}^{\infty} K'_p,$$

where $K'_p = K_p \cup K$. Then $\underline{C}_\alpha(H') = C_\alpha(E)$, since $H \subset H' \subset E$ and (2.2.8) holds.

If the measure γ' is defined with the help of K'_n just as the measure γ was defined with K_n, then

$$\|\gamma'\|^2 = \underline{C}_\alpha(H') = \underline{C}_\alpha(H),$$

and

$$U_\alpha^{\gamma'}(x) \geq 1$$

approximately everywhere in H', i.e. $\gamma' \in \Gamma_{H'} \subset \Gamma_H$. This shows that γ' is a minimal measure in Γ_H and by the uniqueness of the minimal measure it coincides with γ.

Thus $U_\alpha^\gamma(x) \geq 1$ approximately everywhere in H' and hence approximately everywhere on K. The theorem is proved. ▯

Corollary. *At all interior points of E*

$$U_\alpha^\gamma(x) \geq 1.$$

This easily follows from b) of Theorem 2.6 and by Theorem 1.11 (Ch. I).

8. Let us now define and prove the existence of the *outer equlibrium measure* γ_E^* for the set E. Our goal is the proof of the following theorem.

Theorem 2.7. *Suppose that* $\overline{C}_\alpha(E) < \infty$. *Then there exists a unique measure* $\gamma_E^* = \gamma^*$ *with support in* $\overline{E}$ *which satisfies the following conditions:*

(a*) $\|\gamma^*\|^2 = \gamma^*(1) = \overline{C}_\alpha(E)$,

(b*) $U_\alpha^{\gamma^*}(x) \geq 1$ *quasi-everywhere in* E,

(c*) $U_\alpha^{\gamma^*}(x) \leq 1$ *everywhere in* $S(\gamma^*)$.

This measure γ^* will be called the *outer equilibrium* measure for E. Just as in Theorem 2.6, from (b*) and (c*) it follows that

$$U_\alpha^{\gamma^*}(x) = 1 \qquad \gamma^*\text{-almost everywhere},$$

and for $\alpha \leq 2$,

$$U_\alpha^{\gamma^*}(x) = 1 \quad \text{quasi-everywhere in } E;$$

moreover,

$$U_\alpha^{\gamma^*}(x) \leq 1$$

everywhere in R^p.

The basic difficulty is in proving the strong closure of the class Γ_E^* defined as the subset of $\mathscr{E}_\alpha^+$ consisting of those measures μ, for which

$$U_\alpha^\mu(x) \geq 1 \quad \text{quasi-everywhere in } E.$$

To overcome this we need a series of lemmas. Some of these will also be used in succeeding chapters.

Lemma 2.4. *Suppose that* $\nu \in \mathscr{E}_\alpha$ *and*

$$U_\alpha^\nu(x) \geq m \quad (0 < m \leq +\infty)$$

approximately everywhere in the set E. *Then*

$$\underline{C}_\alpha(E) \leq \frac{\|\nu\|^2}{m^2}. \tag{2.2.9}$$

Proof. One must show that for any compact set $K \subset E$, $C_\alpha(K) \leq \|\nu\|^2 m^{-2}$. Suppose γ is the equilibrium measure on K. Then,

$$m\gamma(1) \leq \int_K U_\alpha^\nu(x)\, d\gamma = (\nu, \gamma) \leq \|\nu\| \cdot \|\gamma\|,$$

and by equation (2.1.3) we obtain

$$m\,\|\gamma\|^2 \leq \|\nu\| \cdot \|\gamma\|, \quad m\sqrt{C_\alpha(K)} \leq \|\nu\|,$$

from which

$$C_\alpha(K) \leq \frac{\|\nu\|^2}{m^2}. \quad \square$$

Lemma 2.5. *Suppose* $\mu \in \mathscr{E}_\alpha^+$ *and* $U_\alpha^\mu(x) \geq m$ $(0 < m \leq \infty)$ *quasi-everywhere in* E. *Then*

$$\overline{C}_\alpha(E) \leq \frac{||\mu||^2}{m^2}.$$

Proof. Since the outer capacity is not changed by eliminating from E those points where $U_\alpha^\mu(x) < m$, we may assume that the inequality $U_\alpha^\mu(x) \geq m$ holds everywhere in E. Denote by G_p the open set determined by the inequality $U_\alpha^\mu(x) > p$. Then for $0 < p < m$, $E \subset G_p$ and by the preceding lemma

$$C_\alpha(G_p) \leq \frac{||\mu||^2}{p^2}.$$

Consequently,

$$\overline{C}_\alpha(E) \leq \frac{||\mu||^2}{p^2},$$

and as $p \to m$, we obtain the assertion of the lemma. □

Corollary. *The potential of a signed measure* $\nu \in \mathscr{E}_\alpha$ *is finite quasi-everywhere.*[7]

Indeed, it is sufficient to apply the lemma to ν^+ and ν^-, setting $m = \infty$.

Below we shall establish that the result of Lemma 2.5 holds for any *signed* measure $\nu \in \mathscr{E}_\alpha$. But for this we shall need one lemma on capacitability.

Lemma 2.6. *Every set* A *of type* F_σ, *i.e. representable in the form*

$$A = \bigcup_{n=1}^{\infty} F_n, \quad F_n \text{ closed sets},$$

is capacitable.

Proof. Every set A of type F_σ can be represented in the form

$$A = \bigcup_{n=1}^{\infty} K_n,$$

where K_n are compact sets. It is sufficient, for example, to set $K_n = F_n \cap B_n$ where B_n is the closed ball $|x| \leq n$. The lemma is trivial if $\underline{C}_\alpha(A) = \infty$.

Suppose $\underline{C}_\alpha(A) < \infty$ and that γ is the inner equilibrium measure for A; $\underline{C}_\alpha(A) = ||\gamma||^2$. We will show that $U_\alpha^\gamma(x) \geq 1$ quasi-everywhere in A. Suppose that e_n is the subset of K_n on which we have the inequality

$$U_\alpha^\gamma(x) \leq 1 - \frac{1}{n};$$

---

[7] In Ch.III we shall see that the hypothesis $\nu \in \mathscr{E}_\alpha$ is not necessary.

and e the subset of A for which $U_\alpha^\gamma(x) < 1$. It is easy to see that

$$e = \bigcup_{n=1}^{\infty} e_n,$$

and that the e_n are compact sets of capacity zero. But in that case, $\overline{C}_\alpha(e) = 0$ (cf. Corollary 1 at the end of No. 6), and the assertion is proved. Now applying Lemma 2.5 with $m = 1$, we obtain

$$\overline{C}_\alpha(A) \leq \|\gamma\|^2 = \underline{C}_\alpha(A)$$

that is,

$$\overline{C}_\alpha(A) = \underline{C}_\alpha(A). \quad \square$$

Lemma 2.7. *Suppose that* $\nu \in \mathscr{E}_\alpha$ *and* $U_\alpha^\nu(x) \geq m$, $0 < m < \infty$, *quasi-everywhere in* E. *Then*

$$\overline{C}_\alpha(E) \leq \frac{\|\nu\|^2}{m^2}.$$

Proof. Just as in Lemma 2.5 we may assume that the inequality $U_\alpha^\nu(x) \geq m$ holds everywhere in E. Denote by A_p, $0 < p < m$, the set defined by the inequality $U_\alpha^\nu(x) > p$. We shall show that it is of type F_σ and, consequently, capacitable. For this we note that the inequality $U_\alpha^\nu(x) > p$ or $U_\alpha^{\nu^+}(x) > U_\alpha^{\nu^-}(x) + p$ holds only if there is a rational number r (dependent on x) such that

$$U_\alpha^{\nu^+}(x) > r + p \qquad \text{and} \quad U_\alpha^{\nu^-}(x) \leq r.$$

Therefore, denoting by F_r the closed set defined by the inequality $U_\alpha^{\nu^-}(x) \leq r$, and by G_r the open set defined by the inequality $U_\alpha^{\nu^+}(x) > r + p$, we have

$$A_p = \bigcup_r (F_r \cap G_r);$$

since every G_r is of type F_σ, then this is also true for A_p.

By Lemma 2.4,

$$C_\alpha(A_p) = \underline{C}_\alpha(A_p) \leq \frac{\|\nu\|^2}{p^2},$$

and since $E \subset A_p$, then

$$\overline{C}_\alpha(E) \leq C_\alpha(A_p) \leq \frac{\|\nu\|^2}{p^2}.$$

Now it is sufficient to pass to the limit $p \to m$, and the lemma will be proved. $\square$

Lemma 2.8. *Let $\nu_n \in \mathscr{E}_\alpha$ and $\nu_n \to \nu$ strongly. Then*

$$\underline{\lim}_{n\to\infty} U_\alpha^{\nu_n}(x) \le U_\alpha^{\nu}(x)$$

quasi-everywhere.

Proof. Let us exclude from consideration all points at which even one of the potentials $U_\alpha^{\nu_n}$, U_α^{ν} is infinite or is undefined. By the corollary to Lemma 2.5 the set of such points has outer capacity zero. In the remaining set consider the subset N defined by the inequality

$$U_\alpha^{\nu}(x) < \underline{\lim}_{n\to\infty} U_\alpha^{\nu_n}(x).$$

We must prove that $\overline{C}_\alpha(N) = 0$. It is easy to see that

$$N = \bigcup_{p,q=1}^{\infty} N_{pq},$$

where N_{pq} is the part of N satisfying the inequality

$$U_\alpha^{\nu}(x) = \inf_{n>p} U_\alpha^{\nu_n}(x) - \frac{1}{q}.$$

The set N_{pq} for any $n > p$ is contained in the set A_{nq} defined by

$$U_\alpha^{\nu}(x) \le U_\alpha^{\nu_n}(x) - \frac{1}{q}$$

or

$$U_\alpha^{\nu_n-\nu}(x) \ge \frac{1}{q}.$$

By Lemma 2.7,

$$\overline{C}_\alpha(A_{nq}) \le q^2 \|\nu_n - \nu\|^2.$$

Consequently,

$$\overline{C}_\alpha(N_{pq}) \le q^2 \|\nu_n - \nu\|^2.$$

Letting n tend to infinity, we obtain

$$\overline{C}_\alpha(N_{pq}) = 0$$

and, consequently, $\overline{C}_\alpha(N) = 0$. ☐

Now it is easy to prove

Lemma 2.9. *The set Γ_E^* of measures μ for which*

$$U_\alpha^{\mu}(x) \ge 1 \quad \text{quasi-everywhere in } E$$

is convex and strongly closed.

Convexity is trivial, and the strong closedness follows from the preceding lemma.

We note also, that for $\mu \in \Gamma_E^*$

$$||\mu||^2 \geq \overline{C}_\alpha(E). \tag{2.2.10}$$

This follows immediately from Lemma 2.5 if there we set $m = 1$. From this it follows that the outer equilibrium measure γ^* is unique, if it exists. Indeed, by (a*) and (2.2.10) it is the minimal measure in Γ_E^*.

Proof of Theorem 2.7. First suppose that E is an open set G (or more generally an F_σ set). Then the inner equilibrium measure γ_E satisfies conditions (a*), (b*), (c*). For (a*) and (c*) this is clear; for (b*) it comes from the following consideration. Suppose N is the set of all points in R^p where

$$U_\alpha^\gamma(x) < 1.$$

This is an F_σ set, since $N = \bigcup_{p=1}^{\infty} N_p$, where the closed set N_p is defined by the inequality $U_\alpha^\gamma(x) \leq 1 - 1/p$. Consequently, $G \cap N$ is also an F_σ set and is therefore capacitable. But by (b) of Theorem 2.6

$$\overline{C}_\alpha(G \cap N) = \underline{C}_\alpha(G \cap N) = 0,$$

so that (b*) actually holds.

For arbitrary E, $\overline{C}_\alpha(E) < \infty$, we may choose a decreasing sequence of open sets $G_n \supset G_{n+1} \supset E$ such that

$$\lim_{n\to\infty} C_\alpha(G_n) = \overline{C}_\alpha(E).$$

Moreover, we may assume that $\bigcap_{n=1}^{\infty} \overline{G}_n = \overline{E}$.

Let γ_n be the inner equilibrium measure for G_n. By (b) of Theorem 2.6,

$$\gamma_n \in \Gamma_{G_p} \quad \text{for } p \geq n$$

and since γ_p is the minimal measure in Γ_{G_p}, by Lemma 2.2

$$||\gamma_n - \gamma_p||^2 \leq ||\gamma_n||^2 - ||\gamma_p||^2 = C_\alpha(G_n) - C_\alpha(G_p).$$

Thus, $\{\gamma_n\}$ is a fundamental sequence in $\mathscr{E}_\alpha^+$, and, by the completeness of $\mathscr{E}_\alpha^+$, $\gamma_n \to \gamma^*$ strongly. The measure desired is γ^*. First, $S(\gamma^*) \subset \overline{E}$. This follows from $S(\gamma_p) \subset \overline{G}_n$ $p \geq n$, since

$$S(\gamma^*) \subset \bigcap_{N=1}^{\infty} \left(\overline{\bigcup_{n=N}^{\infty} S(\gamma_n)}\right) \subset \bigcap_{N=1}^{\infty} \overline{G}_N = \overline{E}.$$

Further,

$$||\gamma^*||^2 = \lim_{n\to\infty} ||\gamma_n||^2 = \lim_{n\to\infty} C_\alpha(G_n) = \overline{C}_\alpha(E),$$

i.e. (a*) holds.

From the inclusion $\gamma_n \in \Gamma_{G_n} \equiv \Gamma^*_{G_n} \subset \Gamma^*_E$ and the strong closure of Γ^*_E it follows that $\gamma^* \in \Gamma^*_E$ i.e. (b*).

Finally, (c*) is obtained by Lemma 2.3 and the theorem is proved. ▯

Corollary. *For the capacitability of a set E, $\overline{C}_\alpha(E) < \infty$, it is necessary and sufficient that*

$$\gamma_E = \gamma^*_E.$$

Sufficiency follows from

$$\underline{C}_\alpha(E) = ||\gamma_E||^2, \qquad \overline{C}_\alpha(E) = ||\gamma^*_E||^2.$$

Now suppose that $\overline{C}_\alpha(E) = \underline{C}_\alpha(E)$. Then $||\gamma_E||^2 = ||\gamma^*_E||^2$ and since both measures γ_E and γ^*_E are in Γ_E and minimal, they coincide.

9. In the theory of capacitability, the key lemma is the following

Lemma 2.10. *If $E_n \subset E_{n+1}$ $n = 1, 2, \ldots,$ and $E = \bigcup_{n=1}^{\infty} E_n$ then*

$$\overline{C}_\alpha(E) = \lim_{n\to\infty} \overline{C}_\alpha(E_n).$$

Proof. The lemma is trivial if

$$\lim_{n\to\infty} \overline{C}_\alpha(E_n) = \infty.$$

Therefore we will assume that $\lim_{n\to\infty} \overline{C}_\alpha(E_n) < \infty$. If γ^*_n denotes the outer equilibrium measure for E_n, then since $\gamma^*_n \in \Gamma^*_{E_p}$ for $n > p$, and γ^*_n is minimal in $\Gamma^*_{E_n}$, we obtain, just as in the proof of Theorem 2.7, that the sequence $\{\gamma^*_n\}$ is fundamental. Let γ^* be the strong limit of γ^*_n. Then

$$||\gamma^*||^2 = \lim_{n\to\infty} ||\gamma^*_n||^2 = \lim_{n\to\infty} \overline{C}_\alpha(E_n) \le \overline{C}_\alpha(E). \tag{2.2.11}$$

On the other hand, by the strong closedness of $\Gamma^*_{E_p}$, $\gamma^* \in \Gamma^*_{E_p}$ for any p; hence by Corollary 1 at the end of No. 6, $\gamma^* \in \Gamma^*_E$, and therefore by (2.2.10),

$$||\gamma^*||^2 \ge \overline{C}_\alpha(E).$$

Together with (2.2.11) this proves the lemma. ▯

Remark. If in addition we require the E_n to be *Borel* sets, then, using inner equilibrium measures instead of outer equilibrium measures, we could prove in exactly the same way that

$$\underline{C}_\alpha(E) = \lim_{n\to\infty} \underline{C}_\alpha(E_n).$$

The restriction to Borel sets is connected with the fact that precisely for these sets was established the corollary to Theorem 2.2 (No. 3) necessary to the proof.

Corollary. *The countable union of capacitable Borel sets is again capacitable.*

We proved Lemma 2.10 using the theory of equilibrium measures. If we restrict ourselves to the Robin case $\alpha \leq 2$, then the proof of Lemma 2.10 can be obtained by a shorter way. We will sketch this method.

If in the inequality

$$\overline{C}_\alpha(E) - \overline{C}_\alpha(E \cup E_1) - \overline{C}_\alpha(E \cup E_2) + \overline{C}_\alpha(E \cup E_1 \cup E_2) \leq 0,$$

expressing the convexity of the outer capacity (cf. No. 6) we assume that $E \subset E_1$, then we obtain

$$\overline{C}_\alpha(E_1 \cup E_2) - \overline{C}_\alpha(E_1) \leq \overline{C}_\alpha(E \cup E_2) - \overline{C}_\alpha(E). \tag{2.2.12}$$

Speaking somewhat informally, one can say that when a set is "incremented", the smaller the original set the larger the corresponding increment in the outer capacity. Starting from (2.2.12), it is not difficult to prove by induction the inequality

$$\overline{C}_\alpha\left(\bigcup_1^n E_i\right) - \overline{C}_\alpha\left(\bigcup_1^n e_i\right) \leq \sum_{i=1}^n \{\overline{C}_\alpha(E_i) - \overline{C}_\alpha(e_i)\} \tag{2.2.13}$$

for any $e_i \subset E_i$ $(i = 1, 2, \ldots,)$.[8]

Now it is possible to prove Lemma 2.10. For this it is obviously sufficient to establish the inequality

$$\overline{C}_\alpha(E) \leq \lim_{n\to\infty} \overline{C}_\alpha(E_n).$$

First suppose that all the E_n are open sets. Then for any compact $K \subset E$, one can find a some E_n containing K. Therefore,

$$C_\alpha(K) \leq \lim_{n\to\infty} C_\alpha(E_n),$$

[8] Of course it is assumed that $\overline{C}_\alpha(e_i) < \infty$.

since all this holds for arbitrary $K \subset E$, and E is open and hence capacitable, then,

$$C_\alpha(E) \leq \lim_{n\to\infty} C_\alpha(E_n).$$

Now let us consider arbitrary E_n. Here we assume that all $\overline{C}_\alpha(E) < \infty$. Take $G_n \supset E_n$ such that

$$C_\alpha(G_n) < \overline{C}_\alpha(E_n) + \frac{\varepsilon}{2^n}.$$

If we set $\Omega_n = \bigcup_{i=1}^{n} G_i$, then by (2.2.13),

$$C_\alpha(\Omega_n) - \overline{C}_\alpha(E_n) \leq \sum_{i=1}^{n} \{C_\alpha(G_i) - \overline{C}_\alpha(E_i)\} < \varepsilon.$$

Writing

$$G = \bigcup_{i=1}^{\infty} G_i = \bigcup_{n=1}^{\infty} \Omega_n,$$

by what has been proved earlier we obtain

$$C_\alpha(G) \leq \lim_{n\to\infty} C_\alpha(\Omega_n) \leq \lim_{n\to\infty} \overline{C}_\alpha(E_n) + \varepsilon,$$

and since $E \subset G$, then

$$\overline{C}_\alpha(E) \leq \lim_{n\to\infty} \overline{C}_\alpha(E_n) + \varepsilon.$$

Since $\varepsilon > 0$ was arbitrary, we obtain the required result.

10. The most general result concerning capacitability was obtained in Lemma 2.6. Namely, we established the capacitability of all F_σ sets. Now we come to the proof of the basic theorem on capacitability.

Theorem 2.8. *Every Borel (even amalytic) set in R^p is capacitable.*

Proof. First we show that every $F_{\sigma\delta}$ (in particular G_δ) set is capacitable.

Suppose that $A = \bigcap_{n=1}^{\infty} A_n$, $A_n = \bigcup_{p=1}^{\infty} K_n^p$ where K_n^p are compact sets which may be chosen such that $K_n^p \subset K_n^{p+1}$. The set $a_1^p = A \cap K_1^p$ also increases with p and $A = \bigcup_{p=1}^{\infty} a_1^p$. By Lemma 2.10 $\overline{C}_\alpha(A) = \lim_{p\to\infty} \overline{C}_\alpha(a_1^p)$. Suppose $\overline{C}_\alpha(A) < \infty$. Then we pick a number p_1 such that

$$\overline{C}_\alpha(A) - \overline{C}_\alpha(a_1^{p_1}) < \frac{\varepsilon}{2}.$$

Further, we construct the sets $a_i^{p_i} \subset A$ by induction. Assume that they have been constructed for $i \leq n$. Set $a_{n+1}^p = a_n^{p_n} \cap K_{n+1}^p$. This set increases with p and $a_n^{p_n} = \bigcap_{p=1}^{\infty} a_{n+1}^p$. Therefore $\overline{C}_\alpha(a_n^{p_n}) = \lim_{p\to\infty} \overline{C}_\alpha(a_{n+1}^p)$, and p_{n+1} is chosen such that

$$\overline{C}_\alpha(a_n^{p_n}) - \overline{C}_\alpha(a_{n+1}^{p_{n+1}}) < \frac{\varepsilon}{2^{n+1}}.$$

In this way the $a_n^{p_n}$ form a decreasing sequence of subsets of A, such that for any n

$$\overline{C}_\alpha(A) - \overline{C}_\alpha(a_n^{p_n}) < \varepsilon.$$

Set $a_\varepsilon = \bigcap_{n=1}^{\infty} a_n^{p_n} = A \cap \left(\bigcap_{n=1}^{\infty} K_n^{p_n}\right)$. Since $K_n^{p_n} \subset A_n$, then

$$\bigcap_{n=1}^{\infty} K_n^{p_n} \subset \bigcap_{n=1}^{\infty} A_n = A,$$

and therefore $\alpha_\varepsilon = \bigcap_{n=1}^{\infty} K_n^{p_n}$ is a compact set. By continuity on the right, $C_\alpha(a_\varepsilon) = \lim_{m\to\infty} C_\alpha\left(\bigcap_{n=1}^{m} K_n^{p_n}\right)$, and by $a_\varepsilon \subset a_m^{p_m} = \bigcap_{n=1}^{m} a_n^{p_n} \subset \bigcap_{n=1}^{m} K_n^{p_n}$ we have $C_\alpha(a_\varepsilon) = \lim_{n\to\infty} \overline{C}_\alpha(a_m^{p_m})$. Combining this with the last inequality, we obtain $C_\alpha(a_\varepsilon) \geq \overline{C}_\alpha(A) - \varepsilon$ from which it follows that $\underline{C}_\alpha(A) \geq \overline{C}_\alpha(A)$ i.e. A is capacitable.

In case $\overline{C}_\alpha(A) = \infty$ in the same way one can construct a compact set $a_N \subset A$, for which $C_\alpha(a_N) > N$ and thus the assertion is proved.

Before proceeding further, let us emphasize that in the above reasoning only the following properties of C_α and $\overline{C}_\alpha$ were used:

(i) C_α is an increasing function defined on all compact sets in R^p and continuous on the right.

(ii) For any sets $E_n \subset E_{n+1}$, $n = 1, 2, \ldots$,

$$\overline{C}_\alpha\left(\bigcup_{n=1}^{\infty} E_n\right) = \lim_{n\to\infty} \overline{C}_\alpha(E_n).$$

Therefore, if any set function $f(E)$ satisfies conditions (i) and (ii), and $\overline{f}$ is defined from f just as $\overline{C}_\alpha$ was defined in No. 6 for C_α, then all $F_{\sigma\delta}$ sets will be f-capacitable. It is precisely this fact that opens the way to the completion of the proof of the theorem.

It is known (cf. Hausdorff [1]) that any analytic (in particular, Borel) set $E \subset R^p$ may be viewed as an orthogonal projection of a certain $G_\delta \subset R^{p+1}$.

Let us define for compact sets $K \subset R^{p+1}$ the set function

$$f(K) = C_\alpha (\text{pr. } K),$$

where pr. K denotes the orthogonal projection of K on R^p. Since under projection compact sets and open sets are mapped into corresponding compact sets and open sets, (i) easily follows as well as the formulas[9]

$$\underline{f}(E) = \underline{C}_\alpha (\text{pr. } E), \qquad \bar{f}(E) = \bar{C}_\alpha (\text{pr. } E). \tag{2.2.14}$$

From the last equation it follows that $\bar{f}$ satisfies condition (ii). Consequently, any $G_\delta \subset R^{p+1}$ will be f-capacitable; hence pr. G_δ will be capacitable. Thus, all analytic sets in R^p are capacitable, which was to be proved. ▯

Corollary. *If a Borel set E has a positive capacity, then there is a compact set $K \subset E$ with positive capacity.*

§ 3. Metric Properties of Capacity

11. Let us return to the definition of the capacity of a compact set given in No. 2. There it is clear that the quantity $C_\alpha(K)$ is determined by the values of the function $|x - y|$ for x, y in K. The nature of this dependence is to some extent explained in the following theorem.

Theorem 2.9. *Suppose given on K a function $x' = f(x) \in R^p$, which defines a "contraction" mapping of K, i.e. such that $|x' - y'| \le |x - y|$. Then, if $K' = f(K)$ is the image of K (which is clearly also a compact set), then*

$$C_\alpha(K') \le C_\alpha(K).$$

Proof. We initially establish the validity of the theorem in two special cases. First, suppose that $f(x)$ defines a *one-to-one* mapping of K onto K'. Then the formulas $\mu'(E') = \mu(f^{-1}(E'))$, $E' \subset K'$; $\mu(E) = \mu'(f(E))$, $E \subset K$, establish a one-to-one correspondence between the elements of $\mathfrak{M}_K^+$ and $\mathfrak{M}_{K'}^+$.

The relation

$$\int_{K' \times K'} |x' - y'|^{\alpha-p} \, d\mu'(x') \, d\mu'(y') = \int_{K \times K} |f(x) - f(y)|^{\alpha-p} \, d\mu(x) \, d\mu(y)$$

$$\ge \int_{K \times K} |x - y|^{\alpha-p} \, d\mu(x) \, d\mu(y), \quad \mu' \in \mathfrak{M}_{K'}^+$$

(cf. Halmos, [1]) shows that $W_\alpha(K') \ge W_\alpha(K)$ i.e. $C_\alpha(K') \le C_\alpha(K)$.

---

[9] In these formulas $\underline{f}(E) = \sup_{K \subset E} f(K)$, $\bar{f}(E) = \inf_{G \supset E} f(G)$.

Next, suppose that $K' \subset R^{p-1}$ and that $x' = f(x)$ is an orthogonal projection of R^p onto R^{p-1}. We may assume without loss of generality that K lies to one side of R^{p-1}. For $x' \in K'$ we denote by $f^{-1}(x')$ a point of the complete pre-image of x' which is closest to x'. With each measure $\mu' \in \mathfrak{M}^+_{K'}$ we associate a measure μ which for any set $E \subset f^{-1}(f(K))$ is defined by the equation $\mu(E) = \mu'(f(E))$[10] and for any Borel set E $\mu(E) = \mu(E \cap f^{-1}(f(E)))$. Clearly, $\mu \in \mathfrak{M}^+_K$ and $\mu'(E') = \mu(f^{-1}(E'))$, where by $f^{-1}(E')$ we mean the *complete* pre-image of E'. Now one can literally repeat the previous reasoning and obtain $C_\alpha(K') \leq C_\alpha(K)$.

The above result also holds for orthogonal projections of R^p into R^{p-k} $(0 < k < p)$, since such a projection can be obtained by projecting R^p on R^{p-1}, then R^{p-1} on R^{p-2}, etc.

Considering now the general case, we note first of all that if $K \subset R^p$ is viewed as a set in $R^{p+k} \supset R^p$, then $C_\alpha(K) = \hat{C}_{\alpha+k}(K)$, where $\hat{C}$ denotes capacity in R^{p+k}. Then we consider the space $R^{2p} = R^p \times R^p$, whose points we denote by (x, y), and the set $\hat{K} \subset R^{2p}$ consisting of the points $(x, 0)$, $x \in K$. By the remark just made we have $\hat{C}_{\alpha+p}(\hat{K}) = C_\alpha(K)$. Further we denote by $\hat{K}_\eta \subset R^{2p}$ the set of points $(x, \eta x)$, $x \in K$, $\eta > 0$. $\hat{K}_\eta$ is a homeomorphic image of $\hat{K}$, its projection on R^p is K, and for sufficiently small η, $\hat{K}_\eta$ lies in any given neighborhood of $\hat{K}$. By the above and the continuity of capacity on the right, we have

$$C_\alpha(K) = \hat{C}_{\alpha+p}(\hat{K}) \leq \hat{C}_{\alpha+p}(\hat{K}_\eta) < \hat{C}_{\alpha+p}(\hat{K}) + \varepsilon = C_\alpha(K) + \varepsilon.$$

Further, besides $K' = f(K) \subset R^p$ we consider in R^{2p} the set $\hat{K}'_\eta$, consisting of the points $(f(x), \eta x)$, $x \in K$. The map $(x, \eta x) \to (f(x), \eta x)$ is a one-to-one mapping of $\hat{K}_\eta$ onto $\hat{K}'_\eta$ in which distance is not increased. Therefore, by what has been proved, we have $\hat{C}_{\alpha+p}(\hat{K}'_\eta) \leq \hat{C}_{\alpha+p}(\hat{K}_\eta)$. Moreover, just as above, we have

$$C_\alpha(K') \leq \hat{C}_{\alpha+p}(\hat{K}'_\eta),$$

since K' is an orthogonal projection of $\hat{K}'_\eta$. Combining the above inequalities, we obtain

$$C_\alpha(K') \leq \hat{C}_{\alpha+p}(\hat{K}'_\eta) \leq \hat{C}_{\alpha+p}(\hat{K}_\eta) < C_\alpha(K) + \varepsilon,$$

and since $\varepsilon > 0$ was arbitrary, the theorem is proved. ▯

---

[10] This is a σ-additive function since $f(\bigcup E_i) = \Sigma f(E_i)$, if $E_i \subset f^{-1}(f(K))$ and $E_i \cap E_j = \emptyset$ for $i \neq j$.

Corollary. *If E is a Borel set and $E' = f(E)$ is its image under a contraction mapping, then*

$$C_\alpha(E') \leq C_\alpha(E). \tag{2.3.1}$$

Remark. It is not difficult to see that for a similarly transformation with coefficient m the α-capacity is multiplied by $m^{p-\alpha}$. Therefore, if for a mapping, $x' = f(x)$ we have $|x' - y'| \leq m\,|x - y|$, then $C_\alpha(E') \leq m^{p-\alpha} C_\alpha(E)$.

In particular, under such mappings "with finite dilation" every set with null capacity is mapped again into a set with null capacity.

12. The connection between $C_\alpha(K)$ and the metric properties of the set K is even more closely observed by considering the *generalized transfinite diameter* of the compact set K.

The construction of a transfinite diameter of order α goes as follows: let x_i ($i = 1, 2, \ldots, n$) be n arbitrary points of K. The function

$$\binom{n}{2}^{-1} \sum_{i<j} |x_i - x_j|^{\alpha-p} \quad (0 < \alpha < p),$$

achieves its minimal value on the compact set K at certain points $x_i = \xi_i^{(n)}$. We set

$$D_n^{(\alpha)} = \binom{n}{2} \Bigl(\sum_{i<j} |\xi_i^{(n)} - \xi_j^{(n)}|^{\alpha-p}\Bigr)^{-1} \tag{2.3.2}$$

and we will show that $D_{n-1}^{(\alpha)} \geq D_n^{(\alpha)}$.[11] For this we note that

$$\sum_{i<j} |\xi_i^{(n)} - \xi_j^{(n)}|^{\alpha-p} = \frac{1}{n-2} \sum_{k=1}^{n} \sum_{i<j}{}^{(k)} |\xi_i^{(n)} - \xi_j^{(n)}|^{\alpha-p},$$

where $\Sigma^{(k)}$ denotes the sum in which the terms for $i = k$ or $j = k$ have been omitted. But

$$\sum_{i<j}{}^{(k)} |\xi_i^{(n)} - \xi_j^{(n)}|^{\alpha-p} \geq \binom{n-1}{2} \frac{1}{D_{n-1}^{(\alpha)}},$$

and therefore

$$\binom{n}{2} \frac{1}{D_n^{(\alpha)}} = \sum_{i<j} |\xi_i^{(n)} - \xi_j^{(n)}|^{\alpha-p} \geq \frac{n}{n-2} \binom{n-1}{2} \frac{1}{D_{n-1}^{(\alpha)}} = \binom{n}{2} \frac{1}{D_{n-1}^{(\alpha)}}.$$

This proves the existence of a non-negative limit

$$D^{(\alpha)}(K) = \lim_{n\to\infty} D_n^{(\alpha)} \tag{2.3.3}$$

---

[11] If K consists of only m points, then for $n > m$, we put $D_n^{(\alpha)} = 0$.

which we shall call the generalized transfinite diameter of order α of the compact set K.

Remark. In the classical case, when $p = 3$, $\alpha = 2$, the quantities $D_n^{(2)}$ have a simple physical interpretation. Consider the distribution in K of n point charges of size $1/n$; then the minimum potential energy of such a system of charges is obtained when the charges are located at the points $\xi_i^{(n)}$; moreover, the potential energy will be equal to $\frac{n-1}{2n} \frac{1}{D_n^{(2)}}$.

We now establish the connection between $D^{(\alpha)}(K)$ and the α-capacity $C_\alpha(K)$.

If the inequality

$$\binom{n}{2} \frac{1}{D_n^{(\alpha)}} \leq \sum_{i<j} |x_i - x_j|^{\alpha-p}$$

is multiplied by $d\mu(x_i)\, d\mu(x_j)$, where $\mu \in \mathfrak{M}_K^+$, $\mu(1) = 1$ and is integrated $\binom{n}{2}$ times over $K \times K$, then we obtain

$$\frac{1}{D_n^{(\alpha)}} \leq \int_{K\times K} |x - y|^{\alpha-p}\, d\mu(x)\, d\mu(y) = \frac{1}{A(p, \alpha)} I_\alpha(\mu).$$

Passing to the limit as $n \to \infty$ and to inf over μ we obtain $\frac{1}{D^{(\alpha)}(K)} \leq$ $\leq \frac{1}{A(p, \alpha)} W_\alpha(K)$, whence $D^{(\alpha)}(K) \geq A(p, \alpha)\, C_\alpha(K)$. On the other hand, consider the measure

$$\lambda_n = \frac{1}{n} \sum_{i=1}^{n} \varepsilon_{\xi_i^{(n)}}, \quad \lambda_n \in \mathfrak{M}_K^+, \quad \lambda_n(1) = 1.$$

This measure has infinite energy, but if we use the "truncated" kernel

$$k_\alpha^{(N)}(x) = \begin{cases} A(p, \alpha)\, |x|^{\alpha-p} & \text{for } |x| > N^{-1}, \\ A(p, \alpha)\, N^{p-\alpha} & \text{for } |x| \leq N^{-1}, \end{cases}$$

then we may write the inequality

$$\begin{aligned} &\int_{K\times K} k_\alpha^{(N)}(x - y)\, d\lambda_n(x)\, d\lambda_n(y) \\ &\leq A(p, \alpha) \left[\frac{1}{n^2} \sum_{i\neq j} |\xi_i^{(n)} - \xi_j^{(n)}|^{\alpha-p} + \frac{N^{p-\alpha}}{n} \right] \\ &= A(p, \alpha) \left[\frac{2}{n^2} \binom{n}{2} \frac{1}{D_n^{(\alpha)}} + \frac{N^{p-\alpha}}{n} \right]. \end{aligned}$$

By the vague compactness theorem (Theorem 0.6), we may assume that $\lambda_n \rightharpoonup \lambda^*$. Fixing N, we pass to the limit as $n \to \infty$. Since $k_\alpha^{(N)}$ is a continuous function, as a result we obtain

$$\int_{K\times K} k_\alpha^{(N)}(x-y)\,d\lambda^*(x)\,d\lambda^*(y) \le A(p,\alpha)\,\frac{1}{D^{(\alpha)}(K)}.$$

From this as $N\to\infty$

$$I_\alpha(\lambda^*) \le A(p,\alpha)\,\frac{1}{D^{(\alpha)}(K)}.$$

Since $\lambda^*\in\mathfrak{M}_K^+$, $\lambda^*(1)=1$, together with the above, this shows that $I_\alpha(\lambda^*)=W_\alpha(K)$ and

$$D^{(\alpha)}(K) = A(p,\alpha)\,C_\alpha(K).^{12} \tag{2.3.4}$$

Moreover, by the uniqueness of the minimizing measure (cf. No. 1), we see that $\lambda^*\equiv\lambda_K$.

13. The latter circumstance sometimes allows one to reach certain conclusions about the nature of the minimizing measure λ_K.

Suppose $\alpha\ge 2$. Denote by $G_\infty, G_1, G_2, \ldots$ the connected components of the complement of K, where G_∞ denotes the region containing infinity Let, further, $\overline{\overline{K}} = CG_\infty = K\cup\left(\bigcup_{i\ge 1} G_i\right)$ and $S\subset K$ be the boundary of G_∞. Then the support $S(\lambda)$ of the *minimizing measure λ is a subset of S.* By the compactness of S it is sufficient to establish this for $S(\lambda_n)$. If $S(\lambda_n)\not\subset S$, then for some index i_0 $\xi_{i_0}^{(n)}\notin S$. Since $\xi_{i_0}^{(n)}\notin G_\infty$ then it is an interior point of $\overline{\overline{K}}$. Denote by ω the connected component of the interior of $\overline{\overline{K}}$ containing the point $\xi_{i_0}^{(n)}$. Then boundary of ω is contained in the boundary of $\overline{\overline{K}}$, and the latter coincides with S. The expression

$$\binom{n}{2}^{-1}\left\{\sum_{i<j}{}^{(i_0)} |\xi_i^{(n)}-\xi_j^{(n)}|^{\alpha-p} + \sum_{i\ne i_0} |\xi_i^{(n)}-x|^{\alpha-p}\right\}$$

is a superharmonic function of x because of the inequality $\alpha\ge 2$ (cf. Ch. I, § 2, No. 8); therefore it cannot attain its least value at an interior point of ω and in particular at $x=\xi_{i_0}^{(n)}$.[13] This, however, contradicts the definition of the points $\xi_i^{(n)}$.

From this property of the minimizing measure we have the following important

Corollary.

$$C_\alpha(K) = C_\alpha(\overline{\overline{K}}) = C_\alpha(S) \quad (\alpha\ge 2),$$

i.e. the capacity of a compact set is equal to the capacity of its outer boundary.

---

[12] Using this relation, it is easy to obtain Theorem 2.9.

[13] The restriction $x\in K$ does not matter since $\partial\omega\subset S\subset K$.

As an example we consider the *ball* K: $|x| \leq R$. For $\alpha \geq 2$ the minimizing measure λ will be concentrated on the sphere $|x| = R$, and by symmetry it will coincide with the measure $\varepsilon^{(R)}$ which is distributed uniformly over the surface $|x| = R$. This enables us to compute $W_\alpha(K)$ and $C_\alpha(K)$:

$$W_\alpha(K) = w(p, \alpha)\, R^{\alpha-p},$$

$$C_\alpha(K) = w^{-1}(p, \alpha)\, R^{p-\alpha}. \text{ [14]}$$

For $\alpha > 2$ at all interior points of K the inequality $U_\alpha^\lambda(x) > W_\alpha(K)$ will hold.

For $\alpha < 2$, however, the minimizing measure *will not be concentrated on the surface*. Its support coincides with the entire ball $|x| \leq R$; it is absolutely continuous and has a density equal to

$$\lambda(x) = \frac{A}{(R^2 - |x|^2)^{\alpha/2}},$$

where

$$A = \frac{\Gamma\left(\frac{p-\alpha}{2} + 1\right)}{\pi^{p/2}\, \Gamma\left(1 - \frac{\alpha}{2}\right)} \frac{1}{R^{p-\alpha}}.$$

Computing the potential (cf. Appendix, (A. 1)) we obtain for $|x| \leq R$ the value

$$U_\alpha^\lambda(x) = A(p, \alpha) \frac{\Gamma\left(\frac{\alpha}{2}\right) \Gamma\left(\frac{p-\alpha}{2} + 1\right)}{\Gamma\left(\frac{p}{2}\right)} R^{\alpha-p},$$

from which

$$W_\alpha(K) = w_1(p, \alpha)\, R^{\alpha-p},$$

$$C_\alpha(K) = w_1^{-1}(p, \alpha)\, R^{p-\alpha},$$

where

$$w_1(p, \alpha) = A(p, \alpha) \frac{\Gamma\left(\frac{\alpha}{2}\right) \Gamma\left(\frac{p-\alpha}{2} + 1\right)}{\Gamma\left(\frac{p}{2}\right)}.$$

For the *equilibrium measure* $\gamma = C_\alpha(K)\lambda$ we obtain the density

$$\gamma(x) = \frac{A_1}{(R^2 - |x|^2)^{\alpha/2}},$$

---

[14] The constant $w(p, \alpha)$ occured in Theorem 1.32. Its value is:

$$w(p, \alpha) = \frac{2^{\alpha-2} \pi^{\alpha - \frac{p+1}{2}}\, \Gamma\left(\frac{p-\alpha}{2}\right) \Gamma\left(\frac{\alpha-1}{2}\right) \Gamma\left(\frac{p}{2}\right)}{\Gamma\left(\frac{\alpha}{2}\right) \Gamma\left(\frac{p+\alpha}{2} - 1\right)}.$$

where

$$A_1 = \frac{1}{A(p,\alpha)} \cdot \pi^{-p/2-1} \Gamma\left(\frac{p}{2}\right) \sin\frac{\pi\alpha}{2}$$

does not depend on R.

In the case $\alpha \leq 2$, it is possible to make one more conclusion about the support $S(\lambda)$ of the minimizing measure.

For any compact set K we define the *reduced kernel* $\check{K}$ as the set of all points of K each neighborhood of which intersects K in a set of positive capacity. In that case, from the first maximum principle, it follows that $S(\lambda) \supset \check{S}$. If $\alpha = 2$, then $S(\lambda) = \check{S}$.

Indeed, otherwise we can find $x \in \check{S} \subset K$ such that $x \notin S(\lambda)$, and a neighborhood $V(x)$ which does not intersect $S(\lambda)$. Since $U_\alpha^\lambda(x)$ does not reduce to a constant in the region $G_\infty \cup V(x)$, then $U_\alpha^\lambda(x) < W_\alpha(K)$ at all points of this region. On the other hand, the intersection $V(x) \cap K$ has positive capacity and since $U_\alpha^\lambda(x) = W_\alpha(K)$ approximately everywhere on K (cf. No. 3, (c_1)), then $V(x)$ contains a point at which $U_\alpha^\lambda(x) = W_\alpha$. Thus we have obtained a contradiction.

Let us look more closely at the Newtonian case $\alpha = 2$. The equilibrium measure $\gamma = C_2(K)\lambda$ of the compact set K is by the above the equilibrium measure on $\overline{\overline{K}}$. Since in this case the maximum principle holds, then by No. 3, (c_1) we have the equation $U_2^\gamma(x) = 1$ approximately everywhere on $\overline{\overline{K}}$ (by Theorem 2.8 even quasi-everywhere); moreover $U_2^\gamma(x) \leq 1$ throughout the whole space.

If the exceptional subset of $\overline{\overline{K}}$ where $U_2^\gamma(x) < 1$ is empty (compare further with Ch. IV), then the measure γ solves the classical equilibrium problem of Robin (cf. n^0 3) and $U_2^\gamma(x)$ for $x \in G_\infty$ gives a solution of the *external Dirichlet problem* for the region G_∞ with boundary value equal to 1. The continuity of $U_2^\gamma(x)$ at the boundary points of G_∞ follows from Theorem 1.7 (Ch. I, No. 8).

If Π is a simple smooth closed hypersurface containing K strictly in its interior, then by Gauss's theorem

$$\int_\Pi \frac{\partial U_2^\gamma}{\partial n} dS = 4\pi^2\gamma(1) = 4\pi^2 C_2(K),$$

where n is the internal normal to Π. Indeed, it is easy to prove the asymptotic formulas

$$U_2^\gamma(x) = A(p, 2)\frac{\gamma(1)}{|x|^{p-2}} + o(|x|^{2-p}), \quad |x| \to \infty,$$

$$\frac{\partial U_2^\gamma}{\partial n} = A(p, 2)\,\gamma(1)\frac{\partial}{\partial n}\frac{1}{|x|^{p-2}} + o(|x|^{2-p}), \quad |x| \to \infty.$$

By the harmonicity of $U_2^\gamma(x)$ in G_∞ the integraton over Π may be replaced by integration over a sphere of large radius; then an elementary calculation yields the required result.

If K is a closed, sufficiently smooth surface, then there is a limiting value for the normal derivative $\partial U_2^\gamma/\partial n$ on K and it is equal to, the surface density of the equilibrium measure γ multiplied by $\frac{1}{4\pi^2}$ (cf. Courant [1]).[15]

Suppose, for example, that K is the sphere $|x| = R$, then it is easy to see that

$$U_2^\gamma(x) = \left(\frac{R}{|x|}\right)^{p-2} \qquad (|x| \geq R).$$

From this it follows that

$$C_2(K) = \frac{R^{p-2}}{A(p, 2)}.$$

14. In R^3 it is sometimes possible to find the equilibrium potential and then the Newtonian capacity by separation of variables in the corresponding curvilinear coordinates.

We present several results of such calculations.

K	$C_2(K)$
1. Elongated ellipsoid of revolution with axes a, b $(b < a)$	$\dfrac{2c}{\pi \ln \dfrac{a+c}{a-c}}, \quad c = \sqrt{a^2 - b^2}$
1a. Line segment	0
2. Prolate ellipsoid of revolution with axes a, b $(b < a)$	$\dfrac{c}{\pi \arcsin \dfrac{c}{a}}, \quad c = \sqrt{a^2 - b^2}$
3. Disc of radius a	$\dfrac{2a}{\pi^2}$
4. Triaxial ellipsoid with axes a, b, c	$\dfrac{2}{\pi}\left(\displaystyle\int_0^\infty \frac{ds}{(a^2+s)(b^2+s)(c^2+s)}\right)^{-1}$
5. Elliptic disc with axes a, b	$\dfrac{a}{\pi}\left[K\left(\dfrac{c}{a}\right)\right]^{-1}$, $K(k)$ is the complete elliptic integral of the first kind

[15] From an argument due to Zaremba, [1] it follows that this will hold if K is a Lyapunov surface with Hölder exponent 1.

K	$C_2(K)$
6. Torus; a is the radius of the revolving circle; l is the radius of revolution	$\frac{2c}{\pi^2}\left[\frac{Q_{-1/2}\left(\frac{l}{a}\right)}{P_{-1/2}\left(\frac{l}{a}\right)} + 2\sum_{n=1}^{\infty}\frac{Q_{n-1/2}\left(\frac{l}{a}\right)}{P_{n-1/2}\left(\frac{l}{a}\right)}\right]$, $c = \sqrt{l^2 - a^2}$, $P_\nu(x)$, $Q_\nu(x)$ are Legendre functions of the first and second kind
7. Spherical envelope of radius a with parameter b [16]	$\frac{b}{2\pi^2 a}\sqrt{4a^2 - b^2} + \frac{2a}{\pi^2}\arctan\frac{b}{\sqrt{4a^2 - b^2}}$
8. Semi-ball of radius a	$\frac{2a}{\pi}\left(1 - \frac{1}{\sqrt{3}}\right)$
9. Two balls of radius a_1, a_2. Distance between the centers is $2l$	$\frac{a_1 \sinh\beta_1}{2\pi}\left\{\frac{1}{\sinh\beta_1} + \frac{1}{\sinh\beta_2} + 2\sum_{n=0}^{\infty}\left[e^{-(n+1/2)\beta_1}\cosh\left(n+\frac{1}{2}\right)(\beta_1+\beta_2) - e^{-(n+1/2)\beta_2}\right] \times \frac{e^{-(n+1/2)\beta_1}}{\sinh(n+1/2)(\beta_1+\beta_2)} + 2\sum_{n=0}^{\infty}\left[e^{-(n+1/2)\beta_2}\cosh\left(n+\frac{1}{2}\right)(\beta_1+\beta_2) - e^{-(n+1/2)\beta_1}\right] \times \frac{e^{-(n+1/2)\beta_2}}{\sinh(n+1/2)(\beta_1+\beta_2)}\right\}$, where $\cosh\beta_1 = \frac{4l^2 + a_1^2 - a_2^2}{4la_1}$, $\cosh\beta_2 = \frac{4l^2 - a_1^2 - a_2^2}{4la_2}$
10. Two tangent balls of radius a	$\frac{2\ln 2}{\pi}a$

[16] The parameter b is equal to the generator of the straight circular cone inscribed in the envelope.

K	$C_2(K)$
11. The volume formed by revolving a circle of radius a around a tangent	$\frac{4a}{\pi^2}\int\limits_0^\infty \frac{K_0(x)}{I_0(x)}\,dx$, $I_0(x)$, $K_0(x)$ are Bessel functions with imaginary arguments.

§ 4. Logarithmic Capacity

15. Let us examine the theory of capacity in the case $\alpha = p = 2$. First we assume that the compact set K lies strictly within the unit disc X: $|x| < 1$. Then for any measure $\mu \in \mathfrak{M}_K^+$, by Theorem 1.16, we have

$$I_2(\mu) = \int\limits_{K \times K} \ln \frac{1}{|x-y|}\, d\mu(x)\, d\mu(y) > 0$$

(for convenience here we omit the factor 2π in the logarithmic kernel $k_2(x)$). Moreover, if $\max\limits_{x \in K} |x| = R < 1$, then (cf. Remark to Theorem 1.16)

$$I_2(\mu) > \mu^2(1) \ln \frac{1}{R}.$$

Therefore the variational problem considered in No. 1, i.e. the problem of minimizing the quantity $I_2(\mu)$ in the class of measures $\mu \in \mathfrak{M}_K^+$ with the normalization condition $\mu(1) = 1$, as before, has a solution $\lambda \in \mathfrak{M}_K^+$ and for $W_2(K) = I_2(\lambda)$ we have

$$0 < W_2(K) \le +\infty.$$

In addition, as seen in No. 1, the minimizing measure λ is unique.

The capacity of the compact set K may be defined as above by the formula

$$C_2(K) = W_2^{-1}(K) \tag{2.4.1}$$

(the so-called *Wiener capacity*), but it is more convenient, as will be seen in the future, to introduce the *logarithmic capacity* of K by the formula

$$C_l(K) = e^{-W_2(K)} = e^{-C_2^{-1}(K)}. \tag{2.4.2}$$

Unlike the Wiener capacity, which may take on arbitrarily large values (for instance, for the disc $|x| \le 1 - \varepsilon$), the logarithmic capacity satisfies $C_l(K) \le 1$. On the other hand, the equations $C_2(K) = 0$ and $C_l(K) = 0$ are equivalent.

To define the logarithmic capacity of *any* compact set K we may proceed in a completely analogous way.

The existence of a measure λ, which minimizes $I_2(\mu)$ is established just as in the beginning of No. 1. Of course, $W_2(K)$ may now be negative (we have the bound $W_2(K) > \ln 1/d$, if K is contained in the disc $|x| < d$). But this does not preclude the definition of the logarithmic capacity by the same formula (2.4.2).

The uniqueness of the measure λ can be obtained by replacing the minimization problem for $I_2(\mu)$ by the equivalent problem for

$$||\mu||_d = I_2^{(d)}(\mu) = \int_{K \times K} \ln \frac{2d}{|x-y|} \, d\mu(x) \, d\mu(y) = I_2(\mu) + \ln 2d$$

and applying the argument based on strong convergence (cf. No. 1).

If we define *d-capacity* $C_2^{(d)}(K)$ by the formula

$$C_2^{(d)} = [W_2(K) + \ln 2d]^{-1},$$

then, clearly,

$$C_l(K) = 2d \exp\left\{- \frac{1}{C_2^{(d)}(K)}\right\}.$$

Consider the logarithmic potential

$$U^\lambda(x) = \int_K \ln \frac{1}{|x-y|} \, d\lambda(y)$$

of the minimizing measure λ. The reasoning in No. 3 does not require the positiveness of the quantity $W_2(K)$ and since the first maximum principle holds (cf. Theorem 1.6), we have $U^\lambda(x) = W_2(x)$ approximately everywhere on K and $U^\lambda(x) \leq W_2(K)$ throughout K.

If the Wiener capacity $C_2(K)$ is positive, i.e. $W_2(K) > 0$, then the measure

$$\gamma = \frac{1}{W_2(K)} \lambda = C_2(K) \, \lambda$$

solves the Robin problem since

$$U^\gamma(x) = 1 \quad \text{on} \quad K$$

and

$$U^\gamma(x) \leq 1$$

everywhere.

In this case the Wiener capacity can be defined by the formula

$$C_2(K) = \max \mu(1),$$

where the maximum is taken over all measures μ for which $U^\mu(x) \leq 1$.

Suppose now that $C_2(K) < 0$. Then the negative signed measure

$$\gamma = C_2(K)\,\lambda$$

is easily seen, to have the property that approximately everywhere in K

$$U^{\gamma}(x) = 1$$

and

$$U^{\gamma}(x) \leq 1$$

everywhere.

Also,

$$C_2(K) = \min \nu(1)\,,$$

where the minimum is taken over all negative signed measures for which $U^{\nu}(x) \leq 1$ on K.

Finally, if $W_2(K) = 0$, then the Wiener capacity $C_2(K)$ is not defined and there is no signed measure γ, $S(\gamma) \subset K$, having a non-zero constant potential approximately everywhere in K.

To obtain more precise results on $S(\lambda)$ it is useful to connect the logarithmic capacity $C_l(K)$ with the *transfinite diameter* $d(K)$ of a compact set K in the plane.

To define $d(K)$ consider, as in No. 12, n points x_i $(i = 1, 2, \ldots n)$ in K and the function

$$\binom{n}{2}^{-1} \sum_{i<j} \ln \frac{1}{|x_i - x_j|}\,.$$

Denote its minimum on K by $\ln 1/d_n$ and by $\xi_i^{(n)}$ those points at which this minimum is attained. Thus,

$$d_n = \exp\left\{-\binom{n}{2}^{-1} \sum_{i<j} \ln \frac{1}{|\xi_i^{(n)} - \xi_j^{(n)}|}\right\}$$

$$= \sqrt[\binom{n}{2}]{\prod_{i<j} |\xi_i^{(n)} - \xi_j^{(n)}|}\,.$$

Just as in No. 12, one proves $d_{n+1} \leq d_n$, and defines

$$d(K) = \lim_{n\to\infty} d_n\,.$$

Changing nothing essential in the subsequent reasoning of No. 12, we arrive at the equation

$$d(K) = C_l(K) \tag{2.4.3}$$

and see that λ is the vague limit of the measures $\lambda_n = \frac{1}{n} \sum_{i=1}^{n} \varepsilon_{\xi_i^{(n)}}$; and finally, using the harmonicity of the function $\binom{n}{2}^{-1} \sum \ln \frac{1}{|x_i - x_j|}$ and the first maximum principle, we also conclude that $S(\lambda) = \check{S}$, where $\check{S}$ is the boundary of G_∞. The latter brings us to the equation

$$C_l(K) = C_l(\overline{\overline{K}}) = C_l(S).$$

We also note that for logarithmic capacity we have the theorem on non-increasing capacity under contraction mappings (cf. No. 11). This can be obtained using the equality $C_l(K) = d(K)$ or can be proved directly by the method of No. 11.

16. The logarithmic capacity $C_l(K)$ is, clearly, a non-negative and monotone increasing function of compact sets. To obtain Theorem 2.8 for $C_l(K)$ let us first study the d-capacity $C_2^{(d)}(K)$. We assume that d is sufficiently large to include all the compact sets we will consider in the disc $|x| < d$. For any measure μ in $\mathfrak{M}_K^+$ we introduce the logarithmic d-potential by the formula

$$U^\mu(x; d) = \int_{S(\mu)} \ln \frac{2d}{|x - y|} \, d\mu(y)$$

and note that on $S(\mu)$ $U^\mu(x; d) \geq 0$. Let us examine the d-equilibrium measure of the compact set K defined by the formula

$$\gamma = C_2^{(d)}(K)\, \lambda, \gamma(1) = C_2^{(d)}(K) > 0.$$

Clearly,

$$U^\gamma(x; d) = C_2^{(d)}(K)\, U^\lambda(x; d) = 1,$$

approximately everywhere in K and $U^\gamma(x; d) \leq 1$ everywhere in K. Thus, the d-equilibrium measure γ solves the Robin problem for the d-potential. Also, just as in No. 3, we see that Theorem 2.3, (i) holds for the d-potential $U^\gamma(x; d)$. Now it is not difficult to verify that all the proofs in No. 5 can be repeated verbatim for $C_2^{(d)}(K)$ since $C_2^{(d)}(K)$ satisfies properties (2) and (3) of No. 5.

Therefore for $C_2^{(d)}$ Lemma 2.10 and Theorem 2.8 hold, i.e.

$$\underline{C}_2^{(d)}(E) = \bar{C}_2^{(d)}(E)$$

for any bounded analytic set E.

However

$$\underline{C}_l(E) = 2d \exp\left(-\frac{1}{\underline{C}_2^{(d)}(E)}\right); \quad \bar{C}_l(E) = 2d \exp\left(-\frac{1}{\bar{C}_2^{(d)}(E)}\right),$$

as the function $2d \exp\left(-\frac{1}{t}\right)$ is increasing and hence

$$\underline{C}_l(E) = \overline{C}_l(E).$$

Thus, the basic Theorem 2.8 (No. 10) also holds for logarithmic capacity.

It is also possible to prove analogues of Theorems 2.6 and 2.7 for logarithmic potential. We formulate this result for the case of Borel, sets, when the distinction between inner and outer equilibrium measures disappears.

Theorem 2.6′. *Suppose E is a Borel set in the plane for which $C_2(E) > 0$. Then there exists a unique equilibrium measure γ with support on $\overline{E}$ satisfying the following conditions:*

(a) $\|\gamma\|^2 = \gamma(1) = C_2(E)$,

(b) $U^\gamma(x) = 1$ *approximately everywhere in* E,

(c) $U^\gamma(x) \leq 1$ *everywhere.*

17. The asymptotic behavior of the logarithmic potential $U^\lambda(x)$ of the minimizing measure λ admits an important interpretation of $C_l(K)$ for the theory of analytic functions. Suppose $C_l(K) > 0$. First of all, as $|x| \to \infty$

$$U^\lambda(x) = \int_K \ln \frac{1}{|x-y|}\, d\lambda(x) = \ln \frac{1}{|x|} + \int_K \ln \frac{1}{\left|1 - \frac{y}{x}\right|}\, d\lambda(x)$$

$$= \ln \frac{1}{|x|} + O\left(\frac{1}{|x|}\right).$$

Therefore, the function

$$g(x) = -U^\lambda(x) + W_2(K)$$

harmonic in G_∞, will be equal to zero quasi-everywhere on its boundary S and at infinity will have the form

$$g(x) = \ln |x| - \ln C_l(K) + O\left(\frac{1}{|x|}\right).$$

This function $g(x)$, which is *uniquely*[17] characterized by harmonicity in G_∞, equality to zero quasi-everywhere on S, and by the first asymptotic

[17] Cf. further Ch.IV, § 1.

term (i.e. by the formula $g(x) \sim \ln |x|$), will be called the *Green function* for the region G_∞ with *pole at infinity*. Its construction enables us to find the logarithmic capacity $C_l(K)$. If G_∞ is simply connected, then, defining the conjugate harmonic function $h(x)$ such that for $x \to \infty$ $h(x) = \arg x + O\,(1/|x|)$ we are able to construct the function $f(x) = e^{g(x)+ih(x)}$ analytic in G_∞, which realizes a conformal map of G_∞ onto the exterior of the unit disc and for large $|x|$ has the form

$$f(x) = \frac{1}{C_l(K)}\,x + O(1).$$[18]

This shows that in the present case the logarithmic capacity $C_l(K)$ coincides with the *external conformal radius* of K (cf. Golouzin [1], p. 342).

Therefore, whenever we can write down a function which gives a conformal mapping of G_∞ onto the exterior of a disc, we can also calculate the logarithmic capacity of K. We give a number of results of such calculations.

K	$C_l(K)$
1. Disc of radius a	a
2. Ellipse with axes a, b	$(a+b)/2$
3. Line segment of length $2a$	$a/2$
4. Square with side a	$\frac{\Gamma^2(1/4)}{4\pi^{3/2}}\,a \approx 0{,}59017a$
5. Equilateral triangle of height h	$\frac{\Gamma^3(1/3)}{4\pi^2}\,h \approx 0{,}4870h$
6. Isosceles right triangle with side a	$\frac{3^{3/4}\Gamma^2(1/4)}{4(2\pi)^{3/2}}\,a \approx 0{,}47563a$
7. Right triangle with acute angle 30° and long side h	$\frac{5^{5/12}\Gamma^3\,(i/3)}{2^{10/3}\pi^2}\,h \approx 0{,}37791h$
8. Regular hexagon with side a	$\frac{3\Gamma^3(1/3)}{2^{8/3}\pi^2}\,a \approx 0{,}92032a$

[18] The existence of a function mapping G_∞ conformally onto the unit disc implies that $g(x) = 0$ everywhere on S.

K	$C_l(K)$
9. Regular n-gon with side a	$\dfrac{n\Gamma(1+1/n)}{2^{1+2/n}\,\Gamma(1/2)\,\Gamma(1/2+1/n)}\,a$
10. Rhombus with angle $\pi\alpha$ and side a	$\dfrac{\sqrt{\pi}}{2\Gamma(1/2-\alpha/2)\,\Gamma(1+\alpha/2)}\,a$
11. Half-disc of radius a	$\dfrac{4}{3\sqrt{3}}\,a \approx 0{,}76980a$
12. Star, consisting of n rays: $\arg z = 2\pi(k/n)$ $\|z\| < a;\ k = 1, 2, \ldots, n$	$2^{-2/n}\,a$
13. Disc with arms, consisting of the unit disc $\|z\| < 1$ and n segments: $1 \leq \|z\| \leq 1 + a$, $\arg z = 2\pi(k/n)$	$\left\{\dfrac{(1+a)^{n/2} + (1-a)^{n/2}}{2}\right\}^{-2/n}$
14. "Lemniscate set": $\|z^n + c_1 z^{n-1} + \cdots + c_n\| \leq a^n$	a [19]

From No. 3 of the table and since capacity decreases under projections (cf. No. 15), it follows that any continuum K has positive capacity and moreover, $K \equiv \check{K}$.

From analogous considerations we obtain the following *lower bound for logarithmic capacity* (cf. Golouzin [1], p. 329).

Let K_π denote the projection of the compact set K onto some line, and let $\operatorname{mes} E$ be the Lebesgue linear measure of a set E on the line; then

$$C_l(K) \geq \frac{1}{4}\operatorname{mes} K_\pi. \tag{2.4.4}$$

Indeed, first of all

$$C_l(K) \geq C_l(K_\pi).$$

Then, for convenience placing K_π on the positive half-axis $\xi > 0$, we define by the function

$$\eta = \operatorname{mes}\,[K_\pi \cap (0, \xi)]$$

---

[19] This is so because $g(z) = \ln \sqrt[n]{|z^n + c_1 z^{n-1} + \cdots + c_n|} - \ln a$ is the Green function of the region G_∞.

a continuous mapping of K_π onto the segment $[0, \operatorname{mes} K_\pi]$. It is clear that under this mapping, distance between points are not increased. Therefore, by Theorem 2.9 and the remark in No. 15

$$C_l(K_\pi) \geq C_l([0, \operatorname{mes} K_\pi]) = \frac{1}{4} \operatorname{mes} K_\pi$$

(the latter, by No. 3 of the table).

18. The theory of capacity for the case $\alpha = 2$ can also be based not on the logarithmic (correspondingly, Newtonian) potential but on the Green potential

$$U_g^\mu(x) = \int_\omega G(x, y)\, d\mu(y),$$

where $G(x, y)$ is the Green function for the region ω.[20] As noted in Ch. I, § 3, No. 11 (cf. further Ch. IV, § 1, No. 2), the associated energy integral

$$I_g(\mu) = \int_{\omega \times \omega} G(x, y)\, d\mu(x)\, d\mu(y)$$

is positive. For any compact set $K \subset \omega$ we put

$$W_g(K) = \inf I_g(\mu),$$

where the inf is taken over all $\mu \in \mathfrak{M}_K^+$, normalized by the condition $\mu(1) = 1$; we define the *Green capacity* by the formula

$$C_g(K) = \big(W_g(K)\big)^{-1},$$

and the equilibrium measure γ by the formula

$$\gamma = C_g(K)\lambda, \quad I_g(\lambda) = W_g(K).$$

In the theory developed in Nos. 2—10 nothing is changed. The potentials $U_g^\mu(x)$ and $U_l^\mu(x)$ (correspondingly, $U_2^\mu(x)$ if $p \geq 3$) differ by a bounded term which implies, by the Vallée-Poussin definition of capacity (cf. No. 6) that the equations $C_g(K) = 0$ and $C_l(K) = 0$ $(K \subset \omega)$ (correspondingly $C_2(K) = 0$) are equivalent. The equilibrium potential is a harmonic function in $\omega \setminus K$, which equals 1 quasi-everywhere on K, and is zero on the boundary of ω. Thus, for the region $\omega \setminus \overline{\overline{K}}$ the function $U_g^\gamma(x)$ is the harmonic measure of the boundary S of the compact set $\overline{\overline{K}}$.

The compact set K is called a set of *absolute harmonic measure zero* (cf. Nevanlinna [1]), if for any region $\omega \supset K$ there is no bounded, harmonic function in $\omega \setminus K$ which equals 1 approximately everywhere in K and is zero on the boundary of ω.

---

[20] For ω we may take an arbitrary regular region (cf. Ch. IV, § 1). In this section we assume for simplicity that ω is a ball ($p \geq 3$) or a disc ($p = 2$).

If K has absolute harmonic measure zero, then

$$U_g^\gamma(x) \equiv 0, \qquad \text{i.e. } C_g(S) = 0$$

and

$$C_g(K) = 0$$

From the above, this implies that $C_l(K) = 0$ (correspondingly, $C_2(K)=0$). Below (cf. Theorem 3.2) we shall establish the converse.

The results of Nos. 11 and 12 and the corresponding results for $p = 2$ also carry over the Green capacity. One need only replace the function $|x - y|^{2-p}$ (correspondingly $\ln 1/|x - y|$) everywhere by $G(x, y)$. Also, by a contraction mapping $x' = f(x)$ of a compact set $K \subset \omega$, into a compact set $K' \subset \omega$ one should understand a mapping such that $G(x', y') \geq G(x, y)$, i.e. which does not increase the "pseudodistance" $\varrho(x, y) = [G(x, y)]^{2-p}$ (correspondingly, $\varrho(x, y) = e^{-G(x,y)}$).

We shall examine the case in which ω is a plane *simply connected region* bounded by the curve Γ, and $F \subset \omega$ is a continuum. Then $\omega \setminus \overline{\overline{F}}$ is a doubly-connected region D, bounded by Γ and S and the equilibrium potential $U_g^\gamma(x)$ is closely connected with the conformal mapping of D onto a circurlar annulus.

We shall denote by $V(x)$ the harmonic conjugate to the function $U(x) = U_g^\gamma(x)$. Traversing a contour l, contained in D and encircling F, $V(x)$ is incremented by an amount

$$\int\limits_l \frac{\partial U}{\partial n}\, ds = 2\gamma(1) = 2\pi C_g(F).$$

Consequently, the analytic function

$$z = f(x) = \exp\left\{-\frac{1}{C_g(F)}\,[U(x) + iV(x)]\right\}$$

is single-valued in D and by the usual methods we see that it conformally maps the region D onto the annulus

$$\exp\left\{-\frac{1}{C_g(F)}\right\} < |z| < 1.$$

From this we see the connection between the Green capacity and the *modulus* of the region D:

$$C_g(F) = \frac{1}{\ln M}, \quad M = \frac{r_2}{r_1}. \tag{2.4.5}$$

As an example consider the case when ω is the unit disc and F is the disc $|z - a| < r$, $|a| + r < 1$. An elementary calculation shows that in this case M is equal to the largest root of the quadratic equation

$rM^2 - (1 + r^2 - |a|^2) M + r = 0$, i.e.

$$M = \frac{1}{2r}\Big\{1 + r^2 - |a|^2 + \sqrt{(1 - r - |a|)(1 - r + |a|)(1 + r - |a|)(1 + r + |a|)}\Big\}.$$

When $|a|$ increases in the interval $(0, 1 - r)$, then M decreases and consequently, $C_g(F)$ increases (as it must by the general theorem on contraction mappings).

Now suppose that ω is the unit disc and F the segment $0 \leq \operatorname{Re} x \leq h < 1$. By elementary mappings we construct the function

$$\zeta = \frac{(1 - z)^2}{2(1 + z)^2 - (1 - z)^2},$$

which maps the doubly-connected region $\omega \setminus F$ onto the ζ-plane with two cuts: $[-1, 0]$ and $[\beta, 1]$, where

$$\beta = \frac{(1 - h)^2}{2(1 + h)^2 - (1 - h)^2}.$$

This latter region is mapped by the function

$$\zeta = \Big\{2 \operatorname{sn}^2 \frac{K' \ln w}{\pi} - 1\Big\}^{-1}; \qquad K' = \int_0^1 \frac{dt}{\sqrt{(1 - t^2)(1 - k'^2 t^2)}}$$

(where $k^2 = \dfrac{2\beta}{1 + \beta} = \left(\dfrac{1 - h}{1 + h}\right)^2$ is the modulus of the elliptic sine, and $k'^2 = 1 - k^2$) onto the circular annulus

$$e^{-\pi K/K'} \leq |w| \leq 1$$

(cf. Akhiezer [1], p. 196).

Consequently, $M = e^{\pi K/K'}$ and

$$C_g(F) = \frac{K'}{\pi K}.$$

Since in the neighborhood of the center of the disc $\varrho(x, y) = e^{-g(x,y)} \sim |x - y|$, one might expect that for small h the Green capacity of the segment $[0, h]$ would be asymptotically equal to its Wiener capacity, namely,

$$C_2(K) = \left(\ln \frac{1}{C_l(K)}\right)^{-1} = \left(\ln \frac{4}{h}\right)^{-1}$$

(cf. No. 3 of the table in No. 17). In fact, as $h \to 0$ $k^2 \sim 1 - 4h$; $k'^2 \sim 4h$, $K' \sim \dfrac{\pi}{2}$, $K \sim \ln \dfrac{4}{k'} = \dfrac{1}{2} \ln \dfrac{4}{h}$ (cf. Akhiezer [1], p. 118), where

$$C_g(K) \sim \left(\ln \frac{4}{h}\right)^{-1}.$$

The last example permits one to obtain a lower bound for the Green capacity of a linear closed set F, lying on a radius of the unit disc. For this, we note first that if $0 < h < x < 1$ the quantity

$$\varrho(x, x-h) = \frac{h}{1 - x(x-h)}$$

monotonically decreases with x. Therefore, moving the set along the radius toward the center of the disc is a contraction mapping as is the map of F onto the segment $[0, \operatorname{mes} F]$, defined by the function

$$\eta = \operatorname{mes}\,[F \cap (0, x)].$$

Consequently,

$$C_g(F) \geq \frac{K'}{\pi K}, \tag{2.4.6}$$

where K and K' are complete elliptic integrals with moduli $k^2 = \left(\frac{1 - \operatorname{mes} F}{1 + \operatorname{mes} F}\right)^2$ and $k'^2 = 1 - k^2$, respectively.

Chapter III

Sets of Capacity Zero. Sequences and Bounds for Potentials

§ 1. Polar Sets

1. A set $E \subset R^p$ will be called *polar* (α-polar) if there exists a potential of measure $U_\alpha^\mu(x)$ which assumes the value ∞ at all the points of E and only at those points.

Since $U_\alpha^\mu(x)$ is lower semicontinuous, the sets where $U_\alpha^\mu(x) > n$ are open and E is G_δ set.

It is not difficult to prove that a polar set E has zero α-capacity. Indeed, suppose $C_\alpha(E) > 0$. Consider a compact set $K \subset E$ such that $C_\alpha(K) > 0$, and denote by γ the equilibrium measure on K. Then if $U_\alpha^\mu(x) = \infty$ at all points of K, we have

$$\int_K U_\alpha^\mu(x)\, d\gamma(x) = \int_{S(\mu)} U_\alpha^\gamma(x)\, d\mu(x) = \infty.$$

But as $x \to \infty$, $U_\alpha^\gamma(x) = O\left(\frac{1}{|x|^{p-\alpha}}\right)$, and therefore by the finiteness condition (1.3.10), the latter integral cannot be infinite.

Remark. From this follows a *more precise uniqueness theorem* (cf. Theorem 1.12). Suppose that μ_1 and μ_2 are C-absolutely continuous measures. If

$$U_\alpha^{\mu_1}(x) = U_\alpha^{\mu_2}(x)$$

quasi-everywhere on $S(\mu_1) \cup S(\mu_2)$, then $\mu_1 \equiv \mu_2$.

In fact, consider the signed measure $\nu = \mu_1 - \mu_2$. Since its potential $U_\alpha^\nu(x)$ is defined and finite quasi-everywhere, then this is also true ν-almost everywhere. In addition, the equality $U_\alpha^\nu(x) = 0$ also holds ν-almost everywhere. Consequently,

$$\|\nu\|^2 = \int_{S(\nu)} U_\alpha^\nu(x)\, d\nu = 0,$$

and $\nu \equiv 0$.

In particular, the result holds if the measure μ_1, μ_2 have finite energy (cf. Ch. II, No. 2).

A more precise formulation of the uniqueness theorem will be given in Ch. IV, No. 14.

The two properties mentioned above completely characterize polar sets. In other words, the following theorem holds.

Theorem 3.1. *Suppose $E \subset R^p$ is a G_δ set which has zero α-capacity. Then there exists a measure λ such that $U_\alpha^\lambda(x) = \infty$ for $x \in E$ and $U_\alpha^\lambda(x) < \infty$ for $x \notin E$.*

Proof. The desired measure λ will be constructed as a sum $\lambda = \sum_{i=1}^{\infty} \gamma_i$ of equilibrium measures γ_i for a certain sequence of open sets G_i. We will choose the sequence G_i in such a way that for any point $x \notin E$

$$\sum_{i=1}^{\infty} U_\alpha^{\gamma_i}(x) < \infty, \tag{3.1.1}$$

while for $x \in E$

$$\sum_{i=1}^{\infty} U_\alpha^{\gamma_i}(x) = \infty,$$

by the corollary to Theorem 2.6, if each point $x \in E$ belongs to an infinite number of the G_i.

Let us proceed with the construction. First we select a decreasing sequence of open sets Ω_n such that[1]

$$E = \bigcap_{n=1}^{\infty} \Omega_n, \qquad C_\alpha(\Omega_n) < \frac{w^{-1}(p, \alpha)}{4^n}.$$

Set $F_n = C\Omega_n$, $d_n(x) =$ distance from the point x to F_n and

$$f_{nm} = \left\{x : \frac{1}{2^{m+1}} \leq d_n^{p-\alpha}(x) \leq \frac{1}{2^m}\right\}.$$

The f_{nm} are closed sets and $\bigcup_m f_{nm} = \Omega_n \supset E$. Moreover, if $m < 2n$, then the set f_{nm} will be empty. If f_{nm} were not empty, then Ω_n would contain a ball of radius $\left(\frac{1}{2^{m+1}}\right)^{\frac{1}{p-\alpha}}$ and, consequently, $C_\alpha(\Omega_n) > \frac{w^{-1}(p, \alpha)}{2^{m+1}}$. Comparing this with the inequality $C_\alpha(\Omega_n) < \frac{w^{-1}(p, \alpha)}{4^n}$, we obtain $m \geq 2n$.

---

[1] The definition of the constant $w(p, a)$ is given in Theorem 1.32 and in Ch. II, No. 13.

Now set $E_{nm} = E \cap f_{nm}$ and choose open sets $G_{nm} \supset E_{nm}$ such that

$$C_\alpha(G_{nm}) < \frac{1}{4^m}, \qquad G_{nm} \subset f_{n,m-1} \cup f_{nm} \cup f_{n,m+1}. \tag{3.1.2}$$

We now show that the measure $\lambda = \sum_{n,m} \gamma_{nm}$, where γ_{nm} is the equilibrium measure for G_{nm}, satisfies the conditions of the theorem.

Since for each n

$$E = \bigcup_{m=2n}^{\infty} E_{nm}$$

we have

$$\sum_{n,m} U_\alpha^{\gamma_{nm}}(x) = \infty \qquad \text{for } x \in E.$$

If $x \notin E$, then for all $n > n_0(x)$ $x \in F_n$. We bound from above

$$\sum_{n,m} U_\alpha^{\gamma_{nm}}(x) = \sum_{n \le n_0} \sum_{m=2n}^{\infty} U_\alpha^{\gamma_{nm}}(x) + \sum_{n > n_0} \sum_{m=2n}^{\infty} U_\alpha^{\gamma_{nm}}(x).$$

First of all, for $y \in \bar{G}_{nm}$ and $x \notin \bigcup_{k=m-2}^{m+2} f_{nk}$ and, in particular, for $x \in F_n$

$$|x-y| > \left(\frac{1}{2^{m+2}}\right)^{\frac{1}{p-\alpha}} - \left(\frac{1}{2^{m+3}}\right)^{\frac{1}{p-\alpha}},$$

that is[2]

$$\frac{1}{|x-y|^{p-\alpha}} < A_1(p, \alpha)\, 2^{m+3}.$$

Therefore, outside the set $\bigcup_{k=m-2}^{m+2} f_{nk}$

$$U_\alpha^{\gamma_{nm}}(x) = A(p, \alpha) \int_{\bar{G}_{nm}} \frac{d\gamma_{nm}}{|x-y|^{p-\alpha}} < A_2(p, \alpha)\, 2^{m+3}\, C_\alpha(G_{nm}) < A_3(p, \alpha) \frac{1}{2^m}.$$

Consequently,

$$\sum_{n > n_0} \sum_{m=2n}^{\infty} U_\alpha^{\gamma_{nm}}(x) < A_3(p, \alpha) \sum_{n > n_0} \sum_{m=2n}^{\infty} \frac{1}{2^m} < \infty.$$

Furthermore, if $n \le n_0$, then there is only a finite set of numbers m for

---

[2] $A_k(p, \alpha)$, $k = 1, 2, 3$ denote certain positive constants.

which $x \in \bigcup_{k=m-2}^{m+2} f_{nk}$. This convinces us of the finiteness of the sum

$$\sum_{n \leq n_0} \sum_{m=2n}^{\infty} U_\alpha^{\gamma_{nm}}(x).$$

The theorem is proved. ▯

Note that the measure λ constructed in the proof of Theorem 3.1 is not concentrated on the set E. If E is a compact set K, then it is possible, using atomic measures, to make λ *concentrated on* K. This is done in the following manner.

Suppose $\{x_i\}_1^n$ is an arbitrary system of n points in K and

$$\mu_n = \frac{1}{n} \sum_{i=1}^{n} \varepsilon_{x_i}.$$

Set

$$v_n = \sup_{\mu_n} \left\{ \min_{x \in K} U_\alpha^{\mu_n}(x) \right\}.$$

Because K is compact there is a system of extremal points $\{\eta_i^{(n)}\}$ in K such that, putting $\lambda_n' = \frac{1}{n} \sum_{i=1}^{n} \varepsilon_{\eta_i^{(n)}}$, we shall obtain

$$\min_{x \in K} U_\alpha^{\lambda_n'}(x) = v_n.$$

We evaluate v_n from below. Keeping Ch. II, No. 12 in mind, we can write:

$$\binom{n}{2} \frac{1}{D_n^{(\alpha)}} = \sum_{i<j} |\xi_i^{(n)} - \xi_j^{(n)}|^{\alpha-p} = \frac{1}{2} \sum_{k=1}^{n} \sum_{j \neq k} |\xi_k^{(n)} - \xi_j^{(n)}|^{\alpha-p}.$$

Since

$$\sum_{j \neq k} |\xi_k^{(n)} - \xi_j^{(n)}|^{\alpha-p} = \min_{x \in K} \sum_{j \neq k} |x - \xi_j^{(n)}|^{\alpha-p} \leq \frac{n-1}{A(p,\alpha)} v_{n-1},$$

then

$$\binom{n}{2} \frac{1}{D_n^{(\alpha)}} \leq \frac{1}{A(p,\alpha)} \frac{n(n-1)}{2} v_{n-1},$$

and, thus,

$$v_{n-1} \geq \frac{A(p,\alpha)}{D_n^{(\alpha)}}.$$

Consequently, (cf. (2.3.4))

$$\varliminf_{n\to\infty} v_n \geq \frac{1}{C_\alpha(K)},\text{[3]} \qquad (3.1.3)$$

i.e. in the given situation $\lim\limits_{n\to\infty} v_n = \infty$.

To construct the desired measure λ it is sufficient to take a sequence of positive numbers a_n, satisfying the conditions,

$$\sum_{n=1}^{\infty} a_n = 1, \qquad \sum_{n=1}^{\infty} a_n v_n = \infty,$$

and to put[4]

$$\lambda = \sum_{n=1}^{\infty} a_n \lambda_n'.$$

Then for any point $x \in K$

$$U_\alpha^\lambda(x) \geq \sum_{n=1}^{N} a_n U_\alpha^{\lambda_n'}(x) \geq \sum_{n=1}^{N} a_n v_n,$$

and since this holds for any N, then $U_\alpha^\lambda(x) = \infty$. The finiteness of $U_\alpha^\lambda(x)$ outside of K is obvious since the support of λ is contained in K.

Remark. The theorem also holds for a Green potential and for the logarithmic potential and logarithmic capacity. No essential changes in the proof are necessary for this case.

2. We now give several applications of Theorem 3.1.

Theorem 3.2. *The compact set K has absolutely harmonic measure zero*[5] *if, and only if, its Newtonian (in the planar case — its logarithmic) capacity is zero.*

Proof. The necessity of the condition was shown in Ch. II, No. 18. To prove sufficiency, let us assume that in a certain region $\omega \setminus K$, there exists a harmonic function $h(x)$, equal to 0 on the boundary of ω and equal to 1 on K. If K has zero capacity, then by Theorem 3.1 there is a potential $U_2^\lambda(x)$ (or a Green potential $U_g^\lambda(x)$ relative to the region ω in the planar case) which is infinite at the points of K and only at those points. For any $\varepsilon > 0$ on the entire boundary of $\omega \setminus K$ we have the inequality $h(x) \leq \varepsilon U_2^\lambda(x)$. By the maximum principle for harmonic functions $0 \leq h(x) \leq \varepsilon U_2^\lambda(x)$ at any point $x \in \omega \setminus K$; hence, as $\varepsilon \to 0$ we

---

[3] It can be shown that $\lim\limits_{n\to\infty} v_n$ exists and is equal to $\dfrac{1}{C_\alpha(K)}$.

[4] Convergence of the series is in the sense of vague convergence of measures.

[5] Cf. definition in Ch.II, No.18.

obtain $h(x) \equiv 0$ in $\omega \setminus K$. This shows that K has absolutely harmonic measure zero. ▯

In an analogous way we can prove the following theorem about removable singularities.

Theorem 3.3. *Suppose that the compact set K has Newtonian (or logarithmic, in the case of the plane) capacity zero and ω is any neighbourhood of K. If the function $h(x)$ is harmonic and bounded in $\omega \setminus K$, then there exists a function $h_1(x)$ harmonic in ω and coincident with $h(x)$ in $\omega \setminus K$.*

Proof. Assuming that ω is a regular region[6], let $h_1(x)$ denote the solution of the Dirichlet problem in ω with boundary values $h(x)$. In addition, consider the Green potential $U_g^\lambda(x)$ equal to ∞ exactly at the points of K (cf. No. 1). Then for any $\varepsilon > 0$, the function $\varepsilon U_g^\lambda(x)$ is a harmonic majorant for $\pm[h(x) - h_1(x)]$ in the region $\omega \setminus K$, i.e.

$$|h(x) - h_1(x)| < \varepsilon U_g^\lambda(x), \quad x \in \omega \setminus K.$$

Since ε is arbitrary, this gives $h(x) = h_1(x)$ in $\omega \setminus K$. ▯

We now show how Theorem 3.1 enables us to strengthen the classical maximum principle for subharmonic functions (correspondingly, the minimum principle for superharmonic functions). This principle consists of the following (cf. Theorem 1.1):

Suppose that $f(x)$ is a subharmonic function in the region G, which does not contain the point at infinity[7]. If for each point s on the boundary of G, the inequality

$$\overline{\lim_{x \to s}} f(x) \le M,$$

holds, then

$$f(x) \le M$$

at all points of G and equality is possible only if $f(x) \equiv M$.

With the help of Theorem 3.1 we shall establish the following generalization.

Theorem 3.4. *Suppose that the function $f(x)$ is bounded from above and is subharmonic in the region $G \subset R^p$ $(p \ge 3)$ which does not contain the*

[6] That is, the Dirichlet problem for the region ω has a solution for arbitrary continuous boundary functions.

[7] Here the boundary of G may contain the point at infinity.

point at infinity. Suppose further that the inequality

$$\overline{\lim_{x\to s}} f(x) \leq M$$

holds at all boundary points s with the possible exception of a set E which has null Newtonian capacity and does not contain the infinite point. Then throughout G we have the inequality

$$f(x) \leq M,$$

where equality is possible, even if only at one point, only in the case that $f(x) \equiv M$.

Proof. Since E can be included in an open set of arbitrarily small capacity, there exists a G_δ-set $E_1 \supset E$ having capacity zero. Suppose $U_2^\lambda(x)$ is a Newtonian potential equal to ∞ at all points of the boundary G which belong to E_1 and only at these point[8]. Then for any $\varepsilon > 0$

$$f_1(x) = f(x) - \varepsilon U_2^\lambda(x)$$

will be a function subharmonic in G for which the condition $\overline{\lim_{x\to s}} f_1(x) \leq M$ holds at all points of the boundary of G.

Consequently,

$$f_1(x) = f(x) - \varepsilon U_2^\lambda(x) \leq M, \quad x \in G,$$

and since ε is arbitrary, then

$$f(x) \leq M. \quad \square$$

For the case $p = 2$ we have the following theorem.

Theorem 3.5. *If G is a region in the plane whose boundary contains a continuum, f(x) is subharmonic and bounded in G, and*

$$\overline{\lim_{x\to s}} f(x) \leq M$$

quasi-everywhere on the boundary, then $f(x) \leq M$ *throughout G and equality is possible, even if only at one point, only in the case that* $f(x) \equiv M$.

The proof in the case of a bounded region G is a repetition of the proof of the preceding theorem, except that in place of $U_2^\lambda(x)$ we take the Green potential $U_g^\lambda(x)$ relative to a region $G_1 \supset \overline{G}$.

In the general case, it is possible to map G conformally onto a bounded region G^*. For the mapping function $x^* = \varphi(x)$ we may take

[8] The intersection of E_1 with the boundary of G is a G_δ-set and therefore Theorem 3.1 is applicable.

an elementary function regular in $\bar{G}$ except perhaps for one point x_0 of the boundary of G.[9] If E denotes the exceptional null set on the boundary of G where $\overline{\lim}\, f(x) > M$, then $E^* = \varphi(E)$ will also have capacity zero. This is because in the subset of $\bar{G}$ not in any neighborhood of x_0, the function $\varphi(x)$ realizes a mapping with "finite dilation" (cf. the remark in Ch. II, No. 11). Taking into account that the function $f[\varphi^{-1}(x^*)]$ is subharmonic in G^*, we see that the general case is reduced to the previous one.

§ 2. Continuity Properties of Potentials

3. The classical theorem of Lusin has a remarkable analogue in potential theory.

Theorem 3.6. *Suppose $U^{\mu}_{\alpha}(x)$ is an arbitrary potential. Then for any $\eta > 0$ there exists an open set G_η with α-capacity less than η such that on its complement the potential $U^{\mu}_{\alpha}(x)$ is continuous.*

Proof.[10] Suppose that μ has finite energy. Form the averaged signed measures

$$\mu_n = \mu * m^{(1/n)}.$$

By Theorem 1.11 of Ch. I the $U^{\mu_n}_{\alpha}(x)$ are continuous and $\lim\limits_{n\to\infty} U^{\mu_n}_{\alpha}(x) = U^{\mu}_{\alpha}(x)$; by Lemma 1.2 of Ch. I, No. 12 $\|\mu - \mu_n\| \to 0$. Selecting a subsequence if necessary, we may assume that

$$\sum_{n=1}^{\infty} 4^n \, \|\mu_{n+1} - \mu_n\|^2 < \infty.$$

We shall show then that the potentials $U^{\mu_n}_{\alpha}(x)$ converge uniformly outside a certain open set of arbitrarily small capacity. For this we set

$$\begin{aligned} G_n &= \left\{x : \left| U^{\mu_{n+1}}_{\alpha}(x) - U^{\mu_n}_{\alpha}(x) \right| > \frac{1}{2^n}\right\} \\ &= \left\{x : U^{\mu_{n+1}}_{\alpha}(x) - U^{\mu_n}_{\alpha}(x) > \frac{1}{2^n}\right\} \\ &\quad \cup \left\{x : U^{\mu_n}_{\alpha}(x) - U^{\mu_{n+1}}_{\alpha}(x) > \frac{1}{2^n}\right\}. \end{aligned}$$

---

[9] In addition we may assume that $\overline{\lim\limits_{x\to x_0}}\, f(x) \le M$.

[10] In this proof we suppose $\alpha \lneqq 2$. In the next No. we shall give another proof without this restriction.

By Lemma 2.4 of Ch. II, No. 8 $C_\alpha(C_n) < 2\,\|\mu_{n+1} - \mu_n\|^2 \cdot 4^n$, and therefore, the series $\sum_{n=1}^{\infty} C_\alpha(G_n)$ converges. We now put $G^{(k)} = \bigcup_{n=k}^{\infty} G_n$. This is an open set and by virtue of the inequality

$$C_\alpha(G^{(k)}) \leq \sum_{n=k}^{\infty} C_\alpha(G_n)$$

its capacity for sufficiently large k can be made arbitrarily small. On the other hand, on the complement of $G^{(k)}$ we have the inequalities

$$|U_\alpha^{\mu_{n+1}}(x) - U_\alpha^{\mu_n}(x)| \leq \frac{1}{2^n} \quad (n > k),$$

which proves the uniform convergence of the sequence $\{U_\alpha^{\mu_n}(x)\}$. Consequently,

$$U_\alpha^{\mu}(x) = \lim_{n\to\infty} U_\alpha^{\mu_n}(x)$$

is in fact continuous on the complement of $G^{(k)}$.

Now suppose that μ is any measure with compact support. For any $N > 0$

$$\inf\,[U_\alpha^{\mu}(x), N] = U_\alpha^{\lambda}(x),$$

where λ is a certain measure (cf. Theorem 1.31). We will show that $\lambda \in \mathscr{E}_\alpha^+$. Indeed,

$$\int U_\alpha^{\lambda}(x)\, d\lambda(x) \leq \int U_\alpha^{\mu}(x)\, d\lambda(x) = \int U_\alpha^{\lambda}(x)\, d\mu(x) \leq N\mu(1).$$

Now consider the set

$$\omega = \{x: U_\alpha^{\lambda}(x) \neq U_\alpha^{\mu}(x)\} = \{x: U_\alpha^{\mu}(x) > N\}.$$

It is open and $C_\alpha(\omega) < \frac{\mu(1)}{N}$.

Indeed, if ν is any measure satisfying the conditions $S(\nu) \subset \omega$, $U_\alpha^{\nu}(x) \leq 1$, then

$$\nu(1) = \int_{S(\nu)} d\nu \leq \frac{1}{N} \int_{S(\nu)} U_\alpha^{\mu}(x)\, d\nu = \frac{1}{N} \int U_\alpha^{\nu}(x)\, d\mu \leq \frac{\mu(1)}{N}.$$

Choosing N such that

$$\frac{\mu(1)}{N} < \frac{\eta}{2},$$

and applying the part of the theorem already proved to the potential $U_\alpha^{\lambda}(x)$ and the number $\eta/2$, we obtain the required result.

Let us now remove the assumption that μ has compact support. Denote by μ_n the restriction of μ to the layer $n-1 \leq |x| < n$ and apply the already proved result to each of the measures μ_n, choosing the corresponding numbers $\varepsilon_n > 0$ such that $\Sigma\, \varepsilon_n < \varepsilon$; also, denote by ω_n the open set, $C_\alpha(\omega_n) < \varepsilon_n$, outside of which $U_\alpha^{\mu_n}(x)$ is continuous. Then $\omega = \bigcup_1^\infty \omega_n$ satisfies the requirements of the theorem. This is implied by

$$U_\alpha^\mu(x) = \sum_{n=1}^\infty U_\alpha^{\mu_n}(x),$$

where on any compact set K, the remainder of the series tends uniformly to zero. Indeed, this remainder can be written in the form

$$\int_{|y|>N} \frac{d\mu(y)}{|x-y|^{p-\alpha}},$$

and the assertion follows from condition (1.3.10) of Ch. I.

The transition from measures to signed measures is trivial. $\square$

4. The above proof of Theorem 3.6 is non-elementary in the sense that it relies on Theorem 1.31 which is the result of an extensive analytic argument and holds only for $\alpha \leq 2$. Therefore, the more elementary proof presented below is of some interest. It is based on the following lemma.

Lemma 3.1. *Let μ be a measure with compact support. For any $\varepsilon > 0$, $\eta > 0$ there is a decomposition $\mu = \lambda + \nu$, λ, $\nu \in \mathfrak{M}^+$ and an open set ω, $C_\alpha(\omega) < \varepsilon$, satisfying the conditions:*

(*i*) $\lambda(1) < \eta$,

(*ii*) *$U_\alpha^\nu(x)$ is continuous on the complement of ω,*

(*iii*) *$U_\alpha^\nu(x) \leq \mu(1)/\varepsilon$ on the complement of ω.*

Proof. Set

$$\omega = \left\{x \colon U_\alpha^\mu(x) > \frac{\mu(1)}{\varepsilon}\right\}.$$

This is an open set, and, as we have seen,

$$C_\alpha(\omega) < \varepsilon.$$

Let μ_1 be the restriction of μ to the complement of ω and $\mu_2 = \mu - \mu_1$ the restriction of μ to ω. Since $U_\alpha^{\mu_1}(x) \leq U_\alpha^\mu(x) \leq \mu(1)/\varepsilon$ on the complement of ω, and, in particular, on $S(\mu_1)$, then by Theorem 1.8 we can find a measure $\mu_1' \leq \mu_1$ with a potential continuous in R^p such that $\mu_1(1) - \mu_1'(1) < \eta/2$. Further, in $\omega \cap S(\mu)$ one can find a compact set K such that $\mu(\omega) - \mu(K) = \mu_2(1) - \mu_2(K) < \eta/2$. Denote by μ_2' the restriction of μ_2 to K. Then $\nu = \mu_1' + \mu_2'$ and $\lambda = \mu - \nu$ satisfy all the requirements of the lemma. Indeed, (i) follows from $\mu(1) = \mu_1(1) + \mu_2(1)$, and $\nu(1) = \mu_1'(1) + \mu_2'(1)$; (ii) follows from the fact that $U_\alpha^{\mu_1'}(x)$ is everywhere continuous and $U_\alpha^{\mu_2'}(x)$ is continuous outside an open neighborhood ω of the support $S(\mu_2') \subset K$. Finally, (iii) holds since

$$U_\alpha^\nu(x) \leq U_\alpha^\mu(x) \leq \frac{\mu(1)}{\varepsilon}$$

on the complement of ω. ▯

Proof of Theorem 3.6. We know that it is sufficient to prove the theorem for measures with compact support. Select two sequences $\{\varepsilon_n\}$, $\{\eta_n\}$ of positive numbers subject to the constraints

$$\sum_1^\infty \varepsilon_n < \varepsilon, \quad \sum_2^\infty \frac{\eta_{n-1}}{\varepsilon_n} < \infty.$$

Applying Lemma 3.1 to the measure μ we obtain a decomposition $\mu = \lambda_1 + \nu_1$ where $\lambda_1(1) < \eta_1$, and $U_\alpha^{\nu_1}(x) \leq \mu(1)/\varepsilon_1$ on the complement of ω_1 where $C_\alpha(\omega_1) < \varepsilon_1$. Further, we apply the same lemma to the measure λ_1. We obtain a decomposition $\lambda_1 = \lambda_2 + \nu_2$, such that $\lambda_2(1) < \eta_2$, and $U_\alpha^{\nu_2}(x) < \eta_1/\varepsilon_2$ on the complement of ω_2 where $C_\alpha(\omega_2) < \varepsilon_2$. Continuing to apply Lemma 3.1 iteratively to $\lambda_2, \lambda_3, \ldots$, we construct an open set

$$\omega = \bigcup_{n=1}^\infty \omega_n,$$

satisfying the requirements of Theorem 3.6. Indeed, first of all,

$$C_\alpha(\omega) \leq \sum_1^\infty C_\alpha(\omega_n) < \sum_1^\infty \varepsilon_n < \varepsilon.$$

Furthermore, all the potentials $U_\alpha^{\nu_n}(x)$ are continuous on the complement of ω, since it is contained in the complement of ω_n ($n = 1, 2, \ldots$), and the series

$$\sum_{n=1}^\infty U_\alpha^{\nu_n}(x)$$

converges uniformly. Consequently, it is sufficient to prove that

$$\sum_{n=1}^\infty U_\alpha^{\nu_n}(x) = U_\alpha^\mu(x). \tag{3.2.1}$$

But

$$\mu = \sum_{n=1}^N \nu_n + \lambda_N$$

and

$$\lambda_N(1) \to 0 \quad \text{as } N \to \infty.$$

Therefore, the monotone increasing sequence of measures $\left\{\sum_1^N \nu_n\right\}$ vaguely converges to μ, and this gives (3.2.1) (cf. Corollary to Theorem 1.3). ▯

We point out an important consequence of the above theorem.

Theorem 3.7. *Let $\mu \in \mathscr{E}_\alpha^+$. Then there exists an increasing sequence of measures μ_n such that*

(*i*) *$\mu_n \dot{\to} \mu$ (and even $\|\mu - \mu_n\| \to 0$),*

(*ii*) *the potentials $U_\alpha^{\mu_n}(x)$ are continuous and*

$$\lim_{n\to\infty} U_\alpha^{\mu_n}(x) = U_\alpha^\mu(x).$$

Proof. Suppose that $G^{(n)}$ is a monotone decreasing sequence of open sets of capacity $<1/n$, in whose complements $U_\alpha^\mu(x)$ is continuous. Let μ_n denote the restriction of μ to the complement of $G^{(n)}$; since the sequence of measures monotonically increases, there exists a limiting measure ν (in the vague and the strong sense). It coincides with μ because the support of $\mu - \nu$ must have null capacity (it is contained in $\bigcap_1^\infty G^{(n)}$) and the measure $\mu - \nu$ itself has finite energy. This proves (i), and consequently, the limiting relation (Ch. I, No. 7, Corollary to Theorem 1.3)

$$U_\alpha^\mu(x) = \lim_{n\to\infty} U_\alpha^{\mu_n}(x).$$

It remains to establish the continuity of the potentials $U_\alpha^{\mu_n}(x)$. By Theorem 1.7 we must verify the continuity of $U_\alpha^{\mu_n}(x)$ on the support of the measure μ_n which is a subset of the complement of $G^{(n)}$. We will show that $U_\alpha^{\mu_n}(x)$ is continuous on the complement of $G^{(n)}$. We have

$$U_\alpha^{\mu_n}(x) = U_\alpha^\mu(x) - U_\alpha^{\mu-\mu_n}(x).$$

Since $U_\alpha^\mu(x)$ is continuous on the complement of $G^{(n)}$, and $U_\alpha^{\mu-\mu_n}(x)$ is lower semicontinuous, the difference $U_\alpha^\mu(x) - U_\alpha^{\mu-\mu_n}(x)$ must be upper semicontinuous on the complement of $G^{(n)}$. On the other hand, the left-hand side $U_\alpha^{\mu_n}(x)$ is lower semicontinuous. Consequently, it is continuous on the complement of $G^{(n)}$. ▯

Remark. In the preceding proposition it would not be difficult to require that the measures μ_n also have compact support. For this it would be sufficient to consider, instead of μ_n, its restriction to the closed ball $|x| \leq n$.

Corollary. *If K is a compact set with positive capacity, then there exists a measure $\nu \not\equiv 0$, $S(\nu) \subset K$ with continuous potential.*

Indeed, it is sufficient to apply the preceding theorem to any measure $\mu \in \mathscr{E}_\alpha^+$, $S(\mu) \subset K$ and set $\nu = \mu_n$.

§ 3. Sequences of Potentials of Measures

5. In the Newtonian case, we have already proved one theorem (Ch. I, No. 18, Theorem 1.26) on the convergence of a monotone increasing sequence of potentials.

Here we shall prove similar but more general theorems applicable to any M. Riesz potential $U_\alpha^\mu(x)$.

Theorem 3.8. *Let $\mu_n \in \mathfrak{M}^+$, $\mu_n \dot{\to} \mu$ and*

$$\lim_{r\to\infty} \int\limits_{|x|>r} \frac{d\mu_n(x)}{|x|^{p-\alpha}} = 0 \tag{3.3.0}$$

uniformly with respect to n. Then

$$U_\alpha^\mu(x) = \underline{\lim}_{n\to\infty} U_\alpha^{\mu_n}(x) \qquad \textit{quasi-everywhere.}$$

Proof. By the "principle of descent"

$$U_\alpha^\mu(x) \leq \underline{\lim}_{n\to\infty} U_\alpha^{\mu_n}(x).$$

Assume that the set

$$E = \left\{x : U_\alpha^\mu(x) < \underline{\lim}_{n\to\infty} U_\alpha^{\mu_n}(x)\right\}$$

has positive capacity. By the corollary to Theorem 2.8, there is a compact set $K \subset E$ such that

$$C_\alpha(K) > 0.$$

Consider any measure $\nu \not\equiv 0$ with support in K and with a continuous potential $U_\alpha^\nu(x)$ (by the corollary given at the end of No. 4, such measures exist). By the Fatou lemma

$$\int\limits_K U_\alpha^\mu(x)\, d\nu < \int\limits_K \underline{\lim}_{n\to\infty} U_\alpha^{\mu_n}(x)\, d\nu \leq \underline{\lim}_{n\to\infty} \int\limits_K U_\alpha^{\mu_n}(x)\, d\nu.$$

In addition we will prove that

$$\lim_{n\to\infty} \int\limits_K U_\alpha^{\mu_n}\, d\nu = \int\limits_K U_\alpha^\mu\, d\nu, \tag{3.3.1}$$

and the resulting contradiction will give a proof of the theorem. We rewrite (3.3.1) in the form

$$\lim_{n\to\infty} \int U_\alpha^\nu \, d\mu_n = \int U_\alpha^\nu \, d\mu$$

and will bound the difference

$$\left| \int U_\alpha^\nu \, d\mu_n - \int U_\alpha^\nu \, d\mu \right|.$$

Let $u_r(x)$ be a positive continuous function with compact support which coincides with $U_\alpha^\nu(x)$ for $|x| < r$, is equal to zero for $|x| > r + 1$, and is $\leq U_\alpha^\nu(x)$ everywhere. Then

$$\left| \int U_\alpha^\nu \, d\mu_n - \int U_\alpha^\nu \, d\mu \right| \leq \int (U_\alpha^\nu - u_r) \, d\mu_n$$
$$+ \left| \int u_r \, d\mu_n - \int u_r \, d\mu \right| + \int (U_\alpha^\nu - u_r) \, d\mu. \qquad (3.3.2)$$

Furthermore, since $U_\alpha^\nu(x) < Q\,|x|^{\alpha-p}$, then

$$\int (U_\alpha^\nu - u_r) \, d\mu_n \leq \int_{|x|>r} U_\alpha^\nu \, d\mu_n \leq Q \int_{|x|>r} \frac{d\mu_n(x)}{|x|^{p-\alpha}},$$

and the third integral is bounded in precisely the same way.

By condition (3.3.0) the quantity r may be chosen such that the first and third integrals in (3.3.2) become arbitrarily small. Then, by the vague convergence of μ_n to μ, we may also make the second integral in (3.3.2) arbitrarily small. Consequently, (3.3.1) is established and with that the theorem is proved. □

Remark 1. Of course, if $\mu_n(1) < M$, then condition (3.3.0) is satisfied.

Remark 2. Set

$$V(x) = \varliminf_{n\to\infty} U_\alpha^{\mu_n}(x)$$

and introduce the so-called *lower semicontinuous regularization* of the function $V(x)$:

$$W(x) = \varliminf_{y\to x} V(y).$$

Then $W(x) \equiv U_\alpha^\mu(x)$. Indeed, first $U_\alpha^\mu(x) \leq V(x)$, and therefore

$$U_\alpha^\mu(x) = \varliminf_{y\to x} U_\alpha^\mu(y) \leq \varliminf_{y\to x} V(y) = W(x).$$

On the other hand, by Theorem 3.8 and Theorem 1.11 (Ch. I, No. 9)

$$W(x) = \lim_{r\to 0} \min_{|y-x|<r} V(y) \leq \varliminf_{r\to 0} V * m^{(r)}(x) = \lim_{r\to 0} U_\alpha^\mu * m^{(r)} = U_\alpha^\mu(x).$$

Consequently, $W(x) = U_\alpha^\mu(x)$ for any x.

6. The condition of vague convergence of measures which is involved in Theorem 3.8 is sometimes not made explicit but follows from other conditions. This is apparent in the following two theorems in which such an additional condition is the monotonicity of the sequence of potentials.

Theorem 3.9. *Let $\{U_\alpha^{\mu_n}(x)\}$ be a monotone increasing sequence of potentials majorized by a fixed potential. Then $\lim_{n\to\infty} U_\alpha^{\mu_n}(x)$ is the potential of a measure μ and $\mu_n \dot{\to} \mu$.*

Proof. For the case $\alpha = 2$, this theorem was obtained in Ch. I (Theorem 1.26). In the general case, the proof is inessentially altered.

Let $U_\alpha^\nu(x)$ be the majorizing potential. By Lemma 1.12 we have throughout R^p

$$\lim_{n\to\infty} U_\alpha^{\mu_n}(x) = U_\alpha^\mu(x) + A \leq U_\alpha^\nu(x).$$

But $A = 0$. For the case $\alpha \leq 2$ this easily follows from

$$\lim_{r\to\infty} \varepsilon_\alpha^{(r)}\,(U_\alpha^\nu) = 0$$

(cf. Corollary to Theorem 1.30), and for $\alpha > 2$, it follows from Theorem 1.32.

The last assertion in the theorem, namely, $\mu_n \dot{\to} \mu$ is proved exactly as in Theorem 1.26. ▯

Theorem 3.10. *Let $\{U_\alpha^{\mu_n}(x)\}$ be a monotone decreasing sequence of potentials of measures. Then $\lim_{n\to\infty} U_\alpha^{\mu_n}(x)$ coincides quasi-everywhere with the potential $U_\alpha^\mu(x)$, and $\mu_n \dot{\to} \mu$.*

Proof. From the inequalities

$$U_\alpha^{\mu_n}(x) \leq U_\alpha^{\mu_1}(x), \quad n = 1, 2, \ldots, \tag{3.3.3}$$

it follows that the sequence of measures $\{\mu_n\}$ is weakly bounded. Otherwise for some compact set $K \subset R^p$ and an infinite sequence of numbers $\{n_k\}$ we would have

$$\mu_{n_k}(K) \to \infty.$$

But then

$$\lim_{k\to\infty} U_\alpha^{\mu_{n_k}}(x) \geq \overline{\lim_{k\to\infty}} \int_K \frac{d\mu_{n_k}(y)}{|x-y|^{p-\alpha}} = \infty$$

for all x, which is impossible by (3.3.3). By Theorem 0.6 (Introduction, No. 3) we may select from the sequence $\{\mu_n\}$ a vaguely convergent subsequence $\{\lambda_n\}$, $\lambda_n \dot{\to} \mu$; we show that it satisfies condition (3.3.0).

First suppose that $\alpha \geq 2$. We will show that there exists a positive constant $Q(p, \alpha)$ such that

$$\frac{1}{|x|^{p-\alpha}} \leq Q(p, \alpha)\, U_\alpha^{\varepsilon^{(r)}}(x), \quad |x| \geq r. \tag{3.3.4}$$

Indeed,

$$|x|^{p-\alpha}\, U_\alpha^{\varepsilon^{(r)}}(x) = \frac{|x|^{p-\alpha}}{\omega_p r^{p-1}} \int\limits_{|y|=r} \frac{dS(y)}{|x-y|^{p-\alpha}},$$

where ω_p is the "surface area" of the unit sphere in R^p, and $dS(y)$ is a "surface" element of the sphere $|y| = r$. Setting $x = rx_1$, $y = ry_1$, we obtain

$$|x|^{p-\alpha}\, U_\alpha^{\varepsilon^{(r)}}(x) = \frac{1}{\omega_p} \int\limits_{|y_1|=1} \frac{|x_1|^{p-\alpha}}{|x_1 - y_1|^{p-\alpha}}\, dS(y_1).$$

But the integral on the right is a continuous, positive function for $|x_1| \geq 1$, which tends to 1 at infinity. It has a positive minimum which implies (3.3.4).

Now we may write

$$\int\limits_{|x|>r} \frac{d\lambda_n}{|x|^{p-\alpha}} \leq Q(p, \alpha) \int\limits_{|x|>r} U_\alpha^{\varepsilon^{(r)}}(x)\, d\lambda_n$$

$$\leq Q(p, \alpha) \int\limits_{R^p} U_\alpha^{\varepsilon^{(r)}}(x)\, d\lambda_n$$

$$= Q(p, \alpha) \int U_\alpha^{\lambda_n}(x)\, d\varepsilon^{(r)} = Q(p, \alpha)\, \varepsilon^{(r)}(U_\alpha^{\lambda_n}).$$

By (3.3.3) this yields

$$\int\limits_{|x|>r} \frac{d\lambda_n}{|x|^{p-\alpha}} \leq Q(p, \alpha)\, \varepsilon^{(r)}(U_\alpha^{\mu_1}),$$

and (3.3.0) follows from Theorem 1.32.

Let us now consider the case $\alpha < 2$. By (1.6.1) for $|x| \geq r$

$$\frac{1}{|x|^{p-\alpha}} = \frac{1}{A(p, \alpha)}\, U_\alpha^{\varepsilon_\alpha^{(r)}}(x),$$

and, reasoning just as above, we obtain

$$\int\limits_{|x|>r} \frac{d\lambda_n}{|x|^{p-\alpha}} \leq \frac{1}{A(p, \alpha)}\, \varepsilon_\alpha^{(r)}(U_\alpha^{\mu_1}).$$

Since as $r \to \infty$ the right side tends to zero (cf. Corollary to Theorem 1.30), this proves (3.3.0).

By Theorem 3.8

$$U_\alpha^\mu(x) = \lim_{n\to\infty} U_\alpha^{\lambda_n}(x) = \lim_{n\to\infty} U_\alpha^{\mu_n}(x).$$

quasi-everywhere.

It only remains to show that $\mu_n \dot{\to} \mu$. Using the monotonicity of the sequence $\{U_\alpha^{\mu_n}\}$, this can be obtained just as in the proof of Theorem 1.26. ∎

7. If we investigate potentials of measure with finite energy, then we can obtain some additional results.

First of all, if the sequence $\{U_\alpha^{\mu_n}(x)\}$ is monotone and is majorized by the potential $U_\alpha^\nu(x)$ with finite energy,[11] then the sequenee of corresponding measures $\{\mu_n\}$ *strongly* converges.

Let us prove this. The inequality

$$U_\alpha^{\mu_n}(x) \leq U_\alpha^\nu(x) \tag{3.3.5}$$

implies that $||\mu_n|| \leq ||\nu||$.[12] Indeed,

$$||\mu||^2 = \int U_\alpha^{\mu_n}\, d\mu_n \leq \int U_\alpha^\nu\, d\mu_n = \int U_\alpha^{\mu_n}\, d\nu \leq \int U_\alpha^\nu\, d\nu = ||\nu||^2.$$

Consequently, the sequence of norms $||\mu_n||$ is monotone and bounded. Furthermore,

$$||\mu_{n+k} - \mu_n||^2 = ||\mu_{n+k}||^2 + ||\mu_n||^2 - 2(\mu_{n+k}, \mu_n).$$

Assume for definiteness that the sequence of potentials is monotone increasing. Then

$$(\mu_{n+k}, \mu_n) = \int U_\alpha^{\mu_{n+k}}\, d\mu_n \geq \int U_\alpha^{\mu_n}\, d\mu_n = ||\mu_n||^2,$$

and therefore

$$||\mu_{n+k} - \mu_n||^2 \leq ||\mu_{n+k}||^2 - ||\mu_n||^2.$$

This shows that the sequence $\{\mu_n\}$ is fundamental, and by Theorem 1.18 on the completeness of $\mathscr{E}_\alpha^+$, there is a measure μ to which the μ_n strongly converge.[13]

---

[11] The assumption of the existence of a majorizing potential $U_\alpha^\nu(x)$ may, as is seen below, be replaced by the condition that the norms $||\mu_n||$ be bounded.

[12] It is sufficient that the inequality (3.3.5) holds approximately everywhere.

[13] From the succeeding Theorem 3.11 it follows that

$$U_\alpha^\mu(x) = \lim_{n\to\infty} U_\alpha^{\mu_n}(x)$$

quasi-everywhere.

Further, we have the following theorem about sequences of potentials of signed measures.

Theorem 3.11. *If the signed measures ν_n strongly converge to ν, then from the sequence of potentials $\{U_\alpha^{\nu_n}(x)\}$ one can select a subsequence which converges to $U_\alpha^\nu(x)$ quasi-everywhere.*

The proof is based on Lemma 2.7, Ch. II, No. 8 and is analogous to the first proof of Theorem 3.6. After the selection of the subsequence of signed measures one may assume that

$$\sum_1^\infty 4^n \, ||\nu_{n+1} - \nu_n||^2 < \infty .$$

We set

$$E_n = \{x : |U_\alpha^{\nu_{n+1}}(x) - U_\alpha^{\nu_n}(x)| \geq 2^{-n}\}$$

$$= \{x : U_\alpha^{\nu_{n+1}}(x) - U_\alpha^{\nu_n}(x) \geq 2^{-n}\} \cup \{x : U_\alpha^{\nu_n}(x) - U_\alpha^{\nu_{n+1}}(x) \geq 2^{-n}\}.$$

By Lemma 2.7

$$\bar{C}_\alpha(E_n) \leq 2 \cdot 4^n \, ||\nu_{n+1} - \nu_n||^2 .$$

It is not difficult to see that outside the set

$$e = \bigcap_{k=1}^\infty \bigcup_{n=k}^\infty E_n$$

the sequence $\{U_\alpha^{\nu_n}(x)\}$ converges and that $\bar{C}_\alpha(e) = 0$. ▯

8. In Ch. I (Theorem 1.31) we proved that for $\alpha \leq 2$ (the Robin case) the set of potentials of measures with finite energy is closed relative to the operation inf, i.e. that for $\mu, \lambda \in \mathscr{E}_\alpha^+$

$$\inf (U_\alpha^\mu, U_\alpha^\lambda) = U_\alpha^\nu, \quad \nu \in \mathscr{E}_\alpha^+ . \tag{3.3.6}$$

Relying on this fact, we can prove the following theorem.

Theorem 3.12. *Let $\{\mu_n\}$ be a sequence of measures with uniformly bounded energies. Then there exists a measure $\mu \in \mathscr{E}_\alpha^+$ such that*

$$\varliminf_{n \to \infty} U_\alpha^{\mu_n}(x) = U_\alpha^\mu(x) \quad (0 < \alpha \leq 2).$$

Proof. By (3.3.6)

$$\inf \{U_\alpha^{\mu_n}, U_\alpha^{\mu_{n+1}}, \ldots, U_\alpha^{\mu_{n+k}}\} = U_\alpha^{\nu_{n,k}}(x),$$

where $\nu_{n,k} \in \mathscr{E}_\alpha^+$. For fixed n, $U_\alpha^{\nu_{n,k}}(x)$ form a non-increasing sequence of potentials and therefore (Theorem 3.10)

$$\lim_{k \to \infty} U_\alpha^{\nu_{n,k}}(x) = U_\alpha^{\lambda_n}(x), \quad \lambda_n \in \mathscr{E}_\alpha^+$$

quasi-everywhere. Furthermore, from the inequalities

$$U_\alpha^{\lambda_n}(x) \leq U_\alpha^{\nu_{n,k}}(x) \leq U_\alpha^{\mu_n}(x)$$

(quasi-everywhere) it follows, that $\|\lambda_n\| \leq \|\mu_n\|$ and, thus, the energies of the measures λ_n are bounded. On the other hand, the sequence of potentials $\{U_\alpha^{\lambda_n}(x)\}$ is monotone increasing and we have (Theorem 3.9)

$$\lim_{n\to\infty} U_\alpha^{\lambda_n}(x) = U_\alpha^{\mu}(x), \quad \mu \in \mathscr{E}_\alpha^+ .$$

Consequently,

$$U_\alpha^{\mu}(x) = \lim_{n\to\infty} \left(\lim_{k\to\infty} U_\alpha^{\nu_{n,k}}(x)\right) = \varliminf_{n\to\infty} U_\alpha^{\mu_n}(x)$$

quasi-everywhere, as required. ▯

§ 4. Metric Criteria for Sets of Capacity Zero and Bounds for Potentials

9. Let us introduce *Hausdorff h-measure* which will be useful below in formulating certain characteristics of sets with capacity zero.

Let $h(r)$, $0 \leq r \leq r_0$ be a monotone increasing function and $h(0) = 0$.

To each set $E \subset R^p$ we associate a number $m_h(E)$, $0 \leq m_h(E) \leq \infty$, in the following way. Let $\{A_i\}$ be a covering of the set E by bounded sets A_i whose diameters we denote by d_i. Assume that $d_i < \varepsilon \leq r_0$, and form the sum (finite or infinite) $\sum_i h(d_i/2)$. Then consider

$$\inf_{d_i<\varepsilon} \sum_i h(d_i/2) = m_h(E;\varepsilon),$$

where the inf is taken over all possible coverings of E for which the inequalities $d_i < \varepsilon$ are satisfied. As $\varepsilon \to 0$ the quantity $m_h(E;\varepsilon)$ monotonically increases, and we set

$$m_h(E) = \lim_{\varepsilon\to 0} m_h(E;\varepsilon).$$

In the case $h(r) = r^\alpha$ $(\alpha > 0)$ we obtain the so-called Hausdorff-measure of *dimension* α, and for $h(r) = (\ln 1/r)^{-1}$ the *logarithmic* measure.

The h-measure $m_h(E)$ is, of course, a monotone increasing set function, i.e. $E_1 \subset E_2$ implies $m_h(E_1) \leq m_h(E_2)$. Any contraction mapping defined in a neighborhood of the set E, can, it is easy to see, only decrease the h-measure. In particular, h-measure does not increase under projections.

Theorem 3.13. *Suppose that E has outer α-capacity zero* $(0 < \alpha < p)$ *(or if* $\alpha = p = 2$, *outer logarithmic capacity zero). Then*

$$m_h(E) = 0$$

for any function $h(r)$ for which

$$\int_0 \frac{h(r)}{r^{p+1-\alpha}}\,dr < \infty. \tag{3.4.1}$$

Proof. If $\overline{C}_\alpha(E) = 0$, then $E \subset A$, where A is a G_δ set and $C_\alpha(A) = 0$. It is sufficient therefore to establish that $m_h(A) = 0$ for $h(r)$ satisfying (3.4.1).

By Theorem 3.1, there is a measure λ for which $U_\alpha^\lambda(x) = \infty$ exactly at the points $x \in A$. If we denote by $\lambda(r, x)$ the λ-measure of the ball of radius r with center at the point x, then the equation $U_\alpha^\lambda(x) = \infty$ will be equivalent to the divergence of the integral

$$\int_0 \frac{d\lambda(r; x)}{r^{p-\alpha}}$$

and consequently, to the divergence of the integral

$$\int_0 \frac{\lambda(r; x)}{r^{p+1-\alpha}}\,dr.$$[14]

Hence, for any function $h(r)$ satisfying (3.4.1), and for any point $x \in A$ there is a sequence $r_k \to 0$ such that

$$h(r_k) < \lambda(r_k; x). \tag{3.4.2}$$

The next step is based on a geometric lemma which is of independent interest.

Lemma 3.2 (*Covering lemma*). *Suppose that a set $A \subset R^p$ is covered by balls such that each point $x \in A$ is the center of a certain ball $S(x)$ of radius $r(x)$. If $\sup_{x\in A} r(x) < \infty$,*[15] *then from the system of balls $\{S(x)\}$ it is possible to select a countable system $\{S(x_k)\}$ covering the entire set A and having multiplicity*[16] *not larger than a certain number $N(p)$ depending only on the dimension of the space.*

---

[14] Indeed if one of these integrals converges, then the other converges, too and upon,

$$\int_0^\varepsilon \frac{\lambda(r; x)}{r^{p+1-\alpha}}\,dr = -\frac{\lambda(\varepsilon; x)}{(p-\alpha)\,\varepsilon^{p-\alpha}} + \frac{1}{p-\alpha}\int_0^\varepsilon \frac{d\lambda(r; x)}{r^{p-\alpha}}.$$

[15] This condition is redundant if A is bounded, but is necessary in the general case.

[16] The *multiplicity* of system of sets is the largest number of sets which have a common point. If there is no such largest number then the multiplicity is defined to be infinite.

Proof of the Lemma. Set

$$R_1 = \sup_{x \in A} r(x).$$

Fixing the number q $(2/\sqrt{5} < q < 1)$, we select $S(x_1)$ such that

$$r(x_1) > qR_1.$$

Then we apply transfinite induction. Let us assume that for all ordinals $\alpha < \beta < \Omega$ the balls $S(x_\alpha)$ have been chosen.

If $\bigcup_{\alpha<\beta} S(x_\alpha) \supset A$, then the process of selecting balls is terminated. Otherwise, set

$$R_\beta = \sup_{x \in A \setminus A_\beta} r(x), \quad A_\beta = \bigcup_{\alpha<\beta} S(x_\alpha),$$

and choose $S(x_\beta)$ such that

$$x_\beta \in A \setminus A_\beta, \quad r(x_\beta) > qR_\beta.$$

Since the centers $\{x_\alpha\}$ obviously form an isolated set, the latter is countable, and consequently, the process will terminate for some $\beta_0 < \Omega$, i.e. we will obtain a system $\{S(x_\alpha)\}_{\alpha<\beta_0}$ which covers A.

The system $\{S(x_\alpha)\}_{\alpha<\beta_0}$ has the following property which follows from the construction:

If $S(x_{\alpha_1})$ and $S(x_{\alpha_2})$ are two balls of this system then either

(i) $x_{\alpha_1} \notin S(x_{\alpha_2})$ and $x_{\alpha_2} \notin S(x_{\alpha_1})$, or

(ii) only one of these centers, say x_{α_1} belongs to $S(x_{\alpha_2})$ $(\alpha_1 < \alpha_2)$.

In case (ii)

$$q < \frac{r(x_{\alpha_1})}{r(x_{\alpha_2})} < 1,$$

since $qR_{\alpha_1} < r(x_{\alpha_1}) < r(x_{\alpha_2}) \leq R_{\alpha_1}$.

Let us estimate the multiplicity of a system of balls satisfying condition (i) or (ii). Together with the balls $S(x_\alpha)$ we consider concentric balls $s(x_\alpha)$ of radius $\sqrt{1-q^2}\, r(x_\alpha)$. The balls $s(x_\alpha)$ pairwise do not intersect. Indeed, if $s(x_{\alpha_1}) \cap s(x_{\alpha_2}) \neq \emptyset$, then

$$|x_{\alpha_1} - x_{\alpha_2}| < 2\sqrt{1-q^2} \max[r(x_{\alpha_1}), r(x_{\alpha_2})].$$

But in case (i)

$$|x_{\alpha_1} - x_{\alpha_2}| > \max[r(x_{\alpha_1}), r(x_{\alpha_2})],$$

and in case (ii)

$$|x_{\alpha_1} - x_{\alpha_2}| > \min[r(x_{\alpha_1}), r(x_{\alpha_2})] = r(x_{\alpha_1}) > qr(x_{\alpha_2}) = q\max[r(x_{\alpha_1}), r(x_{\alpha_2})].$$

Since $2\sqrt{1-q^2} < q$, we reach a contradiction and $s(x_{\alpha_1}) \cap s(x_{\alpha_2})$ is empty.

We now note that if

$$y \in S(x_{\alpha_1}) \cap S(x_{\alpha_2})$$

and case (i) holds, then the angle between the segments yx_{α_1} and yx_{α_2} is greater than $\pi/3$. If case (ii) holds, then there is a lower bound on the angle between yx_{α_1} and yx_{α_2}, if $y \notin s(x_{\alpha_1})$. Under this assumption an elementary calculation shows that the angle is greater that

$$\min\left[\arccos\sqrt{1-q^2}, \arccos\frac{1}{2q}\right] = \arccos\frac{1}{2q} = \frac{\pi}{3} - \eta.$$

Now suppose that

$$y \in \bigcap_{k=1}^{N} S(x_{\alpha_k}). \tag{3.4.3}$$

If in this case y belongs to any of the balls $s(x_{\alpha_k})$ (only one!), say $y \in s(x_{\alpha_N})$, then we eliminate the corresponding index in (3.4.3) and write

$$y \in \bigcap_{k=1}^{N-1} S(x_{\alpha_k}).$$

Now to any pair $S(x_{\alpha_k})$, $S(x_{\alpha_i})$ we may apply the above bound on the angle, i.e. the angle between yx_{α_k} and yx_{α_i} is greater than $\pi/3 - \eta$. But the maximal number $n(p, \eta)$ of such segments yx_α emerging from one point y is determined only by the dimension p and the quantity η which is connected with the choice of q and is an absolute constant.[17] Consequently,

$$N \leq n(p, \eta) + 1$$

and the lemma is proved.[18] ▯

Let us return to the proof of the theorem.

We have seen that with each point $x \in A$ there is associated a sequence r_k for which (3.4.2) holds. If we consider balls of radius $r_k(x)$ with centers at the corresponding points, then we will obtain a covering of A to which the above lemma is applicable. According to the lemma and the ine-

---

[17] For sufficiently small η $n(2, \eta) = 6$ and $n(3, \eta) = 12$. The latter fact is the content of a nice theorem due to Schütte and Van der Waerden (cf. Tot [1]). Succeeding values of $n(p, \eta)$ are apparently unknown.

[18] It is useful to note that when the number of segments yx_α is actually equal to $n(p, \eta)$ then the ball $S(x_{\alpha_N})$ is completely covered by the balls $S(x_{\alpha_k})$ $k = 1, 2, \ldots, N-1$, and therefore can be excluded from the basic system $\{S(x_\alpha)\}_{\alpha<\beta_0}$. Therefore, the multiplicity of this system in fact does not exceed $n(p, \eta)$.

quality (3.4.2), for any $\varepsilon > 0$ there exists a covering of A by balls of radius $r_k < \varepsilon$ having multiplicity no greater than $N(p)$, where for each ball $S(x_k)$

$$h(r_k) < \lambda(r_k, x_k).$$

Then

$$\sum_k h(r_k) < N(p)\,\lambda(1)$$

and, consequently,

$$m_h(A) \le N(p)\,\lambda(1).$$

Since this bound holds for any function $h(r)$ with convergent integral (3.4.1), for any positive constant k

$$m_{kh}(A) \le N(p)\,\lambda(1).$$

But obviously

$$m_{kh}(A) = k m_h(A)$$

and therefore,

$$m_h(A) \le \frac{1}{k} N(p)\,\lambda(1).$$

Letting k tend to infinity, we obtain the assertion of the theorem. $\square$

10. One can give sufficient conditions for null α-capacity by using certain functions $h(r)$ for which the integral (3.4.1) diverges.

Theorem 3.14. *If for $h(r) = r^{p-\alpha}$ (correspondingly, $h(r) = (\ln 1/r)^{-1}$) the h-measure of a set E is finite, then its inner α-capacity (correspondingly, logarithmic capacity) is equal to zero.*

Proof. For definiteness we consider the case $\alpha < p$, and suppose that $\underline{C}_\alpha(E) > 0$. On E consider the measure μ with finite energy:

$$I_\alpha(\mu) = \int_{E\times E} \frac{d\mu(x)\,d\mu(y)}{|x-y|^{p-\alpha}} < \infty.$$

Writing this integral in the form

$$I_\alpha(\mu) = \int_0^\infty \frac{d\sigma(r)}{r^{p-\alpha}},$$

where

$$\sigma(r) = \int_E \mu(r; x)\,d\mu(x),$$

we see that there is a decreasing kernel $K(r)$ which at $r = 0$ has a singularity stronger than $r^{\alpha-p}$ (i.e. such that $r^{p-\alpha} K(r) \to \infty$ as $r \to 0$), and such that

$$\int_0^\infty K(r)\, d\sigma(r) < \infty .$$

This means that with respect to the kernel $K(|x-y|)$ the set E has positive capacity. Therefore (cf. Ch. II, § 1, No. 3) there exists on E a measure μ_1 for which the K-potential is bounded on $S(\mu_1)$, i.e.

$$\int_E K(|x-y|)\, d\mu_1(y) \leq M, \quad x \in S(\mu_1).$$

If B is any ball of diameter d, then for $x \in B \cap S(\mu_1)$, we have

$$K(d)\, \mu_1(B) \leq \int_{S(\mu_1)} K(|x-y|)\, d\mu_1(y) \leq M .$$

Therefore, if $B_1, B_2, \ldots, B_n$ is a covering of E by balls of diameters $<\varepsilon$, then

$$\sum_{i=1}^n K^{-1}(d_i) \geq \frac{1}{M} \sum_{i=1}^n \mu_1(B_i) \geq \frac{1}{M} \mu_1(E) .$$

But $K(r)\, r^{p-\alpha} \to \infty$ as $r \to 0$ and therefore for sufficiently small ε the sum $\sum_{i=1}^n d_i^{p-\alpha}$ will be arbitrarily large. Thus, for the function $h(r) = r^{p-\alpha}$, $m_h(E) = \infty$, and this implies the validity of the theorem. ☐

11. The Covering Lemma 3.2 proved in No. 9 enables us to obtain certain *upper bounds* for potentials of measures. In the case of measures with finite energy the corresponding bounds were obtained rather simply in Ch. II (Lemma 2.5).

For an arbitrary measure μ in R^p, $\mu(1) < \infty$ the following theorem holds.

Theorem 3.15. *Suppose that $h(r)$, $0 \leq r \leq r_0$ is a monotone increasing function normalized by the conditions*

$$h(0) = 0,\ h(r_0) = \mu(1)$$

Then an open set G of points on which is satisfied the inequality

$$U_\alpha^\mu(x) > \int_0^{r_0} \frac{dh(r)}{r^{p-\alpha}} \tag{3.4.4}$$

can be covered by a system of balls with radii $r_k \leq r_0$, satisfying the inequality

$$\sum_k h(r_k) < N(p)\, \mu(1) .$$

Proof. Denote by $S(r; x)$ the ball of radius r with center at the point x and set

$$\mu(r, x) = \mu[S(r; x)].$$

Then

$$U_\alpha^\mu(x) = \int_0^\infty \frac{d\mu[(r, x)}{r^{p-\alpha}}.$$

If we assume that for a given value of x, the inequality

$$\mu(r, x) \leq h(r)$$ [19]

holds for all $r > 0$, then using integration by parts we obtain the inequality

$$U_\alpha^\mu(x) \leq \int_0^{r_0} \frac{dh(r)}{r^{p-\alpha}}.$$

Hence for each point x at which the inequality (3.4.4) is satisfied there is an $r = r(x) < r_0$ such that

$$h(r) < \mu(r, x).$$

We have obtained consequently, a covering of G by balls to which we can apply Lemma 3.2. Choosing a covering of multilpicity not greater than $N(p)$ and denoting by r_k the radii of the balls of this covering, we have

$$\sum_k h(r_k) < \sum_k \mu(r_k, x_k) < N(p)\, \mu(1). \quad \square$$

A completely analogous result holds for the logarithmic potential.

Corollary 1. *For any $\beta > p - \alpha$, the point set G where the inequality*

$$U_\alpha^\mu(x) > m > 0$$

is satisfied can be covered by a system of balls with radii r_k satisfying the inequality

$$\sum_k r_k^\beta < eN(p) \left(\frac{M}{m}\right)^{\frac{\beta}{p-\alpha}}, \quad M = \mu(1) \tag{3.4.5}$$

For the proof of (3.4.5) we set

$$h(r) = M\left(\frac{r}{r_0}\right)^\beta.$$

[19] For $r > r_0$ one must set $h(r) \equiv h(r_0) = \mu(1)$.

The conditions $h(0) = 0$, $h(r_0) = M$ are satisfied, and the integral

$$\int_0^{r_0} \frac{dh(r)}{r^{p-\alpha}} = \frac{M\beta}{\alpha + \beta - p} \frac{1}{r_0^{p-\alpha}}$$

is finite. Defining r_0 by the equation

$$m = \frac{M\beta}{\alpha + \beta - p} \frac{1}{r_0^{p-\alpha}}$$

and using Theorem 3.15, we obtain (3.4.5).

The analogous calculation for the logarithmic potential in R^2

$$U^\mu(x) = \int_{R^2} \ln \frac{1}{|x-y|} \, d\mu(y)$$

shows that the set where $U^\mu(x) > m$ for any $\beta > 0$ can be covered by a system of discs with radii r_k, which satisfy the inequality

$$\sum_k r_k^\beta < 6e \left(e^{-\frac{m}{M}}\right)^\beta.$$

Corollary 2. *Suppose $E \subset R^p$ is a set for which $\overline{C}_\alpha(E) = C$, $0 < C < \infty$. Then for any $\beta > p - \alpha$ it can be covered by a system of balls with radii r_k for which*

$$\sum_k r_k^\beta < 3N(p)\, C^{\frac{\beta}{p-\alpha}}.$$

The corresponding result for sets in the plane says: if $0 < C < \infty$ is the outer logarithmic capacity of a set $E \subset R^2$, then for any $\beta > 0$, it can be covered by a system of discs with radii r_k such that

$$\sum_k r_k^\beta < 18 C^\beta.$$

For the proof consider the outer equilibrium measure γ^* on E. By Theorem 2.7

$$U_\alpha^{\gamma^*}(x) \geq 1 \qquad \text{quasi-everywhere on } E.$$

Let G_ε be the set of points of R^p where $U_\alpha^{\gamma^*}(x) > 1 - \varepsilon$ $(\varepsilon > 0)$. Then

$$E \subset G_\varepsilon \cup E_0,$$

where $\overline{C}_\alpha(E) = 0$. By Corollary 1 G_ε is covered by a system of balls with radii r_k such that

$$\sum_k r_k^\beta < eN(p) \left(\frac{C}{1-\varepsilon}\right)^{\frac{\beta}{p-\alpha}},$$

and for sufficiently small $\varepsilon > 0$

$$\sum_k r_k^\beta < 2.9N(p)\, C^{\frac{\beta}{p-\alpha}}. \tag{3.4.6}$$

For E_0, according to Theorem 3.13 it has h-measure zero if $h(r) = r^\beta$, since the integral

$$\int_0 \frac{r^\beta}{r^{p+1-\alpha}}\, dr$$

converges. Hence, for any $\varepsilon > 0$ it can be covered by a system of balls with radii r_k such that

$$\sum_k r_k^\beta < \varepsilon.$$

Combining this with (3.4.6), we obtain the required result. For a set in the plane the proof is completely analogous.

Chapter IV

Balayage, Green Functions, and the Dirichlet Problem

§ 1. Classical Balayage out of a Region

As in the preceding chapters, by the classical case we shall mean either the case of the Newtonian potential ($\alpha = 2$) for $p \geq 3$, or the case of the logarithmic potential for $p = 2$.

1. In R^p ($p \geq 2$) consider a region G with *compact boundary* ∂G. Let ν be a signed measure in $\overline{G}$, i.e. $S(\nu) \subset \overline{G}$. *The problem of balayage* (or "sweeping out") consists of defining a new signed measure ν' on ∂G ($S(\nu') \subset \partial G$) such that

$$U^{\nu}(x) = U^{\nu'}(x) \tag{4.1.1}$$

for all or at least quasi-all points $x \notin G$.

In the case when equation (4.1.1) holds at *all* points $x \notin G$, we shall say that the balayage problem has been solved *in the strict sense.*

We construct the signed measure ν' relying on two fundamental theorems of functional analysis: the Hahn-Banach theorem on the extension of linear functionals and the theorem of F. Riesz concerning the general form of a linear functional in a space of continuous functions[1].

Denote by $C_{\overline{G}}$ the linear space of real functions $f(x)$ continuous in the *closed* region $\overline{G}$, normed in the usual way:

$$\|f\|_{\overline{G}} = \max_{x \in \overline{G}} |f(x)|.$$

Let $H_{\overline{G}}$ denote the linear subspace of $C_{\overline{G}}$ consisting of all functions which are *harmonic* in G. If G contains the point at infinity, then for $p \geq 3$ we require that elements $h \in H_{\overline{G}}$ satisfy the condition $h(\infty) = 0$. We note that if

$$C_2(\partial G) = C_2(CG) = 0,$$

then it follows easily from Theorem 3.3 that the class $H_{\overline{G}}$ is trivial, i.e. consists only of one function identically equal to zero. On the other

[1] Cf. F. Riesz and B. Nagy [1].

hand, if

$$C_2(\partial G) > 0$$

then, as shown by the corollary to Theorem 3.7, the class $H_{\overline{G}}$ will be nontrivial.

In the future, unless it is otherwise explicitly stated, we shall assume that $C_2(\partial G) > 0$.

By the maximum principle, if $h(x) \in H_{\overline{G}}$, then

$$\|h\|_{\overline{G}} = \max_{x \in \overline{G}} |h(x)| = \max_{x \in \partial G} |h(x)|.$$

Finally, $C_{\partial G}$ will denote the linear space of all *functions continuous* in ∂G with the norm

$$\|f\|_{\partial G} = \max_{x \in \partial G} |f(x)|.$$

According to the preceding remark and Harnack's theorem, $H_{\overline{G}}$ is isomorphic to a linear subspace $H_{\partial G}$ of the space $C_{\partial G}$($H_{\partial G}$ coincides with $C_{\partial G}$, if the Dirichlet problem for G is solvable for any continuous boundary function).

Suppose now that ν with $S(\nu) \subset \overline{G}$, is a given finite signed measure[2]. The formula

$$\nu(h) = \int_{\overline{G}} h(x)\, d\nu(x)$$

defines a linear functional on $H_{\overline{G}}$ and, consequently, on the isomorphic space $H_{\partial G}$. Since

$$|\nu(h)| \leq \|h\|_{\overline{G}} \int |d\nu| = \|h\|_{\overline{G}} \, |\nu| \, (\overline{G}),$$

the norm of this functional on H_G does not exceed $|\nu| \, (\overline{G})$.

By the Hahn-Banach theorem this functional can be extended to the whole space $C_{\partial G}$ without increasing the norm; this extension will be denoted by $\nu'(f)$ ($f \in C_{\partial G}$). By the Riesz representation theorem, the functional $\nu'(f)$ has the form

$$\nu'(f) = \int_{\partial G} f(x)\, d\nu'(x), \qquad f \in C_{\partial G},$$

where ν' is a signed measure on ∂G ($S(\nu') \subset \partial G$). Also

$$\|\nu'\| = |\nu'| \, (\partial G) \leq |\nu| \, (\overline{G}).$$

Thus, we have proved the following theorem.

---

[2] That is, $\int_{R^p} |d\nu| < \infty$. For $p \geq 3$, this condition can be made somewhat weaker.

Theorem 4.1. *For any signed measure ν with $S(\nu) \subset \overline{G}$, one can find a corresponding signed measure ν', with $S(\nu') \subset \partial G$, such that for any function $h(x) \in H_{\overline{G}}$*

$$\int_{\partial G} h(x)\, d\nu'(x) = \int_{\overline{G}} h(x)\, d\nu(x). \tag{4.1.2}$$

Moreover $|\nu'|\,(\partial G) \leq |\nu|\,(\overline{G})$.

We shall call such a signed measure ν' *associated with* ν *relative* to G. In case $C_2(\partial G) = 0$ we set $\nu' = 0$.

We investigate especially the case when ν is a *measure* μ. Then the functional $\mu(h)$ is *positive* and by a known theorem[3] it also admits a *positive* extension $\mu'(f)$. This establishes the validity of

Theorem 4.2. *For any measure μ, $S(\mu) \subset \overline{G}$, there is a measure μ', $S(\mu') \subset \partial G$, such that for any function $h(x) \in H_{\overline{G}}$*

$$\int_{\partial G} h(x)\, d\mu'(x) = \int_{\overline{G}} h(x)\, d\mu(x). \tag{4.1.3}$$

We will show that $\mu'(\partial G) \leq \mu(\overline{G})$, where the sign $<$ holds only for $p \geq 3$ and for an infinite region G.

In fact, if we exclude the latter case, then $h(x) \equiv 1 \in H_{\overline{G}}$ and, choosing this function in the equation (4.1.2), we obtain that $\mu'(\partial G) = \mu(\overline{G})$.

For the case $p \geq 3$ and an infinite region G we denote by G_N the part of G inside the sphere $|x| = N$. Let μ_N be the restriction of μ to G_N, μ'_N its associated measure, λ'_N its restriction to ∂G. Since $\lambda'_N(\partial G) \leq \mu'_N(\partial G) \leq \mu_N(\overline{G}_N) \leq \mu(\overline{G})$, then the family of measures $\{\lambda'_N\}$ is vaguely compact (cf. Theorem 0.6), and one can assume that λ'_N vaguely converges to some measure μ' as $N \to \infty$. Moreover it is easy to see that $\mu'(\partial G) = \lim_{N\to\infty} \lambda'_N(\partial G) \leq \mu(\overline{G})$. We show that μ' is a measure associated with μ relative to G. Since $H_{\overline{G}} \subset H_{\overline{G}_N}$, for $h(x) \in H_{\overline{G}}$ we have the equation

$$\int_{\overline{G}_N} h(x)\, d\mu = \int_{\partial G} h(x)\, d\lambda'_N + \int_{|x|=N} h(x)\, d\mu'_N.$$

As $N \to \infty$, the last integral tends to zero, since $\mu'_N(1) \leq \mu(1)$ and $h(x) \to 0$ at infinity. Hence, after passing to the limit we obtain

$$\int_{\overline{G}} h(x)\, d\mu = \int_{\partial G} h(x)\, d\mu'.$$

Let us assume that for any function $f(x) \in C_{\partial G}$ there exists $h(x) \in H_{\overline{G}}$ which coincides with $f(x)$ on ∂G. In other words, let us assume that the

[3] Cf., for example, N. I. Akhiezer [2].

Dirichlet problem for the region G *has a solution* for any continuous boundary conditions. Such regions G will be called *regular*. In this case $H_{\partial G} \equiv C_{\partial G}$ and the associated measure μ' is *uniquely determined* by the measure μ for any μ, $S(\mu) \subset \overline{G}$. If G is not regular, then $H_{\partial G} \subset C_{\partial G}$, and consequently, there is a linear functional $\nu(f)$, $\nu \not\equiv 0$, $f \in C_{\partial G}$, orthogonal to $H_{\partial G}$, i.e. for all $h(x) \in H_{\partial G}$

$$\nu(h) = \int_{\partial G} h(x)\, d\nu = 0 .$$

Putting

$$\mu = \nu^+, \ \mu' = \nu^-, \quad \text{where} \quad \nu = \nu^+ - \nu^-,$$

we have $S(\mu) \subset \partial G \subset \overline{G}$, $S(\mu') \subset \partial G$ and for $h(x) \in H_{\overline{G}}$

$$\int_{\partial G} h(x)\, d\mu' = \int_{\partial G} h(x)\, d\mu .$$

Thus, the region G is regular if, and only if, for any measure μ, $S(\mu) \subset \overline{G}$, the associated measure μ' is uniquely determined.

From this obtains an interesting consequence. Suppose G is regular, $G_1 \subset G$, ∂G_1 the boundary of G_1. If for a measure μ, $S(\mu) \subset \overline{G}_1$, we consider two associated measures: μ' on ∂G and μ_1' on ∂G_1, and if we denote by μ_1'' the measure associated with μ_1' relative to G, then $\mu' \equiv \mu_1''$. Indeed, $H_{\overline{G}} \subset H_{\overline{G}_1}$, and hence for any $h(x) \in H_{\overline{G}}$

$$\int_{\partial G} h(x)\, d\mu' = \int_{S(\mu)} h(x)\, d\mu = \int_{\partial G_1} h(x)\, d\mu_1' = \int_{\partial G} h(x)\, d\mu_1'' .$$

By uniqueness, $\mu' \equiv \mu_1''$.

As we see, this reasoning is based on the uniqueness of the associated measure for a *given* measure μ. Later we shall see that for a rather large class of measures, this uniqueness of the associated measure μ' holds *without the regularity requirement* on the region G (cf. § 4, No. 16).

We now point out several corollaries of Theorem 4.2.

Corollary 1. *Suppose $p \geq 3$. If $x \in C\overline{G}$,*[4] *then*

$$U^{\mu'}(x) = U^{\mu}(x) .$$

Indeed, the function $h(y) = |x - y|^{2-p}$ is in $H_{\overline{G}}$ and by (4.1.2)

$$U^{\mu'}(x) = \int_{\partial G} \frac{d\mu'(y)}{|x-y|^{p-2}} = \int_{\overline{G}} \frac{d\mu(y)}{|x-y|^{p-2}} = U^{\mu}(x) .$$

[4] CE will always denote the complement of the set E.

Corollary 2. *Suppose $p = 2$ and G is a finite region. Then for $x \in C\overline{G}$*

$$U^{\mu'}(x) = U^{\mu}(x).$$

It is sufficient to set $h(y) = \ln |x - y| \in H_{\overline{G}}$ in (4.1.2).

Corollary 3. *Suppose $p = 2$ and G is an infinite region. Then for $x \in C\overline{G}$*

$$U^{\mu'}(x) = U^{\mu}(x) + \text{const.} \tag{4.1.4}$$

Indeed, if $x_1, x_2 \in C\overline{G}$ then

$$h(y) = \ln \frac{1}{|x_1 - y|} - \ln \frac{1}{|x_2 - y|} \in H_{\overline{G}}.$$

Consequently,

$$U^{\mu'}(x_1) - U^{\mu'}(x_2) = U^{\mu}(x_1) - U^{\mu}(x_2),$$

which yields (4.1.4).[5]

In general the constant in (4.1.4) is distinct from zero. For example, consider G to be the exterior of the unit disc: $|x| > 1$ and for μ take the unit mass uniformly distributed on the circle $|x| = 2$. Then for μ' one can take the unit mass uniformly distributed on the circle $|x| = 1$. Indeed, for any function $h(x) \in H_{\overline{G}}$ we have

$$\int_0^{2\pi} h(e^{i\varphi})\, d\varphi = \int_0^{2\pi} h(2e^{i\varphi})\, d\varphi,$$

which is equivalent to (4.1.2). Meanwhile, for $|x| < 1$

$$U^{\mu}(x) = \ln \frac{1}{2}, \quad U^{\mu'}(x) = 0.$$

From Theorem 4.2 we obtain, under certain restrictions on the structure of ∂G, a solution of the balayage problem in the strict sense, and, as a corollary, the regularity of the region G.

Let us assume the following condition of Poincaré:

(P) *Every point $x \in \partial G$ is the vertex of a cone of revolution K_x:*

$$y_1^2 \geq \gamma \sum_{k=2}^{p} y_k^2; \quad y_1 \geq 0 \text{ [6]}$$

which in a certain neighborhood of the poind x lies outside of $\overline{G}$. We shall also say that each point of ∂G *is accessible from outside of G* by a cone.

---

[5] Below (No.4) the constant in the formula (4.1.4) will be expressed through the Green function for the region G.

[6] Here $(y_1, y_2, \ldots, y_p)$ are certain local Cartesian coordinates of the point x.

Theorem 4.3. *If the condition* (P) *is satisfied, the associated measure* μ' *solves the balayage problem in the strict sense.*[7]

Proof. Suppose that $y \to x$ along the axis of the cone K_x. We shall show that

$$\lim_{y \to x} U^\mu(y) = U^\mu(x); \qquad \lim_{y \to x} U^{\mu'}(y) = U^{\mu'}(x).$$

Since the argument is exactly the same in both cases, we shall give the proof for $U^\mu(x)$. If $U^\mu(x) = +\infty$, then the result follows from the lower semicontinuity of $U^\mu(x)$. If

$$U^\mu(x) = \int_{S(\mu)} \frac{d\mu(z)}{|z-x|^{p-2}} < \infty,$$

then, recalling that for $z \notin K_x$ and y on the axis of K_x

$$|z-y| \geq \sqrt{\gamma}\, |z-x|,$$

we obtain

$$\frac{1}{|z-y|^{p-2}} \leq \frac{1}{\gamma^{p/2-1}} \frac{1}{|z-x|^{p-2}}.$$

Thus, the family of functions $\{|z-y|^{2-p}\}$ of the variable z is majorized by a function integrable relative to the measure μ; by Lebesgue's theorem

$$\lim_{y \to x} U^\mu(y) = \int_{S(\mu)} \frac{d\mu(z)}{|z-x|^{p-2}} = U^\mu(x).$$

Using Corollaries 1 and 2, we obtain for $p \geq 3$ or for $p = 2$ and G finite

$$U^{\mu'}(x) = U^\mu(x) \qquad \text{for all } x \notin G$$

and thus, the measure μ' solves the balayage problem for the measure μ.[8] ▯

Remark 1. Suppose $p \geq 3$, K a compact set in R^p, and G_0 that connected component of CK which includes ∞. Assume that G_0 satisfies condition (P), i.e. every point of the boundary of G_0 is accessible by a cone from outside of G_0. Then the theorem just proved implies the solvability (in the strict sense) of the Robin problem for K, i.e. the existence of an equilibrium measure γ on K for which

$$U^\gamma(x) \equiv 1 \qquad \text{throughout } K.$$

---

[7] Excluding the case $p = 2$ and G, infinite. In this regard, cf. Remark 3 below.

[8] By the Jordan decomposition, this also solves (under the same conditions) the balayage problem for signed measures.

To construct γ, consider the sphere Σ_R: $|x| = R$, containing K and on it a measure μ with constant density, for which

$$U^\mu(x) \equiv 1 \quad \text{for} \quad |x| \leq R$$

If γ results from balayage of μ onto the boundary of G_0, then $U^\gamma \equiv 1$ in CG_0 and, in particular on K. It is useful to note that the support of γ lies on the subset of K which is the boundary of G_0; this subset will be called the *external boundary* of K.

Remark 2. In the case $p = 2$, the preceding argument needs only slight modification. As before we consider the circle $|x| = R$, containing K. Let λ be a unit mass uniformly distributed on this circle. Since for $|x| < R$ $U^\gamma(x) = \ln 1/R$ then, by (4.1.4),

$$U^{\lambda'}(x) \equiv \text{const on } K$$

where $\lambda'(1) = \lambda(1) = 1$. The numerical value of the constant is obtained according to (2.4.2), Ch. II, No. 15:

$$U^{\lambda'}(x) = \ln \frac{1}{C_l(K)}, \quad x \in K,$$

where $C_l(K)$ is the logarithmic capacity of K.

Remark 3. Let us return to the problem of balayage out of an infinite region of the plane which satisfies condition (P). Reasoning just as in the proof of Theorem 4.3, we obtain that $U^{\mu'}(x) = U^\mu(x) + \text{const}$ everywhere in K. If we assume that $C_l(K) \neq 1$, then one can define the coefficient α such that on K

$$U^{\mu' + \alpha\lambda'}(x) = U^\mu(x).$$

Hence, balayage out of an infinite plane region G onto its boundary ∂G is possible in the case $C_l(\partial G) \neq 1$; however, the result of sweeping out a *measure* μ turns out, in general, to be the signed measure $\mu' + \alpha\lambda'$. If $C_l(\partial G) = 1$, then balayage may not be possible, as seen from the example in this section.

2. In the next two sections we will show that if the balayage problem for the region G is solvable in the strict sense, then the region G is regular, and consequently, the associated measure μ' is always unique. In particular, from this it will follow that if condition (P) is satisfied at every point of the boundary ∂G, then the region G is regular.

Ignoring for the moment the case of an infinite region in the plane, let us introduce the *Green function* $G(x, y)$ for the region G — a notion which will later be generalized and often utilized.

Thus, suppose that the balayage problem for the region G is solvable in the strict sense.

Consider the Dirac measure ε_y, where $y \in G$. By the hypothesis, there exists a measure ε'_y with support in ∂G such that

$$U^{\varepsilon_y}(x) = U^{\varepsilon'_y}(x) \quad \text{for all } x \notin G.$$

If $p \geq 3$, this means that

$$\frac{1}{|x-y|^{p-2}} = U^{\varepsilon'_y}(x) \quad \text{for all } x \notin G,$$

and if $p = 2$, that

$$\ln \frac{1}{|x-y|} = U^{\varepsilon'_y}(x) \quad \text{for all } x \notin G.$$

This measure ε'_y will be called the *Green measure* (for the point y and the region G).

Note that the potential $U^{\varepsilon'_y}(x)$, continuous on the support of the measure ε'_y, will be everywhere continuous (cf. Theorem 1.7). In particular, this implies the C-absolute continuity and *uniqueness of the Green measure* (cf. Ch. III, No. 1).

Set

$$G(x, y) = U^{\varepsilon_y}(x) - U^{\varepsilon'_y}(x).$$

This function is equal to zero for $x \notin G$, is superharmonic in G and harmonic for $x \neq y$. This implies that within G $G(x, y) > 0$[9]. We will show that

$$G(x, y) = G(y, x).$$

It is sufficient to verify that $U^{\varepsilon'_y}(x) = U^{\varepsilon'_x}(y)$ (for $x, y \in G$). But this can be seen from the following:

$$U^{\varepsilon'_y}(x) = \int_{\partial G} \frac{d\varepsilon'_y(z)}{|x-z|^{p-2}} = \int_{\partial G} U^{\varepsilon_x}(z)\, d\varepsilon'_y(z)$$

$$= \int_{\partial G} U^{\varepsilon'_y}(z)\, d\varepsilon'_x(z) = \int_{\partial G} \frac{d\varepsilon'_x(z)}{|y-z|^{p-2}} = U^{\varepsilon'_x}(y).$$

Suppose now that $y \to y_0 \in \partial G$. We show that

$$\varepsilon'_y \to \varepsilon_{y_1}. \tag{4.1.5}$$

By Theorem 0.4 it is sufficient to verify the relation

$$\varepsilon'_y(\varphi) \to \varphi(y_0)$$

[9] One should keep in mind that for infinite G we assume $p \geq 3$ and $G(\infty, y) = 0$.

for any class of functions $\varphi(x)$ dense in Φ^+. Consider functions $\varphi(x)$ with compact supports having three continuous partial derivatives. These functions admit a representation

$$\varphi(x) = \int_{R^p} \frac{\psi(z)}{|z-x|^{p-2}}\,dz \quad (p \geq 3)$$

or

$$\varphi(x) = \int_{R^2} \psi(z) \ln \frac{1}{|z-x|}\,dz \quad (p = 2)$$

with the smooth function $\psi(z) = \text{const}\,\Delta\varphi$. For such $\varphi(x)$ we have (for definiteness we write out the case $p \geq 3$)

$$\begin{aligned}\varphi(y_0) - \varepsilon_y'(\varphi) &= \int_{R^p} \psi(z) \left[\frac{1}{|z-y_0|^{p-2}} - U^{\varepsilon_y'}(z)\right] dz\\ &= \int_{R^p} \psi(z) \left[\frac{1}{|z-y|^{p-2}} - U^{\varepsilon_y'}(z)\right] dz\\ &\quad + \int_{R^p} \psi(z) \left[\frac{1}{|z-y_0|^{p-2}} - \frac{1}{|z-y|^{p-2}}\right] dz.\end{aligned}$$

As $y \to y_0$ the second integral tends to zero. The first integral may be rewritten in the form

$$\int_G \psi(z) \left[\frac{1}{|z-y|^{p-2}} - U^{\varepsilon_y'}(z)\right] dz = \int_G \psi(z)\, G(y, z)\, dz$$

and since

$$\left|\int_G \psi(z)\, G(y, z)\, dz\right| < \max_G |\psi(z)| \int_G G(y, z)\, dz$$

one must show that

$$\lim_{y \to y_0} \int_G G(y, z)\, dz = 0.$$

However

$$\begin{aligned}\int_G G(y, z)\, dz &= \int_{|y_0-z|<\delta} G(y, z)\, dz + \int_{|y_0-z|>\delta} G(y, z)\, dz\\ &< \int_{|y_0-z|<\delta} \frac{dz}{|z-y|^{p-2}} + \int_{|y_0-z|>\delta} G(y, z)\, dz.\end{aligned}$$

For sufficiently small δ the first integral becomes arbitrarily small uniformly in y, while the second approaches zero as $y \to y_0$.[10]

---

[10] For $|y - y_0| < \delta/2$, $G(y, z) < (2/\delta)^{p-2}$ the theorem on passing to the limit is also applicable.

Remark. In the proof of (4.1.5) we used the convergence of $G(x, y)$ to zero at only *one point*, namely at y_0 (but for any x).

Now it is easy to solve the Dirichlet problem for any continuous function $f(x)$ on ∂G and hence prove the regularity of the region G.

The solution is given by the formula

$$V(y) = \int_{\partial G} f(x)\, d\varepsilon_y'(x) \quad (y \in G).$$

First of all, if $y \to y_0 \in \partial G$, then by (4.1.5)

$$\lim_{y \to y_0} V(y) = f(y_0).$$

We will show that $V(y)$ is a harmonic function in G. Note that if $f_n(x) \to f(x)$ uniformly on ∂G, then

$$V_n(y) = \int_{\partial G} f_n(x)\, d\varepsilon_y'(x)$$

converges to $V(y)$ uniformly in $\overline{G}$. This follows from

$$|V(y) - V_n(y)| \leq \int_{\partial G} |f_n(x) - f(x)|\, d\varepsilon_y'(x) \leq \max_{\partial G} |f_n(x) - f(x)|.$$

Consequently, it is sufficient to establish the harmonicity of $V(y)$ for any class of functions dense in $C_{\partial G}$. Such a class, for example, is the set of functions $\{|x - z|^{2-p}\}$ $(p \geq 3)$ (correspondingly, $\{\ln |x - z|\}$ for $p = 2$), depending on the parameter $z \notin \partial G$.[11] For these functions

$$V_z(y) = \int_{\partial G} \frac{d\varepsilon_y'(x)}{|x - z|^{p-2}} = \begin{cases} \dfrac{1}{|y - z|^{p-2}} & \text{for } z \notin \overline{G} \\ U^{\varepsilon_y'}(z) = U^{\varepsilon_z'}(y) & \text{for } z \in G \end{cases}$$

is, obviously, a harmonic function of y.

It is useful to note that with the help of the Green measure, one can write down the associated measure μ' for an arbitrary measure μ, $S(\mu) \subset \overline{G}$. Namely, we show that

$$\mu' = \int_{S(\mu)} \varepsilon_y'\, d\mu(y).\,^{12} \tag{4.1.6}$$

---

[11] To prove that this set is dense in $C_{\partial G}$ one can, for example, do the following. Thrice differentiable (in a neighborhood of ∂G) functions $\varphi(x)$ with compact support form a set dense in $C_{\partial G}$. Each such function, as we have pointed out, admits a representation

$$\varphi(x) = \int_{R^p} \frac{\psi(z)}{|x - z|^{p-2}}\, dz.$$

From this the required result easily follows.

[12] Of course, the integral (4.1.6) is defined since the family of measures $\{\varepsilon_y'\}$, $y \in \overline{G}$ is Borel (cf. Introduction, No. 3).

or

$$\mu'(\varphi) = \int_{S(\mu)} \varepsilon_y'(\varphi)\, d\mu(y)\,, \quad \varphi \in \Phi. \tag{4.1.6'}$$

Indeed, if $h(x) \in H_{\overline{G}}$, then

$$\int_{\partial G} h(x)\, d\mu'(x) = \int_{S(\mu)} d\mu(y) \int_{\partial G} h(x)\, d\varepsilon_y'(x) = \int_{S(\mu)} h(y)\, d\mu(y)\,,$$

and this shows that the measure μ', defined by formula (4.1.6) is in fact associated with μ relative to G.

Actually, we have proved somewhat more than was formulated at the beginning of this section; namely, it has been established that *if the balayage problem has a solution in the strict sense for any Dirac measure ε_y, $y \in G$, then the region G is regular.*

We now prove that the measure μ' obtained after balayage (in the strict sense) of the measure μ, coincides with the associated measure (and consequently, is unique).

For any function $h(x) \in H_{\overline{G}}$ we have

$$h(x) = \int_{\partial G} h(y)\, d\varepsilon_x'(y)\,.$$

Integrating this equation over the measure μ and using (4.1.6), we obtain

$$\int_{\overline{G}} h(x)\, d\mu(x) = \int_{\partial G} h(y)\, d\mu'(y)$$

as required.

In concluding this section, we establish the *positiveness of the kernel* $G(x, y)$. We shall prove that for any signed measure $\nu = \nu^+ - \nu^-$, $S(\nu) \subset \overline{G}$

$$\iint_{\overline{G} \times \overline{G}} G(x, y)\, d\nu(x)\, d\nu(y) \geq 0\,, \tag{4.1.7}$$

where equality holds only if $S(\nu) \subset \partial G$. The only conditions that we impose on the signed measure are the finiteness conditions

$$\iint_{\overline{G} \times \overline{G}} G(x, y)\, d\nu^{\pm}(x)\, d\nu^{\mp}(y) < \infty\,,$$

which imply that the integral in (4.1.7) is defined and finite.[13] In other words, we are considering signed measures with *finite energy*. We set

$$\lambda = \nu - \nu'$$

---

[13] It is easy to see the sufficiency of the conditions

$$\iint_{\overline{G} \times \overline{G}} G(x, y)\, d\nu^+(x)\, d\nu^-(y) < \infty,$$

analogous to the condition (1.4.2), Ch.I.

and show that

$$\iint_{\overline{G}\times\overline{G}} G(x, y)\, d\nu(x) d\nu(y) = \iint_{\overline{G}\times\overline{G}} \frac{d\lambda(x)\, d\lambda(y)}{|x-y|^{p-2}} = \|\lambda\|^2 \quad (p \geq 3),$$

or correspondingly

$$\iint_{\overline{G}\times\overline{G}} G(x, y)\, d\nu(x)\, d\nu(y) = \iint_{\overline{G}\times\overline{G}} \ln \frac{1}{|x-y|}\, d\lambda(x)\, d\lambda(y) = \|\lambda\|^2 \quad (p = 2)$$

(in the latter case, we are for the moment assuming G to be a finite region).

In fact,

$$\|\lambda\|^2 = (\nu - \nu', \nu - \nu') = \|\nu\|^2 - 2(\nu', \nu) + \|\nu'\|^2.$$

But

$$(\nu', \nu) = \int_{\partial G} U^{\nu}(x)\, d\nu' = \int_{\partial G} U^{\nu'}(x)\, d\nu' = \|\nu'\|^2,$$

so that

$$\|\lambda\|^2 = \|\nu\|^2 - (\nu', \nu).$$

Furthermore, by formula (4.1.6′),

$$(\nu', \nu) = \int_{\overline{G}} U^{\nu}(x)\, d\int_{\overline{G}} \varepsilon_y'(x)\, d\nu(y) = \int_{\overline{G}} d\nu(y) \int_{\overline{G}} U^{\nu}(x)\, d\varepsilon_y'(x)$$

$$= \int_{\overline{G}} d\nu(y) \int_{\overline{G}} U^{\varepsilon_y'}(x)\, d\nu(x),$$

where the validity of all operations follows from the assumed finiteness conditions. Therefore,

$$\|\lambda\|^2 = \iint_{\overline{G}\times\overline{G}} \{U^{\varepsilon_y}(x) - U^{\varepsilon_y'}(x)\}\, d\nu(x)\, d\nu(y) = \iint_{\overline{G}\times\overline{G}} G(x, y)\, d\nu(x)\, d\nu(y).$$

This implies the positiveness of the kernel $G(x, y)$.[14] Moreover, the last integral equals zero only if

$$\lambda \equiv 0, \qquad \nu \equiv \nu'$$

i.e. in case $S(\nu) \subset \partial G$.

3. Consider now the case of an *infinite region G in the plane*. Just as above, for the Dirac measure ε_y, $y \in G$, let ε_y' be the measure on ∂G associated with it according to Theorem 4.2. By Remark 3, No. 1.

$$\ln \frac{1}{|x-y|} = U^{\varepsilon_y}(x) = U^{\varepsilon_y'}(x) + a(y), \quad x \notin G,$$

where $a(y)$ does not depend on x. Also $\varepsilon_y'(1) = 1$.

---

[14] For $p = 2$, recall that $\lambda(1) = 0$ (cf. Ch.I, Theorem 1.16).

The difference

$$G(x, y) = \ln \frac{1}{|x-y|} - U^{\varepsilon'_y}(x) - a(y)$$

will be called the Green function for the region G. As before, the function $G(x, y) = 0$ for $x \notin G$, is superharmonic in G, harmonic for $x \neq y$, and therefore, within G $G(x, y) > 0$. It is not difficult to see that

$$\lim_{x\to\infty} \left[\ln \frac{1}{|x-y|} - U^{\varepsilon'_y}(x)\right] = 0$$ [15]

and consequently,

$$G(x, y) = \ln \frac{1}{|x-y|} - U^{\varepsilon'_y}(x) + G(\infty, y).$$

Furthermore, for $x, y \in G$

$$\begin{aligned}
U^{\varepsilon'_y}(x) &= \int_{\partial G} \ln \frac{1}{|x-z|}\, d\varepsilon'_y(z) \\
&= \int_{\partial G} [U^{\varepsilon'_x}(z) - G(\infty, x)]\, d\varepsilon'_y(z) \\
&= \int_{\partial G} U^{\varepsilon'_y}(z)\, d\varepsilon'_x(z) - G(\infty, x) \\
&= \int_{\partial G} \left[\ln \frac{1}{|y-z|} + G(\infty, y)\right] d\varepsilon'_x(z) - G(\infty, x) \\
&= U^{\varepsilon'_x}(y) + G(\infty, y) - G(\infty, x),
\end{aligned}$$

whence

$$G(x, y) = G(y, x).$$

This leads to a natural definition of the Green function with a pole at infinity:

$$G(x, \infty) = \lim_{y\to\infty} G(x, y) = G(\infty, x).$$

Since $G(y, x)$ is a harmonic function of $x \in G$, $x \neq y$, and $G(y, x) \to G(\infty, x)$ uniformly in any finite region of the x-plane, then $G(x, \infty)$ will be a harmonic function in G. Also, $G(x, \infty) = O$ for $x \notin G$. Now in the equation

$$G(x, \infty) = G(x, y) - \ln \frac{|x|}{|x-y|} + U^{\varepsilon'_x}(y) + \ln |x|$$

suppose that the point $y \in G$ is fixed and $x \to \infty$. Since the first three terms of the right side are bounded, we have

$$G(x, \infty) = \ln |x| + O(1) \quad (x \to \infty).$$

[15] Uniformly in any finite region of the y-plane.

By the theorem on removable singularities, the harmonic (in G) function $G(x, \infty) - \ln |x|$ will have a limit as $x \to \infty$, and this means that

$$\lim_{x \to \infty} \left\{ G(x, y) - \ln \frac{|x|}{|x - y|} + U^{\varepsilon'_x}(y) \right\} = L$$

exists and is independent of y. Consequently,

$$\lim_{x \to \infty} U^{\varepsilon'_x}(y) = L - G(\infty, y) \quad (y \in G).$$

Since the family of measures $\{\varepsilon'_x\}$ is vaguely compact, one may assume that $\varepsilon'_x \dashrightarrow \lambda$, with

$$U^{\lambda}(y) = L - G(\infty, y).$$ [16]

Letting y approach an arbitrary point of the boundary ∂G, we obtain

$$U^{\lambda}(y) \equiv L \quad \text{for} \quad y \in \partial G.$$

Consequently, the measure λ solves (in the strict sense) the Robin problem for ∂G (and, of course, for the complement of G); therefore, it is unique and (cf. Remark 2, No. 1)

$$L = -\ln C_l(\partial G),$$

where C_l is the logarithmic capacity.

Thus,

$$G(x, \infty) = -\ln C_l(\partial G) - U^{\lambda}(x) = \ln |x| - \ln C_l(\partial G) + O(1),$$

and we have obtained for regular regions the results of Ch. II, No. 17.

Finally, with inessential changes we can repeat the argument in No. 2 to find that

$$\varepsilon'_y \dashrightarrow \varepsilon_{y_0} \quad \text{as} \quad y \to y_0 \in \partial G$$

and that the formula

$$V(y) = \int_{\partial G} f(x)\, d\varepsilon'_y(x) \quad (y \in G)$$

solves the Dirichlet problem for the infinite region G.

Returning to the general problem of sweeping out a measure μ from an infinite region G in the plane, we point out the following formula which sharpens Remark 3 of No. 1:

$$U^{\mu'}(x) = U^{\mu}(x) + \int_{S(\mu)} G(y, \infty)\, d\mu(y) \quad (x \in \partial G).$$

---

[16] We shall see that the limiting measure λ is unique; hence no choice is necessary and $\varepsilon'_x \dashrightarrow \lambda$ for arbitrary $x \to \infty$.

Also, by the same method as in No. 2, one can establish the following formula for an infinite region of the plane

$$\iint\limits_{\bar{G}\times\bar{G}} G(x, y)\, d\nu(x)\, d\nu(y) = \iint\limits_{\bar{G}\times\bar{G}} \ln\frac{1}{|x-y|}\, d\lambda(x)\, d\lambda(y) = ||\lambda||^2,$$

where ν is a signed measure with finite energy, and $\lambda = \nu - \nu'$. This formula implies the positiveness of the kernel $G(x, y)$.

4. In order to compare the potentials of measures $U^{\mu}(x)$, $U^{\mu'}(x)$ *within* a regular region G, we shall need the following theorem.

Theorem 4.4. *Let G be a finite regular region of R^p $(p \geq 2)$ and let $f(x)$ be a superharmonic function in $\bar{G}$.*[17] *If μ is a measure in $\bar{G}$, and μ' its associated measure on ∂G, then*

$$\int\limits_{\bar{G}} f(x)\, d\mu(x) \geq \int\limits_{\partial G} f(x)\, d\mu'(x). \tag{4.1.8}$$

Proof. Suppose that $f(x)$ is continuous on ∂G. Since G is regular, there is a function $h(x) \in H_{\bar{G}}$ satisfying the boundary condition $h(x) = f(x)$ on ∂G. Within G we have the inequality $h(x) \leq f(x)$ (cf. Ch. I, No. 4). Using (4.1.2) we obtain

$$\int\limits_{\partial G} f(x)\, d\mu' = \int\limits_{\partial G} h(x)\, d\mu' = \int\limits_{\bar{G}} h(x)\, d\mu \leq \int\limits_{\bar{G}} f(x)\, d\mu.$$

In the general case, $f(x)$ is the limit of the monotone increasing sequence (cf. (1.2.4)) $f_n(x) = f(x) * m^{(1/n)}$ of continuous superharmonic functions. Passing to the limit in the inequality

$$\int\limits_{\bar{G}} f_n(x)\, d\mu \geq \int\limits_{\partial G} f_n(x)\, d\mu'$$

by means of the Levi theorem, we obtain (4.1.8). □

Theorem 4.4′. *Suppose that G is a region in R^p $(p \geq 3)$ containing the point at infinity, and let $f(x)$ be a function superharmonic in $\bar{G}$, non-negative in the neighborhood of infinity. Then again the inequality* (4.1.8) *holds.*

Proof. As in the preceding theorem, it is essentially sufficient to consider the case when $f(x)$ is continuous on ∂G. Let $h \in H_{\bar{G}}$ be a solution of the external Dirichlet problem with boundary condition $h(x) = f(x)$ on ∂G. We will show that $h(x) \leq f(x)$ everywhere in G. Let S_N be a sphere of radius N with center 0. For sufficiently large N it lies within

[17] That is, in a certain region $G_1 \supset \bar{G}$.

G and the inequality $h(x) < \varepsilon$ $(\varepsilon > 0)$ holds on S_N. Suppose that G_N is the subset of G within S_N. Let us apply the minimum principle to G_N and the superharmonic function $f(x) - h(x)$. Since on the whole boundary of G_N we have $f(x) - h(x) > -\varepsilon$, then for any point of G_N, $h(x) < f(x) + \varepsilon$. Fixing the point x, we let N approach infinity and let ε tend to zero. This yields the inequality $h(x) \leq f(x)$. Since $h(x) \in H_{\overline{G}}$ then

$$\int_{\partial G} f(x)\, d\mu' = \int_{\partial G} h(x)\, d\mu' = \int_{\overline{G}} h(x)\, d\mu \leq \int_{\overline{G}} f(x)\, d\mu,$$

as required. ☐

Now suppose $y \in G$ and

$$f(x) = |x - y|^{2-p} \quad (p \geq 3),$$

correspondingly,

$$f(x) = \ln \frac{1}{|x - y|} \quad (p = 2).$$

Since this is a superharmonic function, satisfying for $p \geq 3$ the usual condition at infinity, the preceding theorems are applicable and this yields the inequality

$$U^{\mu}(y) \geq U^{\mu'}(y) \quad (y \in G).$$

Thus, excluding the case of an infinite region in the plane, we can assert the following result.

By the balayage of a measure μ from a region G, satisfying condition (P), *onto its boundary, the potential inside G does not increase.*

We note in conclusion that if we exclude the trivial case when $S(\mu) \subset \partial G$, then it would be easy in the above formulation to replace the statement "does not increase" by "decreases". This is connected with the fact that for the theorems of this section, the equality sign holds either for $S(\mu) \subset \partial G$, or if $f(x)$ is harmonic.

5. Equation (4.1.2), established in No. 1 for functions $h(x) \in H_{\overline{G}}$ holds also for certain harmonic functions which are *discontinuous* on ∂G. Indeed, suppose $h_n(x) \in H_{\overline{G}}$ form a *monotone* sequence. Since by the Levi theorem, on both sides of (4.1.2) one can pass to the limit, then, setting $\omega(x) = \lim_{n \to \infty} h_n(x)$ we will have

$$\int_{\partial G} \omega(x)\, d\mu' = \int_{\overline{G}} \omega(x)\, d\mu. \tag{4.1.9}$$

By Harnack's theorem $\omega(x)$ is harmonic in G (or $\equiv \infty$). On the boundary ∂G $\omega(x)$ is semicontinuous.

Assume that the region G is regular. Then for $\omega(x)$, $x \in \partial G$ in equation (4.1.9) one can take any semicontinuous function, and, in particular, the characteristic function $\chi_E(x)$ of any closed or open set $E \subset \partial G$. As is known, the corresponding harmonic function $\omega(x) = \omega(x, E; G)$ is called the harmonic measure of the set E relative to the region G. Thus

$$\mu'(E) = \int_{\bar{G}} \omega(x, E; G)\, d\mu(x).^{18} \tag{4.1.10}$$

and, in particular,

$$\mu'(\partial G) = \int_{\bar{G}} \omega(x, \partial G; G)\, d\mu(x) = \int_{\bar{G}} U^{\gamma}(x)\, d\mu(x)$$

and

$$\varepsilon_x'(\partial G) = U^{\gamma}(x),$$

where γ is the equilibrium measure of the compact set ∂G (cf. Ch. II, No. 3). If G is an infinite region in R^p ($p \geq 3$), then $\omega(x, \partial G; G) < 1$ inside G, and therefore

$$\mu'(\partial G) < \mu(\bar{G})$$

if only $S(\mu) \cap G \neq \emptyset$. This is true, as we shall see below (No. 7) without the regularity assumption on G.

6. A second interesting application of (4.1.9) is obtained by setting $\omega(x) = |x - y|^{2-p}$ for $y \notin G$. Since this function is harmonic in G and semicontinuous in $\bar{G}$, then, applying (4.1.9), we obtain

$$\int_{\partial G} \frac{d\mu'(x)}{|x - y|^{p-2}} = \int_{\bar{G}} \frac{d\mu(x)}{|x - y|^{p-2}}, \quad y \notin G,$$

i.e.

$$U^{\mu'}(y) = U^{\mu}(y) \quad \text{everywhere outside } G.$$

Hence, for regular regions, the associated measure μ' solves the balayage problem in the strict sense.

Combining this with the results of No. 2, we arrive at the following theorem.

Theorem 4.5. *G is regular if, and only if the balayage problem for the measure μ, $S(\mu) \subset \bar{G}$, has a solution in the strict sense.*

---

[18] By repeated monotone passage to the limit, this formula may be proved for any Borel set $E \subset \partial G$.

Remark. Setting $\mu = \varepsilon_{x_0}$, $x_0 \in G$ in (4.1.10) we obtain

$$\omega(x_0, E; G) = \varepsilon'_{x_0}(E), \quad x_0 \in G,$$

where ε'_{x_0} is the Green measure for the point x_0.

The formula

$$\mu'(E) = \int_G \omega(x, E; G)\, d\mu(x) + \int_{\partial G} \omega(x, E; G)\, d\mu(x)$$

leads to another corollary. Assume that the restriction μ_∂ of the measure μ to the boundary ∂G is C-absolutely continuous. Then the associated measure μ' is also C-absolutely continuous. Indeed, if a compact set $K \subset \partial G$ has capacity zero, then by Theorem 3.2 its absolute harmonic measure is equal to zero and therefore $\omega(x, K; G) \equiv 0$ for $x \in G$ and $\omega(x, K; G) = \chi_K(x)$ for $x \in \partial G$. Thus,

$$\mu'(K) = \int_{\partial G} \chi_K(x) d\mu(x) = \mu_\partial(K) = 0.$$

§ 2. Balayage for Arbitrary Compact Sets

7. The approximation technique presented in this section will allow us to extend the results of the preceding sections to arbitrary compact sets $K \subset R^p$ ($p \geq 2$).

Any compact set $K \subset R^p$ can be viewed as the limit of a monotone decreasing sequence of compact sets

$$K_1 \supset K_2 \supset \cdots \supset K; \quad K = \bigcap_{i=1}^{\infty} K_i$$

such that all of the complements of K_i (i.e. the connected components of the open set CK_i) are regular. Such compact sets K_i we will briefly call *regular*. They may be constructed, for example, in the following way. Enclose each point $x \in K$ in an open ball $S_i(x)$ of radius $1/i$ and set

$$K_i = \overline{\bigcup_{x \in K} S_i(x)}.$$

The regularity of K_i follows since at each boundary point of K_i condition (P) holds. The monotonicity of K_i and the equation $K = \bigcap_i K_i$ are obvious.[19]

We now carry out an argument which applies equally well to the case $p \geq 3$ and the case $p = 2$, if only the infinite region G_0 bounded by K does not contain any mass μ. The case when G_0 contains a mass μ will be examined at the end of this section.

---

[19] It is not difficult to prove that CK_i consists of a finite number of regions which, of course may be unbounded as $i \to \infty$.

Suppose μ is an arbitrary finite measure. Set

$$CK_i = \bigcup_{j=0}^{N_i} G_j^{(i)},$$

where $G_j^{(i)}$ ($j = 0, 1, \ldots, N_i$) are the connected components of CK_i where $G_0^{(i)}$ contains the point at infinity; let $\mu_j^{(i)}$ be the part of the measure μ which lies in $G_j^{(i)}$. Also write

$$\nu_i = \mu - \sum_{j=0}^{N_i} \mu_j^{(i)}$$

and let $\mu_j'^{(i)}$ be the measure associated with $\mu_j^{(i)}$ relative to $G_j^{(i)}$ and

$$\mu_i' = \nu_i + \sum_{j=0}^{N_i} \mu_j'^{(i)}.$$

Then it is not difficult to see that

$$U^{\mu_i'}(x) = U^{\mu}(x), \quad x \in K_i, \tag{4.2.0}$$

and by No. 4 and the corollary at the end of No. 1, we have

$$U^{\mu_{i+1}'}(x) \leq U^{\mu_i'}(x) \leq U^{\mu}(x)$$

for all x. Applying Theorem 3.10 to the monotone decreasing sequence $\{U^{\mu_i'}(x)\}$, we obtain

$$\mu_i' \rightharpoonup \mu'$$

and

$$\lim_{k\to\infty} U^{\mu_k'}(x) = U^{\mu'}(x) \qquad \text{quasi-everywhere.}$$

On the other hand, for $x \in K$, as we have seen,

$$U^{\mu_i'}(x) = U^{\mu}(x).$$

Consequently, quasi-everywhere in K

$$U^{\mu'}(x) = U^{\mu}(x) \tag{4.2.1}$$

and everywhere

$$U^{\mu'}(x) \leq U^{\mu}(x). \tag{4.2.1'}$$

Moreover, it is not difficult to see (cf. Introduction (0.1.10)), that $S(\mu') \subset K$ (and even $S(\mu') \subset \partial K$). Hence, the measure μ' solves the balayage problem for the measure μ on the compact set K.

The constructed measure μ' is also *accociated* with the measure μ *relative* to CK. This means the following. Suppose that $H_{\overline{CK}}$ is the class of functions continuous in $\overline{CK}$ (for $p \geq 3$, equal to zero at infinity) and harmonic in CK. Then for $h(x) \in H_{\overline{CK}}$

$$\int_{\overline{CK}} h(x)\, d\mu = \int_{\partial K} h(x)\, d\mu' \tag{4.2.2}$$

or, what is the same,

$$\int_{CK} h(x)\, d\mu = \int_{\partial K} h(x)\, d(\mu - \mu_K)',$$

where $\operatorname{var} \mu' \leq \operatorname{var} \mu$.

Analogously, one defines the *signed measure* ν' associated with ν relative to CK.

Since $H_{\overline{CK}} \subset H_{\overline{G_j^{(i)}}}$ for all j, i, then

$$\int h(x)\, d\mu_j^{(i)} = \int h(x)\, d\mu_j'^{(i)},$$

and summing on j and adding integrals over the measure ν_i, we obtain

$$\int_{\overline{CK}} h(x)\, d\mu = \int_{K_i} h(x)\, d\mu_i'.$$

But $\mu_i' \rightharpoonup \mu'$, $S(\mu') \subset \partial K$; this implies (4.2.2).

Similarly Theorem 4.4′ is generalized:

Suppose $f(x)$ is a function superharmonic in $\overline{CK}$, non-negative near infinity (for $p \geq 3$); then

$$\int_{\overline{CK}} f(x)\, d\mu \geq \int_{\partial K} f(x)\, d\mu'.$$

Indeed, since $f(x)$ is superharmonic in $\overline{G_j^{(i)}}$ for all j, i, then

$$\int_{\overline{G_j^{(i)}}} f(x)\, d\mu_j^{(i)} \geq \int_{K_i} f(x)\, d\mu_j'^{(i)}.$$

From this, as above,

$$\int_{\overline{CK}} f(x)\, d\mu \geq \int_{\overline{CK}} f(x)\, d\mu_i'.$$

By the lower semicontinuity of $f(x)$, $\mu_i' \rightharpoonup \mu'$ implies that

$$\int_{\partial K} f(x)\, d\mu' = \int_{\overline{CK}} f(x)\, d\mu' \leq \varliminf_{i\to\infty} \int_{\overline{CK}} f(x)\, d\mu_i'$$

(cf. Introduction, Lemma 0.1), and this proves the assertion.

As in the regular case, with each point $y \notin K$ we associate a *Green measure* ε'_y, which is the vague limit of Green measures $\varepsilon'^{(i)}_y$, constructed for regular compact sets K_i[20] (in the next section we shall be convinced of the independence of ε'_y from the choice of the sequence K_i). If we consider a measure μ outside of K, i.e. a measure such that $S(\mu) \cap K = \emptyset$, then formula (4.1.6) of No. 2 still holds. Indeed, for any $\varphi \in \Phi$

$$\mu'_i(\varphi) = \int_{S(\mu)} \varepsilon'^{(i)}_y(\varphi)\, d\mu(y),$$

and since $|\varepsilon'^{(i)}_y(\varphi)| \leq \|\varphi\|$, one can pass to the limit as $i \to \infty$; this yields

$$\mu'(\varphi) = \int_{S(\mu)} \varepsilon'_y(\varphi)\, d\mu(y)$$

or

$$\mu' = \int_{S(\mu)} \varepsilon'_y\, d\mu(y). \tag{4.2.3}$$

In particular, if $\varphi \equiv 1$ in a neighborhood of K, then

$$\mu'(K) = \int_{S(\mu)} \varepsilon'_y(K)\, d\mu(y).$$

The quantity $\varepsilon'_y(K)$, $y \in CK$ is easily expressed in terms of the potential $U^\gamma(x)$ of the equilibrium measure γ of the compact set K. In fact, $U^\gamma(x) = 1$ quasi-everywhere in K; on the other hand, the potential $U^{\varepsilon'_y}$ is bounded, and hence the measures ε'_y and γ are C-absolutely continuous. This enables us to write the equation

$$\begin{aligned}\varepsilon'_y(K) &= \int_K U^\gamma(x)\, d\varepsilon'_y(x) = \int_K U^{\varepsilon'_y}(x)\, d\gamma(x)\\ &= \int_K U^{\varepsilon_y}(x)\, d\gamma(x) = U^\gamma(y),\end{aligned} \tag{4.2.3'}$$

whence

$$\mu'(K) = \int_{S(\mu)} U^\gamma(y)\, d\mu(y). \tag{4.2.4}$$

As above (No. 5), this permits us to conclude that if G is an infinite region of R^p ($p \geq 3$) and $S(\mu) \cap G \neq \emptyset$,then independent of the regularity of G

$$\mu'(K) < \mu(\overline{G}).$$

Concluding this section, let us consider the case which was earlier excluded: $p = 2$ and G_0 includes some mass μ.

---

[20] The family of measure $\{\varepsilon'_y\}$, $y \notin K$ is easily seen to be Borel.

The construction of the measure μ_i' remains unchanged. But in place of the equation (4.2.0), we will have, according to No. 3,

$$U^{\mu_i'}(x) = U^{\mu}(x) + \int_{G_0^{(i)}} G_i(y, \infty)\, d\mu(y), \quad y \in K_i, \tag{4.2.5}$$

where $G_i(y, \infty)$ is the Green function for the infinite region G_0^i with a pole at infinity. As $i \to \infty$, the functions $G_i(y, \infty)$ form, as seen from the maximum principle, a monotone increasing sequence. Therefore,

$$G_0(y, \infty) = \lim_{i \to \infty} G_i(y, \infty)$$

exists. This non-negative function harmonic in G_0 will also be called a Green function (with pole at infinity).

Thus,

$$\lim_{i \to \infty} U^{\mu_i'}(x) = U^{\mu}(x) + \int_{G_0} G_0(y, \infty)\, d\mu(y)$$

exists on K. The set of measures μ_i' is weakly bounded ($\mu_i'(1) = \mu(1)$) and is therefore vaguely compact (Theorem 0.6). Again denoting by μ_i' a vagualy convergent subsequence $\mu_i' \to \mu$, we apply Theorem 3.8 according to which

$$U^{\mu'}(x) = \underline{\lim}_{i \to \infty} U^{\mu_i'}(x)$$

quasi-everywhere. This implies that quasi-everywhere on K

$$U^{\mu'}(x) = U^{\mu}(x) + \int_{G_0} G_0(y, \infty)\, d\mu(y). \tag{4.2.6}$$

Soon we shall be convinced (cf. No. 8) that

$$U^{\mu'}(x) \leq U^{\mu}(x) + \int_{G_0} G_0(y, \infty)\, d\mu.$$

everywhere.

8. Let us turn to the question of *uniqueness* of the measure μ' obtained after balayage of μ on the compact set K.

First of all, the strengthened form of the uniqueness theorem (cf. Ch. III, No. 1) immediately shows that there cannot exist two distinct C-absolutely continuous measures which solve the balayage problem for the measure μ.

On the other hand, we will see now that the construction of the preceding section applied to $\mu - \mu_K$ where μ_K is the restriction of μ to K, always yields a C-absolutely continuous measure.

Indeed, we let

$$\mu - \mu_K = \sum_{i=0}^{\infty} \varkappa_i,$$

where $\varkappa_0$ is the part of μ outside of K_1, and $\varkappa_i$ $(i \geq 1)$ is the part of μ in $K_i \setminus K_{i+1}$. The potential $U^{\varkappa_i}$ is bounded on K; the same holds, consequently, for $U^{\varkappa'_i}$ and therefore, the measure $\varkappa'_i$ is C-absolutely continuous. But then the sum $\sum_{i=0}^{\alpha} \varkappa'_i$ will be C-absolutely continuous. The assertion will be proved if we establish that

$$(\mu - \mu_K)' = \sum_{i=0}^{\infty} \varkappa'_i. \tag{4.2.7}$$

Slightly changing the notation of No. 7, we write $(\mu - \mu_K)^{(j)}$ and $\varkappa_i^{(j)}$, for the result of the balayage of $\mu - \mu_K$ and $\varkappa_i$, respectively on the regular compact set K_j (here, as above, the part of the corresponding measure lying on K_j is not subject to change). It is clear that $\varkappa_i^{(j)} = \varkappa_i$ for $i \geq j$, and

$$(\mu - \mu_K)^{(j)} = \sum_{i=0}^{\infty} \varkappa_i^{(j)}.$$

On the other hand, as $j \to \infty$ by definition

$$(\mu - \mu_K)^{(j)} \stackrel{\cdot}{\to} (\mu - \mu_K)', \quad \varkappa_i^{(j)} \stackrel{\cdot}{\to} \varkappa'_i$$

and moreover,

$$\sum_{i=0}^{\infty} \varkappa_i^{(j)}(1) \leq \sum_{i=0}^{\infty} \varkappa_i(1) = (\mu - \mu_K)\,(1) < \infty$$

and

$$\sum_{i=0}^{\infty} \varkappa'_i(1) \leq \sum_{i=0}^{\infty} \varkappa_i(1) < \infty. \tag{4.2.8}$$

We must prove that

$$\sum_{i=0}^{\infty} \varkappa_i^{(j)} \stackrel{\cdot}{\to} \sum_{i=0}^{\infty} \varkappa'_i,$$

i.e. that for any function $\varphi(x) \in \Phi$

$$\lim_{j\to\infty} \sum_{i=0}^{\infty} \varkappa_i^{(j)}(\varphi) = \sum_{i=0}^{\infty} \varkappa'_i(\varphi). \tag{4.2.9}$$

By (4.2.8) for any fixed $\eta > 0$ we can define $N = N(\eta)$ such that

$$\left|\sum_{N+1}^{\infty} \varkappa_i^{(j)}(\varphi)\right| < \eta, \quad \left|\sum_{N+1}^{\infty} \varkappa'_i(\varphi)\right| < \eta.$$

But

$$\sum_{i=0}^{N} \varkappa_i^{(j)}(\varphi) \to \sum_{i=0}^{N} \varkappa'_i(\varphi)$$

and this together with the preceding inequalities proves (4.2.9).

Thus, we have

Theorem 4.6. *For any finite measure μ the construction of No. 7 does not depend on the choice of the regular compact sets K_i and yields a unique measure μ' which solves the balayage problem and has the property that $\mu' - \mu_K$ is C-absolutely continuous.*

Corollary 1. *If μ_K is C-absolutely continuous, then the balayage of the measure μ is unique in the class of C-absolutely continuous measures.*

Corollary 2. *If the potential $U^\mu(x)$ is bounded on K, then the measure μ which solves the balayage problem is unique.*

Indeed, if $U^\mu(x) < M$ on K, then $U^{\mu'}(x) < M$ quasi-everywhere on K. By lower semicontinuity of the potential $U^{\mu'}(x)$, the last inequality holds everywhere on K and consequently, the measure μ' is C-absolutely continuous. This proves, in particular, the uniqueness of the Green measure ε'_y.

Corollary 3. *If $U^\mu(x) < \infty$ on K, then the balayage is unique in the class of measures μ' which everywhere decrease the potential, i.e. measures such that*

$$U^\mu(x) \geq U^{\mu'}(x)$$

everywhere.

Indeed, under these conditions $U^{\mu'}(x) < \infty$ on K and hence everywhere; therefore, μ' is also C-absolutely continuous.

Until the end of this section, by the balayage of a measure μ we shall mean the measure μ' specified in Theorem 4.6. With this convention, we have

Corollary 4. *Suppose that $K \subset K_1$. Then the measure μ' which is the result of balayage of μ on K, can be obtained first by balayage of μ onto K_1, and then by balayage of the resulting measure μ_1' onto K.*

Formula (4.2.7), which can be rewritten in the form

$$\mu' = \mu_K + \sum_{i=0}^{\infty} \varkappa'_i,$$

enables us to supplement the study of the potential after balayage out of an infinite region of the plane (No. 7). Namely, we prove that

$$U^{\mu'}(x) \leq U^\mu(x) + \int_{G_0} G_0(y, \infty)\, d\mu(y). \tag{4.2.10}$$

everywhere. First, applying (4.2.6) to $\varkappa_i$ we obtain that quasi-everywhere on K

$$U^{\varkappa_i'}(x) = U^{\varkappa_i}(x) + \int_{G_0} G_0(y, \infty)\, d\varkappa_i(y) \tag{4.2.11}$$

and since the potential $U^{\varkappa_i}(x)$ is bounded on K, then $U^{\varkappa_i'}(x)$ is also bounded. This implies that $\varkappa_i'$ is a measure with finite energy (consequently, is C-absolutely continuous) and equation (4.2.11) holds $\varkappa_i'$ almost everywhere. By the second maximum principle (Ch. I, No. 18) the inequality

$$U^{\varkappa_i'}(x) \leq U^{\varkappa_i}(x) + \int_{G_0} G_0(y, \infty)\, d\varkappa_i(y).$$

holds everywhere. Summing these inequalities over i and adding $U^{\mu_K}(x)$ to both sides, we obtain (4.2.10).

Remark 1. Theorem 4.6 and its basic corollaries can be generalized to the case when K is an arbitrary *closed set* $F \subset R^p$. One must look at the compact sets $K_n = F \cap T_n$, where T_n is the closed ball $|x| \leq n$ and define, according to Theorem 4.6, the measures μ_n' which solve the balayage problem on K_n. Then, using the vague compactness of the family $\{\mu_n'\}$ one can find a measure μ' with support on F, which solves the balayage problem on F.

Remark 2. If the compact set K has capacity zero: $C_2(K) = 0$, then Theorem 4.6 implies that $\mu' - \mu_K = 0$. In particular, if $\mu_K = 0$, then $\mu' = 0$.

9. Suppose that $K \subset R^p$ $(p \geq 2)$ is an arbitrary compact set, G_i, $i = 0, 1, 2, \ldots$ the connected components of its complement. For each G_i we define the *generalized Green function* with pole at $y \in G_i$ by the formula

$$G_i(x, y) = U^{\varepsilon_y}(x) - U^{\varepsilon_y'}(x).$$

Only in the case $p = 2$ and $i = 0$ this definition will be slightly changed. In that case we put

$$G_0(x, y) = U^{\varepsilon_y}(x) - U^{\varepsilon_y'}(x) + G_0(\infty, y),$$

where $G(\infty, y)$ was defined in No. 7.

It is not difficult to see, recalling the construction of ε_y' (No. 7), that $G_i(x, y)$ is, inside G_i, the increasing limit of Green functions for regular regions. Using the results of Nos. 2 and 3, it is easy to prove the symmetry and positiveness of $G_i(x, y)$.

In particular, this implies that

$$U^{\varepsilon_y'}(x) = U^{\varepsilon_x'}(y), \quad x, y \in G_i,$$

excluding the case of an infinite region of the plane.[21] This property is called *the reciprocity law for Green measures.*

Furthermore, from the general properties of measures obtained through balayage (cf. No. 7), it follows that $G_i(x, y)$, as a function of the point x, has the properties:

$G_i(x, y) > 0$ in G_i,

$G_i(x, y) \geq 0$ everywhere on ∂G_i,

$G_i(x, y) = 0$ quasi-everywhere on ∂G_i.

Theorem 4.7. *Suppose that for a certain point $x_0 \in \partial G_i$,*

$$G_i(x_0, y) = 0. \tag{4.2.12}$$

Then for any other position of the pole $y' \in G_i$

$$G_i(x_0, y') = 0.$$

Proof. First, since $G_i(x, y)$ is upper semicontinuous near the boundary ∂G_i, condition (4.2.12) implies that

$$\lim_{x \to x_0} G_i(x, y) = 0.$$

Consider an arbitrary sequence of points $x_n \in G_i$, $x_n \to x_0$. The functions $G_i(x_n, y)$ on any closed ball T with center y which is contained in G_i, beginning from some n_0, are harmonic positive functions of y. For any point y' inside T we have the so-called Harnack inequality[22]

$$G_i(x_n, y') \leq \frac{R^{p-2}(R + r)}{(R - r)^{p-1}} G_i(x_n, y),$$

where R is the radius of T and $r = |y' - y| < R$. From this we see that

$$\lim_{n \to \infty} G_i(x_n, y') = 0.$$

---

[21] In this case (cf. No.3)

$$U^{\varepsilon'_y}(x) + G(\infty, x) = U^{\varepsilon'_x}(y) + G(\infty, y).$$

[22] It easily follows from the Poisson integral formula.

$$v(y') = \frac{1}{\omega_p R} \int_{|z-y|=R} v(z) \frac{R^2 - r^2}{\varrho^p} dS_z, \quad \varrho = |z - y'|, \quad r = |y' - y|,$$

if we apply it to the non-negative harmonic function $G_i(x_n, y)$.

Considering a finite chain of balls which connect the point y with any fixed point $y' \in G_i$, we come to the conclusion that

$$\lim_{n\to\infty} G_i(x_n, y') = 0$$

for $y' \in G_i$ and any sequence $x_n \to x_0$, $x_n \in G_i$.

We shall show that this implies the equation

$$G_i(x_0, y') = 0.$$

Indeed, since $G_i(x, y')$ is a subharmonic function of the point x in a neighborhood of x_0, then (Ch. I, No. 4, (1.2.4))

$$G_i(x_0, y') = \lim_{R\to 0} G_i * m^{(R)} = \lim_{R\to 0} \frac{1}{R^{p-1}\omega_p} \int_{T_R} G_i(x, y')\, dx$$

$$= \lim_{R\to 0} \frac{1}{R^{p-1}\omega_p} \int_{T_R \cap G_i} G_i(x, y')\, dx,$$

where T_R is a ball of radius R with center at x_0. Therefore, if it were true that $G_i(x_0, y') > 0$ then we could construct a sequence of points $x_n \in G_i$, $x_n \to x_0$ such that

$$\underline{\lim}_{n\to\infty} G_i(x_n, y') > 0.$$

This proves the theorem. ▯

Utilizing the remark made at the end of No. 8, one can construct the Green function for a region G with non-compact boundary ∂G.

10. In this section we introduce a very important concept in potential theory: the notion of an *irregular point of a compact set*. Later (Ch. V) this concept will be studies in greater detail and will be generalized.

Definition. *A point $x_0 \in K$ is called an irregular point of the compact set K, if*

(*i*) *it belongs to the boundary ∂G_i of one of the connected components G_i of the complement CK.*

(*ii*) *for some $y \in G_i$*

$$G_i(x_0, y) > 0.$$

By the preceding theorem this definition is correct, since the choice of the point $y \in G_i$ is irrelevant.

The *set of irregular points of* K will be denoted by I_K; by the definition

$$I_K = \bigcup_{i=0}^{\infty} I_K^{(i)},$$

where

$$I_K^{(i)} = I_K \cap \partial G_i.$$

Since $G_i(x, y) = 0$ quasi-everywhere on ∂G_i we have the so-called

Kellogg Lemma. *The set I_K of all irregular points of a compact set has capacity zero:*

$$C_2(I_K) = 0 \quad (\text{correspondingly } C_l(I_K) = 0 \quad \text{for} \quad p = 2).$$

The points of the set $K \setminus I_K$ will be called the regular points of K.

We give one nesessary and sufficient condition for the regularity of a boundary point x of the region G. We define a *barrier* for the point $x \in \partial G$ to be a function $V_x(y)$ defined on a part of G belonging to a certain neighborhood of the point x and which has the following properties:

(a) $V_x(y) > 0$;

(b) $V_x(y)$ is superharmonic;

(c) $\lim\limits_{y \to x} V_x(y) = 0$.

Theorem 4.8. *A point $x \in \partial G$ is regular if, and only if it has a barrier.*

Proof. If x is regular, then the Green function $G(y, x_0)$ with pole at an arbitrary point $x_0 \in G$ is a barrier for x. Conversely, suppose there exists a barrier $V_x(y)$ for the point $x \in \partial G$. As pointed out in the preceding section, $G(y, x_0)$ is within G the limit of an increasing sequence of Green functions $G^{(n)}(y, x_0)$ for a sequence of regular regions $G^{(n)} \subset G^{(n+1)}$ which exhaust G and lie strictly within G.

Let D_ϱ be the intersection of G with the interior of the ball $T_\varrho(x) = \{y : |x - y| < \varrho\}$. We choose ϱ small enough so that the barrier $V_x(y)$ is defined in D_ϱ and $x_0 \notin \overline{D_\varrho}$. Then we let $D_\varrho^{(n)}$ be the subset of D_ϱ in $G^{(n)}$, S_ϱ the subset of ∂D_ϱ in G, $S_\varrho^{(n)}$ the subset of $\partial D_\varrho^{(n)}$ in $G^{(n)}$. Let $M = \sup\limits_{y \in \overline{D}_\varrho} G(y, x_0)$ and let $\omega^{(n)}(y)$ be the harmonic measure of $S_\varrho \setminus S_\varrho^{(n)}$ relative to the ball T_ϱ. Since as $n \to \infty$ the $(p - 1)$ dimensional measure of $S_\varrho \setminus S_\varrho^{(n)}$ tends to zero, we can find n_0 such that

$$\omega^{(n_0)}(x) < \varepsilon/M, \quad \varepsilon > 0.$$

Since at all points of the closed set $\overline{S_\varrho^{(n_0)}}$ the function $V_x(y)$ is positive and lower semicontinuous, it has a positive lower bound on $S_\varrho^{(n_0)}$. Therefore, one can find an $m > 0$ such that on $S_\varrho^{(n_0)}$

$$m V_x(y) > M \geq G(y, x_0) > G^{(n)}(y, x_0) \quad (n > n_0).$$

Hence for $n > n_0$ within $D_\varrho^{(n)}$ we have the inequality

$$G^{(n)}(y, y_0) \leq mV_x(y) + M\omega^{(n_0)}(y). \tag{4.2.13}$$

In fact, since the left side is a harmonic function in $D_\varrho^{(n)}$ and the right side is superharmonic, it is sufficient to verify it on points in $\partial D_\varrho^{(n)}$. We have:

on $S_\varrho^{(n_0)}$, $G^{(n)}(y, x_0) < mV_x(y)$,

on $S_\varrho^{(n)} \setminus S_\varrho^{(n_0)} \subset S_\varrho \setminus S_\varrho^{(n_0)}$, $G(y, x_0) \leq M = M\omega^{(n_0)}(y)$,

on $\partial D_\varrho^{(n)} \cap \partial G^{(n)}$, $G^{(n)}(y, x_0) = 0$,

and this establishes inequality (4.2.13). Successively passing to the limit as $n \to \infty$ and $y \to x$, we obtain

$$\overline{\lim_{y \to x}}\, G(y, x_0) \leq \varepsilon.$$

Since ε was arbitrary and $G(y, x_0)$ non-negative

$$\lim_{y \to x} G(y, x_0) = 0.$$

As shown in the proof of Theorem 4.7, this implies that $G(x, x_0) = 0$, i.e. the point x is regular. The theorem is proved. ☐

We point out at once the important

Corollary. *The regularity of a point is a local property of the boundary ∂G, i.e. it depends on the structure of ∂G in an arbitrarily small neighborhood of the given point.*

Regular points x_0 of the boundary ∂G_i, are characterized by the *"concentration" of the Green measure ε_x'*:

$$\varepsilon_x' \to \varepsilon_{x_0} \quad \text{as} \quad x \to x_0, \quad x \in G_i. \tag{4.2.14}$$

Clearly, $G_i(x_0, y) = 0$, i.e. regularity of x_0, is a consequence of (4.2.14). To obtain the converse one need only repeat the proof of (4.1.5) in No. 2 (cf. Remark).

We prove a somewhat more general result.

Theorem 4.9. *For any regular point x_0 belonging to the boundary ∂K of a compact set K, and any sequence $x_n \to x_0$, $x_n \in CK$,*

$$\varepsilon_{x_n}' \to \varepsilon_{x_0}.$$

Proof. Suppose the theorem were not true. Since the family of measures $\{\varepsilon_{x_n}'\}$ is vaguely compact, one may assume that

$$\varepsilon_{x_n}' \to \varkappa \neq \varepsilon_{x_0}.$$

Then any region G_i can contain only a finite number of points of the sequence $\{x_n\}$. Otherwise $x_0 \in \partial G_i$ and the vague convergence of the subsequence $\varepsilon'_{x_n^*}$, $x_n^* \in G_i$ to a measure distinct from ε_{x_0} is impossible, as we have seen above. Select from $\{x_n\}$ a subsequence of points with the property that distinct points of this sequence belong to distinct regions G_i. In order not to complicate the indexing, we shall also denote this sequence by $\{x_n\}$, and the region which contains x_n by G_n. We will show that $\varepsilon'_{x_n} \dot{\to} \varepsilon_{x_0}$ and since this contradicts the supposition, the theorem will be proved. Let T be a ball with center x_0, $m_n = \varepsilon'_{x_n}(T)$. We show that $m_n \to 1$. Since T is arbitrary, this will mean that $\varepsilon'_{x_n} \dot{\to} \varepsilon_{x_0}$.

Denote by G_n^T the subset of G_n within T. Starting with a certain n, $x_n \in T$ and $G_n^T \neq \emptyset$; we shall consider only these n. Suppose $S_n = \partial G_n^T \cap \partial T$. The spherical closed regions S_n are pairwise non-overlapping. The measure ε'_{x_n} can be obtained by first sweeping ε_{x_n} onto ∂G_n^T (denote the resulting measure by ε''_{x_n}), and then sweeping the restriction of ε''_{x_n} to $\mathring{S}_n = S_n \cap G_n$ onto ∂G_n (cf. Corollary 4 to Theorem 4.6). Then $m_n > 1 - \varepsilon''_{x_n}(\mathring{S}_n)$, and one needs to prove that $\varepsilon''_{x_n}(\mathring{S}_n) \to 0$. For this we note that

$$\varepsilon''_{x_n}(\mathring{S}_n) = \omega(x_n, \mathring{S}_n; G_n^T).$$

By the known principle of enlarging a region

$$\omega(x_n, \mathring{S}_n; G_n^T) \leq \omega(x_n, \mathring{S}_n; T).$$

Furthermore, one may assume that $|x_n - x_0| < R/2$, where R is the radius of T. In this case, it is easy to see that there is a constant A such that

$$\omega(x_n, \mathring{S}_n; T) \leq A\,\omega(x_0, \mathring{S}_n; T).$$

But

$$\sum_n \omega(x_0, \mathring{S}_n; T) = \omega\Big(x_0, \bigcup_n \mathring{S}_n; T\Big) \leq 1,$$

which implies that $\lim_{n\to\infty} \omega(x_0, \mathring{S}_n; T) = 0$ and, consequently,

$$\lim_{n\to\infty} \varepsilon''_{x_n}(\mathring{S}_n) = 0. \quad \square$$

Regular points of K possess another characteristic property: in sweeping out *any* measure μ with support outside K onto K the potential remains unchanged at regular points of K:

$$U^{\mu'}(x) = U^{\mu}(x), \quad x \in K \setminus I_K.$$

This follows easily from formula (4.2.3):

$$\mu' = \int_{S(\mu)} \varepsilon_y' \, d\mu(y)$$

by which for $x \in K \setminus I_K$

$$U^{\mu'}(x) = \int_{S(\mu)} U^{\varepsilon_y'}(x) \, d\mu(y) = \int_{S(\mu)} U^{\varepsilon_y}(x) \, d\mu(y) = U^{\mu}(x).$$

11. The notion of a Green measure ε_x' which we introduced for points $x \notin K$ is usefully generalized for points $x \in K$ as well.

If x is a regular point of K, then we set

$$\varepsilon_x' = \varepsilon_x.$$

If x is irregular, then there exists a sequence of points $x_n \notin K$, $x_n \to x$, such that the measures ε_{x_n}' are not concentrated at the point x as $h \to \infty$. By the compactness of the family $\{\varepsilon_{x_n}'\}$ we may assume that

$$\varepsilon_{x_n}' \dot{\to} \varkappa_x \neq \varepsilon_x.$$

Set $m = \varkappa_x(\{x\})$. Clearly, $0 \leq m < 1$ and we can consider the measure

$$\varepsilon_x' = \frac{1}{1-m} (\varkappa_x - m\varepsilon_x), \quad S(\varepsilon_x') \subset K$$

which we shall call the *Green measure for the irregular point* x. The independence of ε_x' from the selection of the sequence $\{x_n\}$ follows from the following property which completely characterizes ε_x' (cf. Ch. III, § 1, the strengthened uniqueness theorem).

Theorem 4.10. *The measure ε_x' is C-absolutely continuous and quasi-everywhere on K*

$$U^{\varepsilon_x'}(y) = U^{\varepsilon_x}(y) \quad (p \geq 3).$$

Proof. In fact, by Theorem 3.8

$$U^{\varkappa_x}(y) = \varliminf_{n \to \infty} U^{\varepsilon_{x_n}'}(y) \quad \text{quasi-everywhere.}$$

But quasi-everywhere on K

$$\varliminf_{n \to \infty} U^{\varepsilon_{x_n}'}(y) = \lim_{n \to \infty} \frac{1}{|x_n - y|^{p-2}} = \frac{1}{|x-y|^{p-2}},$$

so that

$$U^{\varkappa_x}(y) = \frac{1}{|x-y|^{p-2}} = U^{\varepsilon_x}(y) \quad \text{quasi-everywhere on } K$$

and by Theorem 1.3,

$$U^{\varkappa_x}(y) \leq U^{\varepsilon_x}(y) \quad \text{everywhere.}$$

From the boundedness of $U^{\varkappa_x}(y)$ outside any neighborhood of the point x it follows the C-absolute continuity of the restriction of $\varkappa_x$ to $K \setminus \{x\}$, i.e. the C-absolute continuity of the measure ε_x'.

Furthermore

$$U^{\varkappa_x - m\varepsilon_x}(y) = (1 - m)\, U^{\varepsilon_x}(y)$$

quasi-everywhere on K, which implies that

$$U^{\varepsilon_x'}(y) = U^{\varepsilon_x}(y)$$

quasi-everywhere on K. ▯

An analogous proof yields

Theorem 4.10′. *If $p = 2$, the measure ε_x' is C-absolutely continuous and quasi-everywhere on K*

$$U^{\varepsilon_x'}(y) = \ln \frac{1}{|x - y|}, \quad x \notin \partial G_0,$$

$$U^{\varepsilon_x'}(y) = \ln \frac{1}{|x - y|} + G_0(x, \infty), \quad x \in \partial G_0.$$

From what has been proved, it follows that as $x_n \to x$, the set of limit measures for the family $\{\varepsilon_{x_n}'\}$ is contained in the collection of measures

$$\{m\varepsilon_x + (1 - m)\, \varepsilon_x'\}_{0 \leq m \leq 1}. \tag{4.2.15}$$

In addition, it is clear that the points $x_n \to x$ may be chosen arbitrarily and not only outside of K. If x is regular, then the whole collection of limiting measures reduces to ε_x.

The Green measure ε_x' for an irregular point is associated with ε_x, i.e. for any function $h(x) \in H_{\overline{CK}}$ (definition in No. 7)

$$h(x) = \int_K h(y)\, d\varepsilon_x'(y).$$

Indeed, let $x_n \in CK$, $x_n \to x$ and $\varepsilon_{x_n}' \dashrightarrow m\varepsilon_x + (1 - m)\, \varepsilon_x'$ $(m < 1)$. Since ε_{x_n}' is associated with ε_{x_n} (No. 7), we have

$$h(x_n) = \int_K h(y)\, d\varepsilon_{x_n}'(y),$$

and as $n \to \infty$ this gives the required result.

We give another construction for the Green measure ε_x', $x \in K$. Consider the compact set $K \setminus T_\varrho$, where T_ϱ is an open ball of radius ϱ with center x and denote by $\varepsilon_x^{(\varrho)}$ the measure obtained by balayage of ε_x onto $K \setminus T_\varrho$. By (4.2.1) and (4.2.1′) quasi-everywhere on $K \setminus T_\varrho$

$$U^{\varepsilon_x^{(\varrho)}}(y) = \frac{1}{|x-y|^{p-2}}$$

and

$$U^{\varepsilon_x^{(\varrho)}}(y) \leq \frac{1}{|x-y|^{p-2}}$$

everywhere in R^p. Moreover, Corollary 4 to Theorem 4.6 shows that for $\tau < \varrho$

$$U^{\varepsilon_x^{(\tau)}}(y) \geq U^{\varepsilon_x^{(\varrho)}}(y)$$

since the measure $\varepsilon_x^{(\varrho)}$ can be obtained by balayage of the measure $\varepsilon_x^{(\tau)}$. If ϱ ranges over any decreasing sequence of numbers which tends to zero, then $\{U^{\varepsilon_x^{(\varrho)}}(y)\}$ will be a monotone increasing sequence of potentials. By Theorem 1.26 $\varepsilon_x^{(\varrho)}$ vaguely converges to a certain measure $\varkappa_x$, and

$$U^{\varkappa_x}(y) \doteq \lim_{\varrho\to 0} U^{\varepsilon_x^{(\varrho)}}(y).$$

From this it is seen that

$$U^{\varkappa_x}(y) = \frac{1}{|x-y|^{p-2}}$$

quasi-everywhere on K. Consequently, the part of $\varkappa_x$ outside $\{x\}$ is C-absolutely continuous, and, setting $m = \varkappa_x(\{x\})$, we obtain by Theorem 4.10

$$\varkappa_x = m\varepsilon_x + (1-m)\,\varepsilon_x', \quad 0 \leq m \leq 1.$$

It turns out that only the extreme cases are possible here: either $m = 0$ or $m = 1$.

Indeed if the part of $\varkappa_x$ inside T_ϱ is swept out onto $K \setminus T_\varrho$, then we obtain a measure $\varepsilon_x^{(\varrho)}$. Consequently, outside T_ϱ we have the inequality

$$\varepsilon_x^{(\varrho)} \geq \varkappa_x + m\varepsilon_x^{(\varrho)}.$$

Letting ϱ tend to zero, we obtain that in $R^p \setminus \{x\}$

$$\varkappa_x \geq (1+m)\,\varkappa_x.$$

Hence, either $\varkappa_x = 0$ outside $\{x\}$, i.e. $m = 1$, or $m = 0$.

We will show that $m = 0$, i.e. $\varkappa_x = \varepsilon_x'$. For this we establish the validity of "the law of reciprocity"

$$U^{\varkappa_x}(y) = U^{\varkappa_y}(x). \tag{4.2.16}$$

For any positive ϱ and τ we have

$$\int U^{\varepsilon_x^{(\varrho)}}(z)\, d\varepsilon_y^{(\tau)}(z) = \int U^{\varepsilon_y^{(\tau)}}(z)\, d\varepsilon_x^{(\varrho)}(z).$$

Suppose $\varrho \to 0$; then by the monotone convergence and C-absolute continuity of the measure $\varepsilon_y^{(\tau)}$

$$\lim_{\varrho\to 0}\int U^{\varepsilon_x^{(\varrho)}}(z)\,d\varepsilon_y^{(\tau)}(z) = \int U^{\varkappa_x}(z)\,d\varepsilon_y^{(\tau)}(z) = \int \frac{d\varepsilon_y^{(\tau)}(z)}{|x-z|^{p-2}} = U^{\varepsilon_y^{(\tau)}}(x)$$

and

$$\lim_{\tau\to 0}\lim_{\varrho\to 0}\int U^{\varepsilon_x^{(\varrho)}}(z)\,d\varepsilon_y^{(\tau)}(z) = U^{\varkappa_y}(x).$$

Analogously

$$\lim_{\varrho\to 0}\lim_{\tau\to 0}\int U^{\varepsilon_y^{(\tau)}}(z)\,d\varepsilon_x^{(\varrho)}(z) = U^{\varkappa_x}(y).$$

But by the monotone passage to the limit, the left sides of these latter equations are equal; this gives (4.2.16).

We remark that (4.2.16) holds also when $y \notin K$; in this case, as seen from the construction, $\varkappa_y = \varepsilon_y'$.

Now assume that $m = 1$. Then for any point $y \notin K$ we will have

$$U^{\varepsilon_x}(y) = U^{\varepsilon_y'}(x)$$

i.e.

$$G_i(x, y) = 0, \quad y \in G_i.$$

But this contradicts the irregularity of the point x.

Thus, $\varkappa_x = \varepsilon_x'$, and (4.2.16) may be rewritten in the form[23]

$$U^{\varepsilon_x'}(y) = U^{\varepsilon_y'}(x). \tag{4.2.16'}$$

Remark. From the latter construction of the measure ε_x' we can easily obtain a sharped version of Theorem 4.10: the equation

$$U^{\varepsilon_x'}(y) = \frac{1}{|x-y|^{p-2}}$$

holds at all points $y \in K \setminus I_K$.

One can utilize the Green measure ε_x', $x \in I_K$, for the definition of balayage out of the set of irregular points I_K onto K.

First of all, the family of Green measures $\{\varepsilon_x'\}$, $x \in R^p$ is a Borel family. Indeed, for any function $f \in \Phi$ the function $g(x) = \varepsilon_x'(f)$ will be continuous outside of K (and even harmonic, cf. No. 2) and at the regular points of K (Theorem 4.9). Concerning the set I_K of irregular points, we know first, that it has Lebesgue measure zero, since $C_2(I_K) = 0$,

[23] In the case of an infinite region of the plane (4.2.16′) must be modified; cf. first footnote on p.230.

and second, that it is Borel, since

$$I_K = \bigcup_i I_K^{(i)}$$

and each $I_K^{(i)}$ coincides with the set of discontinuities of the Green function $G_i(y, x_0)$, for $y \neq x_0 \in G_i$. Consequently, the function $g(x)$ is Borel.

Now suppose that μ is a measure concentrated on I_K. Set

$$\mu' = \int_{I_K} \varepsilon_x' \, d\mu(x) .$$

It is not difficult to see that

(i) μ' is C-absolutely continuous,

(ii) $S(\mu') \subset K$,

(iii) $U^{\mu'}(x) = U^{\mu}(x)$ on $K \setminus I_K$.

By the uniqueness theorem (Ch. III, No. 1) these properties completely determine μ'.

Below, for any measure (or signed measure) μ, we shall assume without further comment that

$$\mu' = \int_{S(\mu)} \varepsilon_x' \, d\mu(x) . \tag{4.2.17}$$

This measure (correspondingly, signed measure) solves the problem of sweeping μ out onto K, and $\mu'(I_K) = 0$. As will be shown in § 3, No. 14, these properties uniquely determine μ'. It is also easy to verify that

$$U^{\mu'}(y) = \int_{S(\mu)} U^{\varepsilon_x'}(y) \, d\mu(x) = \int_{S(\mu)} U^{\varepsilon_y'}(x) \, d\mu(x) = \int_{S(\mu)} U^{\mu}(x) \, d\varepsilon_y'(x) ,$$

and that μ' is the associated measure (correspondingly, signed measure) for μ relative to CK.

It will be convenient to introduce, for any configuration of the points x, y, the Green function $G(x, y)$ relative to the compact set K:

$$G(x, y) = U^{\varepsilon_y}(x) - U^{\varepsilon_y'}(x)$$

(with the modification indicated in No. 9, p. 229, if the points x, y lie in an infinite region of the plane or on its boundary).

This definition is correct if $x \neq y$ or if $x = y \notin K$. If $x = y \in K \setminus I_K$, then we shall define $G(x, y) = 0$. If $x = y \in I_K$ the function $G(x, y)$ is undefined.

By (4.2.16′)

$$G(x, y) = G(y, x) .$$

If $x, y \in G_i$, where G_i is a connected component of CK, then

$$G(x, y) \equiv G_i(x, y) .$$

If y is a regular point of K, then

$$G(x, y) \equiv 0.$$

If $y \in I_K$, then there exists a point $x \in CK$ such that

$$G(x, y) > 0.$$

§ 3. The Generalized Dirichlet Problem

12. Consider an arbitrary region $G \subset R^p$ with compact boundary ∂G. If G is non-regular, i.e. $H_{\partial G}$ is a proper subspace of $C_{\partial G}$ (cf. No. 1), then for functions $f(x) \in C_{\partial G} \setminus H_{\partial G}$ the Dirichlet problem will have *no solution*. Nevertheless, with each such function we will be able to associate *uniquely* a well-defined bounded function $h_f(x)$ harmonic in G (equal to zero at infinity), which is continuous at all regular points of ∂G and is there equal to $f(x)$. Such a function $h_f(x)$ will be called a *solution of the generalized Dirichlet problem*.

First we show that a solution of the generalized Dirichlet problem is unique. Indeed, if $h_f^{(1)}(x)$ and $h_f^{(2)}(x)$ are two solutions, then the function

$$g(x) = h_f^{(1)}(x) - h_f^{(2)}(x)$$

(which is harmonic in G) quasi-everywhere on the boundary ∂G is continuous and equal to zero. Applying to g the generalized maximum principle (Theorem 3.4), we obtain that in G $g(x) \leq 0$. Analogously $-g(x) \leq 0$, and hence, $g(x) \equiv 0$ in G.

It is not difficult to exhibit the solution of the generalized Dirichlet problem. It is given by the formula

$$h_f(y) = \int_{\partial G} f(x)\, d\varepsilon_y'(x). \tag{4.3.1}$$

The proof of the harmonicity of $h_f(y)$, given in No. 2 for a regular region G, does not depend on the last assumption and can be repeated word-by-word. If x_0 is a regular point of ∂G, then as we saw in No. 10 (cf. (4.2.14)) as $y \to x_0$

$$\varepsilon_y' \dot{\to} \varepsilon_{x_0}.$$

Therefore $h_f(y)$ is continuous at regular points $x_0 \in \partial G$ and

$$h_f(x_0) = f(x_0).$$

Formula (4.3.1) can be used to define the *harmonic measure* $\omega(y, E; G)$ of an arbitrary Borel set $E \subset \partial G$. Namely, we put

$$\omega(y, E; G) = \int_{\partial G} \chi_E(x)\, d\varepsilon_y'(x) = \varepsilon_y'(E),$$

where χ_E is the characteristic function of the set E. We point out the following properties of harmonic measure:

(i) $\omega(y, E; G)$ is a harmonic function within G,

(ii) $0 \leq \omega(y, E; G) \leq 1$.

(iii) $1 - \omega(y, E; G) \geq \omega(y, \partial G \setminus E; G)$, and for the bounded domain G it is an equality.

(iv) If $\omega(y, E; G) = 0$ for even one point inside G, then $\omega(y, E; G) \equiv 0$. In this case E is called a set of *harmonic measure zero*.

(v) If $x \in E$ is regular and is an interior (relative to ∂G) point of E, then $\omega(x, E; G) = 1$; if $x \in \partial G \setminus E$ is regular and is interior to $\partial G \setminus E$, then $\omega(x, E; G) = 0$.

(vi) If E has positive harmonic measure, $x \in \partial G \setminus E$ is irregular and lies in the exterior of E, then $\omega(x, E; G) > 0$. Otherwise, $\omega(x, E; G)$ would be a barrier for the point x and the later would be regular. Analogously, if the point $x \in E$ is irregular and is an interior point (relative to ∂G) of E, then $\omega(x, E; G) < 1$.

From (vi) it follows that for an irregular point $y \in \partial G$ the Green measure ε_y' is positive on any subset of ∂G with positive harmonic measure.

In the case of an infinite region G and $p \geq 3$, by (4.2.3′) $\omega(y, \partial G; G)$ coincides with the equilibrium potential $U^{\gamma}(x)$ of the compact set ∂G. This implies that for points of the external boundary of K, the inequality

$$U^{\gamma}(y) < 1$$

is necessary and sufficient for irregularity.

If we keep in mind the local character of irregularity (No. 10) and the fact that an irregular point belongs to the boundary ∂G_k of one of the regions of the complement CK, then we arrive at the following criterion:

Suppose $y \in K$ and let γ_r denote the equilibrium measure of the compact set

$$K \cap \{x: |x - y| \leq r\}.$$

Then for the irregularity of the point y it is necessary and sufficient that

$$U^{\gamma_r}(y) < 1 \tag{4.3.2}$$

for sufficiently small r.

We note that in the case of an infinite region G in the plane

$$\omega(y, \partial G; G) \equiv 1$$

and does not coincide with the equilibrium potential of the compact set ∂G.

If $C_2(\partial G) > 0$, then, as we have seen in Ch. II, No. 16, there is a measure γ which solves the Robin problem for ∂G; in this case it is not difficult to show that the condition

$$U^{\gamma}(y) < 1$$

is as before necessary and sufficient for the irregularity of the point $y \in \partial G$.

The condition $C_2(\partial G) > 0$ does not appear in the following "local" formulation:

A point $y \in K$ is irregular if, and only if, for small $r > 0$

$$U^{\gamma_r}(y) < 1$$

where γ_r is the equilibrium measure of the compact set

$$K \cap \{x: |x - y| \leq r\}.$$

One should note that for an arbitrary compact set of positive capacity $K \subset R^p$ (and not only the boundary of a region ∂G) and for a function $f(x)$ continuous on K, formula (4.3.1) provides a unique harmonic "quasi-extension" of $f(x)$ to the whole of R^p, i.e. a function $h_f(y)$, harmonic outside of K, continuous at the regular points of K and there equal to $f(x)$. If K has capacity zero, then (cf. Remark 2 at the end of No. 8) $\varepsilon_y' \equiv 0$ and $h_f(y) \equiv 0$.

13. In order that formula (4.3.1) gives a solution of the Dirichlet problem *in the classical sense*, it is necessary and sufficient that for any point $y_0 \in I_{\partial G}$ (i.e. for any irregular point of ∂G) the following equation holds

$$f(y_0) = \int_{\partial G} f(x)\, d\varepsilon_{y_0}'(x). \tag{4.3.3}$$

Necessity is obvious; sufficiency follows from the fact that for $y \in G$, $y \to y_0$, the set of limiting measures for the family $\{\varepsilon_y'\}$ consists, as we saw in No. 11, of measures of the form $m\varepsilon_{y_0} + (1 - m)\, \varepsilon_{y_0}'$, and if condition (4.3.3) is satisfied, then

$$\int_{\partial G} f(x)\, d\{m\varepsilon_{y_0} + (1 - m)\, \varepsilon_{y_0}'\} = f(y_0).$$

This implies that

$$\lim_{y \to y_0} h_f(y) = \lim_{y \to y_0} \int_{\partial G} f(x)\, d\varepsilon_y'(x) = f(y_0).$$

Thus, conditions (4.3.3) for $y_0 \in I_{\partial G}$ are necessary and sufficient for $f(x)$ to be in $H_{\partial G}$, i.e. for the solvability of the Dirichlet problem with boundary function $f(x)$.

From this is obtained, incidentally, the following *criterion for unsolvability* of the Dirichlet problem at the irregular point y_0:

If the continuous function $f(x) \neq \text{const}$, $x \in \partial G$, attains its least (or greatest) value on ∂G at the point y_0, then the Dirichlet problem with boundary function $f(x)$ has no solution at this point.

We show that there is a *countable set* $D_{\partial G}$ of irregular points, having the property, that if condition (4.3.3) is fulfilled by $y_0 \in D_{\partial G}$ then the Dirichlet problem will be solvable.

Since ∂G has positive capacity, we can find an $r > 0$ so small that the subset of ∂G outside any sphere of radius r will have positive harmonic measure[24]. Let y be any irregular point of ∂G, $T_r(y)$ an open ball of radius r with center at y. Then $\varepsilon'_y(\partial G \setminus T_r(y)) > 0$ (cf. No. 12).

Denote by I_n the set of all points of $I_{\partial G}$, for which $\varepsilon'_y(\partial G \setminus T_r(y)) > 1/n$. Take in I_n an arbitrary countable, dense set D_n and set

$$D_{\partial G} = \bigcup_{n=1}^{\infty} D_n .$$

We will show that if the solvability condition (4.3.3) holds at all points $y \in D_{\partial G}$, then the Dirichlet problem has a solution in the classical sense.

Indeed, if $y_0 \in I_{\partial G}$, then for some n_0, $y_0 \in I_{n_0}$, and therefore there exists a sequence of points $y_k \to y_0$, $y_k \in D_{n_0}$. Without loss of generality, we may assume that

$$\varepsilon'_{y_k} \dot{\to} m\varepsilon_{y_0} + (1 - m)\, \varepsilon'_{y_0} .$$

Then from the definition of I_{n_0} it follows that $m < 1 - 1/n_0$, i.e. $1 - m > 1/n_0$. But then, from the conditions

$$f(y_k) = \int f(x)\, d\varepsilon'_{y_k}(x)$$

after passing to the limit an $k \to \infty$, we obtain

$$f(y_0) = \int f(x)\, d\varepsilon'_{y_0}(x) .$$

i.e. the solvability of the Dirichlet problem at the point y_0.

It is clear that the "solvability" set $D_{\partial G}$ that we have constructed is *not uniquely determined*. It is not difficult, however, to specify a set of irregular points which will always be included in the indicated construction of $D_{\partial G}$.

---

[24] Indeed, otherwise there would exist a sequence $r_n \to 0$ and points $y_n \in \partial G$ such that the subset of ∂G outside the sphere of radius r_n and center y_n would have harmonic measure zero. From this it follows that the whole boundary ∂G has harmonic measure zero, which is impossible since the capacity of ∂G is positive.

A point $y_0 \in I_{\partial G}$ will be called a *quasi-isolated* irregular point, if for any sequence $y_n \in \partial G$, which converges to y,

$$\varepsilon'_{y_n} \dashrightarrow \varepsilon_y .$$

The set of all quasi-isolated irregular points will be denoted by $Q_{\partial G}$. Since $Q_{\partial G} = \bigcup_{n=1}^{\infty} (Q_{\partial G} \cap I_n)$, and the points of $Q_{\partial G} \cap I_n$ are, it is easy to see, isolated in the set I_n, then for any choice of D_n we have $D_n \supset Q_{\partial G} \cap I_n$, and this means that $D_{\partial G} \supset Q_{\partial G}$. Clearly, any point of $D_{\partial G} \setminus Q_{\partial G}$ can be eliminated from $D_{\partial G}$, and the remaining set will still be a "solvability" set. It is more difficult to prove that none of the quasi-isolated points can be eliminated from a "solvability" set. For this we must be convinced of the following. If a point $y \in Q_{\partial G}$ does not belong to the set $D \subset I_{\partial G}$, then one can find a function $f(x)$ continuous on ∂G, for which the Dirichlet problem will be solvable at all points of D and will not be solvable at y. We shall not present here the rather tedious construction of such a function.

14. In concluding this section we consider a question related to the Dirichlet problem: the problem of reconstructing the potential $U^\nu(x)$ from its values on the support $S(\nu)$ of the signed measure ν.

Consider an arbitrary point $y \notin S(\nu)$ and denote by $G(x, y)$ the Green function relative to $S(\nu)$. Then

$$\begin{aligned}\int_{S(\nu)} G(x, y)\, d\nu(x) &= \int_{S(\nu)} U^{\varepsilon_y}(x)\, d\nu(x) - \int_{S(\nu)} U^{\varepsilon'_y}(x)\, d\nu(x) \\ &= U^\nu(y) - \int_{S(\nu)} U^\nu(x)\, d\varepsilon'_y(x).\end{aligned}$$

But at the regular points of $S(\nu)$ the function $G(x, y) = 0$; therefore, writing I for the set of irregular points of $S(\nu)$, we obtain the *formula of M. Riesz*

$$U^\nu(y) = \int_{S(\nu)} U^\nu(x)\, d\varepsilon'_y(x) + \int_I G(x, y)\, d\nu(x). \tag{4.3.4}$$

In exactly the same way, for any compact set $K \supset S(\nu)$ and $y \notin K$

$$U^\nu(y) = \int_K U^\nu(x)\, d\varepsilon'_y(x) + \int_I G(x, y)\, d\nu(x), \tag{4.3.4'}$$

where I is the set of irregular points of K, ε'_y is the Green measure for K and $G(x, y)$ is the Green function relative to K.

One should notice that formula (4.3.4′) remains valid (under the condition that U^ν still makes sense) at all regular points of K and those points of I_K where $\nu = 0$, i.e., in any case, *quasi-everywhere* in R^p.

We will point out several important corollaries to this formula.

Corollary 1. Let I be the set of irregular points of the compact set $K \supset S(\mu)$. Denote by ν_I the restriction of the signed measure ν to I.

If we sweep out the signed measure ν_I onto K and set $\nu' = (\nu - \nu_I) + (\nu_I)'$ (cf. No. 11), then $U^{\nu'} = U^{\nu}$ quasi-everywhere on K, and formula (4.3.4′) gives, for all y, the equation

$$U^{\nu'}(y) = \int_K U^{\nu}(x)\, d\varepsilon_y'(x).$$

This result was obtained in No. 11 (for $K = S(\nu)$).

Corollary 2. *If* $\nu_I \equiv 0$, *i.e. then*

$$U^{\nu}(y) = \int_K U^{\nu}(x)\, d\varepsilon_y'(x).$$

This allows us to give the final form of the uniqueness theorem (cf. Ch. I, Theorem 1.12 and Ch. III, No. 1).

Uniqueness Theorem. *Suppose we have two signed measures* ν_1 *and* ν_2 *such that for the set* I *of irregular points of* $S(\nu_1) \cup S(\nu_2)$

$$\nu_{1I} \equiv \nu_{2I} \equiv 0.$$

Then, if $U^{\nu_1}(x) = U^{\nu_2}(x)$ *quasi-everywhere on* $S(\nu_1) \cup S(\nu_2)$, *it follows that* $\nu_1 \equiv \nu_2$.

Indeed, for the signed measure $\nu = \nu_1 - \nu_2$ the condition $\nu(e) \equiv 0$ is satisfied for $e \subset I$, and hence

$$U^{\nu}(y) = \int_{S(\nu_1) \cup S(\nu_2)} U^{\nu}(x)\, d\varepsilon_y'(x).$$

Since ε_y' is C-absolutely continuous, then $U^{\nu}(y) = 0$ quasi-everywhere, and from this, by Theorem 1.12, it follows that $\nu = 0$.

In addition we have proved that the balayage problem for an arbitrary measure (or signed measure) μ onto the compact set K has a unique solution in the class of measures (correspondingly, signed measures), for which $\mu \equiv 0$ on I_K. We obtained just such a solution in No. 11.

Corollary 3. *Suppose that* $U^{\nu}(x) = 0$ *quasi-everywhere on* $K \supset S(\nu)$. *Then, denoting by* I *the set of irregular points of* K, *and by* ν_I *the restriction of* ν *to* I, *we have*

$$\nu = \nu_I - (\nu_I)'.$$

Indeed, by the M. Riesz formula and Corollary 1 it follows that

$$U^{\nu}(y) = \int_I G(x, y)\, d\nu(x) = \int_I U^{\varepsilon_y}(x)\, d\nu - \int_I U^{\varepsilon_y'}(x)\, d\nu$$

$$= U^{\nu_I}(y) - \int_K U^{\nu_I}(x)\, d\varepsilon_y'(x) = U^{\nu_I}(y) - U^{(\nu_I)'}(y)$$

quasi-everywhere; by Theorem 1.12 $\nu = \nu_I - (\nu_I)'$.

§ 4. The Operator Approach to the Dirichlet Problem and the Balayage Problem

15. The solution $h_f(y)$ of the generalized Dirichlet problem may be viewed as the result of applying a certain operator A to an arbitrary function $f(x) \in C_{\partial G}$. Here the range of A belongs to a space of functions which are harmonic and bounded in G (and equal to zero at infinity, if G is not bounded). From formula (4.3.1) it immediately follows that:

(i) the operator A is linear:

$$A[a_1 f_1 + a_2 f_2] = a_1 A[f_1] + a_2 A[f_2];$$

(ii) the operator A is positive: if $f(x) \geq 0$, then $A[f] \geq 0$.

Moreover, it is known that on the subspace $H_{\partial G} \subset C_{\partial G}$ the operator A coincides with the classical solution of the Dirichlet problem, i.e. $A[f] = h_f \in H_{\overline{G}}$.

M. V. Keldysh [3] has proved that the operator A is uniquely determined by the properties indicated above. In proving this theorem, M. V. Keldysh used the existence in $H_{\overline{G}}$ of a function which attains its strict maximum at a prescribed regular point of ∂G. The highly non-trivial construction of such a function was given by him in [2].

The goal of this section is to prove the theorem of M. V. Keldysh. However, we will take a different approach and will not utilize the existence of the function mentioned above.

Essential to this proof will be the description of the annihilator of the subspace $H_{\partial G}$; this is the content of Lemma 4.1 below.

We now recall certain definitions and theorems from the theory of normed linear spaces.

Let L be a normed linear space, and let L^* be its conjugate space, i.e. the space of all linear continuous functionals on L. If $E \subset L$ is any set of elements, then the *annihilator* $E^{\perp}$ of the set E is, by definition, the set of all functionals which equal zero on all elements of E. Analogously, the annihilator of a set $E \subset L^*$ is the set of all elements of L, on which vanish all the functionals of E.

We introduce in L^* the so-called *w^*-topology* (weak* topology), in which the neighborhood basis for an element $f_0 \in L^*$ is given by the sets

$$V(f_0; x_1, x_2, \ldots, x_n, \varepsilon) = \{f\colon |f(x_i) - f_0(x_i)| < \varepsilon, \quad 1 \leq i \leq n\}$$

where $\varepsilon > 0$, $n \geq 1$, and $x_1, x_2, \ldots, x_n$ are arbitrary elements of L. It is not difficult to see that convergence $f_n \to f$ in the w^*-topology is equivalent to pointwise convergence $f_n(x) \to f(x)$ at each element $x \in L$.

To Banach (cf. [1]) belongs the following

Theorem. *If L is separable, then a set $E \subset L^*$ is w^*-closed if, and only if, the w^*-limit of any sequence of elements of E also belongs to E.*

In addition to this theorem, we will make use of one other fact (cf. Day [1], p. 39).

For any $E \subset L^*$, the repeated annihilator $(E^\perp)^\perp$ is the smallest w^*-closed linear manifold containing E.

Let us now turn to the space $C_{\partial G}$ and its conjugate $C^*_{\partial G}$ which consists of all signed measures concentrated on ∂G.

Lemma 4.1. *The annihilator $H^\perp_{\partial G}$ consists of all those signed measures ν which are representable in the form*

$$\nu = \nu_I - (\nu_I)'$$

where ν_I is the restriction of ν to the set I of irregular points of ∂G.

Proof. The statement

$$\nu_I - (\nu_I)' \in H^\perp_{\partial G},$$

is obvious, since, as we have seen, $(\nu_I)'$ is associated with ν_I, i.e. for $h(x) \in H_{\partial G}$

$$\int_{\partial G} h(x)\, d\nu_I(x) = \int_{\partial G} h(x)\, d(\nu_I)'(x).$$

Suppose now that $\nu \in H^\perp_{\partial G}$. We shall show that

$$U^\nu(x) = 0$$

quasi-everywhere on ∂G; then the result follows from Corollary 3 of the M. Riesz formula (cf. No. 14).

Denote by N_0 the set of all signed measures in $C^*_{\partial G}$ whose potential is equal to zero quasi-everywhere on ∂G; we will prove that N_0 is w^*-closed.

Since the space $C_{\partial G}$ is separable, by the Banach theorem above, it is sufficient to verify that the vague limit of any sequence of signed measures in N_0 again belongs to N_0.[25] Thus, suppose that

$$\nu_n \dashrightarrow \nu, \quad \nu_n \in N_0.$$

Since the norms $||\nu_n|| = \operatorname{var} \nu_n = \nu_n^+(\partial G) + \nu_n^-(\partial G)$ are bounded, then, by the theorem on vague compactness of measures, one may assume that

$$\nu_n^+ \dashrightarrow \lambda_1, \quad \nu_n^- \dashrightarrow \lambda_2, \quad \text{and} \quad \lambda_1 - \lambda_2 = \nu.$$

Furthermore, from $\nu_n \in N_0$ it follows that quasi-everywhere on ∂G

$$U^{\nu_n^+}(x) = U^{\nu_n^-}(x).$$

Hence all these equations hold simultaneously quasi-everywhere on ∂G.

By Theorem 3.8 quasi-everywhere we have

$$U^{\lambda_1}(x) = \varliminf_{n\to\infty} U^{\nu_n^+}(x), \qquad U^{\lambda_2}(x) = \varliminf_{n\to\infty} U^{\nu_n^-}(x).$$

Consequently, quasi-everywhere on ∂G

$$U^{\lambda_1}(x) = U^{\lambda_2}(x) \quad \text{and} \quad U^{\nu}(x) = 0.$$

Thus, $\nu \in N_0$ and N_0 is w^*-closed.

We will now show that $H_{\partial G}$ itself is the annihilator of a simply described set of signed measures. In No. 13 we saw that the condition $f(x) \in H_{\partial G}$, denoting the solvability of the Dirichlet problem with boundary values $f(x)$, is equivalent to the validity of the equations:

$$f(z) = \int_{\partial G} f(x)\, d\varepsilon_z'(x), \quad z \in I.$$

If we set $\nu_z = \varepsilon_z - \varepsilon_z'$, then these conditions can be rewritten in the form

$$\int_{\partial G} f(x)\, d\nu_z(x) = 0, \quad z \in I.$$

Therefore, if we denote the set of all signed measures $\{\nu_z\}$, $z \in I$ by N_1, then we have

$$H_{\partial G} = N_1^{\perp}.$$

Consequently,

$$H_{\partial G}^{\perp} = (N_1^{\perp})^{\perp}$$

is the smallest w^*-closed linear manifold containing N_1.

But it is easily seen that $N_1 \subset N_0$, and since N_0 is a w^*-closed linear manifold in $C_{\partial G}^*$, then

$$H_{\partial G}^{\perp} \subset N_0$$

as required[1]. ▯

---

[25] Clearly, in $C_{\partial G}^*$ the concepts of vague and w^*-convergence coincide.

[26] Note that from the proof of Lemma 4.1 it follows that $H_{\partial G}^{\perp} = N_0$.

Remark. The reasoning used to prove Lemma 4.1 enables us to establish a somewhat more general proposition.

Let K be an arbitrary compact set (or, more generally, a closed set with compact boundary), $H_{\partial K} \subset C_{\partial K}$ a linear manifold consisting of all functions continuous on ∂K which can harmonically be extended to CK. Then ν belongs to the annihilator $H_{\partial K}^{\perp}$ if, and only if,

$$\nu = \nu_I - (\nu_I)'$$

where I is the set of irregular points of K, and ν_I is the restriction of ν to I.

This shows, by the way, that for ν to be in $H_{\partial K}^{\perp}$, it is necessary and sufficient that for any region G which is a component of CK, $\nu_{\partial G} \in H_{\partial G}^{\perp}$, where $\nu_{\partial G}$ is the restriction of ν to ∂G.

We now turn to the theorem of M. V. Keldysh.

Theorem 4.11. *An operator A, acting on $C_{\partial G}$ into the space of bounded and harmonic functions in G (equal to zero at infinity), which has the following properties:*

(*i*) $A[a_1 f_1 + a_2 f_2] = a_1 A[f_1] + a_2 A[f_2]$; $f_1, f_2 \in C_{\partial G}$,

(*ii*) $A[f] \geq 0$, *if* $f(x) \geq 0$,

(*iii*) *if* $f(x) \in H_{\partial G}$, *then* $A[f]$ *gives a solution to the Dirichlet problem with boundary function* $f(x)$,

is unique and gives a generalized solution to the Dirichlet problem.

Proof. Let us fix arbitrarily a point $y \in G$; we will consider the value $A[f]_y$ of the function $A[f]$ at this point as a functional on the space $C_{\partial G}$. From (i) and (ii) this functional is continuous (cf. Introduction), and by the F. Riesz theorem we have the representation

$$A[f]_y = \int_{\partial G} f(x)\, d\alpha_y(x), \tag{4.4.1}$$

where α_y is some measure on ∂G. From condition (iii) it follows that for any function $f(x) \in H_{\partial G}$

$$\int_{\partial G} f(x)\, d\alpha_y(x) = \int_{\partial G} f(x)\, d\varepsilon_y'(x),$$

i.e. the signed measure $\nu_y = \alpha_y - \varepsilon_y'$ belongs to $H_{\partial G}^{\perp}$. Now suppose that x_0 is an arbitrary regular point of ∂G. Obviously, the theorem will be proved if we can show that as $y \to x_0$, $\alpha_y \dot{\to} \varepsilon_{x_0}$. Since

$$\alpha_y(\partial G) = A[1]_y \leq \text{const}$$

then, by the vague compactness of the family of measures $\{\alpha_y\}$, we can select a sequence of points $y_n \to x_0$ such that the α_{y_n} vaguely converge to a certain measure α_{x_0}. Since $\varepsilon'_{y_n} \dot{\to} \varepsilon_{x_0}$, then ν_{y_n} also has a vague limit which we denote by ν_{x_0}. Moreover

$$\nu_{x_0} = \alpha_{x_0} - \varepsilon_{x_0}. \tag{4.4.1'}$$

Further, $\nu_{x_0} \in H^{\perp}_{\partial G}$, since the set $H^{\perp}_{\partial G}$ is vaguely closed. But this possible only in the case $\nu_{x_0} = 0$, i.e. $\alpha_{x_0} = \varepsilon_{x_0}$. Indeed, by Lemma 4.1, $\nu_{x_0} = \varkappa_I - (\varkappa_I)'$, while by (4.4.1'), $\nu_{x_0} \geq 0$ outside of $x_0 \notin I$. But then $\varkappa_I \geq 0$ and ν_{x_0} can not be a measure outside of x_0. The resulting contradiction proves that $\nu_{x_0} = 0$. Thus, the only limiting measure for the subsequences chosen from $\{\alpha_y\}$ is ε_{x_0}, and this means that $\alpha_y \dot{\to} \varepsilon_{x_0}$ as $y \to x_0$. The theorem is proved. ☐

16. The lemma proved in the preceding section can be used to describe the *set of all signed measures* $\{\nu'\}$ *associated with a given signed measures* ν relative to a region G (or CK). Suppose that ν_1 and ν_2 are two signed measures associated with ν. Then for any function $h(x) \in H_{\overline{G}}$ (correspondingly, $H_{\overline{CK}}$)

$$\int_{\overline{G}} h(x)\, d\nu(x) = \int_{\partial G} h(x)\, d\nu_1(x) = \int_{\partial G} h(x)\, d\nu_2(x).$$

From this it is apparent that the signed measure $\varkappa = \nu_1 - \nu_1$ belongs to $H^{\perp}_{\partial G}$. By Lemma 4.1

$$\nu_2 = \nu_1 + \varkappa_I - (\varkappa_I)'. \tag{4.4.2}$$

Conversely, this formula, under the conditions that $\nu_1 \in \{\nu'\}$ and var $\nu_2 \leq$ var ν, will yield a signed measure associated with ν.

Suppose that the signed measure ν_1 is associated with ν and $\nu_{1I} \neq 0$; then for the signed measure

$$\nu_2 = \nu_1 - \nu_{1I} + (\nu_{1I})'$$

we can easily prove the inequality var $\nu_2 \leq$ var ν_1, and hence, we obtain a signed measure ν_2, associated with ν and equal to zero on I. From the same formula (4.4.2) it is clear that there exists a unique signed measure associated with ν having this property.

If μ is a measure, then, by (4.2.17), there exists a measure μ' associated with μ and equal to zero on I_K. This implies that every signed measure associated with μ and equal identically to zero on I_K is a measure coinciding with μ'.

Using the uniqueness of the solution to the balayage problem for signed measures (No. 14, p. 245) and recalling that balayage always yields a signed measure associated with the original one (§ 2), we arrive at the following theorem.

Theorem 4.12. *In the class of signed measures (measures) equal to zero on the set I of irregular points of the compact set K, the problem of sweeping out the signed measure (measure) ν onto K and the problem of constructing a signed measure (measure) associated with ν relative to CK, have the same solution.*

It is sometimes possible to characterize the associated signed measure, mentioned in Theorem 4.12, by other means. We point to two such instances.

(I) Suppose that the compact set K is *reduced*, i.e. a subset of K in any neighborhood of each of its points has positive capacity.[27] Moreover, supposed that μ is a measure on $\overline{CK}$, whose potential is bounded near I.

Then there exists a *unique measure* μ' concentrated on K and satisfying the relations

$$\int_{\overline{CK}} h(x)\, d\mu(x) = \int_{\partial K} h(x)\, d\mu'(x), \quad h \in H_{\overline{CK}}. \tag{4.4.3}$$

This measure μ' is associated with μ and is equal to zero on I.

Indeed, suppose, μ' is a measure associated with μ and equal to zero on I. If μ_1' is any measure on ∂K, satisfying (4.4.3) then $\mu_1' - \mu' \in H_{\partial G}^{\perp}$ and by the remark to Lemma 4.1,

$$\mu_1' = \mu' + \nu_I - (\nu_I)'.$$

From this we see that the restriction of μ_1' to I coincides with ν_I and hence $\nu_I \geq 0$. Therefore, $(\nu_I)' \geq 0$ and $\mu' - (\nu_I)'$ as the restriction of μ_1' to CI, is also a measure. Thus

$$U^{(\nu_I)'}(x) \leq U^{\mu'}(x).$$

But, by Theorem 4.12, the measure μ' is obtained by balayage of the measure μ and therefore

$$U^{\mu'}(x) \leq U^{\mu}(x).$$

Consequently, the potential $U^{(\nu_I)'}(x)$ is bounded near I. This, however, is impossible, if $\nu_I \neq 0$. Indeed, in the latter case, by Theorem 2.4, there is a point $x_0 \in I$ at which $U^{\nu_I}(x_0) = +\infty$. Since K is reduced, in any neighborhood of x_0 there are points at which $U^{\nu_I}(x) = U^{(\nu_I)'}(x)$; but since

$$\lim_{x \to x_0} U^{\nu_I}(x) = \infty$$

this contradicts the boundedness of $U^{(\nu_I)'}(x)$.

---

[27] Any compact set K contains a reduced kernel, i.e. a reduced compact set $\check{K}$ such that $C_2(K \setminus \check{K}) = 0$ (cf. p. 164).

Thus, $\nu_I = 0$, and $\mu_1' = \mu$.

(II) With the same conditions on K and the measure μ as in (I), the associated measure μ', equal to zero on I, has a *smallest variation* in the class of all signed measures satisfying (4.4.3).

If the part of μ in an infinite component of CK is equal to zero, then μ' is the only signed measure associated with μ.

We begin with the verification of the last assertion. Suppose that ν is any signed measure, associated with μ relative to CK. Consider the equilibrium potential $U^{\gamma}(x)$ of the compact set K; although it does not belong to $H_{\overline{CK}}$, as shown in Theorem 1.9, there is an increasing, hence uniformly bounded, sequence of positive functions $h_n(x) \in H_{\overline{CK}}$, which tends to $U^{\gamma}(x)$. In the equation

$$\int_{\overline{CK}} h_n(x)\, d\mu(x) = \int_{\partial K} h_n(x)\, d\nu(x)$$

we pass to the limit as $n \to \infty$, and obtain

$$\int_{\overline{CK}} U^{\gamma}(x)\, d\mu(x) = \int_{\partial K} U^{\gamma}(x)\, d\nu(x).$$

But $U^{\gamma}(x) = 1$ quasi-everywhere on $S(\mu)$; moreover, on the subset of $S(\mu)$ where $U^{\gamma}(x) < 1$, the measure μ is equal to zero (since this subset belongs to I and has null capacity, but the potential $U^{\mu}(x)$ is bounded there). Consequently,

$$\int_{\overline{CK}} U^{\gamma}(x)\, d\mu(x) = \mu(1).$$

On the other hand, if we let J denote the set of irregular points of K belonging to the boundary of the infinite component of CK, then,

$$\begin{aligned}\int_{\partial K} U^{\gamma}(x)\, d\nu(x) &= \nu(\partial K \setminus J) + \int U^{\gamma}(x) d\nu_J(x)\\ &= \nu(\partial K \setminus J) + \int_K U^{\nu_J}(x)\, d\gamma(x)\\ &= \nu(\partial K \setminus J) + \int_K U^{(\nu_J)'}(x)\, d\gamma(x)\\ &= \nu(\partial K \setminus J) + \int_{\partial K} U^{\gamma}(x)\, d(\nu_J)'\,(x) = \nu(\partial K \setminus J) + (\nu)'\,(1),\end{aligned}$$

since γ and $(\nu_J)'$ are C-absolutely continuous.

Consequently,

$$\mu(1) = \nu(\partial K \setminus J) + (\nu_J)'\,(1)$$

from which

$$\mu(1) \leq \operatorname*{var}_{\partial K \setminus J} \nu + \operatorname*{var}_{\partial K \setminus J} (\nu_J)' \leq \operatorname*{var}_{\partial K \setminus J} \nu + \operatorname{var} \nu_J = \operatorname{var} \nu.$$

But on the other hand, since ν is assumed associated with μ, we have

$$\operatorname{var} \nu \leqq \operatorname{var} \mu = \mu(1)$$

so that $\operatorname{var} \nu = \mu(1)$. In particular

$$\mu'(1) = \mu(1) = \operatorname{var} \nu. \tag{4.4.4}$$

Just as in (I), we obtain the equation

$$\nu = \mu' + \nu_I - (\nu_I)'.$$

This implies

$$\operatorname{var} \nu = \operatorname{var} [\mu' - (\nu_I)'] + \operatorname{var} \nu_I \geq \operatorname{var} [\mu' - (\nu_I)'] + \operatorname{var} (\nu_I)'.$$

Together with (4.4.4) this yields

$$\operatorname{var} [\mu' - (\nu_I)'] + \operatorname{var} (\nu_I)' \leqq \mu'(1) = \operatorname{var} \mu'.$$

Using the triangle inequality, we can write

$$\operatorname{var} [\mu' - (\nu_I)'] + \operatorname{var} (\nu_I)' = \operatorname{var} \mu'.$$

But this is possible only in the case that

$$(\nu_I)' \geq 0, \quad \mu' - (\nu_I)' \geq 0$$

and then the proof is completed just as in (I). The first assertion in (II) is more easily obtained. Let us suppose that the signed measure ν satisfies (4.4.3) and $\operatorname{var} \nu \leq \operatorname{var} \mu' = \mu'(1)$. Then one can repeat word-for-word the whole argument following formula (4.4.4) and hence obtain $\nu = \mu'$.

The uniqueness of the associated signed measure leads us to the following result.

Theorem 4.13. *Suppose that $h_f(x)$ is the solution to a certain generalized Dirichlet problem. Then there exists a sequence of functions $h_n(x) \in H_{\overline{G}}$ such that*

$$\lim_{n\to\infty} h_n(x) = h_f(x)$$

uniformly on each compact subset of G.

Proof. Without loss of generality we may assume that the boundary ∂G is reduced. Indeed, if $\hat{\partial} G \subset \partial G$ denotes the reduced kernel of the boundary ∂G and $G_1 \supset G$ is the region bounded by $\hat{\partial} G$, then solutions of the generalized Dirichlet problems for G and G_1 for a function $f(x) \in G_{\partial G}$

will, by Theorem 3.3, coincide. By formula (4.3.1)

$$h_f(x) = \int_{\partial G} f(y)\, d\varepsilon'_x(y).$$

We recall that the Green measure ε'_x is associated with the measure ε_x and is obtained as the *positive extension* to the whole of $C_{\partial G}$ of the linear functional

$$\varepsilon_x(h) = h(x)$$

defined on the subspace $H_{\partial G} \subset C_{\partial G}$. By (I), this extension is unique. Recalling the construction for the *positive extension*[28] we come to the conclusion that for any $f \in C_{\partial G}$

$$\varepsilon'_x(f) = \sup_{h \leq f} \varepsilon_x(h) = \inf_{h \geq f} \varepsilon_x(h),$$

where $h \in H_{\partial G}$. Consequently, there is a sequence of functions $h_n \in H_{\overline{G}}$ such that $h_n(y) \geq f(y)$, $y \in \partial G$ and

$$\varepsilon'_x(f) = \lim_{n \to \infty} \varepsilon_x(h_n).$$

But $\varepsilon_x(h_n) = h_n(x)$, and $\varepsilon'_x(f) = h_f(x)$. Moreover, the inequality $h_n \geq f$, satisfied on ∂G, implies, by the generalized maximum principle, the inequality $h_n(y) \geq h_f(y)$ everywhere in G. Consequently, the sequence of non-negative harmonic functions $h_n(y) - h_f(y)$ tends to zero at the point x. By the well-known theorem of Harnack it will tend to zero uniformly on every compact set $K \subset G$. The theorem is proved. ▯

Remark. From the proof it is apparent that the sequence $h_n(x) \in H_{\overline{G}}$, converging to $h_f(x)$, may be chosen such that for all n, either the inequality $h_n(x) > h_f(x)$ holds, or the converse inequality holds.

17. Suppose that K is an arbitrary fixed compact set in R^p. We will view a signed measure ν' on K obtained by balayage as the result of applying a *balayage operator* β to ν: $\nu' = \beta\nu$. The restriction of this operator to the set M_0 of signed measures equal to zero on I will be denoted by β_0.[29] The operator β is linear and bounded, since var $\beta\nu \leq$ var ν (cf. No. 7).

We shall prove that the operator β_0 is *vaguely continuous*, i.e. that

$$\nu_n \dot{\to} \nu; \quad \nu_n, \nu \in M_0$$

implies

$$\beta_0\nu_n \dot{\to} \beta_0\nu.$$

---

[28] Cf., for instance, N. I. Akhiezer [2].

[29] In § 2 the operator β was constructed by extending the operator β_0.

It is sufficient to verify this property for measures. Let $\mu_n \dashrightarrow \mu$ where μ_n, μ are measures from M_0. The set of measures $\{\beta_0\mu_n\}$ is vaguely compact. Suppose that λ is the vague limit of some subsequence $\{\beta_0\mu_{n_i}\}$. Then

$$\lambda \in M_0.$$

To be convinced of this, it is sufficient to establish that $\lambda(F) = 0$ for any closed set $F \subset I$. In turn, for this it is sufficient to verify that for any $\eta > 0$ there is in ∂G a neighborhood V of the set F such that $\beta_0\mu_{n_i}(V) < \eta$ for $i > i_0$.

By Ω_δ we denote the set of points of $\bar{G}$ whose distance from F is less than δ; let V_δ be the intersection of Ω_δ and ∂G. In what follows we can assume that Ω_δ is μ-almost clopen, i.e. that the μ-measure of the boundary of Ω_δ is zero. There are only a countable or a finite number of values δ not satisfying this condition and we eliminate them from the very beginning. Then

$$\mu(\Omega_\delta) = \lim_{n\to\infty} \mu_n(\Omega_\delta).$$

Since $\mu(F) = 0$, Ω_δ may be chosen such that

$$\mu_n(\Omega_\delta) < \eta \quad \text{for} \quad n > n_0(\Omega_\delta).$$

We take $\delta_1 < \delta$ and evaluate the quantity $\beta_0\mu_{n_i}(V_{\delta_i})$. By (4.2.17)

$$\beta_0\mu_{n_i}(V_{\delta_i}) = \int \beta\varepsilon_x(V_{\delta_1})\, d\mu_{n_i}(x).$$

Since $\mu_{n_i} \in M_0$, in integrating we may exclude the points of I and write

$$\beta_0\mu_{n_i}(V_{\delta_1}) = \int_{\Omega_\delta} \beta_0\varepsilon_x(V_{\delta_1})\, d\mu_{n_i}(x) + \int_{\bar{G}\setminus\Omega_\delta} \beta_0\varepsilon_x(V_{\delta_1})\, d\mu_{n_i}(x).$$

But $\beta_0\varepsilon_x(V_{\delta_1}) \leq 1$ and therefore

$$\int_{\Omega_\delta} \beta_0\varepsilon_x(V_{\delta_1})\, d\mu_{n_i}(x) \leq \mu_{n_i}(\Omega_\delta) < \eta$$

fut $i > i_0(\Omega_\delta)$.

On the orher hand, since the distance from any point $x \in \bar{G} \setminus \Omega_\delta$ to V_{δ_1} exceeds $\delta - \delta_1 > 0$, and the capacity of V_δ for sufficiently small δ_1 is arbitrarily small, then we can find δ_1 such that for $x \in \bar{G} \setminus \Omega_\delta$ the inequality $\beta_0\varepsilon_x(V_{\delta_1}) < \eta$ holds. Then

$$\int_{\bar{G}\setminus\Omega_\delta} \beta_0\varepsilon_x(V_{\delta_1})\, d\mu_{n_i}(x) < \eta\mu_{n_i}(\bar{G}).$$

But the measures μ_n are bounded so that $\mu_{n_i}(\bar{G}) < B$, and

$$\beta_0\mu_{n_i}(V_{\delta_1}) < \eta(1 + B)$$

for $i > i_0(\delta_1)$. This proves that $\lambda \in M_0$.

Furthermore, by (4.2.2)

$$\int_{\partial K} h(x)\, d\beta_0\mu_{n_i}(x) = \int_{\overline{CK}} h(x)\, d\mu_{n_i}(x), \quad h \in H_{\overline{CK}}$$

and passing to the limit as $i \to \infty$, we obtain

$$\int_{\partial K} h(x)\, d\lambda(x) = \int_{\overline{CK}} h(x)\, d\mu(x), \qquad h \in H_{\overline{CK}}.$$

This shows that λ is a measure associated with μ. Recalling that $\lambda \in M_0$ and using Theorem 4.12, we see that $\lambda = \beta_0\mu$. Thus $\beta_0\mu_n \dot{\to} \beta_0\mu$ and the assertion is proved.

There is a connection between the balayage operator β and the operator A studied in No. 15. In order to reveal this, let us consider the range of values of the operator A, i.e. the set of all generalized solutions h_f of the Dirichlet problem for the region G. Here we assume that the function $h_f(y)$ is defined at all points of $\bar{G}$ by the formula

$$h_f(y) = \int_{\partial G} f(x)\, d\varepsilon_y'(x).$$

We denote this linear set by Δ_A, $H_{\bar{G}} \subset \Delta_A$. It becomes a Banach space if we set $||h_f|| = ||f||_{C_{\partial G}}$. The operator A is an isometry between $C_{\partial G}$ and Δ_A.

The conjugate operator A^*, mapping Δ_A^* onto $C_{\partial G}^*$, is defined, as is known, by the equation

$$A^*T(f) = T(Af) = T(h_f)$$

which holds for all $f \in C_{\partial G}$, $T \in \Delta_A^*$.

We will show that *A^* is an extension of the operator β.*[30] First of all, from the generalized maximum principle, it follows that Δ_A^* contains all signed measures with support in $\bar{G}$.[31] Furthermore,

$$\nu(h_f) = \int_{\bar{G}} h_f(y)\, d\nu(y) = \int_{\partial G} f(x)\, d\nu'(x),$$

[30] More precisely, of the operator which is the part of β defined on signed measures, with support in G.

[31] More precisely, to each such signed measure there uniquely corresponds an element of Δ_A^* defined by the formula $\nu(h_f) = \int_{\bar{G}} h_f(y)\, d\nu(y)$.

where

$$\nu' = \int_{S(\nu)} \varepsilon'_y d\nu(\gamma) = \beta_0 \nu .$$

In this way,

$$\beta\nu(f) = \nu(h_f) \quad \text{for all } f \in G_{\partial G},$$

and this shows that A^* is an extension of β.

Now note that if

$$\|h_n - h\| \to 0, \quad h_n, h \in \Delta_A$$

then for any natural number p

$$\partial^p h_n \to \partial^p h$$

uniformly on any compact set contained in G. Therefore, to any distribution T with support in G there uniquely corresponds an element of Δ_A^*. By the preceding, A^*T is a certain signed measure $\nu \in C^*_{\partial G}$. Moreover, for any point $z \notin \bar{G}$

$$U^\nu(z) = \nu\,(|z-x|^{2-p}) = T(|z-x|^{2-p})$$

since $h(x) = |z-x|^{2-p} \in H_{\bar{G}}$. The function

$$U^T(z) = T(|z-x|^{p-2})$$

defined outside the support of T, will be called the *potential of the distribution* T.

We will show that the equation $U^T(z) = U^\nu(z)$ holds also at the regular points of ∂G. For the proof we take a region G_1, containing the support of T and such that $\bar{G}_1 \subset G$. If A_1 is the operator for the generalized Dirichlet problem in G_1, then $A_1^* T = \nu_1 \in C^*_{\partial G_1}$. If we now sweep ν_1 out onto ∂G and set $\nu = \beta\nu_1$, then it is not difficult to see that $\nu = A^*T$. In fact, since $\nu_1 \in D_G^*$, then for any function $f(x) \in C_{\partial G}$ we will have

$$\nu(f) = \int_{\partial G} f(x)\, d\nu(x) = \int_{\partial G_1} h_f(x)\, d\nu_1(x) = A_1^* T(h_f) = T(A_1 h_f) = T(h_f).$$

From the equations

$$U^T(z) = U^{\nu_1}(z), \qquad U^{\nu_1}(z) = U^\nu(z)$$

the first of which holds everywhere in ∂G, and the second at all regular points, we obtain the following theorem.

Theorem 4.14. *Suppose that the support of the distribution T is contained in a region G with compact boundary ∂G. Then on ∂G there exists a signed measure ν whose potential coincides with the potential of T at all* *points outside of G, except, possibly, at irregular points of* ∂G. *Moreover, the signed measure* ν *is unique if we also require that it vanish on the set* I *of irregular points.*

18. In concluding this section we examine several general questions connected with the properties of a family of balayage operators $\{\beta_K\}$ where K is a compact set in R^p. However, in contrast with the previous sections we will assume that the operator β_K does not change the signed measures concentrated on I_K.[32]

The procedure of No. 7 showed how to construct the operator β_K for an arbitrary compact set if these operators were defined for regular compact sets. In connection with this we have the following interesting theorem.

Theorem 4.15. *Suppose that* $\{K_\alpha\}$ *is a basis of compact sets in* R^p.[33] *Then all the operators* β_K *can be defined from* $\{\beta_{K_\alpha}\}$ *by means of the following formulas:*

$$\text{(i)} \quad \beta_{K_1 \cap K_2} = \lim_{n\to\infty} (\beta_{K_1}\beta_{K_2})^n = \lim_{n\to\infty} (\beta_{K_2}\beta_{K_1})^n,$$

$$\text{(ii)} \quad \beta_K = \lim_{n\to\infty} \beta_{K_n}, \quad K_n \supset K_{n+1}, \quad \bigcap_1^\infty K_n = K.$$

In both formulas weak convergence of operators is being used, i.e. $\beta = \lim\limits_{n\to\infty} \beta_n$ means that $\beta_n \nu \dashrightarrow \beta\nu$ for any signed measure ν with compact support.

Proof. Let us establish formula (i) which is in essence a short form of writing an alternating process. Since all the operators which appear in (i) act on a signed measure concentrated on $K_1 \cap K_2$ as an identity operator, we need only consider signed measures (or measures) identically zero on $K_1 \cap K_2$. Such a measure can always be represented in the form $\mu = \sum\limits_{k=1}^\infty \mu_k$, where $U^{\mu_k}(x)$ is bounded on $K_1 \cap K_2$.[34] Therefore it is sufficient to verify (i) only for measures concentrated outside of $K_1 \cap K_2$ which have a bounded potential on $K_1 \cap K_2$. If μ is such a measure, then $\beta_{K_1 \cap K_2}\mu = \mu'$ is uniquely determined by the two conditions: $S(\mu') \subset K_1 \cap K_2$ and $U^{\mu'}(x) = U^\mu(x)$ quasi-everywhere on $K_1 \cap K_2$ (cf.

[32] Consequently, the range of values of the operator β_K does not consist only of C-absolutely continuous signed measures.

[33] This means that any compact set $K \subset R^p$ can be represented as $K = \bigcap\limits_{i=1}^\infty K_{\alpha_i}$.

[34] It is sufficient, for example, to take for μ_k the restriction of μ to the set of points in R^p whose distance to $K_1 \cap K_2$ lies in the half-open interval $\left[\frac{1}{k}, \frac{1}{k-1}\right)$

No. 8, Corollary 3). Set

$$\mu_n = (\beta_{K_1}\beta_{K_2})^n \mu.$$

Since

$$U^{\mu_n}(x) \leq U^{\mu_{n-1}}(x)$$

everywhere (No. 7), we can apply Theorem 3.10, according to which there exists a μ_0 which is the vague limit of measures μ_n; also quasi-everywhere

$$U^{\mu_0}(x) = \lim_{n\to\infty} U^{\mu_n}(x).^{35} \tag{4.4.5}$$

We will show that $S(\mu_0) \subset K_1 \cap K_2$.

By the boundedness of the potentials $U^{\mu_n}(x)$, $U^{\mu_0}(x)$ all the measures μ_n, μ_0, are C-absolutely continuous, and hence the operator $\beta_{K_1}\beta_{K_2}$ is vaguely continuous on them (cf. No. 17). In the equation

$$\mu_{n+1} = (\beta_{K_1}\beta_{K_2})\, \mu_n$$

we pass to the limit as $n \to \infty$ and obtain

$$\mu_0 = (\beta_{K_1}\beta_{K_2})\, \mu_0$$

and this is possible only if $S(\mu_0) \subset K_1 \cap K_2$.

On the other hand, since for any n the equation

$$U^{\mu}(x) = U^{\mu_n}(x)$$

holds quasi-everywhere on $K_1 \cap K_2$, then by (4.4.5) we will have

$$U^{\mu}(x) = U^{\mu_0}(x)$$

quasi-everywhere on $K_1 \cap K_2$. Consequently,

$$\mu_0 = \lim_{n\to\infty} (\beta_{K_1}\beta_{K_2})^n\, \mu = \mu' = \beta_{K_1\cap K_2}\, \mu,$$

and (i) is proved.

For (ii), the proof is a repetition of the argument given in No. 7 (where we do not assume the regularity of the compact sets K_i, however). Here it is important that all the operators β_K do not change the signed measures on I_K; we have assumed this.

To conclude the proof it is sufficient to represent an arbitrary compact set $K = \bigcap_i K_i$ in the form

$$K = K_1 \cap (K_1 \cap K_2) \cap (K_1 \cap K_2 \cap K_3) \cap \cdots,$$

[35] If μ has finite energy, then, by Ch.II, No.7, the convergence of μ_n will be strong.

in order to see, with the help of (i) and (ii), that the operators $\{\beta_{K_i}\}$ uniquely define βK. □

Below (§ 6) we shall define an operator β_E for arbitrary Borel sets $E \subset R^p$.

§ 5. Balayage for M. Riesz Kernels

The method used in the preceding sections to study the balayage operator depends essentially on the properties of harmonic functions in a region.

Here we indicate two other approaches to the problem. The first, due to M. Riesz (Nos. 19 and 20), is based on the Kelvin transformation and reduces the balayage problem for a Dirac measure to an equilibrium problem which we studied in detail in Ch. II. The balayage of an arbitrary signed measure is then reduced to balayage for Dirac measures by the principle of superposition.

The second approach was developed by H. Cartan. Here the balayage of a measure with finite energy is viewed as a geometric problem in a Hilbert space. The requirement of finite energy can then be eliminated to deal with the more general problem. We shall present these results in Nos. 22 and 23.

19. Consider an arbitrary point $x_0 \in R^p$ and a point transformation called *inversion with center at* x_0. Namely, each point $x \neq x_0$ is mapped to a point x^* on the ray through x which issues from x_0 such that

$$|x - x_0| \cdot |x^* - x_0| = 1.$$

This is a one-to-one, bicontinuous mapping of $R^p \setminus \{x_0\}$ onto itself. The following formula can be verified in an elementary way:

$$|x^* - y^*| = \frac{|x - y|}{|x_0 - x| \cdot |x_0 - y|}. \tag{4.5.1}$$

If we extend R^p by adjoining the infinite point ω and introduce as a fundamental system of neighborhoods, the exteriors of the balls $|x| \leq r$, then we obtain a compact space $\overline{R^p}$. The inversion transformation obviously determines a one-to-one, bicontinuous map of $\overline{R^p}$ into itself in which the point x_0 goes into ω and conversely. Therefore, every open (closed) set in $\overline{R^p}$ is transformed under inversion to an open (closed) set in $\overline{R^p}$. In particular, every compact set in $R^p \setminus \{x_0\}$ is mapped again onto a compact set in $R^p \setminus \{x_0\}$. A compact set in R^p containing the point x_0 is mapped by the inversion into a closed unbounded set F.

From formula (4.5.1) it follows that if the compact set K does not contain x_0, then there exist two positive constant m_1, m_2 such that for any $x, y \in K$

$$m_1 |x - y| < |x^* - y^*| < m_2 |x - y|.$$

If we use the remark made at the end of No. 11, Ch. II, then we obtain the inequality

$$m_1^{p-\alpha} C_\alpha(K) < C_\alpha(K^*) < m_2^{p-\alpha} C_\alpha(K),$$

where K^* denotes the compact set obtained from K under inversion. Consequently, under inversion with center at the point $x_0 \notin K$, a compact set with capacity zero is transformed again into a compact set with capacity zero. Now it is easy to verify that *any* E with inner capacity zero is mapped under inversion into a set E^* again with inner capacity zero. The same holds if we consider outer capacity.[36]

With each signed measure ν *which does not have an atomic constituent at the point* x_0, we associate another signed measure ν^* by the formula

$$d\nu^*(x^*) = |x - x_0|^{\alpha - p} d\nu(x) \quad (0 < \alpha < p). \tag{4.5.2}$$

ν^* will be called *the Kelvin transform of the signed measure* ν. It is easy to see that this transformation is an involution, i.e. $(\nu^*)^* = \nu$. Moreover, $\nu^*(1) = U_\alpha^\nu(x_0)$ and consequently,

$$\nu(1) = U_\alpha^{\nu^*}(x_0). \tag{4.5.3}$$

The connection between the potentials of the signed measures ν and ν^* immediately follows from (4.5.1) and (4.5.2):

$$U_\alpha^{\nu^*}(x^*) = \int\limits_{S(\nu^*)} \frac{d\nu^*(y^*)}{|x^* - y^*|^{p-\alpha}} = \int\limits_{S(\nu)} \frac{|x_0 - x|^{p-\alpha}}{|x - y|^{p-\alpha}} d\nu(y) = |x_0 - x|^{p-\alpha} U_\alpha^\nu(x). \tag{4.5.4}$$

If the signed measure ν has at x_0 an atomic constituent $a = \nu(\{x_0\})$ and we define ν^* by the same formula (4.5.2), then the connection between $U_\alpha^{\nu^*}$ and U_α^ν is expressed by the formula

$$U_\alpha^\nu(x^*) + a = |x_0 - x|^{p-\alpha} U_\alpha^\nu(x).$$

This leads us to the idea of introducing signed measures in $\overline{R^p} = R^p \cup \omega$. Each signed measure $\bar{\nu}$ in $\overline{R^p}$ is defined by a charge ν in R^p and a number $a = \bar{\nu}(\omega)$ by the formula

$$\bar{\nu}(E) = \begin{cases} \nu(E) & \text{if } \omega \notin E, \\ \nu(E \setminus \omega) + a & \text{if } \omega \in E. \end{cases}$$

[36] For the proof one should represent E as the countable union of bounded sets all of positive distance away from x_0; then use Corollary 2 from Ch.II, No.6.

If we define the potential of $\bar{\nu}$ by the formula

$$U_\alpha^{\bar{\nu}}(x) = U_\alpha^{\nu}(x) + a$$

and if by $\bar{\nu}^*$ we understand the signed measure in $\overline{R^p}$ for which $\bar{\nu}^*(\omega) = \nu(\{x_0\})$ then formula (4.5.4) still holds if in place of ν^* we write $\bar{\nu}^*$.

Furthemore, the following lemma holds.

Lemma 4.3. *Suppose that* $\nu_k \dot{\to} \nu$, *and*

$$\lim_{r\to\infty} \int_{|x|>r} \frac{d\,|\nu_k|\,(x)}{|x|^{p-\alpha}}\,dx = 0$$

uniformly relative to k. Then

$$\nu_k^* \dot{\to} \nu^*.$$

Proof. Let $f^*(x^*)$ be an arbitrary function in Φ. Define the function f by the formula

$$f(x) = f^*(x^*)$$

Then $f(x)$ is continuous, bounded (but, in general, has not a compact support) and equal to zero near x_0. Also

$$\int_{S(f^*)} f^*(x^*)\,d\nu_k^*(x^*) = \int_{S(f)} f(x)\,|x - x_0|^{\alpha-p} d\nu_k(x)\,.$$

Suppose $f_N(x)$ is a function in Φ satisfying the conditions

$$f_N(x) = \begin{cases} f(x) & \text{for } |x| < N, \\ 0 & \text{for } |x| > N+1 \end{cases}$$

and $|f_N(x)| \le |f(x)|$. Then for any $\eta > 0$, one can find $N(\eta)$ such that for all k

$$\left| \int_{S(f)} f(x)\,|x - x_0|^{\alpha-p}\,d\nu_k(x) - \int_{S(f_N)} f_N(x)\,|x - x_0|^{\alpha-p} d\nu_k(x) \right| < \eta$$

and such that an analogous inequality holds also for integrals over the signed measure ν. Furthermore, for sufficiently large k

$$\left| \int f_N(x)\,|x - x_0|^{\alpha-p} d\nu_k(x) - \int f_N(x)\,|x - x_0|^{\alpha-p}\,d\nu(x) \right| < \eta,$$

and hence

$$\left| \int_{S(f)} f(x)\,|x - x_0|^{\alpha-p} d\nu_k(x) - \int_{S(f)} f(x)\,|x - x_0|^{\alpha-p} d\nu(x) \right| < 3\eta,$$

i.e.

$$\left| \int f^*(x^*)\,d\nu_k^*(x^*) - \int f^*(x^*)\,d\nu^*(x^*) \right| < 3\eta,$$

q.e.d. □

20. We will utilize formula (4.5.4) to establish a connection between the equilibrium problem and balayage problem for the Dirac measure ε_{x_0}. The transition from the measure ε_{x_0} to an arbitrary signed measure does not involve any difficulties.

Consider first a closed set $F \subset R^p$ $p \geq 3$. Take an arbitrary point $x_0 \notin F$ and perform an inversion with center at x_0. The set F^* will be compact. Suppose that γ^* is its equilibrium measure, and ε'_{x_0} its Kelvin transform. By (4.5.4)

$$U_\alpha^{\varepsilon'_{x_0}}(x) = \frac{1}{|x_0 - x|^{p-\alpha}} U_\alpha^{\gamma^*}(x^*)$$

and in addition, by (4.5.3)

$$\varepsilon'_{x_0}(1) = U_\alpha^{\gamma^*}(x_0).$$

For $0 < \alpha \leq 2$ (the Robin case) we know that

$U_\alpha^{\gamma^*} = 1$ quasi-everywhere on F^*,

$U_\alpha^{\gamma^*} \leq 1$ everywhere.

Consequently,

$$U_\alpha^{\varepsilon'_{x_0}}(x) = \frac{1}{|x - x_0|^{p-\alpha}} \quad \text{quasi-everywhere on } F,$$

$$U_\alpha^{\varepsilon'_{x_0}}(x) \leq \frac{1}{|x - x_0|^{p-\alpha}} \quad \text{everywhere.}$$

Moreover, since $S(\varepsilon'_{x_0}) \subset F$, the measure ε'_{x_0} solves the balayage problem for the measure ε_{x_0} onto F.

This measure ε'_{α_0} will be called the *Green measure for* F, and the function

$$G^{(\alpha)}(x, y) = U_\alpha^{\varepsilon_y}(x) - U_\alpha^{\varepsilon'_y}(x), \quad y \notin F$$

will be called *the Green function for the open set* CF. Just as in the classical case,

$$G^{(\alpha)}(x, y) = G^{(\alpha)}(y, x), \quad x, y \notin F.$$

Furthermore, it is clear that $G^{(\alpha)}(x, y) \geq 0$ everywhere, and $G^{(\alpha)}(x, y) = 0$ quasi-everywhere on F. In particular, $G^{(\alpha)}(x, y) = 0$ when x is an interior point of F. This follows from Theorem 1.11 (Ch. I, No. 9). The points $x \in F$ where $G^{(\alpha)}(x, y) > 0$ (they must lie on the boundary ∂F), are called the *irregular points* of F. The set of irregular points has α-capacity zero. One should, of course, check that the definition of irregularity does not depend on the position of the pole y in the component of CF whose boundary contains x. We shall do this below in this section.

Then, just as in No. 11, one can define the Green measure ε'_x for $x \in F$, and the inequality $\varepsilon'_x \neq \varepsilon_x$ will characteristic the irregular points of F.

Now for any signed measure we can set

$$\nu' = \int_{S(\nu)} \varepsilon'_x \, d\nu(x) . \tag{4.5.5}$$

The signed measure ν', concentrated on F, is easily seen to solve the problem of sweeping the signed measure ν onto F, i.e.

$$U_\alpha^{\nu'}(x) = U_\alpha^{\nu}(x)$$

quasi-everywhere on F. Moreover, if ν is a measure, then

$$U_\alpha^{\nu'}(x) \leq U_\alpha^{\nu}(x)$$

everywhere in R^p.

The signed measure ν' is identically equal to zero on the set I of irregular points F, and, just as in No. 14, one can see that the solution to the balayage problem with this property, is unique.

From this, by the way, it follows that if $F_1 \supset F_2$ and ν' and ν'' result from sweeping the signed measure ν onto F_1 and F_2, respectively, then

$$\nu'' = \int_{F_1} \varepsilon''_x d\nu'(x) , \tag{4.5.6}$$

where ε''_x is the Green measure for F_2. In other words, balayage can be carried out "with a rest": first sweep the signed measure ν onto the larger set F_1, and then sweep the result ν' onto the smaller set F_2.

We also point out the following easily proved formula:

$$\nu'(K) = \int_{S(\nu)} U_\alpha^{\gamma}(x) \, d\nu(x) , \tag{4.5.6'}$$

where γ is the equilibrium measure on the compact set K, and ν' is the result of balayage of ν onto K.

In the non-Robin case $(2 < \alpha < p)$ the results are less complete. As we saw in Ch. II, § 1, the potential of the equilibrium measure γ^* of the compact set F^* satisfies the equation

$$U_\alpha^{\gamma^*}(x) = 1$$

quasi-everywhere on $S(\gamma^*)$. This condition together with the equation $\gamma^*(1) = ||\gamma^*||^2 = C_\alpha(F^*)$ determine γ^* uniquely. Therefore, for the Green measure ε'_{x_0} obtained from γ^* by the Kelvin transformation, we have

$$U_\alpha^{\varepsilon'_{x_0}}(x) = \frac{1}{|x - x_0|^{p-\alpha}}$$

quasi-everywhere on $S(\varepsilon'_{x_0})$.

These equalities together with the condition $U_\alpha^{\varepsilon'_{x_0}}(x_0) = C_\alpha(F^*)$ *characterize* the Green measure.

Thus, in the non-Robin case ($\alpha > 2$), the balayage problem as well as the Robin problem have no solution. Only by introducing distributions we can obtain a satisfactory result. We shall do this later (Ch. VI, § 1), and in this section we will only study the Robin case ($\alpha \leq 2$).

As an example, take for F the exterior of a sphere: $|x| \geq R$. After inversion with center at x_0, $|x_0| < R$, we will have a ball F^* of radius $R_1 = (R^2 - |x_0|^2)^{-1} R$ with center at the $x_1 = [1 + (R^2 - |x_0|^2)^{-1}]\, x_0$. By No. 13, Ch. II the equilibrium measure γ^* on F^* is defined by the formula

$$d\gamma^*(x^*) = \gamma^*(x^*)\, dx^* = A_1[R_1^2 - |x^* - x_1|^2]^{-\alpha/2}\, dx^*.$$

Hence, for the Green measure ε'_{x_0} we obtain the density

$$\varepsilon'_{x_0}(x) = \begin{cases} A_2(p, \alpha)\, (R^2 - |x_0|^2)^{\alpha/2}\, (|x|^2 - R^2)^{-\alpha/2}\, |x - x_0|^{-p}, & |x| > R, \\ 0, \quad |x| < R, \end{cases} \tag{4.5.7}$$

where

$$A_2(p, \alpha) = \Gamma\left(\frac{p}{2}\right) \pi^{-p/2-1} \sin \frac{\pi\alpha}{2}$$

(cf. Appendix, (A.5)).

Analogously, if F is the ball $|x| \leq R$, and $|x_0| > R$, then the Green measure ε'_{x_0} will have the density

$$\varepsilon'_{x_0}(x) = \begin{cases} A_2(p, \alpha)\, (|x_0|^2 - R^2)^{\alpha/2}\, (R^2 - |x|^2)^{-\alpha/2}\, |x - x_0|^{-p}, & |x| < R, \\ 0, \quad |x| > R. \end{cases}$$

Using the Green measure and the Green function, we can solve, as in No. 14, the problem of reconstructing the potential $U_\alpha^\mu(x)$ from its values on the support of the measure $S(\mu)$. The same M. Riesz formula holds

$$U_\alpha^\mu(y) = \int_{S(\mu)} U_\alpha^\mu(x)\, d\varepsilon'_y(x) + \int_I G^{(\alpha)}(x, y)\, d\mu(x),$$

where I is the set of irregular points of $S(\mu)$. In this formula $S(\mu)$ can be replaced by any closed set $F \supset S(\mu)$.

In particular, if there is no mass μ on I (for instance, if I is empty), then

$$U_\alpha^\mu(y) = \int_F U_\alpha^\mu(x)\, d\varepsilon'_y(x).$$

This holds, for example, if F is the interior or exterior of any sphere. In this case, the explicit form of ε'_y allows one to make certain additional conclusions. Thus, if μ is a measure (or, more generally, $U_\alpha^\mu(x) \geq 0$), then for any two points $y, y_0 \notin F$ we have the inequality

$$\frac{U_\alpha^\mu(y)}{U_\alpha^\mu(y_0)} = \left|\frac{R^2 - |y|^2}{R^2 - |y_0|^2}\right|^{\alpha/2} \frac{\int\limits_F U_\alpha^\mu(x)\, |R^2 - |x|^2|^{-\alpha/2}\, |x - y|^{-p}\, dx}{\int\limits_F U_\alpha^\mu(x)\, |R^2 - |x|^2|^{-\alpha/2}\, |x - y_0|^{-p}\, dx}$$

$$\leq \left|\frac{R^2 - |y|^2}{R^2 - |y_0|^2}\right|^{\alpha/2} \left|\frac{R - |y|}{R + |y_0|}\right|^{-p}$$

since

$$\frac{|x - y|}{|x - y_0|} \geq \frac{|R - |y||}{|R + |y_0||}$$

Thus, we obtain the *generalized Harnack inequality*

$$U_\alpha^\mu(y) \leq \left|\frac{R^2 - |y|^2}{R^2 - |y_0|^2}\right|^{\alpha/2} \left|\frac{R - |y|}{R + |y_0|}\right|^{-p} U_\alpha^\mu(y_0).$$

From this, just as in the classical case $\alpha = 2$, we can obtain the

Uniform Convergence Theorem. *Suppose that the sequence of non-negative potentials $\{U_\alpha^{\mu_n}(x)\}$, $S(\mu_n) \subset F$, where F is the interior or exterior of a sphere, converges to zero at a point $x_0 \notin F$. Then it converges uniformly to zero in any closed region within $R^p \setminus F$.*

It is easy to see that this result also holds in the case when F is any closed subset of R^p.

Now one can prove the

Generalization of Theorem 4.7. *If $G^{(\alpha)}(x, y_0) = 0$ where $x \in \partial F$, $y_0 \notin F$ then $G^{(\alpha)}(x, y) = 0$ for any point y in the connected component CF, including y_0.*

The proof of Theorem 4.7 can be repeated if we also note that the equation $G^{(\alpha)} = \lim\limits_{r \to 0} G^{(\alpha)} * m^{(r)}$ is obtained by Theorem 1.11 (Ch. I, No. 9).

In concluding this section we point out that the Kelvin transformation enables us to answer easily the following question: *under what condition does there exist an equilibrium measure γ for an unbounded closed set F?*

Let us take a point $x_0 \notin F$ and perform an inversion with center at x_0. Then F^* will be a compact set containing x_0. If γ exists, then γ^* will be a C-absolutely continuous measure on F^*, and quasi-everywhere on F^*

$$U_\alpha^{\gamma^*}(x) = \frac{1}{|x - x_0|^{p-\alpha}}.$$

Looking at the definition of an irregular point, we see that x_0 *must be an irregular point for* F^*. This condition is also clearly sufficient for the existence of γ.

21. The Green measure ε'_{x_0} constructed in the preceding section allows one to solve the following problem, which may be viewed as the *Dirichlet problem for α-harmonic functions* ($0 < \alpha < 2$).

Let Ω be a region in R^p with compact boundary $\partial\Omega$, $F = C\Omega$ and $f(x)$ a function, continuous in F and satisfying the condition

$$\int_F \frac{|f(x)|}{1 + |x|^{p+\alpha}}\, dx < \infty \tag{4.5.8}$$

if F contains the point at infinity. One is required to find a function $h_f(x)$, defined in R^p, α-harmonic in Ω which coincidens with $f(x)$ at all regular points of F.

We will show that the solution to this problem is given by the formula

$$h_f(x) = \int_F f(y)\, d\varepsilon'_x(y)\,. \tag{4.5.9}$$

First of all, if x is a regular point of F, then $\varepsilon'_x \equiv \varepsilon_x$ and $h_f(x) = f(x)$. Furthermore, if $x \in \Omega$ the integral in (4.5.9) converges (we assume, of course, that Ω is bounded in this case). In fact, if we let T denote the exterior of a sphere $|y| \geq R$, where R is chosen so large that $T \subset F$, and if we let ε''_x denote the measure obtained by balayage of ε_x onto T, then it is easy to see that for any set $e \subset T$

$$\varepsilon'_x(e) \leq \varepsilon''_x(e)$$

Keeping in mind formula (4.5.7) and condition (4.5.8) we see that $h_f(x)$ is finite.

The continuity of $h_f(x)$ at any point $y_0 \in \partial\Omega$, regular for F, follows from the fact that for $x \in \Omega$ and $x \to y_0$, $\varepsilon'_x \dot{\to} \varepsilon_{y_0}$. Continuity of $h_f(x)$ within Ω is obvious. Therefore, to prove that $h_f(x)$ is α-harmonic, one must show, according to the definition (cf. Ch. I, No. 20), that for any $x_0 \in \Omega$ there exists an $r_0(x_0)$ such that for $r < r_0(x_0)$

$$h_f(x_0) = \varepsilon^{(r)}_{\alpha x_0}(h_f)$$

(for the definition of $\varepsilon^{(r)}_\alpha$, cf. (1.6.2)).

For such an $r_0(x_0)$ one may take the distance from the point x_0 to $\partial\Omega$. Indeed, if r is smaller than this distance, then, denoting by F_1 the exterior of the open ball of radius r with center x_0, we will have that $F_1 \subset F$. We write F_2 for F and denote balayage onto F_2 with two primes, and onto F_1 with one prime. Then, by (4.5.6), we have for any $e \subset F_2$

$$\varepsilon''_{x_0}(e) = \int_{F_1} \varepsilon''_x(e)\, d\varepsilon'_{x_0}(x),$$

and formula (4.5.9) is rewritten in the form

$$h_f(x) = \int_{F_2} f(y)\, d\varepsilon''_x(y).$$

If we also note that $\varepsilon^{(r)}_{\alpha x_0} \equiv \varepsilon'_{x_0}$, then we will have

$$\varepsilon^{(r)}_{\alpha x_0}(h_f) = \int_{F_1} h_f(x)\, d\varepsilon'_{x_0}(x) = \int_{F_1} d\varepsilon'_{x_0}(x) \int_{F_2} f(y)\, d\varepsilon''_x(y)$$

$$= \int_{F_2} f(y) \int_{F_1} d\varepsilon''_x(y)\, d\varepsilon'_{x_0}(x) = \int_{F_2} f(y)\, d\varepsilon''_{x_0}(y) = h_f(x_0)$$

as required.

22. We will now examine the convex cone $\mathscr{E}^+_\alpha$ of measures with finite energy. By Theorem 1.18 $\mathscr{E}^+_\alpha$ is a complete metric space.

Suppose that $K \subset R^p$ is an arbitrary compact set and $\mathscr{E}^+_K$ is the subset of $\mathscr{E}^+_\alpha$ consisting of all measures whose support belongs to K. Since strong convergence implies vague convergence, it is easy to see that $\mathscr{E}^+_K$ is a closed, convex subset of $\mathscr{E}^+_\alpha$.

Hence, for any measure $\mu \in \mathscr{E}^+_\alpha$ there is a unique measure $\beta\mu \in \mathscr{E}^+_K$ satisfying the condition

$$||\mu - \beta\mu|| < ||\mu - \lambda||, \ \lambda \in \mathscr{E}^+_K, \ \lambda \neq \beta\mu. \tag{4.5.10}$$

This is easy to prove using a classical argument similar to that in Ch. II, No. 1.

The measure $\beta\mu$ can be called the *projection* of the measure μ onto $\mathscr{E}^+_K$. We will later show that the projection operator β is the restriction of the balayage operator to measures with finite energy.

Lemma 4.4. *Condition* (4.5.10) *is equivalent to*

$$(\mu - \beta\mu, \lambda - \beta\mu) \leq 0, \quad \lambda \in \mathscr{E}^+_K, \tag{4.5.11}$$

and also to the conditions

$$(\mu - \beta\mu, \lambda) \leq 0, \quad \lambda \in \mathscr{E}^+_K, \tag{4.5.12}$$

$$(\mu - \beta\mu, \beta\mu) = 0. \tag{4.5.12'}$$

Proof. (4.5.10) follows from (4.5.11) since

$$||\mu - \lambda||^2 = ||\mu - \beta\mu||^2 + ||\beta\mu - \lambda||^2$$
$$- 2(\mu - \beta\mu, \lambda - \beta\mu) > ||\mu - \beta\mu||^2.$$

if $\lambda \in \mathscr{E}_K^+$ and $\lambda \neq \beta\mu$. Conversely, (4.5.10) implies that

$$2(\mu - \beta\mu, \lambda - \beta\mu) \leq \|\lambda - \beta\mu\|^2, \quad \lambda \in \mathscr{E}_K^+.$$

By the convexity of $\mathscr{E}_K^+$ we can write

$$\lambda = m\lambda_0 + (1-m)\,\beta\mu, \quad 0 < m < 1,$$

where λ_0 is any measure in $\mathscr{E}_K^+$. Then

$$2m\,(\mu - \beta\mu, \lambda_0 - \beta\mu) \leq m^2\,\|\lambda_0 - \beta\mu\|^2,$$

from which we obtain (4.5.11) as $m \to 0$.

Then from (4.5.11) we obtain (4.5.12) by replacing λ with $\lambda + \beta\mu \in \mathscr{E}_K^+$; for $\lambda \equiv 0 \in \mathscr{E}_K^+$ we obtain

$$(\mu - \beta\mu, \beta\mu) \geq 0.$$

But $\beta\mu \in \mathscr{E}_K^+$ and hence this inequality, together with (4.5.12) gives (4.5.12′). It is trivial to obtain (4.5.11) from (4.5.12) and (4.5.12′). □

Theorem 4.16. *For $\alpha \leq 2$ the measure $\beta\mu$ solves the balayage problem for the measure $\mu \in \mathscr{E}_\alpha^+$ onto K, i.e.*

$$U_\alpha^{\beta\mu}(x) = U_\alpha^\mu(x) \qquad \text{quasi-everywhere on } K$$

$$U_\alpha^{\beta\mu}(x) \leq U_\alpha^\mu(x) \qquad \text{everywhere in } R^p.$$

Proof. By (4.5.12) for any measure $\lambda \in \mathscr{E}_K^+$

$$\int_K [U_\alpha^\mu(x) - U_\alpha^{\beta\mu}(x)]\,d\lambda(x) \leq 0.$$

Set

$$e = \{x : x \in K,\ U_\alpha^\mu(x) > U_\alpha^{\beta\mu}(x)\}.$$

Since the restriction of λ to e also belongs to $\mathscr{E}_K^+$, then

$$\int_e [U_\alpha^\mu(x) - U_\alpha^{\beta\mu}(x)]\,d\lambda(x) \leq 0.$$

This shows, that $\lambda(e) = 0$ for any $\lambda \in \mathscr{E}_K^+$, and by Theorem 2.1, $C_\alpha(e) = 0$.

Thus, quasi-everywhere on K

$$U_\alpha^\mu(x) \leq U_\alpha^{\beta\mu}(x). \tag{4.5.13}$$

On the other hand, from (4.5.12′) we obtain

$$\int_K [U_\alpha^\mu(x) - U_\alpha^{\beta\mu}(x)]\,d\beta\mu(x) = 0.$$

By the preceding, the difference $U_\alpha^\mu(x) - U_\alpha^{\beta\mu}(x) \leq 0$ $\beta\mu$-almost everywhere. Consequently,

$$U_\alpha^\mu(x) = U_\alpha^{\beta\mu}(x)$$

$\beta\mu$-almost everywhere. By the second maximum principle (cf. (1.5.12)), which by Ch. I, No. 25, holds for $\alpha \leq 2$, we obtain that everywhere in R^p

$$U_\alpha^{\beta\mu}(x) \leq U_\alpha^\mu(x).$$

Together with (4.5.13) this shows that

$$U_\alpha^\mu(x) = U_\alpha^{\beta\mu}(x)$$

quasi-everywhere on K. ▯

23. It is desirable to free ourselves of the assumption made in No. 22 that the measure μ has finite energy.

This can partially be achieved in the following way. Assuming for the moment that $\mu \in \mathscr{E}_\alpha^+$ we note that the measure $\beta\mu$ solves the problem of minimizing the so-called *Gaussian integral*

$$J(\lambda) = \int_K [U_\alpha^\lambda(x) - 2U_\alpha^\mu(x)]\, d\lambda(x) \tag{4.5.14}$$

in the class of all measures $\lambda \in \mathfrak{M}_K^+$. Indeed, one can assume from the very start that $\lambda \in \mathscr{E}_K^+$, since for $\lambda \notin \mathscr{E}_\alpha^+$ it is clear that $J(\lambda)$ cannot attain its minimum. In that case

$$J(\lambda) = (\lambda - 2\mu, \lambda),$$

and this quantity differs from

$$||\mu - \lambda||^2 = ||\lambda||^2 - 2(\mu, \lambda) + ||\mu||^2$$

only by the constant $||\mu||^2$.

The Gauss integral $J(\lambda)$ is still defined for $\mu \notin \mathscr{E}_\alpha^+$. Hence there arises the idea of posing the variational problem directly for $J(\lambda)$ with the hope that its solution will also solve the balayage problem for μ onto K.

This can indeed be done if it is assumed that the potential $U_\alpha^\mu(x)$ is *continuous* on K.

Let us describe briefly this argument.

Since $U_\alpha^\mu(x)$ is bounded on K, then $J(\lambda)$ has a finite lower bound. Suppose that $\{\lambda_n\}$ is a minimizing sequence. Then it is vaguely compact, and therefore it may be assumed that $\lambda_n \rightharpoonup \beta\mu$. Using the semi-continuity of $J(\lambda)$ (here it is important that the potential $U_\alpha^\mu(x)$ be continuous) we obtain

$$J(\beta\mu) = \inf_{\lambda \in \mathfrak{M}_K^+} J(\lambda). \tag{4.5.15}$$

The minimizing measure $\beta\mu \in \mathscr{E}_K^+$ has the following properties:

(a) Quasi-everywhere on K

$$U_\alpha^{\beta\mu}(x) \geq U_\alpha^\mu(x).$$

In fact, if

$$e = \{x : x \in K,\ U_\alpha^{\beta\mu}(x) < U_\alpha^\mu(x)\}$$

and $C_\alpha(e) > 0$, then there is a measure $\varkappa$, $S(\varkappa) \subset e$ with bounded potential (cf. Corollary to Theorem 2.8).

But then, for sufficiently small $\eta > 0$

$$J(\beta\mu + \eta\varkappa) - J(\beta\mu) = 2\eta \int_e [U_\alpha^{\beta\mu}(x) - U_\alpha^\mu(x)]\, d\varkappa(x) + \eta^2 \int_e U_\alpha^\varkappa(x)\, d\varkappa(x) < 0$$

which contradicts (4.5.15)

(b) $\beta\mu$-almost everywhere

$$U_\alpha^{\beta\mu}(x) \le U_\alpha^\mu(x).$$

Indeed, if the $\beta\mu$-measure of the set

$$e_1 = \{x : U_\alpha^{\beta\mu}(x) > U_\alpha^\mu(x)\}$$

is positive, then, denoting by π the restriction of $\beta\mu$ to e_1, we would have for $\eta < 1$, that $\beta\mu - \eta\pi \in \mathfrak{M}_K^+$, and for sufficiently small $\eta > 0$

$$J(\beta\mu - \eta\pi) - J(\beta\mu) = -2\eta \int_{e_1} [U_\alpha^{\beta\mu}(x) - U_\alpha^\mu(x)]\, d\pi(x) + \eta^2 \int_{e_1} U_\alpha^\pi(x)\, d\pi(x) < 0.$$

This again contradicts (4.5.15).

Using the second maximum principle which holds for $\alpha \le 2$ (cf. Ch. I, No. 25), we conclude that the inequality

$$U_\alpha^{\beta\mu}(x) \le U_\alpha^\mu(x)$$

holds throughout R^p. But then, combining this with (a) we obtain that

$$U_\alpha^{\beta\mu}(x) = U_\alpha^\mu(x), \qquad \text{quasi-everywhere on } K,$$

i.e. that $\beta\mu$ actually solves the balayage problem for μ onto K.

The method above enables us, in particular, to obtain the Green measure $\beta\varepsilon_x$ for any position of the point x outside of K.

Furthermore, one could define the Green measure $\beta\varepsilon_x$ for any point x (as was done for $\alpha = 2$ in No. 11) and complete the construction of the theory of balayage with formula (4.5.5).

However, there is another approach which achieves the goal more quickly.

From Theorem 4.16 it follows that for any measure $\lambda \in \mathscr{E}_K^+$

$$(\mu - \beta\mu, \lambda) = \int_K [U_\alpha^\mu(x) - U_\alpha^{\beta\mu}(x)]\, d\lambda(x) = 0.$$

Consequently, for any $\lambda, \mu \in \mathcal{E}_\alpha^+$

$$(\mu - \beta\mu, \beta\lambda) = 0, \quad (\lambda - \beta\lambda, \beta\mu) = 0,$$

i.e.

$$(\mu, \beta\lambda) = (\lambda, \beta\mu)$$

or

$$\int_K U_\alpha^\mu(x)\, d\beta\lambda(x) = \int_K U_\alpha^\lambda(x)\, d\beta\mu(x) = \int U_\alpha^{\beta\mu}(x)\, d\lambda(x).$$

We now prove the following theorem.

Theorem 4.16′. *Fore any measure $\mu \in \mathfrak{M}^+$ there is a unique measure $\beta\mu \in \mathfrak{M}_K^+$ satisfying the equation*

$$\int U_\alpha^{\beta\mu}(x)\, d\lambda(x) = \int_K U_\alpha^\mu(x)\, d\beta\lambda(x) \tag{4.5.16}$$

for any measure $\lambda \in \mathcal{E}_\alpha^+$.

The measure $\beta\mu$ solves the balayage problem, i.e.

$$U_\alpha^{\beta\mu}(x) = U_\alpha^\mu(x) \quad \text{quasi-everywhere on } K, \tag{4.5.16′}$$
$$U_\alpha^{\beta\mu}(x) \leq U_\alpha^\mu(x) \quad \text{everywhere in } R^p.$$

Proof. First we prove the existence of $\beta\mu$.

We construct a sequence of measures $\mu_n \in \mathcal{E}_\alpha^+$ such that the potentials $U_\alpha^{\mu_n}(x)$, increasingly tend to $U_\alpha^\mu(x)$. This can be done by taking any fixed measure $\gamma \in \mathcal{E}_\alpha^+$, for example, the equilibrium measure for the ball $|x| \leq 1$, and setting

$$U_\alpha^{\mu_n}(x) = \inf\,[U_\alpha^\mu(x), nU_\alpha^\gamma(x)]$$

(cf. Theorem 1.31).

We note that the potentials $U_\alpha^{\beta\mu_n}(x)$ are also monotone increasing. Indeed, quasi-everywhere on K, and consequently $\beta\mu_n$-almost everywhere

$$U_\alpha^{\beta\mu_n}(x) = U_\alpha^{\mu_n}(x) \leq U_\alpha^{\mu_{n+1}}(x) = U_\alpha^{\beta\mu_{n+1}}(x).$$

By the second maximum principle,

$$U_\alpha^{\beta\mu_n}(x) \leq U_\alpha^{\beta\mu_{n+1}}(x) \qquad \text{everywhere in } R^p.$$

Then, at all points of R^p

$$U_\alpha^{\beta\mu_n}(x) \leq U_\alpha^{\mu_n}(x) \leq U_\alpha^\mu(x).$$

Applying Theorem 3.9, we obtain

$$\lim_{n\to\infty} U_\alpha^{\beta\mu_n}(x) = U_\alpha^{\beta\mu}(x)$$

and $\beta\mu_n \dot{\rightarrow} \beta_\mu$. The latter shows that $\beta\mu \in \mathfrak{M}_K^+$. Finally, having written the equation (4.5.16) for $\mu_n \in \mathscr{E}_\alpha^+$, we may, by the Levi theorem, pass to the limit as $n \to \infty$ and obtain (4.5.16) for the original measure μ.

The uniqueness of $\beta\mu$ follows from the fact that the equation

$$\int U_\alpha^{\beta\mu}(x)\, d\lambda(x) = \int U_\alpha^{\varkappa}(x)\, d\lambda(x)$$

for any $\lambda \in \mathscr{E}_\alpha^+$ implies in particular that

$$U_\alpha^{\beta\mu} * m^{(r)} = U_\alpha^{\varkappa} * m^{(r)}$$

whence, by Theorem 1.11, it follows that $U_\alpha^{\beta\mu}(x) \equiv U_\alpha^{\varkappa}(x)$ and $\beta\mu = \varkappa$.

Finally, writing the relations

$$U_\alpha^{\beta\mu_n}(x) = U_\alpha^{\varkappa_n}(x) \qquad \text{quasi-everywhere on } K,$$

$$U_\alpha^{\beta\mu_n}(x) \leq U_\alpha^{\mu_n}(x) \qquad \text{everywhere}$$

and letting $n \to \infty$, we obtain (4.5.16′). ☐

One should be reminded that the relations (4.5.16′) do not by themselves uniquely determine the measure $\beta\mu$. For example, if x_0 is an irregular point of K and $\mu = \varepsilon_{x_0}$, then any measure

$$m\varepsilon_{x_0} + (1 - m)\, \varepsilon'_{x_0} \quad (0 < m < 1),$$

where ε'_{x_0} is the Green measure considered in No. 20, satisfies conditions (4.5.16′).

§ 6. Balayage onto Borel Sets

In this section we will assume that $0 < \alpha \leq 2$ and $\alpha < p$. Suppose E is an arbitrary Borel set. Our aim is to construct the balayage operator β_E. The attempt to define it by the conditions

$$\begin{aligned} &U_\alpha^{\beta_E\mu}(x) = U_\alpha^\mu(x) && \text{quasi-everywhere on } E, \\ &U_\alpha^{\beta_E\mu}(x) \leq U_\alpha^\mu(x) && \text{everywhere in } R^p, \\ &\beta_E\mu && \text{is concentrated on } E, \end{aligned} \tag{4.6.1}$$

fails, since there are simple examples (an open ball in R^p for $\alpha = 2$) which show that these requirements are incompatible. We will require less, namely, that $S(\beta_E\mu) \subset \overline{E}$, i.e. that $\beta_E\mu$ is concentrated on $\overline{E}$. With this approach, however, it turns out that $\beta_E\mu$ is not uniquely defined. One can get around this difficulty by demanding that the operator β_E as a set function of E, be weakly continuous on the left; this means that we want to define β_E as the weak limit of a generalized sequence $\{\beta_K\}$ where $K \subset E$ are compact sets ordered by inclusion.

We use the equilibrium measure $\gamma = \gamma_E$ constructed in Ch. II, and, applying the Kelvin transformation, arrive, at our goal. Note that $\beta_E\mu$ can also be characterized by the fact that the potential $U_\alpha^{\beta_E\mu}(x)$ is the least potential of a measure satisfying conditions (4.6.1).

24. Suppose that $E \subset R^p$ is an arbitrary *Borel* set satisfying the condition

$$C_\alpha(E) < \infty$$

By the results of Ch. II (Theorems 2.6, 2.7, and 2.8) there exists a unique measure γ, which we called the equilibrium measure for E, which satisfies the following conditions:

(a) $\gamma(1) = \|\gamma\|^2 = C_\alpha(E)$,

(b) $U_\alpha^\gamma(x) = 1$ quasi-everywhere on E,

(c) $U_\alpha^\gamma(x) \leq 1$ everywhere in R^p,

(d) $S(\gamma) \subset \overline{E}$.

This measure was obtained as the vague (and also the strong) limit of equilibrium measure γ_n of compact sets K_n, satisfying the conditions

$$K_n \subset K_{n+1} \subset E; \quad C_\alpha(K_n) \to C_\alpha(E). \tag{4.6.2}$$

For certain arguments in this and the succeeding chapter it will be more convenient to consider, instead of the sequence of compact sets K_n satisfying (4.6.2), the *generalized sequence* (*filter*) $\mathfrak{K}$ of all compact subsets of E, ordered by inclusion. Then the generalized sequence of equilibrium measures γ_K will converge on this filter (strongly, and hence vaguely) to the measure γ.[37] This situation, which will be denoted by

$$\lim_{\mathfrak{K}} \gamma_K = \gamma \quad \text{or } \gamma_K \xrightarrow{\mathfrak{K}} \gamma$$

immediately follows from the considerations in the proof of Theorem 2.6.[38]

---

[37] Here is the definition of vague convergence for the generalized sequence γ_K: for any function $f \in \Phi$ and any $\eta > 0$ there is a compact set $K_0(f, \eta)$ such that for $K \supset K_0$

$$|\gamma_K(f) - \gamma(f)| < \eta.$$

To become familiar with the properties of generalized sequences and their limits one can read. Dunford and Schwartz [1], Ch.1, General theory, I.7.

We point out that the theorems of Ch.III, § 3 carry over without any difficulty to generalized sequences of potentials.

[38] Here it should be noted that $\underline{C}_\alpha(E) = \lim_{\mathfrak{K}} C_\alpha(K)$ and that Theorem 1.18 (on the completeness of $\mathscr{E}_\alpha^+$) remains valid if instead of fundamental sequences we consider fundamental generalized sequences.

From this we obtain the following extremal property of the equilibrium potential $U_\alpha^\gamma(x)$.

Lemma 4.5. *The potential $U_\alpha^\gamma(x)$ is the lower bound of the potentials of measure $U_\alpha^\mu(x)$ satisfying the condition*

$$U_\alpha^\mu(x) \geq 1 \tag{4.6.3}$$

quasi-everywhere on E.

Proof. Since, according to (b), $U_\alpha^\gamma(x)$ satisfies condition (4.6.3), we need only show that (4.6.3) implies the inequality

$$U_\alpha^\gamma(x) \leq U_\alpha^\mu(x), \quad x \in R^p. \tag{4.6.4}$$

Let us consider the equilibrium potential $U_\alpha^{\gamma_n}(x)$ of the compact set K_n. Since $U_\alpha^\mu(x) \geq 1$ quasi-everywhere on K_n, then

$$U_\alpha^{\gamma_n}(x) \leq U_\alpha^\mu(x)$$

quasi-everywhere on the support $S(\gamma_n) \subset K_n$ of the measure γ_n i.e. γ_n-almost everywhere. According to the second maximum principle this implies the inequality

$$U_\alpha^{\gamma_n}(x) \leq U_\alpha^\mu(x)$$

everywhere in R^p. But $U_\alpha^\gamma(x)$ is the limit of a monotone increasing sequence of potentials $\{U_\alpha^{\gamma_n}(x)\}$ (cf. Theorem 3.9), and hence as $n \to \infty$, we obtain (4.6.4). □

Corollary. *If the potential of the measure μ satisfies the inequality*

$$U_\alpha^\mu(x) \geq 1 \qquad \text{quasi-everywhere on } E,$$

then $\mu(1) \geq C_\alpha(E)$.

We now turn to the construction of $\beta_E \varepsilon_{x_0}$, i.e. balayage for the Dirac measure ε_{x_0} for $x_0 \notin \overline{E}$.

As in No. 20 we use the inversion transformation with pole at x_0 and denote by E^* the image of the set E under this inversion. E^* is bounded and hence

$$C_\alpha(E^*) < \infty. \tag{4.6.5}$$

As we have seen, there exists an equilibrium measure γ for E^*; consider its Kelvin transform ε'_{x_0}. By (4.5.4) and (4.5.3)

$$U_\alpha^{\varepsilon'_{x_0}}(x) = \frac{1}{|x - x_0|^{p-\alpha}} U_\alpha^{\gamma^*}(x^*); \quad \varepsilon'_{x_0}(1) = U_\alpha^{\gamma^*}(x_0).$$

From this one sees that the measure $\varepsilon'_{x_0} = \beta_E \varepsilon_{x_0}$ satisfies the conditions

$$U_\alpha^{\varepsilon'_{x_0}}(x) = \frac{1}{|x - x_0|^{p-\alpha}} \quad \text{quasi-everywhere on } E,$$

$$U_\alpha^{\varepsilon'_{x_0}}(x) \leq \frac{1}{|x - x_0|^{p-\alpha}} \quad \text{everywhere in } R^p.$$

It actually solves the balayage problem for ε_x onto E, since by Lemma 4.3, which trivially extends to generalized sequences, we can conclude that

$$\beta_K \varepsilon_{x_0} \xrightarrow{\mathfrak{K}} \beta_E \varepsilon_{x_0}. \tag{4.6.6}$$

Here, by monotone convergence (cf. Theorem 3.9)

$$U_\alpha^{\beta_E \varepsilon_{x_0}}(x) = \lim_{\mathfrak{K}} U_\alpha^{\beta_K \varepsilon_{x_0}}(x).$$

From this or directly from Lemma 4.5, it follows that the potential $U_\alpha^{\beta_E \varepsilon_{x_0}}$ can be characterized as the lower bound of potentials of measures $U_\alpha^\mu(x)$ which satisfy the condition

$$U_\alpha^\mu(x) \geq \frac{1}{|x - x_0|^{p-\alpha}} \quad \text{quasi-everywhere on } E.$$

Remark. Suppose that

$$I_{x_0} = \{x \in E : U_\alpha^{\beta_E \varepsilon_{x_0}}(x) < U_\alpha^{\varepsilon_{x_0}}(x)\}, \quad x_0 \in C\overline{E}.$$

According to the foregoing, the set I_{x_0} has capacity zero. We prove that this set does not depend on the position of the point x_0 when the latter is varying within any component G of the open set $C\overline{E}$.

Suppose $x \in E \setminus I_{x_0}$, so that

$$U_\alpha^{\varepsilon_{x_0}}(x) - U_\alpha^{\beta_E \varepsilon_{x_0}}(x) = 0, \quad x_0 \in G.$$

We denote $\mathfrak{K}_x$ the generalized sequence of the compact sets contained in $E \setminus \{x\}$. According to (4.6.6)

$$\beta_K \varepsilon_{x_0} \xrightarrow{\mathfrak{K}_x} \beta_{E \setminus \{x\}} \varepsilon_{x_0}.$$

Taking into account (4.6.2) and the equality

$$C_\alpha(E \setminus \{x\}) = C_\alpha(E),$$

we conclude that

$$\beta_{E \setminus \{x\}} \varepsilon_{x_0} = \beta_E \varepsilon_{x_0},$$

from which

$$U_\alpha^{\beta_E \varepsilon_{x_0}} = \lim_{\mathfrak{K}_x} U_\alpha^{\beta_K \varepsilon_{x_0}}.$$

Since $x, x_0 \notin K$, then by the symmetry of the Green's function (see p. 263) we can write

$$U_\alpha^{\varepsilon_{x_0}}(x) - U_\alpha^{\beta_K \varepsilon_{x_0}}(x) = U_\alpha^{\varepsilon_x}(x_0) - U_\alpha^{\beta_K \varepsilon_x}(x_0) \geq 0 .$$

Now, on the basis the theorem of uniform convergence (p. 266) we can conclude that the equality

$$U_\alpha^{\varepsilon_{x_0}}(x) - U_\alpha^{\beta_E \varepsilon_{x_0}}(x) = 0 , \qquad x_0 \in G ,$$

implies for any point $x_1 \in G$, that

$$U_\alpha^{\varepsilon_{x_1}}(x) - U_\alpha^{\beta_E \varepsilon_{x_1}}(x) = 0 ,$$

i.e. $x \in E \setminus I_{x_1}$. This proves that $I_{x_0} = I_{x_1}$.

The definition of $\beta_E \varepsilon_{x_0}$ for $x_0 \in \overline{E}$ encounters the following difficulty. The set E^* may not be bounded and may have infinite capacity. In this case the equilibrium measure γ^* may not exist. The consideration of this question is intimately connected with the theory of irregular points of the set E; we defer this to the next chapter.

25. We now turn to the definition of the operator β_E for any measure μ concentrated on the complement of $\overline{E}$. Define

$$\beta_E \mu = \int_{C\overline{E}} \beta_E \varepsilon_x \, d\mu(x) . \tag{4.6.7}$$

Then $\beta_E \mu$ will be concentrated on $\overline{E}$ and by means of the remark at the end of No. 24 it is easy to verify the validity of (4.6.1).

We will show that

$$\beta_K \mu \xrightarrow{\mathfrak{K}} \beta_E \mu . \tag{4.6.6'}$$

According to (4.6.6) for any function $f \in \Phi$ and any $x \in C\overline{E}$

$$\lim_{\mathfrak{K}} \beta_K \varepsilon_x(f) = \beta_E \varepsilon_x(f) .$$

To justify the formula

$$\lim_{\mathfrak{K}} \int \beta_K \varepsilon_x(f) \, d\mu(x) = \int \beta_E \varepsilon_x(f) \, d\mu(x) ,$$

equivalent to (4.6.6'), we will show that

$$\beta_K \varepsilon_x(f) = O(|x|^{\alpha - p})$$

uniformly relative to $K \in \mathfrak{K}$. For this we note that

$$|\beta_K \varepsilon_x(f)| \leq \int |f(y)| \, d\beta_K \varepsilon_x(y) \leq M \beta_K \varepsilon_x(S(f)) ,$$

where $M = \max |f(y)|$. Write

$$K_1 = K \cap S(f) .$$

Then by (4.5.6) we obtain

$$\beta_K \varepsilon_x(S(f)) = \beta_K \varepsilon_x(K_1) \leq \beta_{K_1} \varepsilon_x(K_1) \leq \beta_{S(f)} \varepsilon_x(S(f)).$$

Furthermore (4.5.6′) yields

$$\beta_{S(f)} \varepsilon_x(S(f)) = U_\alpha^\gamma(x) \sim \frac{\gamma(1)}{|x|^{p-\alpha}},$$

where γ is the equilibrium measure of the compact set $S(f)$, q.e.d.

Using the second maximum principle, it is easy to show that the generalized sequence of potentials $\{U_\alpha^{\beta_K \mu}(x)\}$ is increasing. In that case, as in Lemma 4.5, we see that the potential $U_\alpha^{\beta_E \mu}(x)$ is the lower bound of potentials of measures λ which satisfy the condition

$$U_\alpha^\lambda(x) \geq U_\alpha^\mu(x) \quad \text{quasi-everywhere on } E. \tag{4.6.8}$$

This condition completely characterizes the measure $\beta_E \mu$.

We point out in conclusion that the following reciprocity law holds:

$$U_\alpha^{\beta_E \varepsilon_x}(y) = U_\alpha^{\beta_E \varepsilon_y}(x), \quad x, y \notin \overline{E}. \tag{4.6.9}$$

It can be obtained by writing according to (4.2.16) (No. 11) the equation

$$U_\alpha^{\beta_K \varepsilon_x}(y) = U_\alpha^{\beta_K \varepsilon_y}(x)$$

and then passing to the limit through the filter $\mathfrak{K}$.

If (4.6.9) is integrated over the two measures $d\mu_1(x)$ and $d\mu_2(y)$, concentrated outside of $\overline{E}$, then by (4.6.7) we obtain the formula

$$\int U_\alpha^{\beta_E \mu_1}(y)\, d\mu_2(y) = \int U_\alpha^{\beta_E \mu_2}(x)\, d\mu_1(x). \tag{4.6.10}$$

In the next chapter it will shown that this formula remains valid for any measures μ_1, μ_2.

Chapter V

Irregular Points

§ 1. Irregular Points of Borel Sets. Criteria for Irregularity

From now until No. 6 we shall assume that $0 < \alpha \leq 2$, $\alpha < p$.

1. The connections between the balayage problem and the equilibrium problem can be used to define and study irregular points; with this in mind, we consider the following question.

Suppose there is given an arbitrary Borel set $E \subset R^p$. *Under what conditions does there exist an equilibrium measure* γ_E? In Ch. II we saw that $C_\alpha(E) < \infty$ is a sufficient condition for this. By Theorems 2.6, 2.7, and 2.8 this measure $\gamma = \gamma_E$ can be characterized by the conditions

(a) $\gamma(1) = C_\alpha(E)$,

(b) $U_\alpha^\gamma(x) = 1$ quasi-everywhere on E,

(c) $S(\gamma) \subset \overline{E}$.

Moreover, it has finite energy and $\|\gamma\|^2 = C_\alpha(E)$. In making the transition to sets of infinite capacity we must more precisely define what is meant by an equilibrium measure, since conditions (b) and (c) *do not determine* (in the general case of a non-closed E) the measure γ_E uniquely. We will require continuity on the left, i.e. the relations

$$\gamma_{En} \dashrightarrow \gamma_E$$

for any increasing sequence of Borel sets E_n, which satisfy the conditions

$$C_\alpha(E_n) < \infty, \quad \lim E_n = E.$$

As will be shown below, the equilibrium measure γ can also be characterized by the following extremal property: *its potential $U_\alpha^\gamma(x)$ is the greatest lower bound of potentials of measures μ, which satisfy the condition*

$$U_\alpha^\mu(x) \geq 1$$

quasi-everywhere on E (cf. Lemma 4.5)

Theorem 5.1. *Suppose that the positive numbers r_n monotonically tend to infinity. Set*

$$E_n = E \cap \{x: r_n \leq |x| < r_{n+1}\}, \quad n = 0, 1, 2, \ldots; \; r_0 = 0.$$

Then for the existence of the equilibrium measure γ_E it is sufficient that

$$\sum_{n=1}^{\infty} \frac{C_\alpha(E_n)}{r_n^{p-\alpha}} < \infty. \tag{5.1.1}$$

If in addition we require that

$$1 < q \leq \frac{r_{n+1}}{r_n} \leq Q \quad (Q < q^2), \tag{5.1.2}$$

then condition (5.1.1) *will also be necessary for the existence of γ_E.*[1]

Proof. We introduce the sets

$$E^{(n)} = \bigcup_{k=0}^{n} E_k = E \cap \{x: |x| < r_{n+1}\}$$

and their equilibrium measures γ_n. The potentials $U_\alpha^{\gamma_n}(x)$ form a monotone increasing sequence, as is easily shown by Lemma 4.5. We will show that under condition (5.1.1) there is a potential $U_\alpha^\mu(x)$ which majorizes this sequence. Indeed, denoting by $\varkappa_n$ the equilibrium measure of E_n, we set

$$\mu = \sum_{n=1}^{\infty} \varkappa_n$$

and assert that $U_\alpha^\mu(x) \not\equiv \infty$. For this, by (1.3.10), one must verify that

$$\int\limits_{|x| \geq r_1} \frac{d\mu(x)}{|x|^{p-\alpha}} < \infty.$$

But

$$\int\limits_{|x| \geq r_1} \frac{d\mu(x)}{|x|^{p-\alpha}} = \sum_{n=1}^{\infty} \int\limits_{E_n} \frac{d\mu(x)}{|x|^{p-\alpha}} = \sum_{n=1}^{\infty} \int\limits_{E_n} \frac{d\varkappa_n(x)}{|x|^{p-\alpha}}$$

$$\leq \sum_{n=1}^{\infty} \frac{1}{r_n^{p-\alpha}} \int\limits_{E_n} d\varkappa_n = \sum_{n=1}^{\infty} \frac{C_\alpha(E_n)}{r_n^{p-\alpha}} < \infty.$$

Now on the basis of Theorem 3.9, we conclude that $\gamma_n \dot{\to} \gamma$ and $U_\alpha^\gamma(x) = \lim_{n\to\infty} U_\alpha^{\gamma_n}(x)$. The measure γ is the desired one. Indeed, since

[1] It will be clear from what follows (cf. No. 3, (5.1.9″)) that the condition $Q < q^2$ is not essential.

$S(\gamma_n) \subset \bar{E}$, then $S(\gamma) \subset \bar{E}$. Also $U_\alpha^\gamma(x) = 1$ quasi-everywhere on E, since this is true quasi-everywhere on $E^{(n)}$ for each n.

Complicating the construction somewhat, it is possible to replace the sets $E^{(n)}$ with an increasing sequence of compact sets $K^{(n)}$. Even more convenient is to consider the generalized sequence $\mathfrak{K}$ of all compact subsets $\mathfrak{K} \subset E$, ordered by inclusion.

Finally, we show that

$$U_\alpha^\gamma(x) = \inf_\mu U_\alpha^\mu(x)$$

under the condition that

$$U_\alpha^\mu(x) \geq 1$$

quasi-everywhere on E. In fact, it follows from the latter condition and Lemma 4.5 that for any n

$$U_\alpha^{\gamma_n}(x) \leq U_\alpha^\mu(x]$$

and therefore,

$$U_\alpha^\gamma(x) \leq U_\alpha^\mu(x).$$

We turn to the proof of the second part of the theorem. Suppose that γ is the equilibrium measure for E. We will show that under condition (5.1.2)

$$\sum_{n=1}^\infty \frac{C_\alpha(E_{2n})}{r_{2n}^{p-\alpha}} < \infty, \quad \sum_{n=1}^\infty \frac{C_\alpha(E_{2n-1})}{r_{2n-1}^{p-\alpha}} < \infty. \tag{5.1.3}$$

Since the convergence of both series is established in a completely analogous way, we confine ourselves with the first of these. Set

$$E' = \bigcup_{n=1}^\infty E_{2n},$$

and denote by γ' the restriction of γ to $\bar{E}'$, and let $\gamma'' = \gamma - \gamma'$. Suppose that $\beta_E\gamma''$ is the result of sweeping γ'' out onto E' and

$$\bar{\gamma}' = \gamma' + \beta_{E'}\gamma''.$$

Then quasi-everywhere on E'

$$U_\alpha^{\bar{\gamma}'}(x) = 1.$$

From the inequality

$$\int\limits_{|x| \geq r_2} \frac{d\bar{\gamma}'(x)}{|x|^{p-\alpha}} < \infty$$

it follows that

$$\sum_{n=1}^{\infty} \frac{\bar{\gamma}'(\bar{E}_{2n})}{r_{2n+1}^{p-\alpha}} \leq \sum_{n=1}^{\infty} \int_{r_{2n} \leq |x| < r_{2n+1}} \frac{d\bar{\gamma}'(x)}{|x|^{p-\alpha}} = \int \frac{d\bar{\gamma}'(x)}{|x|^{p-\alpha}} < \infty,$$

and by (5.1.2)

$$\sum_{n=1}^{\infty} \frac{\bar{\gamma}'(\bar{E}_{2n})}{r_{2n}^{p-\alpha}} < \infty. \tag{5.1.4}$$

Therefore, the theorem will be proved if we can establish the inequality

$$C_\alpha(E_{2n}) \leq M \bar{\gamma}'(\bar{E}_{2n}),$$

where M is a certain constant.

Denote by $\varkappa_n$ the restriction of $\bar{\gamma}'$ to $\bar{E}_{2n}$. Then for $i > n$

$$U_\alpha^{\varkappa_i}(x) \leq \frac{\varkappa_i(\bar{E}_{2i})}{(r_{2i} - r_{2n+1})^{p-\alpha}} = \frac{\bar{\gamma}'(\bar{E}_{2i})}{(r_{2i} - r_{2n+1})^{p-\alpha}}, \quad x \in E_{2n},$$

while for $i < n$

$$U_\alpha^{\varkappa_i}(x) \leq \frac{\varkappa_i(\bar{E}_{2i})}{(r_{2n} - r_{2i+1})^{p-\alpha}} = \frac{\bar{\gamma}'(\bar{E}_{2i})}{(r_{2n} - r_{2i+1})^{p-\alpha}}, \quad x \in E_{2n}.$$

Hence, quasi-everywhere on E_{2n} we have

$$U_\alpha^{\varkappa_n}(x) = U_\alpha^{\bar{\gamma}'}(x) - \sum_{i \neq n} U_\alpha^{\varkappa_i}(x)$$

$$\geq 1 - \sum_{i<n} \frac{\bar{\gamma}'(\bar{E}_{2i})}{(r_{2n} - r_{2i+1})^{p-\alpha}} - \sum_{i>n} \frac{\bar{\gamma}'(\bar{E}_{2i})}{(r_{2i} - r_{2n+1})^{p-\alpha}}.$$

The corollary to Lemma 4.5 shows that

$$\varkappa_n(1) = \bar{\gamma}'(\bar{E}_{2n})$$

$$\geq C_\alpha(E_{2n}) \left[1 - \sum_{i<n} \frac{\bar{\gamma}'(\bar{E}_{2i})}{(r_{2n} - r_{2i+1})^{p-\alpha}} - \sum_{i>n} \frac{\bar{\gamma}'(\bar{E}_{2i})}{(r_{2i} - r_{2n+1})^{p-\alpha}}\right],$$

and we have to find a lower bound to the last factor. For $i > n$

$$\frac{r_{2i} - r_{2n+1}}{r_{2i}} = 1 - \frac{r_{2n+1}}{r_{2i}} \geq 1 - \frac{r_{2n+1}}{r_{2n+2}} \geq 1 - \frac{1}{q},$$

and hence

$$\sum_{i>n} \frac{\bar{\gamma}'(\bar{E}_{2i})}{(r_{2i} - r_{2n+1})^{p-\alpha}} \leq (1 - q^{-1})^{\alpha-p} \sum_{i>n} \frac{\bar{\gamma}'(\bar{E}_{2i})}{r_{2i}^{p-\alpha}}.$$

If $i < n$, then

$$\frac{r_{2n} - r_{2i+1}}{r_{2i}} \geq \frac{r_{2n}}{r_{2n-2}} - \frac{r_{2i+1}}{r_{2i}} \geq q^2 - Q$$

and therefore

$$\sum_{i<n} \frac{\bar{\gamma}'(\bar{E}_{2i})}{(r_{2n} - r_{2i+1})^{p-\alpha}} \leq (q^2 - Q)^{\alpha-p} \sum_{i<n} \frac{\bar{\gamma}'(\bar{E}_{2i})}{r_{2i}^{p-\alpha}}.$$

Let

$$\theta = \max\,[(1 - q^{-1})^{\alpha-p},\ \ (q^2 - Q)^{\alpha-p}].$$

Then

$$1 - \sum_{i<n} \frac{\bar{\gamma}'(\bar{E}_{2i})}{(r_{2n} - r_{2i+1})^{p-\alpha}} - \sum_{i>n} \frac{\bar{\gamma}'(\bar{E}_{2i})}{(r_{2i} - r_{2n+1})^{p-\alpha}} \geq 1 - \theta \sum_{i=1}^{\infty} \frac{\bar{\gamma}'(\bar{E}_{2i})}{r_{2i}^{p-\alpha}}.$$

But the series (5.1.4) converges, and without loss of generality, we may assume that

$$\sum_{i=1}^{\infty} \frac{\bar{\gamma}'(\bar{E}_{2i})}{r_{2i}^{p-\alpha}} < \frac{1}{2\theta}.$$

Indeed, if it were not so, then we could replace the set E' with the set $\bigcup_N^\infty E_{2n}$ having chosen N appropriately and then we could repeat the same reasoning. Consequently,

$$C_\alpha(E_{2n}) \leq 2\bar{\gamma}(\bar{E}_{2n}),$$

and the theorem is proved. ▯

Corollary. *For the existence of an equilibrium measure γ_E for the Borel set E it is necessary and sufficient that*

$$\sum_{n=1}^{\infty} \frac{C_\alpha(E_n)}{q^{n(p-\alpha)}} < \infty, \tag{5.1.5}$$

where $q > 1$, and

$$E_n = E \cap \{x : q^n \leq |x| < q^{n+1}\}.$$

Remark. In Theorem 5.1 and in formula (5.1.5) for E_n one can take the set

$$E \cap \{x : r_n \leq |x - x_0| < r_{n+1}\}$$

and correspondingly

$$E \cap \{x: q^n \leq |x - x_0| < q^{n+1}\}.$$

If γ_E exists, then using the established extremal property, it is easy to prove that

$$\gamma_{E_i} \dot{\to} \gamma_E,$$

if the E_i are monotone increasing and $\cup E_i = E$.

2. Now we will be able to define the balayage $\beta_E \varepsilon_{x_0}$ for any point $x_0 \in \bar{E}$ and to introduce the notion of an irregular point for any Borel set E.

Under inversion with pole at x_0, the set E is mapped into an *unbounded* set E^*. If there is an equilibrium measure for E^*, then looking at its Kelvin transform $\beta_E \varepsilon_{x_0}$ we shall see, just as in Ch. IV, § 6, that $\beta_E \varepsilon_{x_0}$ actually solves the balayage problem for ε_{x_0} onto E. We note that $\beta_E \varepsilon_{x_0}$ as well as γ^* is a C-absolutely continuous measure.

From the extremal property of the potential $U_\alpha^{\gamma^*}(x)$ (cf. proof of Theorem 5.1) it follows that $U_\alpha^{\beta_E \varepsilon_{x_0}}(x)$ also has the corresponding extremal property, namely,

$$U_\alpha^{\beta_E \varepsilon_{x_0}}(x) = \inf_\alpha U_\alpha^\mu(x)$$

under the condition

$$U_\alpha^\mu(x) \geq \frac{1}{|x - x_0|^{p-\alpha}}$$

quasi-everywhere on E. This property uniquely determines the potential $U_\alpha^{\beta_E \varepsilon_{x_0}}(x)$ in the whole space and, consequently, uniquely defines the measure $\beta_E \varepsilon_{x_0}$. We also note that

$$U_\alpha^{\beta_E \varepsilon_{x_0}}(x) = \lim\nolimits_{\mathfrak{R}} U_\alpha^{\beta_K \varepsilon_{x_0}}[x).$$

If γ^* does not exist, then it is not hard to see that there is no measure distinct from ε_{x_0} whose potential is equal to $|x - x_0|^{\alpha - p}$ quasi-everywhere on E. In this case we set $\beta_E \varepsilon_{x_0} = \varepsilon_{x_0}$ and call the point x_0 *regular for* E.

An irregular point for E *is a point* $x_0 \in \bar{E}$ *for which* $\beta_E \varepsilon_{x_0} \neq \varepsilon_{x_0}$.

Only the points of the boundary ∂E can be irregular. Indeed, if $x_0 \in \mathring{E}$, then E^* contains the exterior of a certain ball and by Theorem 5.1 the equilibrium measure γ^* can not exist.

Now, supplementing Ch. IV, No. 25 we are able to define the balayage operator β_E for any finite measure μ. We set

$$\beta_E\mu = \int\limits_{S(\mu)} \beta_E\varepsilon_x d\mu(x).^2$$

We list the properties of the measure $\beta_E\mu$ which follow from this definition and the corresponding properties of $\beta_E\varepsilon_{x_0}$.

(1) The measure $\beta_E\mu$ is concentrated on $\bar{E}$.

(2) Quasi-everywhere on E

$$U_\alpha^{\beta_E\mu}(x) = U_\alpha^\mu(x).$$

(3) Everywhere in R^p

$$U_\alpha^{\beta_E\mu}(x) \leq U_\alpha^\mu(x).$$

(4) $U_\alpha^{\beta_E\mu}(x) = \lim_{\mathfrak{K}} U_\alpha^{\beta_K\mu}(x)$.

(5) $U_\alpha^{\beta_E\mu}(x) = \inf\limits_\lambda U_\alpha^\lambda(x)$, where $U_\alpha^\lambda(x) \geq U_\alpha^\mu(x)$ quasi-everywhere on E.

(6) $U_\alpha^{\beta_E\varepsilon_x}(y) = U_\alpha^{\beta_E\varepsilon_y}(x)$ for any configuration of the points x, y. From this, as in Ch. IV, § 6 we obtain the formula

$$\int U_\alpha^{\beta_E\mu_1}(x)\, d\mu_2(x) = \int U_\alpha^{\beta_E\mu_2}(x)\, d\mu_1(x) = \int U_\alpha^{\mu_1}(x)\, d\beta_E\mu_2(x)$$

for any measures.

(7) If $E_1 \supset E_2$, then $\beta_{E_2}\mu = \beta_{E_2}(\beta_{E_1}\mu)$.
In particular

$$\beta_E\mu = \beta_E(\beta_E\mu),$$

i.e. $\beta_E^2 = \beta_E$.

Properties (1)—(6) are proved just as the analogous properties in Ch. IV, § 6. Property (7) follows from (5).

(8) Let $E_{\mathrm{I}} \subset \partial E$ denote the set of all irregular points of E. Then

$$\beta_E\mu(E_{\mathrm{I}}) = 0.$$

Otherwise, setting

$$\beta_E\mu = \lambda_1 + \lambda_2,$$

[2] It is easy to prove that the family of measure $\{\beta_E\varepsilon_x\}$, $x \in R^p$ is Borel, if one takes into consideration that

$$\beta_K\varepsilon_x \overset{\mathfrak{K}}{\to} \beta_E\varepsilon_x.$$

where $\lambda_2 \neq 0$ is the restriction of $\beta_E \mu$ to E_{I}, we would have

$$\beta_E^2 \mu = \beta_E \lambda_1 + \beta_E \lambda_2 = \lambda_1 + \beta_E \lambda_1 \neq \lambda_1 + \lambda_2 = \beta_E \mu,$$

which contradicts property (7).

(9) Denote by $E_{\mathrm{II}} \subset \bar{E}$ the set of all points in R^p, regular for E, and by $\mathfrak{M}^+_{E_{\mathrm{II}}} \equiv \mathfrak{M}^+_{\mathrm{II}}$ the set of all measures concentrated on E_{II}. Then

$$\bar{E} = E_{\mathrm{I}} \cup E_{\mathrm{II}},$$

and by (1) and (8) $\mathfrak{M}^+_{\mathrm{II}}$ coincides with the range of values of the operator β_E.

(10) $\mu \in \mathfrak{M}^+_{\mathrm{II}}$ if, and only if, for any measure λ

$$\int U^\lambda_\alpha(x)\, d\mu(x) = \int U^{\beta_E \lambda}_\alpha(x)\, d\mu(x).$$

Indeed, by (6) this equation is equivalent to:

$$\int U^\lambda_\alpha(x)\, d\mu(x) = \int U^\lambda_\alpha(x)\, d\beta_E \mu(x),$$

which, by the arbitrariness of λ, is equivalent to $\mu \equiv \beta_E \mu$ (cf. Lemma 1.1), i.e. $\mu \in \mathfrak{M}^+_{\mathrm{II}}$.

(11) In (10) setting $\mu \equiv \varepsilon_{x_0}$, we obtain that in order for a point x_0 to be regular it is necessary and sufficient that for any measure λ

$$U^{\beta_E \lambda}_\alpha(x_0) = U^\lambda_\alpha(x_0).$$

Consequently, if x_0 is irregular, then there exists a measure λ for which

$$U^{\beta_E \lambda}_\alpha(x_0) < U^\lambda_\alpha(x_0).$$

(12) If $U^\mu_\alpha(x) \le U^\lambda_\alpha(x)$ quasi-everywhere in E, then $U^\mu_\alpha(x) \le U^\lambda_\alpha(x)$ on E_{II}.

In fact, (5) implies that everywhere in R^p

$$U^{\beta_E \mu}_\alpha(x) \le U^\lambda_\alpha(x).$$

But on E_{II} by (11)

$$U^{\beta_E \mu}_\alpha(x) = U^\mu_\alpha(x),$$

which implies (12).

(13) If $\mu \in \mathfrak{M}^+_{\mathrm{II}}$ and

$$U^\mu_\alpha(x) \le U^\lambda_\alpha(x) \qquad \text{quasi-everywhere on } E,$$

then

$$U^\mu_\alpha(x) \le U^\lambda_\alpha(x) \qquad \text{everywhere in } R^p.$$

This follows from (5) if one takes into account that $\beta_E\mu \equiv \mu$. Thus, the following *uniqueness theorems* hold:

(14) If $\mu_1, \mu_2 \in \mathfrak{M}_{\mathrm{II}}^+$ and

$$U_\alpha^{\mu_1}(x) = U_{\alpha_1}^{\mu_2}(x) \qquad \text{quasi-everywhere on } E$$

then $\mu_1 \equiv \mu_2$.

(15) The measure $\beta_E\mu$ is the unique measure in $\mathfrak{M}_{\mathrm{II}}^+$ whose potential is equal to $U_\alpha^\mu(x)$ quasi-everywhere on E.

3. It is important to formulate necessary and sufficient conditions for the irregularity of the point $x_0 \in \partial E$ in terms of the set E itself, without relying on E^*. This is achieved by the following theorem.

Theorem 5.2. (*Wiener Criterion*). *Suppose that* $0 < q < 1$ *and*

$$E_n = E \cap \{x : q^{n+1} \le |x - x_0| < q^n\}.$$

Then for the irregularity of the point $x_0 \in \partial E$ *it is necessary and sufficient that*

$$\sum_{n=1}^{\infty} \frac{C_\alpha(E_n)}{q^{n(p-\alpha)}} < \infty. \tag{5.1.6}$$

Proof. Let us perform an inversion with pole at x_0; E goes over into a set E^* and the E_n into the sets

$$E_n^* = E^* \cap \{x : q^{-n} < |x - x_0| \le q^{-n-1}\}.$$

From (4.5.1) it can be seen that for any two points $x, y \in E_n$

$$q^{-2n} |x - y| \le |x^* - y^*| \le q^{-2n-2} |x - y|.$$

But then, on the basis of Ch. II, No. 11

$$q^{-2n(p-\alpha)} C_\alpha(E_n) \le C_\alpha(E_n^*) \le q^{-(2n+2)(p-\alpha)} C_\alpha(E_n).$$

Condition (5.1.5), necessary and sufficient for the existence of the equilibrium measure γ_E, can be written in the form

$$\sum_{n=1}^{\infty} C_\alpha(E_n^*)\, q^{n(p-\alpha)} < \infty.$$

By the preceding inequality, it is equivalent to (5.1.6) and the theorem is proved.

Example. Consider in R^3 a closed volume of revolution K which in spherical coordinates is given by the conditions

$$\theta \le f(\varrho), \quad \varrho \le 1,$$

where $f(\varrho)$ is a monotone increasing function and $f(0) = 0$, $f(1) < \pi$.

Suppose $0 < q < 1$ and $\theta_n = f(q^n)$. Then the set

$$K_n = K \cap \{x : q^{n+1} \leq |x| \leq q^n\}$$

is included in the right circular cylinder of height q^n and base radius $q^n \sin \theta_n$ and moreover in the ellipsoid with semi-axes $2q^n \sin \theta_n$, $2q^n \sin \theta_n$, $2q^n$.

Using the table in Ch. II, No. 14 we obtain

$$C_2(K_n) < \frac{4q^n \cos \theta_n}{\pi \ln \cot \theta_n/2}.$$

On the other hand, K_n contains a right circular cylinder of height $q^n - q^{n+1}$ with base radius $q^{n+1} \sin \theta_{n+1}$ and, hence, the ellipsoid with semi-axes $q^{n+1} \sin \theta_{n+1}$, $q^{n+1} \sin \theta_{n+1}$, $q^n(1-q)/2$.

Replace it with a smaller ellipsoid with semi-axes

$$\frac{1}{2} q^{n+1}(1-q) \sin \theta_{n+1}, \quad \frac{1}{2} q^{n+1}(1-q) \sin \theta_{n+1}, \quad \frac{1}{2} q^{n+1}(1-q);$$

then

$$C_2(K_n) > \frac{q^{n+1}(1-q) \cos \theta_{n+1}}{\pi \ln \cot \theta_{n+1}/2}$$

and thus,

$$\frac{q(1-q) \cos \theta_{n+1}}{\pi \ln \cot \theta_{n+1}/2} < \frac{C_2(K_n)}{q^n} < \frac{4 \cos \theta_n}{\pi \ln \cot \theta_n/2}.$$

This shows, that a cusp 0 of the volume K will be a regular point of K if, and only, if the series

$$\sum_{n=1}^{\infty} \frac{\cos \theta_n}{\ln \cot \theta_n/2}.$$

diverges. Since $\theta_n \to 0$ the divergence of this series is equivalent to the divergence of the series

$$\sum_{n=1}^{\infty} \frac{1}{\ln \theta_n} = \sum_{n=1}^{\infty} \frac{1}{\ln f(q^n)}.$$

Therefore, an "algebraic" cusp of the function $f(\varrho) = A\varrho^m$ is a regular point, while the function $f(\varrho) = Ae^{-m/\varrho}$ generates a cusp which is an irregular point of K (Lebesgue cusp).

The Wiener criterion can be reduced to other forms sometimes more convenient. Set

$$E^{(n)} = E \cap \{x : |x - x_0| < q^n\}, \quad 0 < q < 1.$$

Then for the irregularity of the point x_0 it is necessary and sufficient that

$$\sum_{n=1}^{\infty} \frac{C_\alpha(E^{(n)})}{q^{n(p-\alpha)}} < \infty. \tag{5.1.7}$$

Indeed,

$$C_\alpha(E_n) \le C_\alpha(E^{(n)}).$$

Consequently, (5.1.7) implies (5.1.6) and the irregularity of x_0. Also

$$C_\alpha(E^{(n)}) \le C_\alpha(E_n) + C_\alpha(E^{(n+1)}),$$

from which

$$\sum_{n=1}^{N} \frac{C_\alpha(E_n)}{q^{n(p-\alpha)}} \ge \sum_{n=1}^{N} \frac{C_\alpha(E^{(n)})}{q^{n(p-\alpha)}} - q^{p-\alpha} \sum_{n=1}^{N} \frac{C_\alpha(E^{(n+1)})}{q^{(n+1)(p-\alpha)}}$$

$$= (1 - q^{p-\alpha}) \sum_{n=1}^{N} \frac{C_\alpha(E^{(n)})}{q^{n(p-\alpha)}} + C_\alpha(E^{(1)}) - \frac{C_\alpha(E^{(N+1)})}{q^{N(p-\alpha)}}.$$

But $E^{(N+1)}$ is contained in the ball $\{x\colon |x - x_0| < q^{N+1}\}$ whose capacity. according to Ch. II, No. 13 is proportional to $q^{(N+1)(p-\alpha)}$. Hence

$$\frac{C_\alpha(E^{(N+1)})}{q^{N(p-\alpha)}} \le A,$$

and

$$\sum_{n=1}^{N} \frac{C_\alpha(E_n)}{q^{n(p-\alpha)}} \ge (1 - q^{p-\alpha}) \sum_{n=1}^{N} \frac{C_\alpha(E^{(n)})}{q^{n(p-\alpha)}} - A.$$

Consequently, (5.1.6) implies (5.1.7) and the assertion is proved.

Criterion (5.1.7) can be reduced to the form of an integral. For this let us denote by $c_\alpha(\varrho)$ the capacity of the set

$$E \cap \{x\colon |x - x_0| < \varrho\}.$$

This is a non-decreasing function of the variable ϱ, $\varrho > 0$, and we will now show that irregularity criterion (5.1.7) is equivalent to the following:

$$\int_0^1 \frac{c_\alpha(\varrho)}{\varrho^{p-\alpha+1}}\, d\varrho < \infty. \tag{5.1.8}$$

Indeed, for $q < 1$

$$\int_{q^{n+1}}^{q^n} \frac{c_\alpha(\varrho)}{\varrho^{p-\alpha+1}}\, d\varrho \le \frac{c_\alpha(q^n)}{p-\alpha} \left(\frac{1}{q^{(n+1)(p-\alpha)}} - \frac{1}{q^{n(p-\alpha)}} \right) = \frac{1 - q^{p-\alpha}}{(p-\alpha)\, q^{p-\alpha}} \frac{C_\alpha(E^{(n)})}{q^{n(p-\alpha)}}$$

and

$$\int_{q^{n+1}}^{q^n} \frac{c_\alpha(\varrho)}{\varrho^{p-\alpha+1}}\, d\varrho \leq \frac{c_\alpha(\varrho^{n+1})}{p-\alpha}\left(\frac{1}{q^{(n+1)(p-\alpha)}} - \frac{1}{q^{n(p-\alpha)}}\right) = \frac{1-q^{p-\alpha}}{p-\alpha}\, \frac{C_\alpha(E^{(n+1)})}{q^{(n+1)(p-\alpha)}}.$$

Hence

$$\frac{1-q^{p-\alpha}}{p-\alpha} \sum_{n=0}^{\infty} \frac{C_\alpha(E^{(n+1)})}{q^{(n+1)\,(p-\alpha)}} \geq \int_0^1 \frac{c_\alpha(\varrho)}{\varrho^{p-\alpha+1}}\, dp \leq \frac{1-q^{p-\alpha}}{(p-\alpha)\, q^{p-\alpha}} \sum_{n=0}^{\infty} \frac{C_\alpha(E^{(n)})}{q^{n(p-\alpha)}}$$

and conditions (5.1.8) and (5.1.7) are actually equivalent.

If the same calculation is repeated for any monotone decreasing sequence of positive numbers ϱ_n satisfying the condition

$$1 < a < \frac{\varrho_n}{\varrho_{n+1}} < b,$$

then we will obtain the irregularity criterion in the form

$$\sum_{n=1}^{\infty} \frac{c_\alpha(\varrho_n)}{\varrho_n^{p-\alpha}} < \infty. \tag{5.1.9}$$

If instead of $c_\alpha(\varrho)$ we introduce the relative capacity $\mathfrak{C}_\alpha(\varrho)$ of the set $E \cap \{x: |x - x_0| < \varrho\}$, i.e. the ratio of $c_\alpha(\varrho)$ to $\varrho^{p-\alpha}$,[3] then the criteria (5.1.7)—(5.1.9) can be written in the form

$$\sum_{n=1}^{\infty} \mathfrak{C}_\alpha(q^n) < \infty, \tag{5.1.7'}$$

$$\int_0^1 \frac{\mathfrak{C}_\alpha(\varrho)}{\varrho}\, d\varrho < \infty, \tag{5.1.8'}$$

$$\sum_{n=1}^{\infty} \mathfrak{C}_\alpha(\varrho_n) < \infty, \quad 1 < a < \frac{\varrho_n}{\varrho_{n+1}} < b. \tag{5.1.9'}$$

Analogously, the Wiener criterion (5.1.6) could also be rewritten.

In criterion (5.1.9) or (5.1.9′) one could replace the quantity $c_\alpha(\varrho_n)$ with $C_\alpha(E_n)$, where

$$E_n = E \cap \{x: \varrho_{n+1} \leq |x - x_0| < \varrho_n\},$$

i.e. write it in the form

$$\sum_{n=1}^{\infty} \frac{C_\alpha(E_n)}{\varrho_n^{p-\alpha}} < \infty. \tag{5.1.9''}$$

---

[3] The latter quantity differs only by a constant factor from the capacity of the ball $\{x: |x - x_0| < \varrho\}$ (cf. p. 163).

Finally, in place of $C_\alpha(E_n)$ one could introduce the values $U_\alpha^{\gamma_n}(x_0)$, where γ_n is the equilibrium measure of the set E_n. Indeed,

$$\frac{C_\alpha(E_n)}{\varrho_n^{p-\alpha}} \leq U_\alpha^{\gamma_n}(x_0) \leq \frac{C_\alpha(E_n)}{\varrho_n^{p-\alpha}} < \frac{C_\alpha(E_n)}{\varrho_n^{p-\alpha}} b^{p-\alpha}.$$

Thus, a necessary and sufficient condition for the irregularity of the point $x_0 \in \partial E$ takes on yet another form

$$\sum_{n=1}^{\infty} U_\alpha^{\gamma_n}(x_0) < \infty. \tag{5.1.10}$$

4. Now consider several sufficient conditions for regularity of a point.

Theorem 5.3. *If for any sequence $\varrho_n > 0$ monotonically tending to zero*

$$\mathfrak{C}_\alpha(\varrho_n) > m > 0, \tag{5.1.11}$$

then the point x_0 is regular.

Proof. Select from $\{\varrho_n\}$ a subsequence (which we again call $\{\varrho_n\}$) such that

$$\frac{\varrho_n}{\varrho_{n+1}} > a > 1.$$

Now note that

$$\int_{\varrho_{n+1}}^{\varrho_n} \frac{c_\alpha(\varrho)}{\varrho^{p-\alpha+1}} d\varrho \geq \frac{1}{p-\alpha} \frac{c_\alpha(\varrho_{n+1})}{\varrho_{n+1}^{p-\alpha}} \left(1 - \frac{\varrho_{n+1}^{p-\alpha}}{\varrho_n^{p-\alpha}}\right)$$

$$> \frac{1-a^{\alpha-p}}{p-\alpha} \frac{c_\alpha(\varrho_{n+1})}{\varrho_{n+1}^{p-\alpha}} = Q\mathfrak{C}_\alpha(\varrho_{n+1}),$$

and hence by the condition of the theorem we have

$$\int_0^1 \frac{c_\alpha(\varrho)}{\varrho^{p-\alpha+1}} d\varrho = \infty.$$

In view of (5.1.8) we see that the theorem is proved. ▯

The next characterization concerns the *classical* case: $p = 3$, $\alpha = 2$.

Theorem 5.4. *Let $K \subset R^3$ be a compact set, $x_0 \in \partial K$ and*

$$S_\varrho = K \cap \{x : |x - x_0| = \varrho\}.$$

Let us suppose that for $\varrho \leq \varrho_0$, S_ϱ contains a continuum Q_ϱ, where

$$\operatorname{diam} Q_\varrho > \varrho^k \quad (k \geq 1). \tag{5.1.12}$$

Then the point x_0 is regular for K.

Proof. Suppose that $x_0\xi$ and $x_0\eta$ are two perpendicular lines through x_0. Let us define circular projection operators P_ξ, P_η in the following way: for any point $x \in R^3$ $P_\xi x$ and $P_\eta x$ are points in the plane $\xi x_0 \eta$ which lie in the half-planes $\eta \geq 0$ and $\xi \geq 0$ respectively and which have the property that the distance from $P_\xi x$ (respectively, $P_\eta x$) to the axis $x_0\xi$ (respectively, $x_0\eta$) is equal to the distance from x to the same axis. Since the point transformations P_ξ and P_η do not increase the distances between points, then by the corollary to Theorem 2.9 (Ch. II, § 3), for any Borel subset $e \subset K$

$$C_2[P_\xi(e)] \leq C_2(e), \quad C_2[P_\eta(e)] \leq C_2(e).$$

Hence,

$$C_2[P_\xi(e) \cup P_\eta(e)] \leq 2C_2(e),$$

and it will be sufficient to verify that the point x_0 is regular for the plane compact set

$$K_1 = P_\xi(K) \cup P_\eta(K).$$

We now note that $P_\xi(Q_\varrho)$ and $P_\eta(Q_\varrho)$ are arcs of the circle $\xi^2 + \eta^2 = \varrho^2$, where it is not difficult to show that

$$\operatorname{diam} P_\xi(Q_\varrho) + \operatorname{diam} P_\eta(Q_\varrho) \geq \operatorname{diam} Q_\varrho.$$

Consequently,

$$\max [\operatorname{diam} P_\xi(Q_\varrho), \quad \operatorname{diam} P_\eta(Q_\varrho)] \geq \frac{1}{2} \operatorname{diam} Q_\varrho,$$

and this shows that the intersection of K_1 with any circle of radius $\varrho \leq \varrho_0$ with center at x_0 contains an arc whose diameter is greater than $\varrho^k/2$. This intersection, denoted by L_ϱ, is a compact set, and $-\pi/2 \leq \arg z \leq \pi$ for any point zL_ϱ. Suppose that $A_\varrho \in L_\varrho$ is the point with least argument which is an endpoint of an arc $s_\varrho \subset L_\varrho$ of diameter equal to $\varrho^k/2$. Since L_ϱ is closed, such a point exists and the function $\varphi(\varrho) = \arg A_\varrho$ will be lower semi-continuous (and therefore measurable). The other end of the arc s_ϱ will be denoted by B_ϱ,

$$\arg B_\varrho > \arg A_\varrho.$$

In proving the regularity of the point x_0, we will use the Wiener criterion (5.1.6) where for definiteness we set $q = 1/2$.

Denote by K_{1n} the subset of K between the circles of radius 2^{-n-1} and 2^{-n}; we will bound the capacity $C_2(K_{1n})$ from below. For this, we consider the measure μ which on the Borel set

$$H = \bigcup_{2^{-n-1}<\varrho<2^{-n}} s_\varrho$$

coincides with Lebesgue measure in the plane, and outside of it, is equal to zero (this is the restriction of Lebesgue measure to the set H). First of all,

$$\mu(1) = \operatorname{mes} H = \int_{2^{-n-1}}^{2^{-n}} d\varrho \int_{A_\varrho}^{B_\varrho} ds > \frac{1}{2} \int_{2^{-n-1}}^{2^{-n}} \varrho^k \, d\varrho = \alpha 2^{-n(k+1)},$$

where α depends only on k.

Now let us bound the potential

$$U_2^\mu(x) = \int_{2^{-n-1}}^{2^{-n}} d\varrho \int_{A_\varrho}^{B_\varrho} \frac{ds(y)}{|x-y|}.$$

from above.

For this we note that if x' is the closest point to x on the circle containing s_ϱ, then

$$|x-y| > \frac{1}{4}(|x-x'| + |y-x'|),$$

and hence

$$\int_{A_\varrho}^{B_\varrho} \frac{ds(y)}{|x-y|} < 4 \int_{A_\varrho}^{B_\varrho} \frac{ds(y)}{|x-x'|+|y-x'|} < 6 \int_{-\varrho^{k/2}}^{\varrho^{k/2}} \frac{dt}{|x-x'|+|t|}$$

$$< 12 \ln \left| 1 + \frac{\varrho^k}{|x-x'|} \right|.$$

Further

$$|x-x'| = ||x-x_0| - |x'-x_0|| = ||x-x_0| - \varrho|$$

and consequently,

$$U_2^\mu(x) < 12 \int_{2^{-n-1}}^{2^{-n}} \ln \left| 1 + \frac{\varrho^k}{|x-x_0| - \varrho} \right| d\varrho$$

$$< 12 \int_{2^{-n-1}}^{2^{-n}} \ln \left| 1 + \frac{2^{-nk}}{||x-x_0| - \varrho|} \right| d\varrho$$

$$< 24 \int_0^{2^{-n}} \ln \left| 1 + \frac{2^{-nk}}{t} \right| dt \le 24 \, \frac{1 + [1 + n(k-1)] \ln 2}{2^{nk}}.$$

Thus,

$$C_2(K_{1n}) \geq \frac{\mu(1)}{\sup U_2^\mu(x)} \geq \frac{\beta \cdot 2^{-n}}{1 + [1 + n(k-1)] \ln 2},$$

where β depends only on k,

$$\sum_{n=1}^{\infty} \frac{C_2(K_{1n})}{2^{-n}} = \infty,$$

and the point x_0 is regular. □

The assumption that K is compact was used essentially at that point in the proof when it was established that the function $\varphi(\varrho) = \arg A_\varrho$ was lower semicontinuous (and hence measurable). It is not difficult to see, however, that if K is assumed to be open, then the corresponding function $\varphi(\varrho)$ will be upper semicontinuous and, consequently, will also be measurable. This shows the validity of the following theorem.

Theorem 5.4′. *Suppose $G \subset R^3$ is an open set, $x_0 \in \partial G$ and*

$$S_\varrho = G \cap \{x: |x - x_0| = \varrho\}.$$

Assume that for $\varrho \leq \varrho_0$, S_ϱ contains a continuum Q_ϱ with

$$\operatorname{diam} Q_\varrho > \varrho^k \quad (k \geq 1).$$

Then the point x_0 is regular for G.

Theorems 5.4 and 5.4′ can be generalized to any Borel set E, but we will not dwell on that here.

Remark. The conditions of Theorems 5.4 and 5.4′ can be slightly weakened by requiring that (5.1.12) be satisfied for almost all values $\varrho \leq \varrho_0$. The proof remains unchanged.

A similar theorem is

Theorem 5.5. *Suppose that M is a set of positive numbers ϱ containing an infinite sequence of intervals $b_i \leq \varrho \leq a_i$ $(i = 1, 2, \ldots)$ such that $\frac{a_i}{b_i} > \lambda > 1$.*

Preserving the notation of Theorem 5.4, let us assume that for $\varrho \leq M$ the set S_ϱ contains a continuum Q_ϱ such that

$$\operatorname{diam} Q_\varrho > \eta\varrho \quad (\eta > 0). \tag{5.1.13}$$

Then the point x_0 is regular for K.

The proof is basically the same as the preceding except that instead of the Wiener criterion (5.1.6), the sufficiency criterion (5.1.11) is used.

Considering as above the plane set K_1, we look for a lower bound of the capacity of

$$K_{1n} = K \cap \{x: a_n \leq |x - x_0| \leq b_n\}.$$

We obtain, for the restriction of the Lebesgue measure μ, that

$$\mu(1) > \frac{\eta}{4}(a_n^2 - b_n^2) > \frac{\eta}{4} a_n^2 (1 - \lambda^{-2}) = \alpha_1 a_n^2,$$

and for the corresponding potential, wo obtain the bound

$$U_2^\mu(x) < 6 \int_{b_n}^{a_n} \ln \left| 1 + \frac{\eta a_n}{||x - x_0| - \varrho|} \right| d\varrho < \alpha_2 a_n,$$

where α_1, α_2 depend only on η and λ.

Thus

$$C_2(K_{1n}) \geq \alpha_3 a_n \quad \text{and} \quad \mathfrak{C}_2(a_n) \geq \alpha_3,$$

where $\alpha_3 > 0$ depends only on η and λ. By Theorem 5.3, the point x_0 is regular. ▯

5. We now consider some of the consequences of the above criteria.

Here are two propositions which follow directly from the Wiener criterion (5.1.6):

(1) If $E_1 \subset E_2$ and $x_0 \in \partial E_1 \cap \partial E_2$, then if x_0 is irregular for E_2, it is also an irregular point for E_1.

(2) The irregularity of the point x_0 ∂E is a local property, i.e. it is determined by the nature of the set

$$E \cap \{x: |x - x_0| < \eta\} \qquad (\eta > 0 \text{ is arbitrary}).$$

Also the following lemma holds.

Lemma 5.1. *For the irregularity of the point x_0 $E \cap \partial E$ it is necessary and sufficient that it be irregular for any compact set K, $x_0 \in K \subset E$. For the irregularity of the point $x_0 \in \partial E \setminus E$, it is necessary and sufficient that it be irregular for some open set $G \supset E$.*

Proof. The necessity in the first part of the lemma and the sufficiency in the second part follows directly from proposition (1). For the proof of sufficiency in the first part of the lemma, suppose that

$$\sum_{n=1}^{\infty} \frac{C_\alpha(E_n)}{q^{n(p-\alpha)}} = \infty,$$

$$E_n = E \cap \{x: q^{n+1} \leq |x - x_0| < q^n\}, \quad 0 < q < 1,$$

i.e. that x_0 is regular for E. We will show that in this case, there is a compact set $K \subset E$, for which x_0 is also regular. It can be constructed as follows. Take a sequence of numbers $\eta_n > 0$ such that

$$\sum_{n=1}^{\infty} \frac{\eta_n}{q^{n(p-\alpha)}} < \infty,$$

and find a compact set $K_n \subset E_n$ satisfying the condition

$$C_\alpha(K_n) > C_\alpha(E_n) - \eta_n.$$

Then the set $\bigcup_{n=1}^{\infty} K_n \cup \{x_0\}$ will clearly be the required compact set.

Necessity in the second part of the lemma is proved in a completely similar way: if

$$\sum_{n=1}^{\infty} \frac{C_\alpha(E_n)}{q^{n(p-\alpha)}} < \infty,$$

then we can find open sets $G_n \supset E_n$ such that

$$C_\alpha(G_n) < C_\alpha(E_n) + \eta_n.$$

Then the point x_0 will be irregular for the open set

$$G = \bigcup_{n=1}^{\infty} G_n.$$

This follows from

$$G_n' = G \cap \{x: q^{n+1} \leq |x - x_0| < q^n\} \subset G_n,$$

and hence

$$\sum_{n=1}^{\infty} \frac{C_\alpha(G_n')}{q^{n(p-\alpha)}} < \infty. \quad \Box$$

Corollary. *Since the nature of the point $x_0 \in \partial E$ is the same for the sets $E \setminus \{x_0\}$ and $E \cup \{x_0\}$, from Lemma 5.1 it follows that $x_0 \in \partial E$ is irregular if, and only if, one of the following two conditions is fulfilled:*

(*a*) *x_0 is irregular for any compact set K, $x_0 \in K \subset E \cup \{x_0\}$.*

(*b*) *x is irregular for some open set $G \supset E \setminus \{x_0\}$.*

This implies the following generalization of the Kellogg Lemma.

Lemma 5.2. *The set $E \cap E_I$ of irregular points for E, which belong to this set, has capacity zero.*[4]

---

[4] In § 4 it will be shown that the set E_I can have positive capacity.

Proof. Let us assume that $C_\alpha(E \cap E_I) > 0$. Then there exists a compact set $K \subset E \cap E_I$ such that $C_\alpha(K) > 0$. By the Kellogg Lemma there is a point $x_0 \in K$, regular for K. But then from (a) it will follow that x_0 is regular for $E \supset K$, which is impossible. ▯

Finally, we point out the natural generalization of Theorem 4.9.

Lemma 5.3. *If the point $x_0 \in \partial E$ is regular for E and $x_n \to x_0$ then*

$$\beta_E \varepsilon_{x_n} \dot{\to} \varepsilon_{x_0}.$$

We sketch the proof. If this were not true, then there would exist a sequence of points $x_n \to x_0$, a ball $T = \{x: |x - x_0| < r\}$ and a number $\eta > 0$ such that

$$\beta_E \varepsilon_{x_n}(CT) > \eta, \quad n \geqq n_0.$$

Consider the compact set $K \subset E \cup \{x_0\}$. for which x_0 is regular, and the capacity $K \cap CT$ is positive. Then by Theorem 4.9

$$\beta_K \varepsilon_{x_n} \dot{\to} \varepsilon_{x_0}.$$

On the other hand, for any e (cf. (4.5.6))

$$\beta_K \varepsilon_{x_n}(e) = \int_E \beta_K \varepsilon_x(e) \, d\beta_E \varepsilon_{x_n}(x).$$

Consequently,

$$\beta_K \varepsilon_{x_n}(CT) \geq \int_{CT} \beta_K \varepsilon_x(CT) \, d\beta_E \varepsilon_{x_n}(x),$$

and it is not difficult to see that this quantity tends to zero as $n \to \infty$. But this contradicts Theorem 4.9.

6. We now briefly investigate the results concerning the case of a logarithmic potential in the plane, i.e. the case $\alpha = p = 2$. Here we first limit ourselves to compact subsets K in R^2.

The definition of an irregular point of a compact set in the plane was given in Ch. IV, No. 10. There we also saw the local nature of this concept (Corollary to Theorem 4.8), and then in No. 12 we obtained a necessary and sufficient condition for irregularity of the point $x_0 \in K$ in the form

$$U^{\gamma_r}(x_0) < 1 \quad (r < r_0), \tag{5.1.14}$$

where γ_r is the equilibrium measure of the compact set

$$K \cap \{x: |x - x_0| \leq r\}.$$

In this case the Kelvin transformation is not available to us and hence neither is Theorem 5.1.

Nevertheless, it turns out that the following theorem is valid.

Theorem 5.6. *Suppose that*

$$K_n = K \cap \{x : \varrho_{n+1} \leq |x - x_0| \leq \varrho_n\} \quad (1 < a < \varrho_n/\varrho_{n+1} < b)$$

and that γ_n is the equilibrium measure for K_n. Then a necessary and sufficient condition for the irregularity of the point x_0 is that

$$\sum_{n=1}^{\infty} U^{\gamma_n}(x_0) < \infty.$$

Now it is easy to obtain the irregularity criterion in the form

$$\sum_{n=1}^{\infty} C_2(K_n) \ln \frac{1}{\varrho_n} < \infty \quad (1 < a < \varrho_n/\varrho_{n+1} < b). \tag{5.1.15}$$

If we go from the Wiener capacity C_2 to logarithmic capacity C_l (cf. Ch. II, No. 15) by the formula

$$C_2 = \left(\ln \frac{1}{C_l}\right)^{-1},$$

then this criterion may be rewritten as

$$\sum_{n=1}^{\infty} \frac{\ln \varrho_n}{\ln C_l(K_n)} < \infty. \tag{5.1.15'}$$

If in (5.1.15) we set $\varrho_n = q^n$, $0 < q < 1$ then this criterion will take the form

$$\sum_{n=1}^{\infty} n C_2(K_n) < \infty, \tag{5.1.16}$$

where

$$K_n = K \cap \{x : q^{n+1} \leq |x - x_0| \leq q^n\}.$$

The condition $1 < a < \varrho_n/\varrho_{n+1} < b$ can be replaced by

$$1 < a < \frac{\ln \varrho_{n+1}}{\ln \varrho_n} < b$$

Taking $\varrho_n = e^{-q^{-n}}$, $0 < q < 1$ we obtain the necessary and sufficient condition for irregularity in the form

$$\sum_{n=1}^{\infty} \frac{C_2(K_n)}{q^n} < \infty, \tag{5.1.17}$$

where

$$K_n = K \cap \left\{x : q^{-n} < \ln \frac{1}{|x - x_0|} \leq q^{-n-1}\right\}.$$

We point out one consequence of these criteria: if the point x_0 belongs to any continuum contained in K, then it is regular. Indeed the Wiener capacity of the set $K_n = K \cap \{x: q^{n+1} \leq |x - x_0| \leq q^n\}$ for sufficiently large n is not less than the capacity of the interval of length $q^n - q^{n+1} = q_n (1 - q)$.

By Ch. II, § 4, for such an interval

$$C_2 = \left(\ln \frac{4}{q^n(1 - q)}\right)^{-1} \sim \frac{A}{n},$$

and criterion (5.1.16) proves the regularity of the point.

However, this can also be obtained by a conformal mapping of the exterior of the continuum (which one can assume does not separate the plane) onto the unit disc.

This result brings us to the following conclusion: if G is a finitely-connected region of the plane, then the irregular points of ∂G must be isolated.

In addition, the basic results of the theory of irregular points of Borel sets and the theory of balayage can be extended with certain modifications to the case of the logarithmic potential. These modifications deal exclusively with the balayage of a measure concentrated in an infinite component of the complement of a given set (cf. Vallée-Poussin [5]).

The irregularity criteria (5.1.15)—(5.1.17) hold also for any Borel set.

Note, especially, that the uniqueness theorem (14) from No. 2 also holds for a logarithmic potential.

§ 2. The Characteristics and Types of Irregular Points

7. From the results of the preceding chapter and § 1 of this one, one observes that the irregularity of the point $x_0 \in \partial E$ is connected with the smallness as $\varrho \to 0$ of one of the following functions:

(a) the relative capacity $\mathfrak{C}(\varrho)$ of the subset of E within the sphere $|x - x_0| = \varrho$;

(b) the quantity $\omega(\varrho) = U_\alpha^{\gamma_\varrho}(x_0)$, where γ_ϱ is the equilibrium measure of the set $E \cap \{x: |x - x_0| < \varrho\}$;

(c) the size $\Gamma(\varrho)$ of the Green measure $\beta_E \varepsilon_{x_0}$ of the ball $|x - x_0| < \varrho$.

If the point x_0 is regular then $w(\varrho) \equiv 1$ and $\Gamma(\varrho) \equiv 1$ while $\mathfrak{C}(\varrho)$ tends to zero rather slowly. But if x_0 is irregular the hehavior of all three above functions is very similar.

The number θ_1 will be called an *exponent of convergence* for the function $\mathfrak{C}(\varrho)$ if it has the property

$$\int\limits_0 \frac{\mathfrak{C}(\varrho)}{\varrho^{\eta}}\, d\varrho < \infty \quad \text{for} \quad \eta < \theta_1,$$
$$\int\limits_0 \frac{\mathfrak{C}(\varrho)}{\varrho^{\eta}}\, d\varrho = \infty \quad \text{for} \quad \eta > \theta_1. \tag{5.2.1}$$

Furthermore, if

$$\int\limits_0 \frac{\mathfrak{C}(\varrho)}{\varrho^{\theta_1}}\, d\varrho < \infty,$$

then we will say that $\mathfrak{C}(\varrho)$ belongs to the convergence class (of the exponent θ_1), and otherwise, to the divergence class.

The quantities θ_2, θ_3 will denote the exponents of convergence for $w(\varrho)$ and $\Gamma(\varrho)$ respectively.

It is clear that $\theta_1, \theta_2, \theta_3 \geq 1$, and, if x_0 is regular, then $\theta_1 = \theta_2 = \theta_3 = 1$.

By the Wiener criterion, for irregularity of x_0 it is necessary and sufficient that either $\theta_1 > 1$, or $\theta_1 = 1$ and $\mathfrak{C}(\varrho)$ belongs to the convergence class.

Theorem 5.7. *The equation $\theta_1 = \theta_2 = \theta_3$ always holds. If this common value is >1, then all three functions $\mathfrak{C}(\varrho)$, $w(\varrho)$, $\Gamma(\varrho)$ belong either to the convergence class or to the divergence class. If the common value is equal to* 1, *but the point is irregular, then*

$$\lim_{\varrho\to 0} w(\varrho) = \lim_{\varrho\to 0} \Gamma(\varrho) = 0.$$

Proof. It is sufficient to consider the case of an irregular point x_0. By Theorem 5.3 $\lim\limits_{\varrho\to 0} \mathfrak{C}(\varrho) = 0$. We set

$$E^{(\varrho)} = E \cap \{x : |x - x_0| < \varrho\}$$

and write $c(\varrho)$ for the capacity of this set. Then, suppose that for $r \leq \varrho$

$$\gamma_\varrho(r) = \gamma_\varrho(\overline{E^{(r)}}),$$

so that $\gamma_\varrho(r)$ is the γ_ϱ-measure of the ball $|x - x_0| \leq r$. Then

$$\gamma_\varrho(\varrho) = c(\varrho), \qquad w(\varrho) = \int\limits_0^{\varrho} \frac{d\gamma_\varrho(r)}{r^{p-\alpha}},$$

and

$$w(\varrho) = \mathfrak{C}(\varrho) + (p - \alpha) \int_0^{\varrho} \frac{\gamma_\varrho(r)}{r^{p-\alpha+1}} dr > \mathfrak{C}(\varrho). \tag{5.2.2}$$

On the other hand, it is not difficult to see (cf. property (7), No. 2), that $\gamma_\varrho(r) \leq \gamma_r(r) = c(r)$ and hence

$$\mathfrak{C}(\varrho) < w(\varrho) < \mathfrak{C}(\varrho) + (p - \alpha) \int_0^{\varrho} \frac{\mathfrak{C}(r)}{r} dr. \tag{5.2.3}$$

From this follows $\lim_{\varrho \to 0} w(\varrho) = 0$ and the equation $\lim_{\varrho \to 0} \Gamma(\varrho) = 0$ follows from the C-absolute continuity of the Green measure $\beta_E \varepsilon_{x_0}$; thus the last assertion of the theorem is proved.

From (5.2.3) it follows that for any $\eta > 1$, the statements

$$\text{(I)} \int_0 \frac{w(\varrho)}{\varrho^\eta} d\varrho < \infty \quad \text{and} \quad \text{(II)} \int_0 \frac{\mathfrak{C}(\varrho)}{\varrho^\eta} d\varrho < \infty$$

are equivalent. It is obvious that (I) implies (II). But the converse also holds since

$$\int_0^1 \frac{d\varrho}{\varrho^\eta} \int_0^{\varrho} \frac{\mathfrak{C}(r)}{r} dr$$

$$= \frac{1}{\eta - 1} \lim_{\varrho \to 0} \frac{1}{\varrho^{\eta-1}} \int_0^{\varrho} \frac{\mathfrak{C}(r)}{r} dr - \frac{1}{\eta - 1} \int_0^1 \frac{\mathfrak{C}(r)}{r} dr + \frac{1}{\eta - 1} \int_0^1 \frac{\mathfrak{C}(r)}{r^\eta} dr, \tag{5.2.4}$$

and

$$\lim_{\varrho \to 0} \frac{1}{\varrho^{\eta-1}} \int_0^{\varrho} \frac{\mathfrak{C}(r)}{r} dr \leq \lim_{\varrho - 0} \int_0^{\varrho} \frac{\mathfrak{C}(r)}{r^\eta} dr = 0.$$

Thus, the theorem is proved as it relates to θ_1 and θ_2.

We now look for bounds on $\Gamma(\varrho)$. We first assume that E is a compact set K. Then $\beta_K \varepsilon_{x_0}$ will be concentrated on K and γ_ϱ on

$$K^{(\varrho)} = K \cap \{x : |x - x_0| \leq \varrho\}.$$

Then,

$$\int_K U_\alpha^{\gamma_\varrho}(x)\, d\beta_K \varepsilon_{x_0}(x) = \int_{K^{(\varrho)}} U_\alpha^{\beta_K \varepsilon_{x_0}}(x)\, d\gamma_\varrho(x)$$

$$= \int_{K^{(\varrho)}} \frac{d\gamma_\varrho(x)}{|x - x_0|^{p-\alpha}} = U_\alpha^{\gamma_\varrho}(x_0) = w(\varrho).$$

But $U_\alpha^{\gamma_\varrho}(x) = 1$ quasi-everywhere on $K^{(\varrho)}$, hence

$$w(\varrho) = \Gamma(\varrho) + \int_{K \setminus K^{(\varrho)}} U_\alpha^{\gamma_\varrho}(x)\, d\beta_K \varepsilon_{x_0}(x)$$

and

$$\Gamma(\varrho) \le w(\varrho)$$

We bound $\Gamma(\varrho)$ from below. Since outside the ball $|x - x_0| \le 3\varrho$

$$|x - y| \ge 2\varrho, \quad y \in K^{(\varrho)},$$

we have

$$U_\alpha^{\gamma_\varrho}(x) = \int_{K^{(\varrho)}} \frac{d\gamma_\varrho(y)}{|x - y|^{p-\alpha}} \le 2^{\alpha-p}\,\mathfrak{C}(\varrho),$$

and hence

$$\int_{K \setminus K^{(\varrho)}} U_\alpha^{\gamma_\varrho}(x)\, d\beta_K \varepsilon_{x_0}(x) = \int_{K \setminus K^{(3\varrho)}} + \int_{K^{(3\varrho)} \setminus K^{(\varrho)}}$$

$$\le 2^{\alpha-p}\,\mathfrak{C}(\varrho) \int_{K \setminus K^{(3\varrho)}} d\beta_K \varepsilon_{x_0}(x) + \int_{K^{(3\varrho)} \setminus K^{(\varrho)}} U_\alpha^{\gamma_\varrho}(x)\, d\beta_K \varepsilon_{x_0}(x)$$

$$\le 2^{\alpha-p}\mathfrak{C}(\varrho)\,[1 - \Gamma(3\varrho)] + \Gamma(3\varrho) - \Gamma(\varrho).$$

Consequently,

$$w(\varrho) \le \Gamma\,(3\varrho) + 2^{\alpha-p}\mathfrak{C}(\varrho)\,[1 - \Gamma\,(3\varrho)],$$

and since by (5.2.3) $\mathfrak{C}(\varrho) < w(\varrho)$ then

$$w(\varrho) \le \Gamma\,(3\varrho) + 2^{\alpha-p} w(\varrho)$$

or

$$\Gamma\,(3\varrho) \ge (1 - 2^{\alpha-p}) w\,(\varrho).$$

Thus, for any compact set $K \subset E \cup \{x_0\}$, containing the point x_0 we have proved the inequalities

$$(1 - 2^{\alpha-p})\, w\,(\varrho/3) \le \Gamma\,(\varrho) \le \omega\,(\varrho). \tag{5.2.5}$$

We can now pass to the limit along the generalized sequence $\mathfrak{K}$ of compact sets K. Since in that case

$$\beta_K \varepsilon_{x_0} \overset{\mathfrak{K}}{\to} \beta_E \varepsilon_{x_0}, \ \lim_{\mathfrak{K}} w_K\,(\varrho) = \omega_E\,(\varrho)$$

(the latter follows from Theorem 3.9 for generalized sequences), we have the validity of (5.2.5) for the set $E \cup \{x_0\}$ and hence for E.

After this, it is clear that for any $\eta > 1$, the statements

$$\text{(I)} \int_0 \frac{w(\varrho)}{\varrho^\eta}\, d\varrho < \infty, \quad \text{(II)} \int_0 \frac{\Gamma(\varrho)}{\varrho^\eta}\, d\varrho < \infty$$

are equivalent, and the theorem is proved. ☐

8. If the point x_0 is a positive distance away from the set E, then the measure $\beta_E \varepsilon_{x_0}$ will have finite energy, since its potential is bounded and the measure itself is finite. The same holds if for sufficiently small ϱ the set

$$E \cap \{x : |x - x_0| < \varrho\}$$

has capacity zero.

Let us find a necessary and sufficient condition for $\beta_E \varepsilon_{x_0} \in \mathscr{E}_\alpha^+$. This is easy to do in terms of the set E^* obtained from E by inversion with pole at x_0. In Ch. IV, § 6 and in No. 2 of this chapter we have seen that the Kelvin transform γ^* of the measure $\beta_E \varepsilon_{x_0}$ is the equilibrium measure on E^*. On the other hand, formulas (4.5.2) and (4.5.4) directly imply that for any two signed measures ν and ν^* related by the Kelvin transformation,

$$\|\nu\| = \|\nu^*\|.$$

Consequently, the condition $\beta_E \varepsilon_{x_0} \in \mathscr{E}_\alpha^+$ is equivalent to the condition $\gamma^* \in \mathscr{E}_\alpha^+$, i.e. $\gamma^*(1) < \infty$ or

$$C_\alpha(E^*) < \infty. \tag{5.2.6}$$

We now wish to express condition (5.2.6) in terms of the set E itself. We prove two preliminary lemmas.

Lemma 5.4. *If E_1 and E_2 are two Borel sets which aré separated by a positive distance d, then*

$$C_\alpha(E_1) + C_\alpha(E_2) \leq C_\alpha(E_1 \cup E_2) \left[1 + \frac{\max[C_\alpha(E_1), C_\alpha(E_2)]}{d^{p-\alpha}}\right]. \tag{5.2.7}$$

Proof. Suppose that $K_1 \subset E_1$, $K_2 \subset E_2$ are compact sets and that γ_1, γ_2 are the corresponding equilibrium measures. Then $\gamma_1 + \gamma_2$ is a measure concentrated on $E_1 \cup E_2$, and for any point $x \in E_1 \cup E_2$,

$$U_\alpha^{\gamma_1+\gamma_2}(x) \leq 1 + \frac{\max[\gamma_1(1), \gamma_2(1)]}{d^{p-\alpha}} \leq 1 + \frac{\max[C_\alpha(E_1), C_\alpha(E_2)]}{d^{p-\alpha}},$$

since

$$\gamma_1(1) = C_\alpha(K_1), \gamma_2(1) = C_\alpha(K_2).$$

Consequently,

$$C_\alpha(E_1 \cup E_2) \geq (\gamma_1(1) + \gamma_2(1)) \left\{ 1 + \frac{\max [C^\alpha(E_1), C^\alpha(E_2)]}{d^{p-\alpha}} \right\}^{-1},$$

and since $\gamma_1(1), \gamma_2(1)$ can be taken arbitrarily close to $C_\alpha(E_1)$ and $C_\alpha(E_2)$ respectively, this implies (5.2.7). ▯

Lemma 5.5. *Set*

$$E_n = E \cap \{x : r_n \leq |x - x_0| < r_{n+1}\};\ n = 1, 2, \ldots,$$

where the numbers r_n are chosen so that

$$\sum_{n=1}^{\infty} \frac{1}{(r_{n+1} - r_n)^{p-\alpha}} < \infty. \tag{5.2.8}$$

Then the condition

$$C_\alpha(E) < \infty \tag{5.2.9}$$

is equivalent to the condition

$$\sum_{n=1}^{\infty} C_\alpha(E_n) < \infty. \tag{5.2.10}$$

Proof. It is clear that (5.2.10) implies (5.2.9). We will prove the converse. For this we establish that

$$\sum_{k=1}^{\infty} C_\alpha(E_{2k-1}) < \infty. \tag{5.2.10'}$$

and

$$\sum_{k=1}^{\infty} C_\alpha(E_{2k}) < \infty. \tag{5.2.10''}$$

Since the proofs of (5.2.10′) and (5.2.10″) are exactly the same, we will only verify (5.2.10′). Set $C_\alpha(E) = M$. Applying Lemma 5.4, we obtain

$$C_\alpha(E_1) + C_\alpha(E_3) \leq C_\alpha(E_1 \cup E_3) \left[+1 \frac{M}{(r_3 - r_2)^{p-\alpha}} \right].$$

Repeated application yields

$$\begin{aligned} C_\alpha(E_1) + C_\alpha(E_3) + C_\alpha(E_5) &\leq C_\alpha(E_1 \cap E_3) \left[1 + \frac{M}{(r_3 - r_2)^{p-\alpha}} \right] + C(E_5) \\ &\leq [C_\alpha(E_1 \cup E_3) + C_\alpha(E_5)] \left[1 + \frac{M}{(r_3 - r_2)^{p-\alpha}} \right] \\ &\leq C_\alpha(E_1 \cup E_2 \cup E_3) \left[1 + \frac{M}{(r_3 - r_2)^{p-\alpha}} \right] \left[1 + \frac{M}{(r_5 - r_4)^{p-\alpha}} \right]. \end{aligned}$$

By induction we obtain for any natural number N, the inequality.

$$\sum_{k=1}^{N} C_\alpha(E_{2k-1}) \leq C_\alpha\left(\bigcup_{1}^{N} E_{2k-1}\right) \prod_{k=1}^{N-1}\left[1+\frac{M}{(r_{2k+1}-r_{2k})^{p-\alpha}}\right]$$

$$\leq M \prod_{k=1}^{N-1}\left[1+\frac{M}{(r_{2k+1}-r_{2k})^{p-\alpha}}\right].$$

But by (5.2.8)

$$\sum_{k=1}^{\infty} \frac{1}{(r_{2k+1}-r_{2k})^{p-\alpha}} < \infty,$$

which is equivalent to the convergence of the infinite product

$$\prod_{k=1}^{\infty}\left[1+\frac{M}{(r_{2k+1}-r_{2k})^{p-\alpha}}\right]$$

and thus, (5.2.10′) is established. ▯

Note that condition (5.2.8) will be fulfilled if we put $r_n = q^n$ $(q > 1)$.

Theorem 5.8. *$\beta_E \varepsilon_{x_0} \in \mathscr{E}_\alpha^+$ if, and only if*

$$\sum_{n=1}^{\infty} \frac{C_\alpha(E_n)}{q^{2n(p-\alpha)}} < \infty \quad (0 < q < 1), \tag{5.2.11}$$

where

$$E_n = E \cap \{x : q^{n+1} \leq |x - x_0| < q^n\}.$$

Proof. At the beginning of the section we obtained a necessary and sufficient condition for $\beta_E \varepsilon_{x_0} \in \mathscr{E}_\alpha^+$ in the form (5.2.6). Applying Lemma 5.5, we obtain another necessary and sufficient condition

$$\sum_{n=1}^{\infty} C_\alpha(E_n^*) < \infty,$$

where

$$E_n^* = E^* \cap \{x : q^{-n} \leq |x - x_0| < q^{-n-1}\} \quad (0 < q < 1).$$

By using the inequalities

$$q^{-2n(p-\alpha)} C_\alpha(E_n) \leq C_\alpha(E_n^*) \leq q^{-(2n+2)(p-\alpha)} C_\alpha(E_n)$$

(cf. proof of Theorem 5.2) it can be given the form (5.2.11) and the theorem is proved. ▯

Just as for the Wiener criterion, condition (5.2.11) can be given an alternate form. If again, as in No. 3, we set

$$E^{(n)} = E \cap \{x : |x - x_0| < q^n\} \quad (0 < q < 1),$$

then (5.2.11) turns out to be equivalent to the condition

$$\sum_{n=1}^{\infty} \frac{C_\alpha(E^{(n)})}{q^{2n(p-\alpha)}} < \infty. \tag{5.2.12}$$

In fact, (5.2.12) obviously implies (5.2.11). To obtain the converse, we use the formula

$$\sum_{n=1}^{N} \frac{C_\alpha(E_n)}{q^{2n(p-\alpha)}} \geq (1 - q^{2(p-\alpha)}) \sum_{n=1}^{N} \frac{C_\alpha(E^{(n)})}{q^{2n(p-\alpha)}} + C_\alpha(E^{(1)}) - \frac{C_\alpha(E^{(N+1)})}{q^{2N(p-\alpha)}},$$

which is obtained just as the analogous formula in No. 3. Now it is sufficient to show that if (5.2.11) holds, then

$$C_\alpha(E^{(N+1)}) = O(q^{2N(p-\alpha)}).$$

But

$$E^{(N+1)} = \bigcup_{k=1}^{\infty} E_{N+k},$$

from which

$$C_\alpha(E^{(N+1)}) \leq \sum_{k=1}^{\infty} C_\alpha(E_{N+k}) \leq q^{2N(p-\alpha)} \sum_{k=1}^{\infty} \frac{C_\alpha(E_{N+k})}{q^{2(N+k)(p-\alpha)}},$$

and the right side is even $o(q^{2N(p-\alpha)})$.

Then, just as in No. 3, we obtain the analogue of condition (5.1.8) in the form

$$\int_0^1 \frac{c_\alpha(\varrho)}{\varrho^{2(p-\alpha)+1}}\, d\varrho < \infty \tag{5.2.13}$$

or

$$\int_0^1 \frac{\mathfrak{C}_\alpha(\varrho)}{\varrho^{p-\alpha+1}}\, d\varrho < \infty,$$

where $\mathfrak{C}_\alpha(\varrho)$ is the relative capacity of the set

$$E \cap \{x: |x - x_0| < \varrho\}.$$

These factes together with Theorem 5.7 allow us to make the following conclusion.

Theorem 5.9. *$\beta_E \varepsilon_{x_0} \in \mathscr{E}_\alpha^+$ if, and only if the exponent of convergence θ of any of the functions $\mathfrak{C}(\varrho)$, $w(\varrho)$, $\Gamma(\varrho)$ is either greater than $p - \alpha + 1$, or is equal to $p - \alpha + 1$ but under the condition that these functions belong to the convergence class.*

§ 3. The Fine Topology

9. In the preceding sections we have seen that the irregularity of a point x_0 relative to a set E was connected with a certain "rareness" of E near x_0. This "rareness" was characterized by the rate of decrease of the functions $\mathfrak{C}(\varrho)$, $w(\varrho)$, or $\Gamma(\varrho)$ as $\varrho \to 0$.

However, one can characterize it in a completely different way, relying on the specific behavior of functions which are superharmonic near x_0. This new point of view will lead us to the thought of *strengthening the topology* of the space R^p in such a way that all superharmonic functions will be continuous; in a certain sense this will be the most economical way of strengthening that topology.

The following definition is basic.

Definition. *A set E will be called thin (α-thin) at the point x_0 in the following two cases:*

(1) x_0 is not a limit point of E, i.e. $x_0 \notin E'$, where E' is the derived set of E.

(2) $x_0 \in E'$, and there exists a function $v(x)$, superharmonic (α-superharmonic) in R^p such that

$$v(x_0) < \varliminf_{x \to x_0} v(x), \quad x \in E \setminus \{x_0\}. \tag{5.3.1}$$

Theorems 1.22 and 1.30 imply that in this definition in place of the superharmonic function $v(x)$ we can consider the potential $U_\alpha^\mu(x)$ of a measure μ.

Thus, the thinness of E at x_0 is equivalent (if we ignore case (1)) to the following condition: there exists a ball T with center at x_0, a number $\eta > 0$, and a potential of measure $U_\alpha^\mu(x)$ such that

$$U_\alpha^\mu(x_0) < \infty, \quad U_\alpha^\mu(x) \geq U_\alpha^\mu(x_0) + \eta, \quad x \in E \cap T \setminus \{x_0\}. \tag{5.3.2}$$

From these definitions it follows immediately that if the sets E_1, $E_2, \dots, E_n$ are thin at x_0, then the union $E = \bigcup_{i=1}^{n} E_i$ will also be thin at x_0. Indeed, if $v_i(x)$ is the superharmonic function connected with E_i by the relation (5.3.1), then the superharmonic function

$$v(x) = \sum_{i=1}^{n} v_i(x)$$

will by semicontinuity of $v_i(x)$ satisfy (5.3.1).

From the definition of thinness, we have the following remark.

If E is not thin at the point x_0 and the potentials $U_\alpha^{\mu_1}(x)$, $U_\alpha^{\mu_2}(x)$ of the two measures μ_1, μ_2 coincide on $E \setminus \{x_0\}$, then

$$U_\alpha^{\mu_1}(x_0) = U_\alpha^{\mu_2}(x_0).$$

Theorem 5.10. *The set E will be thin at x_0 if, and only if, one of the two conditions below is satisfied:*

(*1*) $x \notin \bar{E}$.

(*2*) *x_0 is an irregular point of E.*

Proof. We will consider the non-trivial case when $x_0 \in E'$.

Suppose that E is thin at x_0. Then, by (5.3.2), it is easy to construct a measure μ such that

$$U_\alpha^\mu(x_0) < 1, \quad U_\alpha^\mu(x) \geq 1 \quad \text{for} \quad x \in E \cap T \setminus \{x_0\}.$$

Now we take a number $a > 0$ sufficiently small, so that outside T

$$U_\alpha^{a\varepsilon_{x_0}}(x) = \frac{a}{|x - x_0|^{p-\alpha}} \leq U_\alpha^\mu(x).$$

By Theorem 1.31 the functions

$$\inf\,[U_\alpha^\mu(x), \quad U_\alpha^{a\varepsilon_{x_0}}(x)] \quad \text{and} \quad \inf\,[1, U_\alpha^{a\varepsilon_{x_0}}(x)]$$

are potentials of certain measures, say λ_1 and λ_2. It is not difficult to see that throughout $E \setminus \{x_0\}$

$$U_\alpha^{\lambda_1}(x) \geq U_\alpha^{\lambda_2}(x).$$

By property (12) of No. 2 we have

$$U_\alpha^{\lambda_1}(x) \geq U_\alpha^{\lambda_2}(x).$$

everywhere on the set E_{II} of regular points for E. But

$$U_\alpha^{\lambda_1}(x_0) < U_\alpha^{\lambda_2}(x_0) = 1,$$

and hence $x_0 \notin E_{\text{II}}$, i.e. x_0 is an irregular point for E.

Now suppose that $x_0 \in E'$ is an irregular point for E. Denoting by γ the equilibrium measure for $E \cap T$ ($T = \{x\colon |x - x_0| < r\}$), we have the inequality

$$U_\alpha^\gamma(x_0) < 1$$

for sufficiently small r (cf. Theorem 5.7). This shows that quasi-everywhere on $E \cap T$

$$1 = U_\alpha^\gamma(x) \geq U_\alpha^\gamma(x_0) + \eta,$$

where $\eta > 0$ is sufficiently small.

We construct a measure μ for which the inequality

$$U^{\mu}_{\alpha}(x) \geq U^{\mu}_{\alpha}(x_0) + \frac{1}{2}\eta \qquad (5.3.2')$$

will hold on the entire set $E \cap T \setminus \{x_0\}$; this shows that E is thin at x_0. Suppose e is that subset of $E \cap T$ where

$$U^{\gamma}_{\alpha}(x) < U^{\gamma}_{\alpha}(x_0) + \eta .$$

By the preceding, $C_{\alpha}(e) = 0$. Let us assume that there is a measure ν satisfying the conditions

$$\begin{aligned} &U^{\nu}_{\alpha}(x) \geq U^{\gamma}_{\alpha}(x_0) + \eta \quad \text{for} \quad x \in e \setminus \{x_0\}, \\ &U^{\nu}_{\alpha}(x_0) < \eta/2. \end{aligned} \qquad (5.3.3)$$

Then the measure $\mu = \gamma + \nu$ will be the desired one, since for $x \in e\setminus\{x_0\}$

$$U^{\mu}_{\alpha}(x) = U^{\gamma}_{\alpha}(x) + U^{\nu}_{\alpha}(x) \geq U^{\gamma}_{\alpha}(x_0) + \eta > U^{\mu}_{\alpha}(x_0) + \eta/2,$$

and for $x \in E \cap T \setminus (e \cup \{x_0\})$

$$U^{\mu}_{\alpha}(x) > U^{\gamma}_{\alpha}(x) \geqq U^{\gamma}_{\alpha}(x_0) + \eta > U^{\mu}_{\alpha}(x_0) + \eta/2.$$

We will show how to construct the measure ν. Suppose that e_n is the subset of e in the layer $2^{-n}r \leq |x - x_0| < 2^{-n+1}r$. For any $\delta_n > 0$ one can find an open set $G_n \supset e_n$ lying in the layer $2^{-n-1}r < |x - x_0| < 2^{-n+1}r$ and such that $C_{\alpha}(G_n) < \delta_n$.

Suppose that γ_n is the equilibrium measure for G_n and

$$\nu = [U^{\gamma}_{\alpha}(x_0) + \eta] \sum_{n=1}^{\infty} \gamma_n .$$

Then the first of the inequalities in (5.3.3) will obviously be satisfied while

$$U^{\nu}_{\alpha}(x_0) < [U^{\gamma}_{\alpha}(x_0) + \eta] \sum_{n=1}^{\infty} \frac{\delta_n \cdot 2^{(n+1)(p-\alpha)}}{r^{p-\alpha}} .$$

By the choice of the numbers δ_n, one can also make the second inequality of (5.3.3) gold. The theorem is proved. □

10. We present for the convenience of the reader certain notions from general topology.

To introduce a *topology* $\mathfrak{T}$ on a certain set E, say the set of points in R^p, means to specify a system of subsets $\{G\}$, called the *open* sets of this topology, satisfying following axioms:

(I) $\emptyset \in \{G\}$, $E \in \{G\}$.

(II) Any union of open sets is open.

(III) The intersection of a finite number of open sets is open.

If E has been given two topologies $\mathfrak{T}$ and $\mathfrak{T}'$, and $\{G\} \subset \{G'\}$, then the topology $\mathfrak{T}'$ is called *stronger* than the topology $\mathfrak{T}$. Thus the set $\{\mathfrak{T}\}$ of all topologies on E is partially ordered. The *weakest* topology is the topology defined by the system $\{\emptyset, E\}$, and the *strongest* is the so-called *discrete* topology, for which $\{G\}$ consists of all subsets of E.

Let $\{A\}$ be an arbitrary family of subsets of E. There is a weakest topology in which all the sets of this family are open. It can be obtained in the following way: add the two sets $\emptyset$ and E to $\{A\}$; form the system $\{A_1\}$ of all finite intersections of these sets, and then the system $\{A_2\}$ of all unions of sets in $\{A_1\}$. The system $\{A_2\}$ satisfies the axioms (I), (II), (III) and defines the desired topology. It is called the topology *generated* by the system of sets $\{A\}$.

A set V is called a neighborhood of the point $x \in E$, if $x \in G \subset V$. The system $\mathfrak{V}(x)$ of neighborhoods of a point x possesses the following properties:

(a) If $V_1 \in \mathfrak{V}(x)$ and $V \supset V_1$, then $V_2 \in \mathfrak{V}(x)$.

(b) If $V \in \mathfrak{V}(x)$, then $x \in V$.

(c) If $V_k \in \mathfrak{V}(x)$, $k = 1, 2, \ldots, n$, then $\bigcap_1^n V_k \in \mathfrak{V}(x)$.

(d) If $V \in \mathfrak{V}(x)$, then there exists a "sufficiently small" neighborhood $W \in \mathfrak{V}(x)$, such that $V \in \mathfrak{V}(y)$ for any $y \in W$.

It can be shown that if to each point $x \in E$ there corresponds a system of sets $\mathfrak{V}(x)$ which satisfies the axioms (a)—(d), then there exists a unique topology $\mathfrak{T}$, in which the set of neighborhoods of any point $x \in E$ coincides with $\mathfrak{V}(x)$. Thus, a topology may be specified by a system of neighborhoods.

In order to strengthen the topology, one must enrich at least one of the systems $\mathfrak{V}(x)$. In other words, if $\mathfrak{T}'$ is stronger than $\mathfrak{T}$, then $\mathfrak{V}'(x) \supset \mathfrak{V}(x)$ and $\mathfrak{V}'(x) \neq \mathfrak{V}(x)$ for at least one point x. The converse is also true.

Suppose $f(x)$ is a function on E with values in $[-\infty, +\infty]$. If it is continuous in the topology $\mathfrak{T}$, then it will be continuous in any stronger topology $\mathfrak{T}'$. To see this most simply, recall that the continuity of $f(x)$ is equivalent to the requirement that the inverse image $f^{-1}(I)$ of any interval I of the extended real line[5] be an open set.

If $\{f_\alpha(x)\}$ is an arbitrary family of functions defined on E, then there exists a weakest topology in which all the functions $f_\alpha(x)$ are continuous. It is easy to see that this topology is generated by the family of sets

[5] By intervals of the extended real line we mean sets of the form $[-\infty, b)$, $(a, +\infty]$, (a, b).

$\{f_\alpha^{-1}(I)\}$ where I ranges over all the intervals of the extended real line, and f_α over all the functions in the given family.

11. The weakest topology on R^p in which all the potentials $U_\alpha^\mu(x)$ of measures μ are continuous will be called the *fine topology* on R^p. This topology is stronger than the usual (Euclidean) topology on R^p.

Indeed, first of all, any open set (in the usual sense) is open in the fine topology. It is sufficient to verify this for the balls $|x - x_0| < r$. Every such ball is the inverse image of the interval $(r^{\alpha-p}, +\infty]$ under the mapping defined by the potential $U_\alpha^{\varepsilon_{x_0}}(x) = \frac{1}{|x - x_0|^{p-\alpha}}$, and, hence, is open in the fine topology. On the other hand, the fine topology does not coincide with the usual topology, since for the latter there exist discontinuous potentials.

Thus, the system $\mathfrak{V}(x)$ of neighborhoods in the fine topology (we will call them *fine neighborhoods*) must be richer than the usual system of neighborhoods. It turns out that the structure of the fine neighborhoods is intimately connected with the notion of thinness, and consequently, with irregular points. This connection is established by the following theorem.

Theorem 5.11. *The system $\mathfrak{V}(x_0)$ of fine neighborhoods of the point x_0 consists of all those sets V containing x_0 and having the property that CV is thin at x_0.*

Proof. Suppose that V is a thin neighborhood of x_0. Then V contains a set G which is the finite intersection of inverse images $I_1, I_2, \ldots, I_k$ under the mappings realized by the functions $U_\alpha^{\mu_1}(x), U_\alpha^{\mu_2}(x), \ldots, U_\alpha^{\mu_k}(x)$. Since any finite interval (a, b) can be represented in the form $[-\infty, b) \cap (a, +\infty]$ without loss of generality we may assume that all the intervals I are semi-infinite.

Now note that if the interval I has the form $(a, +\infty]$ then by the lower semi-continuity of the potential $U_\alpha^\mu(x)$ its inverse image is an open (in the usual topology) set (containing x_0). Suppose that $I \equiv [-\infty, b)$, $b > 0$ and $U_\alpha^\mu(x_0) = c \in [-\infty, b)$. Since $U_\alpha^\mu(x) > 0$, then the inverse image of I coincides with the inverse image of $I_0 = (0, b)$. If this inverse image V_0 does not contain a usual neighborhood of the point x_0, then the set CV_0 will be thin at the point x_0. Indeed, in CV_0 $U_\alpha^\mu(x) \geq b$, while $U_\alpha^\mu(x_0) = c < b$. Since $CV \subset CG$, and CG is a finite union of sets of the type CV_0, this implies that CV is thin.

Now suppose that $x_0 \in V$ and CV is thin at x_0. If x_0 is an exterior point of CV, then V is a usual and also a fine neighborhood of x_0. Therefore, we shall assume that $x_0 \in \partial V$. Then there exists a potential $U_\alpha^\mu(x)$ which satisfies (5.3.2) with $E \equiv CV$. In that case the inverse image of the interval $(0, U^{\mu}_{\alpha}(x_0) + \frac{1}{2}\eta)$ under the map $U^{\mu}_{\alpha}(x)$ does not contain any points of the set $CV \cap T \setminus \{x_0\}$ and is therefore contained in the set $V \cup CT$. But the above inverse image is an open set in the fine topology; accordingly, $V \cup CT$ is a fine neighborhood of x_0. Since the ball T is also a fine neighborhood, then by property (c) in No. 10, V will be a fine neighborhood of the point x_0 and the theorem is proved. $\square$

Remark. A region G whose boundary ∂G contains irregular points (in the classical sense, i.e. for $\alpha = 2$), is an open set in the fine topology. It remains open after the adjunction of all the irregular points of ∂G, which in the fine topology are isolated points of ∂G.

§ 4. Properties of Set of Irregular Points

12. The basic property of the set E_{I} of irregular points for E is expressed in the Kellogg Lemma and its generalization (cf. Lemma 5.2). Its conclusion is that $E \cap E_{\mathrm{I}}$ has capacity zero. In this connection it is useful to note that the set E_{I} itself may have positive capacity arbitrarily large in comparison with $C_{\alpha}(E)$. An extreme example of this type is the set E of rational points within the unit ball T: $|x| \leq 1$. It is clear that $C_{\alpha}(E) = 0$ while $E_{\mathrm{I}} \equiv T$ and $C_{\alpha}(E_I) = 1$.

A similar effect is possible for an open set and even for a region. Let us look at some examples.

Example 1. Suppose $\{r_k\}$ is the set of all rational points within the ball T: $|x| \leq 1$, ω_k an open ball with radius δ_k and center at r_k and

$$G = \bigcup_{k=1}^{\infty} \omega_k.$$

The radius δ_k is chosen in the following way. Let $\eta_k > 0$ and $\sum_1^{\infty} \eta_k = \eta < 1$. We choose the number δ_k such that first, $C_2(\omega_k) < \eta_k$ and second, such that the equilibrium potential of ω_k on the sphere $|x| = 1$ is less than η_k. Then $C_2(G) < \eta$, and all points on the sphere $|x| = 1$ will be irregular for G. Consequently, $C_2(G_I) = 1$. Note that G_I includes more than just the sphere $|x| = 1$. It is not difficult to show that the subset of G_I lying within T and belonging to any neighborhood of any point on $|x| = 1$, has positive capacity.

Example 2. Suppose that Y is a cylinder in R^3 determined by conditions $x^2 + y^2 \leq 1$, $0 \leq z \leq 1$. Let $\{r_k\}_0^{\infty}$ be all the rational points of the open disc $x^2 + y^2 < 1$; encircle r_k with an open disc Γ_{α} of radius δ_k and denote by Y_k the cartesian product of T_k with the semi-interval $0 \leq z < 1$ $(k = 0, 1, \ldots)$. Here the radii δ_k are chosen so small that the equilibrium potential of Y_k on the cylindrical surface S: $x^2 + y^2 = 1$,

$0 \leq z \leq 1$, is less than η_k, and $\sum_{k=0}^{\infty} \eta_k < 1$. Finally, let Y_{-1} denote the open cylinder $x^2 + y^2 < 1$, $-1 < z < 0$. Form the simply connected region

$$G = \bigcup_{k=-1}^{\infty} Y_k .$$

It is clear that G_I includes all the points of S, and therefore, $C_2(G_I) > 0$. As in Example 1, the points of G_I lying within Y form a set which has positive capacity in the neighborhood of any point of S.

Example 3. Let us modify somewhat the region G of Example 2, by closing some of the "organic tubes" Y_k on the bottom and opening them at the top. More precisely, decompose the set $\{Y_k\}_0^\infty$ of all tubes into two subsets $\{Y'_k\}$ and $\{Y''_k\}$ such that $\bigcup_k Y'_k$ and $\bigcup_k Y''_k$ have the same boundary in the layer $0 < z < 1$.

Then the simply-connected region

$$G' = \bigcup_k Y'_k \cup Y_{-1}$$

will have the property that both the set G'_I and the set $(C\bar{G'})_I$ of irregular points of the exterior of G' will have positive capacity. Moreover, within Y any point of the set G_I of Example 2 belongs to both G'_I as well as $(C\bar{G'})_I$.

13. We now consider compact sets in R^3 and will treat the classical case $\alpha = 2$ exclusively. In addition we assume that the compact set K is the boundary ∂G of a region G. The Kellogg Lemma asserte that

$$C_2(K_I) = 0 .$$

However, there may be many different ways of distributing the set K_I on the boundary K of a region. Using the "principle of accumulating singularities" it is possible to construct an example where K_I will be everywhere dense in K. It is also easy to construct a simply-connected region, whose irregular points form a linear continuum.

Example 4. Consider the broken line, consisting of the segments

$$s_{2i} \ : 0 \leq x \leq 1, \quad y = \frac{1}{i} - \frac{x}{i\,(i+1)},$$

$$s_{2i+1} : 0 \leq x \leq 1, \quad y = \frac{1}{i+1} \qquad (i = 1, 2, \ldots),$$

which have the segment AB: $0 \leq x \leq 1$, $y = 0$ as a limiting continuum.

Taking into account that the capacity of a circular cylinder of finite height tends to zero with the radius, we can construct cylinders Z_i with axes s_i whose radii as $i \to \infty$ approach zero so fast that $\Sigma_\varrho C_2(Z_i) < \varrho^2$, where the sum is taken over all cylinders Z_i containing points which are a distance not larger than ϱ from the segment AB. Let Z be the set consisting of all the cylinders Z_i and the limiting segment AB. Then denoting by $c(x, \varrho)$ the capacity of the intersection of Z with the ball of radius ϱ centered at x on AB, we will have

$$\int_0 \frac{c\,(x, \varrho)}{\varrho^2}\, d\varrho < \infty.$$

This shows that for the simply-connected region $G = CZ$, all points x on the segment AB will be irregular.

In this example it is interesting to note that each point of the "irregular line" is inaccessible from outside the constructed region.

One can, however, construct an example in which all points of the "irregular line" will be accessible from outside the region.

Example 5. In the (x, y)-plane consider the parallel segments $L_n: 0 \leq x \leq 1$, $y = \frac{1}{n}$ $(n = 1, 2, \ldots)$ and on each L_n the 2^n points P_{ni} $(i = 0, 1, \ldots, 2^n - 1)$ with abscissas $x_i = i/2^n$.

Denote by $l_{i_1 i_2 \cdots i_n \cdots}$ the infinite broken line passing through the points $P_{n i_n}$, $2i_{n-1} \leq i_n \leq 2i_{n-1} + 1$, $n \geq 2$. Each such broken line has a unique limit point on the segment $y = 0$, $0 \leq x \leq 1$ and, conversely, to each point on this segment there corresponds one broken line leading to this point.

Let Λ be the set consisting of all these broken lines and the segment $0 \leq x \leq 1$, $y = 0$. It has capacity zero, since that is the case for each subset of Λ lying between L_n and L_{n+1}, and Λ is the countable union of these subsets and the segment $0 \leq x \leq 1$, $y = 0$.

We enclose each such link of every broken like $l_{i_1 i_2 \ldots}$ with a cylinder whose base lies in the plane perpendicular to $0y$, and whose axis is the link itself. Besides it is easy to satisfy the following requirements:

(a) Between L_n and L_{n+1} non-neighboring cylinders do not have any points in common, hence the tubes enclosing $l_{i_1 \ldots i_n i_{n+1} \ldots}$ and $l_{i_1 \ldots i_n i'_{n+1} \ldots}$ $(i_{n+1} \neq i'_{n+1})$, below L_{n+1} do not have any points in common;

(b) $c_n < 1/n^2$, where c_n is the capacity of all cylinders situated below L_n.

The complement of the resulting set clearly yields the desired example.

The above examples show that if we wish to distinguish in which cases the irregular points do not form a continuum, we must confine our attention to a narrower class of regions than those in which all boundary points are accessible from the exterior.

We shall assume the following condition.

Every open Jordan arc L, lying on the boundary ∂G of a region, is accessible from the exterior through a Jordan surface.

This is to be understood in the sense that in the complement of the given region, there is a Jordan surface S (a homeomorphic image of the closed disc), whose intersection with ∂G is the curve L. Here we may assume that L is on the border of S, i.e. under the homeomorphism from the closed disc into S, L is the image of some arc on the circumference.[6]

Simply-connected regions which satisfy the above condition will be called regions of class (Γ).

The following conjecture seems highly probable.

Conjecture. *For regions of class (Γ) the irregular points do not form a Jordan curve.*

In other words, on every Jordan arc L which lies on the boundary ∂G of a region G of class (Γ), there is at least one regular point.

We will show the validity of this conjecture for the case of a rectifiable curve (No. 14) and the case of a plane curve (No. 15).

For the moment we note the following. If the above conjecture is actually true, then it immediately implies the following, more precise assertion.

On a Jordan curve L, the irregular points form a set of the *first category*. This situation follows from the next lemma.

Lemma 5.6. *Suppose that $x \in \partial G$. The function*

$$J(x) = \int_0^1 \frac{c(x, \varrho)}{\varrho^2} \, d\varrho \leq +\infty,$$

where

$$c(x, \varrho) = C_2[CG \cap \{y : |y - x| \leq \varrho\}]$$

is lower semicontinuous on ∂G.

Proof. Since

$$J(x) = \lim_{\eta \to +0} \int_\eta^1 \frac{c(x, \varrho)}{\varrho^2} \, d\varrho,$$

---

[6] Indeed, if L is the image of a Jordan arc L^* which belongs to the closed disc K (the inverse image of S), and L^* is not an arc of the circumference, then if necessary we extend L^* to a Jordan arc whose endpoints are on the circumference and consider the region $K_1 \subset K$, whose boundary includes the whole arc $\bar{L}$, and in place of S we take its subset homeomorphic to K_1.

it is sufficient to establish that the function

$$J_\eta(x) = \int_\eta^1 \frac{c(x, \varrho)}{\varrho^2}\, d\varrho .$$

is continuous on ∂G. Let $|x - x_1| < \delta < \eta$; then

$$c(x, \varrho - \delta) \le c(x_1, \varrho) \le c(x, \varrho + \delta),$$

and

$$\int_\eta^1 \frac{c(x, \varrho - \delta)}{\varrho^2}\, d\varrho \le J_\eta(x_1) \le \int_\eta^1 \frac{c(x, \varrho + \delta)}{\varrho^2}\, d\varrho .$$

But

$$\int_\eta^1 \frac{c(x, \varrho + \delta)}{\varrho^2}\, d\varrho = \int_{\eta+\delta}^{1+\delta} \frac{c(x, \varrho)}{(\varrho - \delta)^2}\, d\varrho ,$$

and since the capacity $c(x, \varrho)$ is bounded, this expression is a continuous function of δ for $|\delta| < \eta$. Consequently, for $0 < \delta < \delta_0(\varepsilon)$

$$|J_\eta(x_1) - J_\eta(x)| \le \max\left\{ \left| \int_\eta^1 \frac{c(x, \varrho + \delta)}{\varrho^2}\, d\varrho - \int_\eta^1 \frac{c(x, \varrho)}{\varrho^2}\, d\varrho \right| , \left| \int_\eta^1 \frac{c(x, \varrho - \delta)}{\varrho^2}\, d\varrho - \int_\eta^1 \frac{c(x, \varrho)}{\varrho^2}\, d\varrho \right| \right\} < \varepsilon,$$

q.e.d. □

Corollary. *The set I of irregular points of ∂G is a set of type F_σ.*

Indeed, the set

$$I_m = \{x : x \in \partial G,\ J(x) \le m\}$$

is closed, and by (5.1.8)

$$I = \bigcup_{m=1}^{\infty} I_m .$$

In addition, if we assume the truth of the above conjecture, then I_m will be nowhere dense in L, so that $L \cap I$ will in fact be of the first category in L.

14. We have the following theorem.

Theorem 5.12. *One every rectifiable curve L which lies on the boundary ∂G of a region G of class (Γ), the irregular points form a set of the first category.*

Proof. First of all we make the following remark which will also be used in the future.

Instead of considering a region G of class (Γ) with an arc L on ∂G, we can simply look at the *complement* CS of a certain Jordan surface S, whose boundary contains the arc L. This follows from the local nature of irregularity and property (1), No. 5.

Thus, suppose, that S is a Jordan surface whose boundary contains the rectifiable arc L. Denote by A and B the endpoints of the arc L and as a parameter on L take the arc length s measured from the endpoint A. We assume that s varies from 0 to 1.

Consider any homeomorphism $z\ (= x + iy) = f(Q)\ \ (Q \in S)$ of the surface S onto the half-disc $y \geq 0$, $(x - 1/2)^2 + y^2 \leq 1/4$, where L is mapped into the diameter $0 \leq x \leq 1$. Map this half-disc homeomorphically onto itself using the function $\zeta = h(z)$ such that the diameter is mapped into itself, and the distance between the points $\zeta_1 = h(z_1)$ and $\zeta_2 = h(z_2)$ on the diameter is equal to the length of the subset of L bounded by the points P_1 and P_2: $z_1 = f(P_1)$, $z_2 = f(P_2)$. Then $\zeta = h[f(Q)] = \varphi(Q)$ will be a homeomorphism of S onto the half-disc in which L is mapped into the diameter $0 \leq x \leq 1$ and the point $P(s) \in L$ is mapped into the point with abscissa s.

Assume now that all points of the arc L are irregular.

Let us encircle the interior point $P(s) \in L$ with a sphere $\sigma(P, r)$ of radius r. The intersection $\sigma(P, r) \cap L$ is a closed set whose image on the diameter $0 \leq s \leq 1$ we denote by L_r. Consider the contiguous interval $(\underline{s}, \overline{s})$ of this set, containing the point s. To this corresponds an arc of L which lies entirely (except for the endpoints) within $\sigma(P, r)$ and goes through its center P. The intersection of $\sigma(P, r)$ with the surface S is also a closed set whose image under the mapping φ will be denoted by S_r. Obviously, $S_r \supset L_r$ and for sufficiently small r it does not contain any points of the semi-circle $y \geq 0$, $(x - 1/2)^2 + y^2 = 1/4$. (Moreover, by the continuity of the mapping φ, as $r \to 0$, all points of S_r uniformly approach the point s on the diameter.) Considering only such r, we can assert that S_r contains a continuum K_r which connects a point $s_1 \leq \underline{s}$ with a point $s_2 \geq \overline{s}$. Otherwise, in the half-disc we would have a curve γ connecting the point s with a point of the semicircle and having no points in common with S_r. To this curve γ there would correspond a curve Γ on the surface S, connecting the center P of the sphere $\sigma(P, r)$ with a point of its exterior and not intersecting $\sigma(P, r)$, which is absurd.

Thus, to each point s and for each sufficiently small number r one can associate an interval (s_1, s_2) whose length tends to zero as $r \to 0$, which contains s inside, and whose endpoints are connected by a continuum K_r lying in the half-disc. To this continuum K_r there also corresponds a continuum M_r in $S \cap \sigma(P, r)$.

By our assumption of the irregularity of all points of L and by Theorem 5.4 to each point P on L there corresponds a sequence $r_1 > r_2 > \cdots, r_n \to 0$, such that $S \cap \sigma(P, r_i)$ contains no connected subset with diameter greater than or equal to r_i^2. Hence, for these r_i, the diameter of M_{r_i} will be less than r_i^2. For each point $P(s)$ considering only these values of r and their corresponding intervals (s_1, s_2) we obtain a covering of the segment $0 \leq s \leq 1$ in the sense of Vitali. By a well-known theorem, from this covering one can select a countable sequence of non-intersecting intervals $(s_1^{(i)}, s_2^{(i)})$, $i = 1, 2, \ldots,$ such that the set $e = [0,1] \setminus \bigcup_{i=1}^{\infty} (s_1^{(i)}, s_2^{(i)})$ will be of measure zero. Also, the lengths of the intervals $(s_1^{(i)}, s_2^{(i)})$ may be chosen so small, that the corresponding radii r_i will be less than a pre-determined number $\varepsilon > 0$. Then $\sum_{i=1}^{\infty} r_i^2 < \varepsilon \sum_{i=1}^{\infty} r_1 \leq \varepsilon$.

Now notice the following. The endpoints of each arc L which corresponds to the interval $(s_1^{(i)}, s_2^{(i)})$ under the inverse mapping φ^{-1}, are separated by a distance less than r_i^2 (since this distance is not greater than the diameter of M_{r_i}). Connecting the endpoints of these arcs by rectilinear segments and replacing the arcs by these segments, we obtain a "reduced" curve $\hat{L}$. Its subset consisting of intervals has linear measure less than ε. But subset consisting of the image of the set e has linear measure zero, since the mapping $\varphi^{-1}(s)$ is absolutely continuous on the segment $0 \leq s \leq 1$.

Thus, the length of $\hat{L}$ is less than ε. Since the arc $\hat{L}$ has the same endpoints as L and $\varepsilon > 0$ is arbitrary, we evidently have a contradiction.

Thus, the arc L has at least one regular point, and, by a remark in the preceding section, we see that the theorem is proved. ▯

15. It is significantly more difficult to prove

Theorem 5.13. *One every plane Jordan curve L lying on the boundary ∂G of a region G of class (Γ), the irregular points form a set of first category.*

The proof of this theorem will be divided into several stages; in the process we shall prove several lemmas.

I. As in No. 14 we again consider a Jordan surface S whose boundary includes the plane curve L. Introducing coordinates in the plane containing the arc L, we describe it with the equations $\xi = \xi(t)$, $\eta = \eta(t)$ $(0 \leq t \leq 1)$. Extend L in the (ξ, η)-plane without bound in both directions[7], and the resulting Jordan curve L^+ will again be described by

[7] The possibility of such an extension can easily be proved if one uses the fact that L does not separate the plane.

$\xi = \xi(t)$, $\eta = \eta(t)$ $(-\infty < t < +\infty)$. This curve L^+ separates the plane into two regions, which we denote by R_1 and R_2.

Lemma 5.7. *On the curve L there exists a point which is simultaneously regular for R_1 and for R_2.*

Proof. A point of L, accessible from R_1 through an angle, will be regular for R_1. This follows from the fact that the capacity of a disc sector is proportional to the radius. It is easy to see that the set of such points is dense on L, and by a remark in No. 13, the set of irregular points of R_1 is of first category in L. The same holds for the irregular points of R_2. The assertion of the lemma now follows since a Jordan curve is a set of second category in itself. ▯

In what follows we aim to prove that a point $P \in L$, regular for R_1 and R_2 is also regular for any Jordan surface whose border includes L.

II. Consider the homeomorphism $x + iy = f(Q)$, $Q \in S$, of the surface S onto the half-disc K: $x^2 + y^2 \leq 1$, $y \geq 0$, for which $f(P) = 0$.

Again, let $\sigma(P, r)$ denote a sphere of radius r centered at the point P, $S_r = f[\sigma(P, r) \cap S]$ and assume that r is sufficiently small so that S_r contains no points of the semicircle $x^2 + y^2 = 1$, $y \geq 0$.

Also, let T_r denote the set of points of K consisting of the image of the subset of S lying within $\sigma(P, r)$, and let T_{re} be the set of points of K consisting of the image of the subset of S lying outside of $\sigma(P, r)$. Both of these sets are open.

Consider that component T_r^0 of the set T_r whose boundary contains the point 0. The external boundary of T_r^0 is a continuum containing the point 0 together with a certain interval of the x axis. If we denote by a_r the leftmost point of this continuum belonging to the x axis, and by b_r the rightmost point, then a_r and b_r divide this continuum into two continua k_r and k_r' where k_r consists only of interior points of K. Let A_r, B_r be the inverse images of a_r, b_r in L, and M_r, the inverse image of k_r. M_r is a continuum on the sphere $\sigma(P, r)$ and connects the points A_r and B_r.

Let us construct the circular cone N with vertex at P and axis $P\zeta$, perpendicular to the (ζ, η)-plane in which lies the curve L. Denote by $2\gamma_0$ the angle of incidende of the axis with this cone (the quantity $\gamma_0 < \pi/4$ will be specified later).

Consider the set A_0 of those values of r in the interval $0 < r < r_0$ (r_0 sufficiently small), for which $\sigma(P, r) \cap S$ contains a continuum Q_r, which satisfies the condition

$$\operatorname{diam} Q_r > \vartheta r \quad \big(\vartheta = \min(\gamma_0, \pi/2 - \gamma_0)\big).$$

Let us assume that A_0 contains a sequence of intervals (b_i, a_i), $i = 1, 2, \ldots$ $(b_i < a_i < b_{i+1})$, for which $a_i/b_i > \lambda > 1$.

Then, by Theorem 5.5, the point P will be regular for S.

III. If the above assumption of sufficient "thickness" of the set A_0 does not hold, then in its complement, which we write A_1, one can find a sequence of numbers $r_1 > r_2 > \cdots$, tending to zero for which $r_i/r_{i+1} <$ const. Selecting, if necessary, a subsequence from it, we may assume that

$$1 < a < \frac{r_i}{r_{i+1}} < b\text{[8]}$$

Further we may assume that for these chosen r_i the set $\sigma(P, r_i) \cap S$ has no interior points, nor any components separating σ into more than two regions.

Indeed, $\sigma(P, r) \cap S$ contains interior points only for a countable set of values r, since to each such r there corresponds a neighborhood in S and two distinct r correspond to non-intersecting neighborhoods.

Similarly, there are only countably many values of r for which $\sigma(P, r) \cap S$ contains a continuum which separates σ and S into more than two regions. To prove this assertion we can argue as follows. If $\sigma(P, r) \cap S$ for some r separates σ and S into more than two regions, then the closed set $S_r = f[\sigma(P, r) \cap S] \subset K$ contains a continuum which decomposes the half-disc K into at least three regions. But $S_{r_1} \cap S_{r_2} = \emptyset$ for $r_1 \neq r_2$, and hence if the above assertion were not true, then in the half-disc one could find an uncountable set of continua, mutually non-intersecting, each of which would decompose K into at least three regions. It is not difficult to see that this is impossible.

Thus, for the values r_i we have chosen in A_1, we have:

(a) $\sigma(P, r_i) \cap S$ has no interior points, and each of its components decomposes σ and S into no more than two regions;

(b) $\sigma(P, r_i) \cap S$ does not contain a continuum enveloping the cone N;

(c) every continuum of $\sigma(P, r) \cap S$ is mapped, by a circular projection P_ς with axis $P\zeta$, into an arc with radian measure not greater than θ.

In particular:

(c_1) a component of $\sigma(P, r_1) \cap S$ having points in common with the cone N does not have any points in common with the coaxial cone N_1 which is incident with the axis at an angle equal to $4\gamma_0$;

---

[8] Indeed, if $x_i = -\ln r_i$, then $x_i \to \infty$ monotonically and $x_{i+1} - x_i < \beta$. Subdividing the x axis into intervals of length β and in each interval with odd index selecting one x_i, we obtain a sequence $x_1' < x_2' < \cdots$ for which $\beta \leq x_{i+1}' - x_i' < 3\beta$ Then $e^\beta < r_i'/r_{i+1}' < e^{3\beta}$.

(c_2) the continuum M_{r_i} constructed in II has no points in common with N.

As in II consider the image $S_{r_i} = f[\sigma(P, r_1) \cap S]$ in the half-disc K and the region $T^0_{r_i}$ whose external boundary was there denoted by $k_{r_i} \cup k'_{r_i}$. The remaining components of the boundary of T^0_{ri} will be denoted by

$$l_2^{(i)}, l_2^{(i)}, \ldots,; l_k^{(i)} \subset S_{r_i}.$$

We will be interested only in those $l_k^{(i)}$ which are simultaneously a part of the boundary of $T_{r_i e}$. By (a) each such $l_k^{(i)}$ is a continuum, which decomposes K into two regions.

First consider the simplest case when for all i, k, $f^{-1}(l_k^{(i)})$ is outside the cone N. In this case it is not difficult to obtain a lower bound to the capacity of the subset Σ_{r_i} of the surface S in the sphere $\sigma(P, r_i)$, which is sufficient for asserting the regularity of the point P. First of all, $f^{-1}(T^0_{r_i}) \subset \Sigma_{r_i}$ implies that

$$C_2(\Sigma_{r_i}) \geq C_2[f^{-1}(T^0_{r_i})].$$

We now orthogonally project the subset $f^{-1}(T^0_{r_i})$ of the surface S onto the (ξ, η) plane. Consider any segment pq, perpendicular to the (ξ, η) plane, whose endpoints p and q lie on the sphere $\sigma(P, r_i)$ and inside the cone N; connect the points p and q by a curve lying outside the sphere $\sigma(P, r_i)$. The resulting closed curve, by the simplifying assumption above, will not be incident with any of the continua $f^{-1}(l_k^{(i)})$. Therefore, if it is incident with $f^{-1}(k_{r_i} \cup k'_{r_i})$ then the segment pq will intersect $f^{-1}(T^0_{r_i})$ and the corresponding point in the (ξ, η) plane will be included in the projection of $f^{-1}(T^0_{r_i})$.

By (c_2) the continuum $M_{r_i} = f^{-1}(k_{r_i})$ can be continuously deformed, without intersecting N, either in an arc belonging to R_1 or in an arc belonging to R_2. Suppose, for example, that the first statement holds (the other case is analogous). Then if the segment pq intersects R_1 in a point belonging to the disc $\xi^2 + \eta^2 < r_i^2 \sin^2 \gamma_0$ then the closed curve constructed above will be incident with $f^{-1}(k_{r_i} \cup k'_{r_i})$ and hence the segment pq will intersect the surface $f^{-1}(T^0_{r_i})$.

Consequently, for each i, the projection of $f^{-1}(T^0_{r_i})$ onto the (ξ, η) plane contains either a subset of R_1, or a subset of R_2 which lies in the above disc.

Denoting by $c_1(r)$, respectively $c_2(r)$, the capacity of the subset of R_1, respectively R_2, in the disc $\xi^2 + \eta^2 \leq r^2$, we will have

$$C_2(\Sigma_{r_i}) \geq \min\{c_1(r_i \sin^2\gamma_0),\quad c_2(r_i \sin^2\gamma_0)\} = c_0(r_i \sin^2\gamma_0).$$

If, starting with a certain r, $c_0(r) = c_1(r)$ or $c_0(r) = c_2(r)$ then by criterion (5.1.9) and the regularity of the point P for R_1 and R_2, it follows that P is regular for S.

Otherwise, by the inequality

$$c_1(r) + c_2(r) \geq \frac{2}{\pi} r^9$$

and the easily proved continuity of $c_1(r)$ and $c_2(r)$, there exists an infinite sequence of numbers $\varrho_1 > \varrho_2 > \cdots$, tending to zero, for which

$$c_0(\varrho_i) = c_1(\varrho_i) = c_2(\varrho_i) \geq \frac{1}{\pi}\varrho_i.$$

Since $c_0(r)$ is monotone decreasing, the integral $\int_0 \frac{c_0(\varrho)}{\varrho^2} d\varrho$ diverges.

But then, by criterion (5.1.9) the series

$$\sum_{i=1}^{\infty} \frac{c_0(r_i \sin^2\gamma_0)}{r_i}$$

diverges, and hence, the series

$$\sum_{i=1}^{\infty} \frac{C_2(\Sigma_{r_i})}{r_i}$$

also diverges.

This proves that the point P is regular for the surface S.

IV. The study of the general case will be based on the following considerations. The subset of the surface S, delimited by the portion L between A_{r_i} and B_{r_i} and the continuum M_{r_i} may within the cone N protrude beyond the sphere $\sigma(P, r_i)$. This implies that under the projection which we used in III, we would have no assurance of completely covering either the subset of R_1, or the subset of R_2 in the disc

$$\xi^2 + \eta^2 \leq r_i^2 \sin^2\gamma_0.$$

Therefore, before projecting we shall transform the subset of S of interest to us in a certain way. Roughly speaking, the parts of S lying outside of $\sigma(P, r_i)$ but within $\sigma(P, 2r_i)$ will be replaced by parts of the surface of the same sphere $\sigma(P, r_i)$ with the same boundary. Here one must

[9] This inequality follows from $C_2(E_1 \cup E_2) \leq C_2(E_1) + C_2(E_2)$ and the fact that the capacity of a disc of radius r is equal to $\frac{2}{\pi} r$.

carefully determine which parts of S are to be replaced and show that with this replacement, the capacity of the modified surface will not be greatly increased. This is the goal of the two lemmas proved below.

Lemma 5.8. *Let σ be a sphere with center P, $\{\mathfrak{G}_i\}_1^\infty$ mutually non-intersecting connected Jordan surfaces*[10] *not containing the point P, lying within σ, whose entire border lies on σ such that each component of this border separates σ into two parts. Denote by v_i a three-dimensional region not containing P and bounded by the surface $\mathfrak{G}_i$ and a certain open subset, G_i, of the sphere.*[11] *If the spherical radian measure of G_i is less than $4\pi - m$ $(m > 0)$ and F is any compact set, then*

$$C_2(\overline{\bigcup_i G_i \cup F}) < \alpha C_2(\overline{\bigcup_i \mathfrak{G}_i \cup F}),$$

where $\alpha > 0$ depends only on M.

Proof. Any homothetic transformation with center at the point P does not affect the statement to be proved. Therefore, we may assume that the radius of the sphere σ is equal to one.

Consider the equilibrium measure γ on $\overline{\bigcup G_j \cup F}$. Its potential (cf. Ch. II, § 1 and Ch. IV, § 3) does not exceed one and is equal to one at all regular points of $\overline{\bigcup G_i \cup F}$. Let us sweep out this measure onto the set $\overline{\bigcup \mathfrak{G}_i \cup F}$. As a result we obtain a new measure μ on $\overline{\bigcup \mathfrak{G}_i \cup F}$ whose potential does not exceed one, since under balayage, potential is not increased. Therefore,

$$\mu(\overline{\bigcup \mathfrak{G}_i \cup F}) \leq C_2(\overline{\bigcup \mathfrak{G}_i \cup F}).$$

Let us find a lower bound for the total mass μ. By (4.2.4) it is equal to

$$\int_{\overline{\cup G_i \cup F}} W(x)\, d\gamma(x),$$

where $W(x)$ denotes the equilibrium potential of $\overline{\cup \mathfrak{G}_i \cup F}$. Now it is sufficient to show that quasi-everywhere on $\overline{\cup G_i \cup F}$ we have

$$W(x) > 1/\alpha.$$

Since $\overline{\cup G_i \cup F} \setminus \cup G_i \subset \overline{\cup \mathfrak{G}_i \cup F}$, then quasi-everywhere on this set $W(x) = 1$.

Suppose therefore that $x \in G_i$. If $w_i(x)$ is the equilibrium potential of $\mathfrak{G}_i$, then obviously, $W(x) \geq w_i(x)$. Then, evaluating $w_i(x)$ we may assume, without loss of generality that the contours which bound G_i are sufficiently smooth curves having capacity zero.

[10] That is, homeomorphic images of closed plane regions.

[11] The boundary of G_i coincides with the border of $\mathfrak{G}_i$.

Let H_i be the complement of G_i with respect to the sphere σ. By the assumption just made, mes $H_i > m$.

Now construct a bounded harmonic function $V_i(x)$ defined in the infinite region bounded by H_i and $\mathfrak{G}_i$ (i.e. outside the complement of v_i with respect to the unit ball) with the following boundary conditions: $V_i = 1$ on $\mathfrak{G}$, $V_i = 0$ on H_i[12].

Obviously,

$$w_i(x) \geq V_i(x).$$

Suppose, also, that $u_i(x)$ is a harmonic function which is a solution of the external Dirichlet problem for the region bounded only by H_i with boundary values equal to zero as one approaches points of H_i from the outside of the sphere, and equal to one as they are approached from inside σ.

Using the principle of extending regions, it is easy to see that $V_i \geq u_i$ wherever the function V_i is defined. Thus, for $x \in G_i$

$$W(x) \geq w_i(x) \geq V_i(x) \geq u_i(x),$$

and we have only to find a lower bound for $u_i(x)$.

For this we examine the Gaussian integral

$$I(x) = \frac{1}{4\pi} \iint_{H_i} \frac{\cos \varphi}{|x - y|^2} \, dS_y.$$

where φ is the angle between the vector yx and the interior normal to the sphere. Recalling the geometric significance of $I(x)$, we observe that the limiting values of $I(x)$ on H_i from within σ are less than one. Since the jump of $I(x)$ at interior points of H_i is equal to one, by the maximum principle,

$$I(x) < u_i(x),$$

Enclose the point x on the sphere by a small disc ω of area $m/2$, and denote by H_i^- the subset of H_i outside ω. Obviously, mes $H_i^- > m/2$, and

$$I(x) \geq \frac{1}{4\pi} \iint_{H_i^-} \frac{\cos \varphi}{|x - y|^2} \, dS_y > \frac{1}{16\pi} \iint_{H_i^-} \cos \varphi \, dS_y > m \cdot \text{const}.$$

Thus, the lemma is proved. ☐

V. **Lemma 5.9.** *Suppose again that σ is a sphere of radius r with center at P, $\mathfrak{H}_i$ $(i = 1, 2, \ldots)$ are mutually non-intersecting non-closed connected*

---

[12] That V_i is not defined on the boundary of $\mathfrak{G}$ is not essential, since the capacity of that set is equal to zero.

Jordan surfaces, where the border of $\mathfrak{H}_i$ *consists of one continuum* $\varkappa_i$, *lying on* σ *and separating* σ *into two regions, one of whose diameters, say* B_i, *is less than* $r/2$. *Suppose, further, that near* $\varkappa_i$, $\mathfrak{H}_i$ *lies outside* σ *and all the regions* B_i *belong to a disc* δ *(of the sphere* σ*) whose angular radius is less than* γ_0 *(this number will be determined more precisely in what follows). Denote by* $\varkappa_i^{(k)}$ $(k = 1, 2, \ldots)$ *all components of the intersection of* $\mathfrak{H}_i$ *with* σ *and assume that each continuum* $\varkappa_i^{(k)}$ *has no interior points of* σ *and separates* σ *into two regions one of which* $G_i^{(k)}$ *satisfies the condition of Lemma* 5.8 *(where the role of* $\mathfrak{G}_i$ *is played by* $\mathfrak{H}_i$*).*

Then, if F *is any compact set, and* H_i *is the subset of* $\mathfrak{H}_i$ *which lies within the sphere* σ_1 *of radius* kr $(k = 4\gamma_0 + 1)$, *then*

$$C_2(\overline{\bigcup B_i} \cup F) < \alpha C_2(\overline{\bigcup H_i} \cup F),$$

where α *depends only on* m *and* γ_0.

Proof. As in the proof of Lemma 5.8, we shall assume that r is equal to one. The regions $G_i^{(k)}$ for fixed i are decomposed into finite or countable subdivisions

$$\begin{array}{l} G_i^{(k_{11})}, G_i^{(k_{12})}, \ldots \\ G_i^{(k_{21})}, G_i^{(k_{22})}, \ldots \\ \ldots\ldots\ldots\ldots\ldots\ldots \end{array}$$

according to the following rule: the open set

$$G_{im} = \bigcup_{s=1}^{\infty} G_i^{(k_{ms})}$$

together with a certain connected subset $\mathfrak{G}_{im}$ of the surface $\mathfrak{H}_i$, which lies within σ, bounds the three-dimensional region v_{im}.

By hypothesis, we may use the result of Lemma 5.8, taking for G_i the sets G_{im}, and for F the set

$$\overline{\bigcup_i H_i \setminus \bigcup_{i,m} \mathfrak{G}_{im}} \cup F.$$

This yields

$$C_2\left(\overline{\bigcup_{i,m} G_{im}} \cup \overline{\bigcup_i H_i \setminus \bigcup_{i,m} \mathfrak{G}_{im}} \cup F\right) < \beta C_2\left(\overline{\bigcup_i H_i} \cup F\right),$$

where $\beta > 0$ depends only on the constant m.

The surface $\mathfrak{H}_i$, after passing to H_i and replacing $\bigcup_{lim} \mathfrak{G}_{im}$, by $\bigcup_{lim} G_{im}$, becomes, in general, a disconnected set. Denote by $\mathfrak{R}_i$ that component of this set which contains $\varkappa_i$. $\mathfrak{R}_i$ is a Jordan surface having no points inside σ with border $\varkappa_i \cup \lambda_i$ where λ_i lies on the sphere σ_1 of radius kr. From the preceding, we have

$$C_2(\overline{\bigcup_i \mathfrak{K}_i \cup F}) < \beta C_2(\overline{\bigcup_i H_i \cup F}).$$

Now we will show that

$$C_2(\overline{\bigcup_i B_i \cup F}) < \beta_1 C_2(\overline{\bigcup_i \mathfrak{K}_i \cup F}),$$

where β_1 depends only on γ_0. Clearly, this will imply the conclusion of the lemma with the constant α equal to $\beta\beta_1$.

We now make the following construction. Take a disc δ_1 (on the sphere σ) concentric with δ but with angular radius $2\gamma_0$, and project it from the center P of the sphere σ on the sphere σ_1. The subset of the projecting cone between (and including) δ_1 and its projection on σ_1 bounds a volume D. Denote by h_i the connected subset of $\mathfrak{K}_i$ which lies in D and contains $\varkappa_i$; we shall prove the stronger inequality

$$C_2(\overline{\bigcup_i B_i \cup F}) < \beta_1 C_2(\overline{\bigcup_i h_i \cup F}).$$

We can repeat verbatim the reasoning in the proof of Lemma 5.8 replacing the interior of the sphere σ with the interior of the region D until we come to finding a lower bound for the equilibrium potential $w_i(x)$ of the surface h_i at the point x of the region B_i.

To evaluate $w_i(x)$ we proceed as follows. Let B_i' denote the set on $\partial D \setminus h_i$ whose points can be connected with B_i by a polygonal line not intersecting h_i and lying entirely (except the endpoints) within D. The complement of $B_i \cup B_i'$ with respect to ∂D will be denoted by F_i. Since the subset of the boundary h_i which belongs to the sphere σ consists only of $\varkappa_i$ which lies within δ and, possibly, a certain set on the boundary of δ_1, then F_i contains the spherical annulus $\delta_1 \setminus \delta$. Consequently,

$$\operatorname{mes} F_i > A\gamma_0.$$

Now we can again repeat the corresponding reasoning in Lemma 5.8 and reduce the problem to finding a lower bound at points of B_i of a function $u_i(x)$ harmonic outside of F_i which is the solution of the Dirichlet problem for the infinite region bounded by F_i with boundary values equal to one as the points of F_i are approached from within D and equal to zero as they are approached from without.

The Gaussian integral

$$I(x) = \frac{1}{4\pi} \iint_{F_i} \frac{\cos\varphi}{|x-y|^2} dS_y \tag{5.4.1}$$

is a minorant for $u_i(x)$ as before, but due to the change in sign of $\cos\varphi$ does not give the necessary bound for $u_i(x)$. Therefore, we will have to find a better minorant for $u_i(x)$.

For this we set

$$\Theta = \min_{x\in F_i} \frac{1}{4\pi} \iint\limits_{B_i \cup B_i'} \frac{\cos\varphi}{|x-y|^2}\, dS_y .$$

If we denote by $I^+(x)$ the limiting value of the Gaussian integral (5.4.1) at the point $x \in F_i$ as x is approached from inside D, then, clearly,

$$I^+(x) + \Theta \leq 1. \tag{5.4.2}$$

We now consider the equilibrium potential

$$U^{\gamma_i}(x) = \int\limits_{F_i} \frac{d\gamma_i(y)}{|x-y|}$$

of the set F_i. Recall, that this is a function harmonic outside F_i which on F_i is equal to one. Then the function

$$I^*(x) = I(x + \Theta U^{\gamma_i}(x)$$

will also be a minorant for $u_i(x)$. This follows from inequality (5.4.2) and the fact that the jump in I^* at points of F_i is equal to the jump in I, i.e. one.

Before we evaluate $I^*(x)$ from below for $x \in B_i$, we make the following remark.

If γ_0 is sufficiently small, then for the three-dimensional region D we have the *Neumann inequality*[13]

$$\frac{1}{4\pi}(I_A^\alpha + I_B^\beta) > \lambda > 0,$$

where I_A^α denotes the solid angle within which the subset α of the boundary of D can be seen from the point A also located on this boundary, I_B^β denotes the solid angle within which from the point B the complementary subset β of the boundary can be seen, and λ depends only on γ_0.

Indeed for small values of γ_0 the region can be obtained from a right circular cylinder which has a square cross section (side equal to $4\gamma_0$) by means of a very small (in the sense of distance) first order deformation. Therefore, if we denote by λ_0 the Neumann constant of such a cylinder, then it is clear that for sufficiently small γ_0 the Neumann inequality will hold with constant $\lambda > \lambda_0/2$.

---

[13] Cf. Goursat [1], No 533.

Since a further decrease of γ_0 in no case decreases the constant λ, we may also require that $2\pi \sin 2\gamma_0 < \varepsilon$ ($\varepsilon > 0$ is fixed below). An elementary calculation shows that for any $\alpha \subset \partial D$, $A \in \partial D$

$$I_A^\alpha > 2\pi \sin 2\gamma_0 > -\varepsilon.$$

Turning to the evaluation of $I^*(x)$ for $x \in B_i$, we con. ider two cases. First, suppose that $I(x) > 0$. Then

$$I^*(x) = I(x) + \Theta U^{\gamma_i}(x) > [I(x) + \Theta]\, U^{\gamma_i}(x) > \lambda U^{\gamma_i}(x) > \lambda\omega(x),$$

where by $\omega(x)$ we mean the equilibrium potential of the spherical annulus $\delta_1 \setminus \delta$.

Denoting by ν the equilibrium measure on $\delta_1 \setminus \delta$, we have

$$\omega(x) = \int\limits_{\delta_1\setminus\delta} \frac{d\nu(y)}{|x-y|} > \frac{1}{2\sin\frac{3}{2}\gamma_0} \int\limits_{\delta_1\setminus\delta} d\nu(y) = \frac{C_2(\delta_1\setminus\delta)}{2\sin\frac{3}{2}\gamma_0}.$$

But for small γ_0 $C_2(\delta_1 \setminus \delta)$ is a quantity on the order of γ_0, and hence

$$\omega(x) > a > 0,$$

where a is an absolute constant.

Thus,

$$I^*(x) > a\lambda$$

and this, as in Lemma 5.8, proves the desired result.

Now suppose that $I(x) < 0$. By the Neumann inequality and the above choice of γ_0 we have

$$\Theta > \lambda, \quad I(x) > -\varepsilon.$$

Consequently,

$$I^*(x) > -\varepsilon + \lambda U^{\gamma_i}(x) > -\varepsilon + \lambda\omega(x) > -\varepsilon + a\lambda > a\lambda/2,$$

if we take $\varepsilon < a\lambda/2$.

Thus, in this case, too, the lemma is proved. ▯

VI. Let us return to the proof of the theorem as indicated at the beginning of IV.

In III we denoted by

$$l_1^{(i)}, l_2^{(i)}, \ldots$$

continua which decompose the half-disc K into two regions and which are the common boundary of $T_{r_i}^0$ and T_{r_ie}.

From this sequence we select those $l_k^{(i)}$ for which $f^{-1}(l_k^{(i)})$ is either within the cone N, or has points in common with it. Denote them by

$$l_1'^{(i)}, l_2'^{(i)}, \ldots,$$

and the remaining continua $l_k^{(i)}$ will be denoted by

$$l_1''^{(i)}, l_2''^{(i)}, \ldots.$$

The subsets of K bounded by these continua will be denoted, respectively, by $d_k'^{(i)}$ and $d_k''^{(i)}$.

By (c_1) all of the $f^{-1}(l_k'^{(i)})$ will lie "inside" the cone N_1 and none of them, by (b) will enclose the cone N.

Fixing γ_0 and k as in Lemma 5.9, we shall evaluate the capacity $C_2(\Sigma_{kr_i})$ from below. Consider the subset of Σ_{kr_i} which belongs to the set $f^{-1}(\bigcup_k d_k'^{(i)} \cup T_{r_i}^0)$ and denote it by Σ_{kr_i}'. If $\delta_i^{(1)}$ and $\delta_i^{(2)}$ are spherical discs cut out of σ by the cone N_1, then we apply Lemma 5.9 twice, taking the set $f^{-1}(T_{r_i}^0)$ for F, $f^{-1}(\delta_k'^{(i)})$ for $\mathfrak{H}_k$, and for δ — $\delta_i^{(1)}$ and $\delta_i^{(2)}$, successively. It is not difficult to see that due to the special precaution in the choice of the r_i, all the conditions for applicability of Lemma 5.9, and Lemma 5.8 on which it depends, will be fulfilled; thus, the constants m and α will depend only on the choice of γ_0.

In this way

$$C_2(\Sigma_{kr_i}) \geq C_2(\Sigma_{kri}') > \frac{1}{\alpha^2} C_2[\overline{\bigcup_k B_k^{(i)} \cup f^{-1}(T_{r_i}^0)}],$$

where the $B_k^{(i)}$ are subsets of the sphere with diameter less than $r_i/2$, bounded by the continua $f^{-1}(l_k'^{(i)})$.

Now we project the set $\overline{\bigcup_k B_k^{(i)} \cup f^{-1}(T_{r_i}^0)}$ onto the (ξ, η) plane and note that the reasoning we employed in III will be valid here, since the continua $f^{-1}(l_k''^{(i)})$ lie outside the cone N_1.

Consequently,

$$C_2(\Sigma_{kr_i}) > \frac{1}{\alpha^2} c_0\Big(r_i \sin\frac{\gamma_0}{2}\Big),$$

and the divergence of the series

$$\sum_i \frac{1}{r_i} c_0\left(r_i \sin\frac{\gamma_0}{2}\right)$$

implies the divergence of the series

$$\sum_i \frac{C_2(\Sigma_{kr_i})}{kr_i}$$

and hence, the regularity of the point P. The theorem is proved. ▯

§ 5. Stability of the Dirichlet Problem. Approximation of Continuous Functions by Harmonic Functions

16. We confine ourselves to the classical case $\alpha = 2$, and for simplicity assume that $p = 3$.

Suppose a continuous function $f(x)$ is specified on the compact boundary ∂G of a region $G \subset R^3$. In Ch. IV, § 3 we obtained the solution of the generalized Dirichlet problem in the form

$$h_f(y) = \int_{\partial G} f(x) d\varepsilon'_y(x), \quad y \in \overline{G},$$

where ε'_y is obtained from the Dirac measure ε_y by applying the balayage operator: $\varepsilon'_y = \beta_{\partial G} \varepsilon_y$.

By a known extension theorem, the function $f(x)$ can be continuously extended to all of R^3. We shall denote this extension by the same symbol $f(x)$, $x \in R^3$. We shall be interested in the behavior of the solution $h_f(y)$ under variations of the boundary ∂G. Here variations of the boundary "into" the region G will in general be essentially distinct from variations "away" from the region G.

Variations of the first type are always stable. This is seen from the following theorem.

Theorem 5.14. *Suppose that $\{G_n\}$ is a monotone increasing sequence of regions and*

$$G = \bigcup_{n=1}^{\infty} G_n.$$

All the boundaries ∂G_n, ∂G are assumed compact. Denoting by $h_f^{(n)}(y)$, $h_f(y)$ the solutions of the generalized Dirichlet problem for G_n and G, respectively, we will have

$$h_f(y) = \lim_{n \to \infty} h_f^{(n)}(y), \tag{5.5.1}$$

and convergence is uniform on any compact set $K \subset G$.

Proof. Recall that by Theorem 4.15, for compact sets the balayage operator β_K has a property which might be called continuity on the right. This means that if $\{K_n\}$ is a monotone decreasing sequence of compact sets and $K = \bigcap_{1}^{\infty} K_n$, then

$$\beta_K = \lim_{n \to \infty} \beta_{K_n}$$

in the sense of weak convergence of operators.

Set

$$K_n = \bigcup_{m=n}^{\infty} \partial G_m, \quad K_1 \supset K_2 \supset \cdots$$

It is not difficult to see that $\bigcap_1^\infty K_n = \partial G$ and for any signed measure ν concentrated within G_n,

$$\beta_{K_n}\nu = \beta_{\partial G_n}\nu.$$

In particular, for any point $y \in G_n$

$$\beta_{K_n}\varepsilon_y = \beta_{\partial G_n}\varepsilon_y$$

and consequently,

$$\beta_{\partial G_n}\varepsilon_y \dot{\to} \beta_{\partial G}\varepsilon_y.$$

Thus,

$$\lim_{n\to\infty} \int_{\partial G_n} f(x)\, d\beta_{\partial G_n}\varepsilon_y(x) = \int_{\partial G} f(x)\, d\beta_{\partial G}\varepsilon_y(x),$$

i.e.

$$\lim_{n\to\infty} h_f^{(n)}(y) = h_f(y), \quad y \in G.$$

Moreover, the family of harmonic functions $\{h_f^{(n)}(y)\}$ is uniformly bounded in each region G_m. This implies, as is known, the uniform convergence on any compact set $K \subset G$. ▯

If for $\{G_n\}$ one takes a sequence of regular regions, then we obtain the construction of a solution of the generalized Dirichlet problem first presented by Wiener.

17. Turning to the study of the variation of ∂G ,,away“ from the region G, we subject the region G to one natural constraint. Namely, we shall assume that

$$\partial G = \partial \overline{G} \tag{5.5.2}$$

or, what is the same thing,

$$\partial G = \partial(C\overline{G}).$$

In this case it is sometimes said that G has no *internal boundary points* or G is called a region of *Caratheodory type*.

Consider an arbitrary sequence of regions $\{G_n\}$ satisfying the conditions

$$G_n \supset \overline{G}, \quad \partial G_n \to \partial G. \tag{5.5.3}$$

The latter means that for any $\delta > 0$ one can specify $N(\delta)$ such that for $n > N(\delta)$ each of the sets ∂G_n and ∂G lies in a δ-neighborhood of another. By (5.5.2) such a sequence exists.

Can we assert that in this case

$$h_f(y) = \lim_{n\to\infty} h_f^{(n)}(y), \quad y \in \overline{G}?$$

The answer turns out to be negative. It is not difficult to understand the reason for this.

Consider a sequence of compact sets

$$K_n = CG_n \cap \{x : |x| \leq n\},$$

contained in the open set $C\overline{G}$. For any position of the point y in G and for sufficiently large n

$$\beta_{K_n}\varepsilon_y = \beta_{\partial G_n}\varepsilon_y,$$

since the measure $\beta_{K_n}\varepsilon_y$ is concentrated on ∂G_n.

In Ch. IV, § 6 we saw (cf. (4.6.6)) that

$$\beta_K\varepsilon_y \overset{\mathfrak{K}}{\dashrightarrow} \beta_{C\overline{G}}\varepsilon_y,$$

where $\mathfrak{K}$ denotes generalized sequence of compact sets belonging to $C\overline{G}$, ordered by inclusion. The above sequence $\{K_n\}$ is cofinal with $\mathfrak{K}$, i.e. for any $K \in \mathfrak{K}$ there is a number $n_0(K)$ such that $K \subset K_{n_0}$. In fact, K is a positive distance from ∂G and then (5.5.3) implies for a sufficiently large n that $K \subset CG_n$. Increasing n if necessary, one finally obtains that $K \subset K_n$.

As a consequence of this, we will have

$$\beta_{K_n}\varepsilon_y = \beta_{\partial G_n}\varepsilon_y \dashrightarrow \beta_{C\overline{G}}\varepsilon_y$$

and hence

$$\lim_{n=\infty} h_f^{(n)}(y) = \lim_{n=\infty} \int f(x)\, d\beta_{\partial G_n}\varepsilon_y(x) = \int f(x)\, d\beta_{C\overline{G}}\varepsilon_y(x), \quad x \in \overline{G}.$$

The function

$$h_f^*(y) = \int_{\partial G} f(x)\, d\beta_{C\overline{G}}\varepsilon_y(x), \quad y \in \overline{G}, \tag{5.5.4}$$

may be called the *external solution* of the Dirichlet problem for the region G with boundary conditions $f(x)$. Just like $h_f(y)$, this is a harmonic and bounded function in G. At the regular points of $C\overline{G}$, by Lemma 5.3, $h_f^*(y)$ is continuous and equal to $f(x)$. This coincidence will in general not occur at the irregular points of $C\overline{G}$.

The function

$$G^*(x, y) = U^{\varepsilon_y}(x) - U^{\beta_{C\overline{G}}\varepsilon_y}(x); \quad y \in \overline{G},$$

is called the external Green function for the region G. This is a function of the point x, which is harmonic in $G \setminus \{y\}$ and positive within G.

By property (6), No. 2

$$G^*(x, y) = G^*(y, x).$$

This implies that $G^*(x, y) > 0$ for $y \in G$ and $x \in (C\bar{G})_I$. On the other hand, if $y \in G$ and x is a regular point of $C\bar{G}$, then $G^*(x, y) = 0$. This follows from property (11), No. 2.

It is clear, that

$$G^*(x, y) \geq G(x, y),$$

and

$$G^*(x, y) > G(x, y)$$

for $y \in G$ and $x \in (C\bar{G}_I) \setminus (CG)_I$.

Theorem 5.15. *Suppose $C\bar{G}$ has no irregular points. Then the region G is regular, $h_f^*(y) \equiv h_f(y)$, and the sequence of functions $h_f^{(n)}(y)$ constructed for the regions G_n, satisfying conditions* (5.5.3) *converges to $h_f^*(y)$ uniformly in $\bar{G}$.*

Proof. The first assertion follows since a point irregular for CG, will also be irregular for $C\bar{G}$ (cf. Proposition (1) in No. 5). From this and the uniqueness of the solution of the classical Dirichlet problem follows the coincidence of $h_f^*(y)$ and $h_f(y)$.

For the proof of the last assertion it is sufficient to verify the uniform convergence of $h_f^{(n)}(y)$ to $f(y)$ on ∂G. We shall assume that G is a bounded region. This does not inhibit generality, since otherwise one could invert the region G with pole at an external point of G (by (5.5.2) such a point exists) and by well-known methods transform the "exterior" Dirichlet problem into the "interior" problem.

Then, it is sufficient to verify the assertion in the case that the function $f(x)$ is superharmonic in a finite region $G_0 = \bigcup_1^\infty G_k$. Indeed, if

$$h_f^{(n)}(y) \to f(y)$$

in the mertic of the space $C_{\partial G}$ (of course, by $h_f^{(n)}(y)$ we mean the restriction of this function to ∂G) for any superharmonic function f, then the same will hold for $f(x) = U^\nu(x)$ where ν is any signed measure with continuous potential. But for any function $f(x)$ continuous in $\bar{G}_0$ and for any $\delta > 0$, by Lemma 1.1 (cf. Remark), one can find a potential $\varphi(x) = U^\nu(x)$ satisfying the condition

$$|f(x) - \varphi(x)| < \delta, \quad x \in \bar{G}_0.$$

In this case for any n

$$|h_f^{(n)}(y) - h_\varphi^{(n)}(y)| < \delta, \quad y \in G_n.$$

By hypothesis

$$|h_\varphi^{(n)}(y) - \varphi(y)| < \delta, \quad y \in \partial G,$$

for $n > n_0(\delta)$, from which it follows that for these n

$$|h_f^{(n)}(y) - f(y)| < 3\delta, \quad y \in \partial G.$$

Thus, suppose that $f(x)$ is a superharmonic function in G_0. Since pointwise convergence $h_f^{(n)}(y) \to f(y)$ holds on ∂G, it is sufficient to show that the family of functions $\{h_f^{(n)}\}$ restricted to ∂G is compact in $C_{\partial G}$. Consider any sequence of this family which as before we denote by $\{h_f^{(n)}\}$. Pick a subsequence of indices $\{n_k\}$ such that the regions $\{G_{n_k}\}$ form a decreasing sequence. This can be done as follows. Suppose $n_1 = 1$ and G_{n_1}, G_{n_2}, ..., G_{n_k} have already been chosen. Denote by $\delta_k > 0$ the distance between the compact sets ∂G_{n_k} and ∂G. By (5.5.3) there is an index n_{k+1} such that $\partial G_{n_{k+1}}$ will lie in a δ_k-neighborhood of ∂G. Therefore $G_{n_{k+1}} \subset G_{n_k}$, and the sequence $\{G_{n_k}\}$ is constructed inductively.

By the superharmonicity of $f(x)$, the restriction of the functions $\{h_f^{(n_k)}\}$ to ∂G will form a monotone increasing sequence of continuous functions. This sequence converges to a continuous function $f(y)$. By the Dini theorem convergence will be uniform. Thus, the family $\{h_f^{(n)}\}$ is compact and the theorem is proved. ▯

We introduce the following terminology. If the sequence $\{h_f^{(n)}\}$ constructed for arbitrary regions $\{G_n\}$ satisfying condition (5.5.3), converges uniformly to $f(x)$ on ∂G, then we shall call the corresponding Dirichlet problem *stable* in $\bar{G}$.

From Theorem 5.15 and the remarks preceding it, it follows that *any Dirichlet problem is stable in $\bar{G}$ if, and only if, the open set $C\bar{G}$ has no irregular points.*

There arises the question of conditions for stability in $\bar{G}$ of a Dirichlet problem with *specified* function $f(x)$ on ∂G. The answer is given by the following lemma.

Lemma 5.10. *The Dirichlet problem with boundary conditions $f(x)$, $x \in \partial G$ will be stable in $\bar{G}$ if, and only if*

$$h_f^*(x) = f(x)$$

i.e.

$$\int_{\partial G} f(x)\, d\mathring{v}_z(x) = 0, \tag{5.5.4'}$$

where $\mathring{v}_z = \varepsilon_z - \beta_{C\bar{G}}\varepsilon_z$, *and* z *is any irregular point of* $C\bar{G}$.

Proof. The necessity of condition (5.5.4') is obvious. We will show its sufficiency, assuming, as in Theorem 5.15 that the region G is bounded.

We show, first of all, that if the condition holds, then

$$h_f^*(x) \equiv h_j(x).$$

For this it is sufficient to verify that for any point $y \in \partial G$

$$\lim_{x\to y} h_f^*(x) = f(y).$$

Consider a sequence of points $x_n \to y$ for which there exists vague limit of measures $\beta_{C\bar{G}}\varepsilon_{x_n}$. Call it α_y. At all points of the open set $C\bar{G}$ we have

$$U^{\alpha_y}(x) = \frac{1}{|x-y|^{p-2}}, \quad x \in C\bar{G}.$$

If we denote by $(\alpha_y)_I$ the restriction of the measure α_y to the set I of irregular points of $C\bar{G}$ then, by the uniqueness theorem (14) No. 2, § 1, we obtain the equation

$$\alpha_y - (\alpha_y)_I + \beta_{C\bar{G}}(a_y)_I = \beta_{C\bar{G}}\varepsilon_y.$$

From equation (5.5.4') integration with respect to z yields be relation

$$\int_{\partial G} f(x)\, d[\beta_{C\bar{G}}(\alpha_y)_I - (\alpha_y)_I] = 0.$$

Therefore,

$$\lim_{n\to\infty} h_f^*(x_n) = \int_{\partial G} f(x)\, d\alpha_y(x) = \int_{\partial G} f(x)\, d\beta_{C\bar{G}}\varepsilon_y(x) = f(y),$$

and from this, by the usual compactness argument, we obtain

$$\lim_{x\to y} h_f^*(x) = f(y).$$

Now we must obtain an estimate of the quantity

$$|h_f^{(n)}(y) - f(y)| = |h_f^{(n)}(y) - h_f(y)|,$$

uniform relative to $y \in \partial G$. For this, we fix y and consider the function

$$\Phi_n(x) = h_f^{(n)}(x) - h_f(y),$$

harmonic in G_n; also consider the region $\Delta_{r,n}$ which is a connected subset, containing the point y, of the intersection $G_n \cap \{x: |x-y| < r\}$. It will be shown that for all sufficiently large n, the number r may be chosen such that the function $|\Phi_n(x)|$ is very small on the whole boundary $\Gamma_{r,n}$ of the region $\Delta_{r,n}$, except for a set of small harmonic measure. This will then imply the smallness of

$$|\Phi_n(y)| = |h_f^{(n)}(y) - h_f(y)|$$

uniformly relative to y.

Consider the function $F(x)$ equal to $h_f(x)$ in $\bar{G}$ and $f(x)$ in $C\bar{G}$. Since it is uniformly continuous in $\bar{G}_0$, where $G_0 = \bigcup_1^\infty G_n$, then, given the number $\varepsilon > 0$, we will be able to find $\delta > 0$, such that the oscillation of $F(x)$ in any ball of radius δ with center in $\bar{G}$ is less than ε. Then we select a number $N_0(\varepsilon)$ so large that for $n > N_0$ the following two conditions hold:

$$\text{mes}\,(G_n \setminus \bar{G}) < \varepsilon\delta^p,$$

$$\text{mes}\, e_n(\varepsilon) < \varepsilon\delta^p,$$

where

$$e_n(\varepsilon) = \{x: x \in \bar{G},\quad |h_f^{(n)}(x) - h_f(x)| > \varepsilon\}.$$

The first condition will be fulfilled for sufficiently large n, since

$$G_n \setminus \bar{G} \subset \omega(\eta),$$

where $\omega(\eta)$ is the subset of an η-neighborhood of $\bar{G}$, lying outside $\bar{G}$ and $\bigcap_{n>0} \omega(\eta) = \emptyset$.

The possibility of satisfying the second condition follows since the sequence $h_f^{(n)}(x)$ converges in measure to $h_f^*(x) = h_f(x)$.

Finally, we take a number r, $\delta/2 < r < \delta$ such that the intersection H_r of the sphere $S_r = \{x: |x-y| = r\}$ with the set $e_n(\varepsilon) \cup (G_n \setminus \bar{G})$ will have $(p-1)$-dimensional measure

$$\text{mes}_{p-1} H_r < 4\varepsilon\delta^{p-1}.$$

The possibility of choosing such an r follows from the inequality

$$\int_{\delta/2}^{\delta} \text{mes}_{p-1} H_r\, dr < \text{mes}\,[e_n(\varepsilon) \cup (G_n \setminus \bar{G})] < 2\varepsilon\delta^p.$$

Now let us return to the region $\Delta_{r,n}$ defined above. The part of the boundary $\Gamma_{r,n}$ belonging to H_r has a harmonic measure (relative to $\Delta_{r,n}$)

which at the point y is less than $\frac{4\varepsilon}{\omega_p}\left(\frac{\delta}{r}\right)^{p-1}$.[14] This easily follows from the principle of extending a region (one must replace $\Delta_{r,n}$ by the ball $\{x: |x-y| < r\}$). For the remainder of $\Gamma_{r,n}$ we have

$$|\Phi_n(x)| \leq |F(x) - F(y)| < \varepsilon.$$

Also, setting

$$M = \max |F(x)|$$

we obtain the inequality

$$|\Phi_n(y)| < \varepsilon + \frac{16}{\omega_p} M\varepsilon,$$

which holds for $n > N_0(\varepsilon)$. The lemma is proved. □

Corollary. *The set $\mathfrak{H}_{\partial G}$ of those continuous functions on ∂G for which the Dirichlet problem is stable in $\bar{G}$, is a closed linear manifold in $C_{\partial G}$.*

The construction of the external solution h_f^*, may in contrast with h_f, be generalized to the case of any compact set $K \subset R^p$.

If for G_n we take a $1/n$-neighborhood of K, then clearly, we will have $\partial G_n \to \partial K$. Furthermore, if $f(x)$ is continuous on K, then the reasoning described at the beginning of this section will lead to the function

$$h_f^*(y) = \int_{\partial K} f(x)\, d\beta_{CK}\varepsilon_y(x), \quad y \in K,$$

which is harmonic in $\mathring{K}$.

If CK has no irregular points, then, just as above, we see that $h_f^*(x) = f(x)$ on ∂K, and on this set the convergence $h_f^{(n)}(x) \to f(x)$ is uniform. If the original function $f(x)$ is harmonic in $\mathring{K}$ (and, obviously, only in this case), the convergence

$$h_f^{(n)}(x) \to f(x).$$

will be uniform on K.

18. In addition to the notion of stability in a closed region $\bar{G}$, we introduce the definition of *stability in an open region* G. The Dirichlet problem with boundary conditions $f(x)$ will be called *stable* (in the open region G), if

$$h_f^*(x) = h_f(x), \quad x \in G. \tag{5.5.5}$$

If is not hard to find a sufficient condition for stability in G of any Dirichlet problem. Namely, assume that the set $(C\bar{G})_I$ of irregular points

[14] Here, ω_p is the "area" of the unit sphere in R^p.

for $C\bar{G}$ coincides with the set $(CG)_I$ of irregular points of the closed set CG. In that case the balayage operators $\beta_{C\bar{G}}$ and β_{CG} are identical. This follows from the uniqueness theorem (15), No. 2. In particular, for any position of the point y in $\bar{G}$

$$\beta_{C\bar{G}}\varepsilon_y = \beta_{CG}\varepsilon_y = \beta_{\partial G}\varepsilon_y.$$

Taking into account (5.5.4) and (4.3.1) we see that equation (5.5.5) holds also in $\bar{G}$. In connection with this we prove the following theorem.

Theorem 5.16. *The condition $(C\bar{G})_I = (CG)_I$ is necessary and sufficient in order that every Dirichlet problem solvable in G be stable in $\bar{G}$.*

Proof. Sufficiency follows since under the specified condition $h_f^*(y) = h_f(y)$ in $\bar{G}$ and this means that on ∂G

$$h_f^*(x) = h_f(x) = f(x).$$

By Lemma 5.10 the problem will be stable in $\bar{G}$.

To prove necessity, assume that

$$(C\bar{G})_I \supset (CG)_I, \qquad (C\bar{G})_I \neq (CG)_I.$$

Let $x_0 \in (C\bar{G})_I \setminus (CG)_I \subset \partial G$.

In Ch. IV, § 4 we saw that the subspace $H_{\partial G} \subset C_{\partial G}$ of functions for which the Dirichlet problem is solvable is characterized by its annihilator: the set N of signed measures on ∂G, which have the form

$$\nu = \nu_I - \beta_{\partial G}\nu_I,$$

where ν_I is the restriction of ν to $(CG)_I$. From the structure of N it is seen that the signed measure $\varepsilon_{x_0} - \beta_{C\bar{G}}\varepsilon_{x_0}$ which is concentrated on ∂G, does not belong to N. Consequently, in $H_{\partial G}$ there is a function $g(x)$ for which

$$g(x_0) \neq \int_{\partial G} g(x)\, d\beta_{C\bar{G}}\varepsilon_{x_0}(x),$$

i.e. $h_g^*(x_0) \neq g(x_0)$. For this function, the Dirichlet problem is solvable but in $\bar{G}$ it is not stable (cf. Lemma 5.10). ▯

Theorem 5.16 can also be obtained from the following

Proposition. *Suppose $\overset{\circ}{N}$ denotes the set of all signed measures on ∂G orthogonal to $\mathfrak{H}_{\partial G}$. Then for $\nu \in \overset{\circ}{N}$ it is necessary and sufficient that*

$$\nu = \nu_I - \beta_{C\bar{G}}\nu_I,$$

where I denotes here the set of irregular points of $C\bar{G}$, i.e. the set designated by $(C\bar{G})_I$ in Theorem 5.16.

This statement is the analogue of Lemma 4.1, and its proof can be obtained in a similar manner. We shall point out the basic steps of the proof.

(a) For any $f(x) \in \mathfrak{H}_{\partial G}$, as we have seen,

$$f(z) = \int f(x)\, \partial\beta_{C\bar{G}}\varepsilon_z(x);$$

this implies that

$$\int f(z)\, d\nu_I(z) = \int f(z)\, d\beta_{C\bar{G}}\nu_I(z),$$

and sufficiency has been proved.

(b) If $U^\nu(x) = 0$ for $x \in C\bar{G}$ and $S(\nu) \in \partial G$, then $\nu = \nu_I - \beta_{C\bar{G}}\nu_I$. Indeed, from No. 2 it follows that $U^\nu(x) = 0$ at all regular points of $C\bar{G}$. Consequently, for any point $y \in \bar{G}$

$$\int_{\partial G} G^*(x, y)\, d\nu(x) = U^\nu(y) - \int U^\nu(x)\, d\beta_{C\bar{G}}\varepsilon_y(x) = U^\nu(y),$$

from which

$$U^\nu(y) = \int_I G^*(x, y)\, d\nu(x) = U^{\nu_I}(y) - \int U^{\nu_I}(x)\, d\beta_{C\bar{G}}\varepsilon_y(x)$$

$$= U^{\nu_I}(y) - U^{\beta_{C\bar{G}}\nu_I}(y).$$

Since for $y \in C\bar{G}$ both parts of the last equation are equal to zero, we have

$$\nu = \nu_I - \beta_{C\bar{G}}\nu_I.$$

(c) The remaining argument proceeds just as in the proof of Lemma 4.1.

Remark. Let K be an arbitrary compact set and $\mathfrak{H}_{\partial K}$ the set of all those functions continuous on ∂K for which $h_f^* = f$ (i.e. for which the Dirichlet problem is stable on K). With inessential changes in the preceding proof, one can see that the set $\mathring{N}$ of those signed measures on ∂K which arre orthogonal to $\mathfrak{H}_{\partial K}$ consists of those and only those signed measures ν, which are representable in the form

$$\nu = \nu_I - \beta_{CK}\nu_I,$$

where I is the set of irregular points of CK.

Returning to now stability in an open region G, we prove the following theorem giving necessary and sufficient conditions for such stability.

Theorem 5.17. *The Dirichlet problem with arbitrary boundary conditions $f(x) \in C_{\partial G}$ will be stable in G if, and only if, the set $(C\bar{G})_I \subset \partial G$ has harmonic measure zero.*

Proof. The difference

$$g(x) = h_f(x) - h_f^*(x) \tag{5.5.6}$$

is a bounded harmonic function in G, continuous and equal to zero at all points of $\partial\bar{G}$, except for the set $(CG)_I$. Therefore, if $\omega(x)$ denotes the harmonic measure of this set and $M = \sup|g(x)|$ on $(C\bar{G})_I$, then

$$|g(x)| \leq M\omega(x), \quad x \in G.$$

Consequently, if $\omega(x) \equiv 0$ then the Dirichlet problem is stable in G.

Now suppose that $\omega(x) > 0$. We will show that it is possible to find a function $f(x) \in C_{\partial C}$ for which the difference (5.5.6) will be distinct from zero in G. For this fix a point $x_0 \in G$ and set

$$f(x) = \frac{1}{|x - x_0|} \in C_{\partial G}.$$

Then

$$h_f(x) = U^{\beta_{\partial G}\varepsilon_{x_0}}(x),$$

$$h_f^*(x) = U^{\beta_{C\bar{G}}\varepsilon_{x_0}}(x),$$

and

$$g(x) = h_f(x) - h_f^*(x) = G^*(x, x_0) - G(x, x_0)$$

will be positive at all points of $(C\bar{G})_I \setminus (CG)_I$ (cf. No. 17). Hence, there is a number $\alpha > 0$ and a compact set $K \subset (C\bar{G})_I \setminus (CG)_I$ of positive harmonic measure, on which $g(x) > \alpha$. Recall now that $h_f^*(x)$ is the limit of a sequence of functions $h_f^{(n)}(x)$ harmonic in $\bar{G}$ constructed for the regions $\{G_n\}$ of type (5.5.3). We shall assume that this sequence of regions is monotone decreasing; then, by the superharmonicity of the function $\frac{1}{|x-x_0|}$, the sequence $\{h_f^{(n)}\}$ will be monotone increasing. Therefore,

$$h_f(x) - h_f^{(n)}(x) \geq g(x), \quad x \in \bar{G}.$$

In contrast with $g(x)$, the limiting values of the difference $h_f(x) - h_f^{(n)}(x)$ at any point of ∂G can be estimated in an elementary way, and from the maximum principle it follows that

$$h_f(x) - h_f^{(n)}(x) \geq \alpha\omega(x, K, G), \quad x \in G.$$

Since this holds for any n, then

$$g(x) = \lim_{n\to\infty} [h_f(x) - h_f^{(n)}(x)] \geq \alpha\omega(x, K, G), \quad x \in G,$$

and the theorem is proved. ▯

19. The results of No. 17 have a close connection with the problem of *uniform approximation* of a function $f(x)$ continuous on a compact set K by means of functions harmonic on K.[15]

Denote, as in Ch. IV, by H_K the space of functions continuous on K and harmonic in $\mathring{K}$. The metric in H_k is assumed uniform. This space, by the maximum principle, can be isometrically imbedded in the space $C_{\partial K}$. From the results of Ch. IV, § 1 and 3, it follows that if $C\mathring{K}$ has no irregular points (in particular, if $\mathring{K}$ is empty), then H_K is isometric to $C_{\partial K}$ (correspondingly, C_K).

Suppose also that $\mathfrak{H}_K \subset H_K$ denotes the linear manifold of functions harmonic on K.

Then from No. 18 and the remark made at the end of Ch. IV, No. 15, one can extract the following theorem.

Theorem 5.18. *$\mathfrak{H}_K$ is dense in H_K if, and only if,*

$$(CK)_I \equiv (C\mathring{K})_I.$$

In particular, *if K has no interior points, then this condition reduces to the condition that the open set CK have no irregular points.*

We present here still another proof of the last assertion, relying on simple facts from functional analysis and the uniqueness theorem (14), No. 2.

Denote by $L_{CK} \subset \mathfrak{H}_K$ the linear manifold generated by the family of functions $\left\{\frac{1}{|x-x_0|}\right\}$, $x_0 \in CK$. It is not difficult to show that L_{CK} is dense in $\mathfrak{H}_K$. Indeed, if $f(x) \in \mathfrak{H}_K$, then there is a neighborhood of the compact set K which is bounded by a smooth surface S in which the function $f(x)$ is harmonic. We may assume that it is also harmonic on S. Then by Green's formula

$$f(x) = \frac{1}{4\pi}\iint\limits_S \left\{\frac{1}{|x-x_0|}\frac{\partial f}{\partial n_{x_0}} - f(x_0)\frac{\partial}{\partial n_{x_0}}\frac{1}{|x-x_0|}\right\} dS_{x_0},$$

where $x \in K$, n_{x_0} denotes the normal to S at the point x_0, and dS_{x_0} is an element of area of the surface S, connected with the point x_0.

[15] Recall, that a function $h(x)$ is called *harmonic on* K, if it is harmonic in some neighborhood of K.

Replacing the integral by an appropriate integral summation, and the derivative $\partial/\partial n_{x_0}\, 1/|x - x_0|$ by a difference quotient, we obtain a function from L_{CK} which uniformly approximates on K the function $f(x)$ with any desired accuracy.

Thus, the density of $\mathfrak{H}_K$ in $H_K \equiv C_K$ is equivalent to the density of L_{CK} in C_K.

By a known theorem of functional analysis L_{CK} will not be dense in C_K if, and only if, there exists a signed measure $\nu \neq 0$, concentrated on ∂K such that

$$U^\nu(x_0) = \int\limits_{\partial K} \frac{d\nu(x)}{|x - x_0|} = 0, \quad x_0 \in CK.$$

In other words, there must exist two distinct measures ν^+ and ν^- concentrated on the boundary ∂K of the open set CK which have the same potential at all points of CK. As we saw in No. 2 this is possible only if the set CK has irregular points. The theorem is proved. ☐

Remark. Suppose that $\mathring{K} = \emptyset$ and

$$CK = \bigcup_{j=0}^{\infty} G_j,$$

where the G_j are the connected components of CK. Select from $\{G_j\}$ an arbitrary subsequence $\{G_{n_j}\}$ in such a way that the boundary $\partial\omega$ of the open set

$$\omega = \bigcup_j G_{n_j}$$

coincides with ∂K. Form the linear manifold L_ω generated by the family of functions

$$\left\{\frac{1}{|x - x_0|}\right\}, \quad x_0 \in \omega.$$

Repeating the same reasoning as above, we arrive at the conclusion that L_ω is dense in C_k if, and only if, ω has no irregular points.

Consider one more linear manifold in $\mathfrak{H}_K$. In each component of the set CK (or ω) fix one point x_0 and denote by Y_j the set of so-called *spherical functions* with pole x_j. This is the set of functions representable in the form

$$\frac{\partial^n}{\partial l_1^{\alpha_1} \partial l_2^{\alpha_2} \cdots \partial l_m^{\alpha_m}} \frac{1}{|x - x_j|} \quad (\alpha_1 + \alpha_2 + \cdots + \alpha_m = n),$$

where $l_1, l_2, \ldots, l_m$ are arbitrary directions; equivalently (cf. Courant-Hilbert [1]), they are representable as

$$\frac{P_n(x)}{|x - x_j|^{2n+1}},$$

where $P_n(x)$ is the homogeneous harmonic polynomial of degree n. In case $x_j = \infty$ the set Y_j consists, by definition, of all homogeneous harmonic polynomials.

Denote by Z_{CK} (or Z_ω) the linear manifold generated by all the sets Y_j.

It turns out that Z_ω is dense in H_K if, and only if, L_ω is dense in H_K. This is because every signed measure ν on ∂K orthogonal to L_ω will be orthogonal to Z_ω and conversely.

Indeed, if

$$U^\nu(x) = \int \frac{d\nu(y)}{|x-y|} = 0, \quad x \in \omega, \tag{5.5.7}$$

then

$$\frac{\partial^n}{\partial l_1^{\alpha_1} \partial l_2^{\alpha_2} \cdots \partial l_m^{\alpha_m}} U^\nu(x_j) = 0 \quad (j = 1, 2, \ldots) \tag{5.5.8}$$

for any $n \geq 0$ and directions $l_1, l_2, \ldots, l_m$. This then means that the signed measure ν is orthogonal to Z_ω. Conversely, (5.5.8) together with the analyticity of the function $U^\nu(x)$ harmonic in ω, implies (5.5.7).

We point two simple consequences of Theorem 5.18 and the above remarks.

Theorem 5.19. *If K has spatial Lebesgue measure null, then L_{CK} and hence Z_{CK}, is dense in C_K.*

Proof. Since K has no interior points, $H_K \equiv C_K$. Therefore, it is sufficient to prove that any point $x_0 \in K$ is regular for CK. This becomes obvious if one notes that

$$c_2(\varrho) = \varrho,$$

where $c_2(\varrho)$ is the Newtonian capacity of the set $CK \cap \{x: |x - x_0| < \varrho\}$. By the monotonicity of $c_2(\varrho)$ the last equation need only be verified on a dense set of values ϱ. Consider those ϱ for which the set $K \cap \{x: |x - x_0|, = \varrho\}$ has two-dimensional measure zero. These are almost all ϱ. For these values of ϱ the measure $\varrho\varepsilon_{x_0}^{(\varrho)}$ is an equilibrium measure for the ball $\{x: |x - x_0| < \varrho\}$ and for the set $CK \cap \{x: |x - x_0| < \varrho\}$, too. This implies that $c_2(\varrho) = \varrho$. ▯

Theorem 5.20. *Suppose that S is a simple Jordan surface (a homeomorphic image of the closed disc). Then the linear manifold of harmonic polynomials is dense in C_S.*

Proof. The set CS consists of one region. Taking the point at infinity contained in CS as a pole, we obtain the linear manifold Z_{CS} consisting of all harmonic polynomials. Therefore, for the proof of the theorem, it is sufficient to establish that any point $x_0 \in S$ is regular for CS. We shall use Theorem 5.4′ and will show that the intersection of CS with the sphere $\Sigma_p = \{x: |x - x_0| = \varrho\}$ contains for all ϱ, except for a countable set, a continuum Q_ϱ such that

$$\operatorname{diam} Q_\varrho > \varrho. \tag{5.5.9}$$

In the proof of Theorem 5.13, (III), it was established that the set $\Sigma'_\varrho = Z_\varrho \cap S$ can contain continua which separate the sphere Σ_ϱ into more than two connected components or it can have interior points only for a countable set of values ϱ. Suppose that ϱ does not belong to this exceptional countable set. If Σ'_ϱ does not separate Σ_ϱ at all, then in $\Sigma_\varrho \setminus \Sigma'_\varrho = CS \cap \Sigma_\varrho$ there of course exists a continuum Q_ϱ with the property (5.5.9). Otherwise, suppose that $K \subset \Sigma'_\varrho$ is a continuum which decomposes Σ_ϱ into two regions. At least one of these has diameter greater than ϱ and hence contains a continuum Q_ϱ satisfying condition (5.5.9). This completes the proof of the theorem. ▯

20. In certain cases, the regularity of the open set CK, which under the condition $\mathring{K} = \emptyset$ is equivalent to the density of $\mathfrak{H}_K$ in H_K, can be formulated in another way.

Theorem 5.21. *Suppose that the compact set $K \subset R^3$ has no interior points. Denote by $c(x_0, \varrho)$ the capacity of the set*

$$CK \cap \{x: |x - x_0| < \varrho\} \quad x_0 \in K.$$

Assume that the condition

$$\overline{\lim_{\varrho \to 0}} \frac{c(x_0, \varrho)}{\varrho^3} = \infty \tag{5.5.10}$$

is fulfilled for almost all points $x_0 \in K$. Then the set CK is regular and moreover, there exists a constant $A > 0$ such that for all points $x_0 \in K$

$$c(x_0, \varrho) > A\varrho. \tag{5.5.11}$$

Conversely, if CK is regular, then condition (5.5.10) *is satisfied at all points of K.*

Proof. The second part of the theorem is a trivial consequence of the Wiener criterion:

$$\int_0^1 \frac{c(x_0, \varrho)}{\varrho^2}\, d\varrho = \infty,$$

which by the condition is fulfilled at all points of K.

We will therefore prove the first part. Fix an arbitrary point $x_0 \in K$ and a number $\varrho > 0$. Let $T(x_0, \varrho)$ denote the open ball of radius ϱ with center at x_0. For almost all points of the set $K \cap \overline{T}(x_0, \varrho/2)$ (5.5.10) holds. Since outside K condition (5.5.10) is certainly satisfied, we may assume that it holds on the compact set

$$K_0 \subset \overline{T}\,(x_0, \varrho/2)\,,$$

whose measure differs by an arbitrarily small amount from the volume of $\overline{T}(x_0, \varrho/2)$, for example mes $K_0 > \pi\varrho^3/7$.

With each point $x \in K_0$ we associate the ball $T(x, \varrho_x)$ of radius $\varrho_x < \varrho/2$, satisfying the condition

$$c(x, \varrho_x) > N\varrho_x^3. \tag{5.5.12}$$

Here $N > 0$ is a number not dependent on x, which we will sepcify later. Suppose that $\{T(x_k, \varrho_k)\}$ is any finite covering of K_0 by the balls indicated. Now make the following construction. Let $T(x_1, \varrho_1)$ be a ball of largest radius (in the given covering). Eliminate from $\{T(x_k, \varrho_k)\}$ those balls which intersect with $T(x_1, 3\varrho_1)$ and from the remaining system select a ball of greatest radius; let this be $T(x_2, \varrho_2)$. Then from the remaining system eliminate those balls which intersect with $T(x_2, 3\varrho_2)$, and continue the process.

As a result we shall arrive at a new finite system $\{T(x_k, \varrho_k)\}_1^n$ possessing the following properties:

(1) for any i, $1 \le i \le n$, the ball $T(x_i, 3\varrho_i)$ does not intersect any other ball $T(x_j, \varrho_j)$;

(2) the system $\{T(x_k, 5\varrho_k)_i^n\}$ is a covering of K_0;

(3) the balls $T(x_i, 2\varrho_i)$ and $T(x_j, 2\varrho_j)$ for $i \neq$ j do not intersect.

Property (1) follows directly from the construction, property (3) follows from (1), and (2) holds because at the ith step in the construction we excluded only those balls which were contained in $T(x_i, 5\varrho_i)$.

From (2) it follows that

$$\frac{4}{3}\,\pi \sum_{i=1}^{n} (5\varrho_i)^3 > \text{mes}\, K_0 > \pi\varrho^3/7\,,$$

i.e.

$$\sum_{i=1}^{n} \varrho_i^3 > A_1 \varrho^3\,, \tag{5.5.13}$$

where A_1 is an absolute constant.

By (5.5.12), in $CK \cap T(x_i, \varrho_i)$ we can find a compact set K_i such that

$$C_2(K_i) = N\varrho_i^3 \quad (i = 1, 2, \ldots, n).$$

The compact set

$$F = \bigcup_{i=1}^{n} K_i$$

belongs to $CK \cap T(x_0, \varrho)$. We shall bound its capacity from below.

Let γ_i denote the equilibrium measure on K_i and set $\gamma = \sum_{i=1}^{n} \gamma_i$; then for any point $x \in K_j$, $1 \leq j \leq n$, the inequalities

$$U^{\gamma}(x) \leq 1 + \sum_{i \neq j} \frac{C_2(K_i)}{\varrho(x, K_i)} \leq 1 + 2N \sum_{i \neq j} \frac{\varrho_i^3}{|x - x_i|}$$

will hold.

The latter inequality is implied by

$$\varrho(x, K_i) \geq |x - x_i| - \varrho_i,$$

and by (3)

$$\varrho_i < \frac{1}{2} |x - x_i|.$$

Furthermore, for $x \in K_j$ and $y \in T(x_i, \varrho_i)$

$$|x - x_i| \geq |x - y| - \varrho_i > |x - y| - \frac{1}{2} |x - x_i|,$$

so that

$$\frac{\varrho_i^3}{|x - x_i|} \leq \frac{2}{3} \int\limits_{T(x_i, \varrho_i)} \frac{dy}{|x - y|},$$

and hence

$$U^{\gamma}(x) \leq 1 + \frac{4}{3} N \sum_{i \neq j} \int\limits_{T(x_i, \varrho_i)} \frac{dy}{|x - y|} < 1 + \frac{4}{3} N \int\limits_{T(x_0, \varrho)} \frac{dy}{|x - y|}$$

$$< 1 + \frac{16}{3} N\pi \int\limits_0^{2\varrho} r\, dr = 1 + A_2 N \varrho^2,$$

where A_2 is an absolute positive constant.

Now take $N = 1/\varrho^2$. Then

$$U^{\gamma}(x) < 1 + A_2;$$

consequently,

$$C_2(F) \geq \frac{\gamma(1)}{1+A_2} = \frac{1}{1+A_2}\sum_{i=1}^{n}\gamma_i(1) = \frac{1}{1+A_2}\sum_{i=1}^{n} C_2(K_i)$$

$$= \frac{N}{1+A_2}\sum_{i=1}^{n}\varrho_i^3 = \frac{1}{1+A_2}\sum_{i=1}^{n}\frac{\varrho_i^3}{\varrho^2},$$

and by (5.5.13),

$$C_2(F) > \frac{A_1}{1+A_2}\varrho = A\varrho.$$

This implies (5.5.11) and thus, the theorem is proved. □

21. Consider now a compact set K in the plane. The method which we used in No. 19 to prove Theorem 5.18, carries over without change to the case of the plane; one need only use for L_{CK} the linear manifold generated by the family of functions $\left\{\ln \frac{1}{|z-z_0|}\right\}$, $z_0 \in CK$, and take into consideration the remark made at the end of No. 6.

If K has no interior points, then one can give a sufficient condition, which is to a certain extent "geometric", for the regularity of CK and, hence for the density of L_{CK} in C_K.

We shall call a continuum l *g-decomposable* if it is representable as

$$l = l_1 \cup l_2,$$

where $l_1 \neq \emptyset$ is a continuum which does not separate the plane and $l_2 \neq l$ (although it may be empty). If such a decomposition is impossible, then l will be called a *g-indecomposable* continuum.

We note that any continuum which does not separate the plane is g-decomposable. An example of a g-indecomposable continuum can be obtained from a closed disc by eliminating an everywhere dense family of open, non-intersecting discs.

Now we can formulate the criterion mentioned above.

Theorem 5.22. *Suppose that the compact set K has no interior points and does not contain a g-indecomposable continuum. Then L_{CK} is dense in C_K and, consequently, the set CK has no irregular points.*

Proof. From the proof of Theorem 5.18, it is clear that to prove the density of L_{CK} in C_K it is sufficient to show that the values outside the set K of the potential

$$U^\mu(z) = \int_K \ln \frac{1}{|z-\zeta|}\, d\mu(\zeta)$$

of the measure μ concentrated on K, uniquely define U^μ on the set K. Indeed, in this case, according to the uniqueness theorem (Theorem 1.12), the measure μ will also be uniquely defined.

Let G_k ($k = 0, 1, \ldots$) denote, as in No. 19, the connected components of CK. Since K has no interior points,

$$K = \partial K = \overline{\bigcup_{i=0}^{\infty} \partial G_i}.$$

Let $z_0 \in \partial G_i$; we will show that G_i is not thin at the point z_0. Then, by § 3, No. 9, the potential U^μ will be uniquely defined at the point z_0. By Theorem 5.10 we need to verify that z_0 is a regular point for G_i. Note that the set

$$G_i^{(n)} = G_i \cap \{z : 2^{-n-1} \le |z - z_0| \le 2^{-n}\}$$

contains, for sufficiently large n, a broken line of diameter 2^{-n-1}. It can be obtained by connecting any point of the set $G_i \cap \{z: |z - z_0| = 2^{-n-1}\}$ by a broken line within G_i to a point on the set $G_i \cap \{z: |z - z_0| = 2^{-n}\}$ and taking the subset of this broken line in $G_i^{(n)}$. Since the logarithmic capacity of such a line is not less than 2^{-n-3} (cf. Ch. II, No. 17), we have

$$C_l(G_i^{(n)}) \ge 2^{-n-3},$$

and applying criterion (5.1.15′) we will have

$$\sum_n \frac{\ln 2^{-n}}{\ln C_l(G_i^{(n)})} \ge \sum_n \frac{\ln 2^n}{\ln 2^{n+3}} = \infty.$$

This proves the regularity of the point z_0.

Thus, the potential $U_\mu(z)$ is already defined at all points of the set

$$r = \bigcup_{i=1}^{\infty} \partial G_i.$$

It is not difficult to see also, that the indicated construction is applicable not only to points of r, but also to those points of $K \setminus r$ which have a sufficiently small neighborhood which does not completely contain any region G_i. Denote this set of points by e and consider the compact set

$$K_1 = \overline{K \setminus (r \cup e)}.$$

Since

$$CK_1 \subset C(K \setminus (r \cup e)) = CK \cup r \cup e,$$

the potential $U^\mu(z)$ is everywhere defined outside K_1.

We will show that K_1 is a proper subset of K. Let l be any connected component of K. Since this is a g-decomposable continuum, we have

$$l = l_1 \cup l_2,$$

where the continuum $l_1 \neq \emptyset$ and does not separate the plane, and the continuum $l_2 \neq l$. Consider a point $z_1 \in l_1$ which is a positive distance (denote this by δ_1) from l_2 (if $l_2 = \emptyset$, then z_1 is any point of l_1). For $\delta < \delta_1$ a δ-neighborhood of the point z_1 does not contain any region G_i and hence $z_1 \in e$. Thus, e contains all points of l which are a positive distance from l_2, i.e. an entire portion of the set l. Consequently, K_1 is a proper subset of K.

Further, the compact sets K_α are constructed by transfinite induction: $K_{\alpha+1}$ is defined from K_α just as K_1 was defined from K, and, if β is a limit ordinal, then we set $K_\beta = \bigcap_{\alpha<\beta} K_\alpha$. From the above, it follows that if K_α is not empty, then K_β for $\beta > \alpha$ is a proper subset of K_α.

By the well-known Baire theorem,[16] among the sets $\{K_\alpha\}$ there is only a countable number of distinct sets. Consequently, after a countable number of steps we shall arrive at a set K_α which is empty, and the potential $U^\mu(z)$ will be defined throughout the plane. This proves the theorem. ▯

22. Together with the linear manifold L_{CK} generated by the harmonic functions $\left\{\ln \frac{1}{|z-z_0|}\right\}$, $z_0 \in CK$, consider yet another linear manifold, belonging to $\mathfrak{H}_K$; namely, we denote by $\mathfrak{R}_{CK} \subset \mathfrak{H}_K$ the linear manifold generated by the rational fractions

$$\left\{\mathrm{Re}\frac{1}{z-z_0},\quad \mathrm{Im}\frac{1}{z-z_0}\right\},\quad z_0 \in CK.$$

We saw (cf. p. 341) that L_{CK} is dense in $\mathfrak{H}_K$ and, therefore, dense relative to $\mathfrak{R}_{CK}$ in C_K. The converse, however, is false; hence, the question of the density of $\mathfrak{R}_{CK}$ in C_K can not be resolved in terms of irregular points.

We now give one sufficient criterion for $\mathfrak{R}_{CK}$ to be dense in C_K.

We call the compact set, l, *singular* if any neighborhood of any of its points intersects an infinite set of connected components of Cl.

It can be shown that a singular continuum is g-indecomposable. On the other hand, a singular compact set may not contain any g-indecomposable continua. Here is a simple example: the set of circles with center at the point 0, drawn through all the points of the Cantor set on the segment [0,1].

---

[16] Cf. Hausdorff [1].

Theorem 5.23. *Suppose the compact set K has no interior points and does not contain a singular compact set. Then $\mathfrak{R}_{CK}$ is dense in C_K.*

Proof. If $\mathfrak{R}_{CK}$ is not dense in C_K, then there exists a signed measure $\nu \neq 0$, concentrated on K and such that

$$\int_K \operatorname{Re} \frac{1}{z-\zeta}\, d\nu(z) = 0, \int_K \operatorname{Im} \frac{1}{z-\zeta}\, d\nu(z) = 0, \quad \zeta \in CK.$$

But in that case

$$\int_K \frac{d\nu(z)}{z-\zeta} = 0, \quad \zeta \in CK,$$

from which, integrating with respect to ζ, we obtain in each connected component G_i of the open set CK

$$\int_K \ln \frac{1}{z-\zeta}\, d\nu(z) = A_i,$$

where A_i is a certain constant. Determining the real part, we obtain

$$U^\nu(\zeta) = a_i, \quad \zeta \in G_i,$$

where $a_i = \operatorname{Re} A_i$.

If G_0 is an infinite component of CK, then $a_0 = 0$. Indeed, $U^\nu(\zeta)$ is naturally bounded in G_0, and therefore $\nu(1) = 0$. But then

$$-a_0 = \int_K \ln|z-\zeta|\, d\nu(z) = \int_K \ln\left|1 - \frac{z}{\zeta}\right| d\nu[z),$$

and forcing $\zeta \to \infty$ we obtain that $a_0 = 0$.

We now show that the equation

$$U^\nu(\varrho) = a_i$$

holds at all points of ∂G_i. Indeed, as was established in the proof of Theorem 5.22, every point $\varrho_0 \in \partial G_i$ is regular for G_i. Hence, by the definition of thinness (5.3.1), we can write

$$U^{\nu+}(\zeta_0) = \varliminf_{G_i \ni \zeta \to \zeta_0} U^{\nu+}(\zeta),$$

$$U^{\nu-}(\zeta_0) = \varliminf_{G_i \ni \zeta \to \zeta_0} U^{\nu-}(\zeta).$$

Subtracting these equations, we obtain

$$U^\nu(\zeta_0) \geq \varliminf_{G_i \ni \zeta \to \zeta_0} [U^{\nu+}(\zeta) - U^{\nu-}(\zeta)] = a_i.$$

Applying the same reasoning to the potential $U^{-\nu}(\zeta)$, we discover that

$$U^{-\nu}(\zeta_0) \geq -a_i,$$

and hence $U^\nu(\zeta_0) = a_i$.

Consequently, if $\partial G_i \cap \partial G_j \neq \emptyset$, then $a_i = a_j$.

The theorem will be proved, if we are convinced that $U_\nu(\zeta) \equiv 0$ throughout the plane, since this would imply that $\nu \equiv 0$ and hence that $\mathfrak{R}_{CK}$ is dense in C_K.

Let π denote the operation which attaches to any set ω all those regions G_j, for which $\partial\omega \cap \partial G_j \neq \emptyset$ together with the sets $\partial\omega \cap \partial G_j$. Then $\pi(G_i)$ is a set which contains G_i and $U^\nu(\zeta) = a_i$ throughout $\pi(G_i)$. If $\pi(G_i) = G_i$, then we will say that the operation π does not apply to G_i.

Set $\pi^{\alpha+1}(G_0) = \pi(\pi^\alpha(G_0))$ and $\pi^\beta(G_0) = \bigcup_{\alpha<\beta} \pi^\alpha(G_0)$. By the Baire theorem, after a countable number of steps we will obtain a region $\Omega_0 = \pi^\beta(G_0)$ to which the operation π does not apply; and in Ω_0, $U^\nu(\zeta) = 0$.

If $\Omega = R^2$, then the theorem is proved. Otherwise, consider the countable system of regions, consisting of Ω_0 and all the G_i not included in Ω_0. Iterate the operation π beginning with some one of the specified regions G_i. Denote by Ω_1 that region of the type $\pi^\beta(G_i)$ to which the operation π does not apply. If $\Omega_1 = R^2$, then the theorem is proved. Otherwise, we repeat the same reasoning, beginning with a region G_j not included in $\Omega_0 \cup \Omega_1$.

After a countable number of steps either $U^\nu(\zeta) \equiv 0$ throughout the whole plane, or we shall have a countable system of regions $D_0, D_1, D_2, \ldots$, to each of which the operation π does not apply. In the latter case the set $l = C(\bigcup_0^\infty D_i) \subset K$ as it is easy to see, is singular, which contradicts the condition of the theorem. Thus only the first possibility can hold, and the theorem is proved. $\square$

A complete solution of the question of the density of $\mathfrak{R}_{CK}$ in C_K (i.e. finding necessary and sufficient conditions) requires the introduction of such notions as the analytic capacity of a set or the minimal boundary of a function algebra and therefore goes beyond the limits of this book.

Chapter VI

Generalizations

§ 1. Distributions with Finite Energy and Their Potentials

1. In Ch. I, § 4 we saw that the space $\mathscr{E}_\alpha$ $(1 < \alpha < p)$, of signed measures in R^p with finite energy, together with the Hilbert structure given by the formula

$$(\nu_1, \nu_2) = I[\nu_1, \nu_2] = \int U_\alpha^{\nu_1}(x)\, d\nu_2(x)$$

is not complete[1] (cf. Theorem 1.19).

Our goal is to describe the Hilbert space $\overline{\mathscr{E}}_\alpha$, obtained by *completion* of $\mathscr{E}_\alpha$. Our description will follow *J.* Deny [4] and will utilize the Fourier transform. For this, it will be more convenient to view $\mathscr{E}_\alpha$ as a space of *complex* measures, in which the scalar product of complex measures $\nu_1 = \alpha_1 + i\beta_1$ and $\nu_2 = \alpha_2 + i\beta_2$ is defined by the formula

$$(\nu_1, \nu_2) = (\alpha_1, \alpha_2) + (\beta_1, \beta_2) + i[(\beta_1, \alpha_2) - (\alpha_1, \beta_2)] = \int U_\alpha^{\nu_1}(x)\, \overline{d\nu_2(x)}.$$

Consider first the linear manifold $\mathscr{E}_\alpha^{(0)}$ of complex measures with compact supports in $\mathscr{E}_\alpha$. By Corollary 2 to Lemma 1.2′ it is dense in $\mathscr{E}_\alpha$. If $\nu \in \mathscr{E}_\alpha^{(0)}$, then $\tilde{\nu}(x)$ will denote the Fourier transform of the complex measure ν (cf. Introduction, § 3), which is an infinitely differentiable function in R^p.

Lemma 6.1. *If* $\nu \in \mathscr{E}_\alpha^{(0)}$, *then* $k_\alpha * \nu * \check{\bar{\nu}}$ *is a continuous function* R^p,[2] *and*

$$||\nu||^2 = k_\alpha * \nu * \check{\bar{\nu}}(0) = \int \frac{|\tilde{\nu}(x)|^2}{|x|^\alpha}\, dx. \tag{6.1.1}$$

Proof. By Lemma 0.6, the double convolution

$$k_\alpha * \nu * \check{\bar{\nu}},$$

is defined and is a locally integrable function. The latter follows since

---

[1] In this case, it is said that the space is a pre-Hilbert space.

[2] More precisely, it is equivalent to a continuous function.

it is the potential of the complex measure $\bar{\check{\nu}} * \nu$ with compact support Furthermore, by the Corollary to Theorem 0.12,

$$\mathfrak{F}[k_\alpha * \nu * \bar{\check{\nu}}] = \frac{|\tilde{\nu}(x)|^2}{|x|^\alpha},$$

whence

$$k_\alpha * \nu * \bar{\check{\nu}} = \mathfrak{F}^{-1}\left[\frac{|\tilde{\nu}(x)|^2}{|x|^\alpha}\right].$$

The assertions of the lemma will clearly follow from

$$\int \frac{|\tilde{\nu}(x)|^2}{|x|^\alpha}\, dx < \infty. \tag{6.1.2}$$

To prove (6.1.2), consider the smoothed complex measure[3]

$$\nu = \nu * m^{(1/n)}.$$

It is easy to see that $k_\alpha * \nu_n * \bar{\check{\nu}}_n$ is a bounded continuous function in R^p, and

$$\mathfrak{F}[k_\alpha * \nu_n * \bar{\check{\nu}}] = \frac{|\tilde{\nu}(x)|^2}{|x|^\alpha} \cdot |h_n(x)|^2 \geq 0,$$

where $h_n(x) = \mathfrak{F}[m^{(1/n)}]$. Consequently[4]

$$\int \frac{|\tilde{\nu}(x)|^2}{|x|^\alpha} |h_n(x)|^2\, dx = k_\alpha * \nu_n * \bar{\check{\nu}}_n(0)$$

$$= \int k_\alpha (x - y)\, d\nu_n(x)\, \overline{d\nu_n(y)} = ||\nu_n||^2.$$

However, by Lemma 1.2, $||\nu_n||^2 \to ||\nu||^2$ and since at each point x $h_n(x) \to 1$ as $n \to \infty$, then the Fatou lemma immediately implies (6.1.2). ▯

Lemma 6.2. *The formula*

$$||\nu||^2 = \int \frac{|\tilde{\nu}(x)|^2}{|x|^\alpha}\, dx \tag{6.1.1}$$

holds for any complex measure $\nu \in \mathscr{E}_\alpha$.

Proof. Suppose $\nu \in \mathscr{E}_\alpha$ and $\nu_n \in \mathscr{E}_\alpha^{(0)}$ and $||\nu - \nu_n|| \to 0$. Then $||\nu_n - \nu_m|| \to 0$, and by Lemma 6.1

$$\int \frac{|\tilde{\nu}_n(x) - \tilde{\nu}_m(x)|^2}{|x|^\alpha}\, dx \to 0.$$

---

[3] For the definition of $m^{(r)}$, cf. Introduction, No.5, (d).

[4] To be convinced of this, consider the functions

$$\varphi_1(x) = \frac{|\tilde{\nu}(x)|^2}{|x|^\alpha} |h_n(x)|^2, \qquad \varphi_2(x) = \begin{cases} \prod_1^p \left(1 - \frac{|x_i|}{\lambda}\right), & |x_i| < \lambda, \\ i = 1, 2, \dots, p, & \\ 0 \text{ at the other points of } R^p, & \end{cases}$$

apply the convolution theorem (cf. Corollary to Theorem 0.12), and then let $\lambda \to \infty$.

By the completeness theorem, there exists a function $\tilde{v}(x)$ such that

$$\int \frac{|\tilde{v}(x) - \tilde{v}_n(x)|^2}{|x|^\alpha} dx \to 0.$$

It is easy to see that this implies (6.1.1). Moreover, since $v_n \dot{\to} v$ and the Fourier operator is weakly continuous (cf. Theorem 0.11′), then $v(x) = \mathfrak{F}(v)$. ▯

If we denote by $L^2_{(\alpha)}$ the Hilbert space of functions in R^p, whose modulus is square-integrable with weight $|x|^{-\alpha}$, then Lemma 6.2 shows that $v \to \tilde{v}(x)$ is an isometry of $\mathscr{E}_\alpha$ into a certain linear manifold E of the space $L^2_{(\alpha)}$. Now to obtain the completion of $\mathscr{E}_\alpha$ is very simple: it will be isometric to the closure $\bar{E}$ of the manifold E in the space $L^2_{(\alpha)}$. From what follows, it will turn out that $\bar{E} = L^2_{(\alpha)}$.

2. **Lemma 6.3.** *Any function $t(x) \in L^2_\alpha$ belongs to S^* and consequently,*[5] *is the Fourier transform of some distribution $T \in S^*$.*

Proof. By Theorem 0.10 we must show that the function $t(x)$ is slowly increasing, i.e. there exists an $m > 0$, such that

$$\int \frac{|t(x)|}{(1 + |x|^2)^m} dx < \infty.$$

Using the Bunyakovsky-Schwartz inequality, for sufficiently large m we obtain

$$\left(\int \frac{|t(x)|}{(1 + |x|^2)^m} dx\right)^2 \leq \int \frac{|t(x)|^2}{|x|^\alpha} dx \int \frac{|x|^\alpha}{(1 + |x|^2)^{2m}} dx,$$

which proves the lemma. ▯

Definition. *Distributions $T \in S^*$ for which $\tilde{T}(x) \in L^2_\alpha$ will be called distributions with finite energy, and the quantity*

$$||T||^2 = \int \frac{|\tilde{T}(x)|^2}{|x|^\alpha} dx \tag{6.1.3}$$

will be called the energy of T.

The set of all such distributions will be denoted by S^*_α.

For $T_1, T_2 \in S^*_\alpha$, if we set

$$(T_1, T_2) = \int \tilde{T}_1(x) \overline{\tilde{T}_2(x)} \frac{dx}{|x|^\alpha}, \tag{6.1.3′}$$

then, by the preceding, S^*_α becomes a Hilbert space, and the mapping $T \leftrightarrow \tilde{T}(x)$ is an *isometry* between S^*_α and $L^2_{(\alpha)}$. By Lemma 6.2 $\mathscr{E}_\alpha \subset S^*_\alpha$.

[5] Cf. Theorem 0.11′.

Now we show that $\mathscr{E}_\alpha$ is dense in S^*. This will imply that E is dense in $L^2_{(\alpha)}$.

Lemma 6.4. *If* $T \in S^*_\alpha$ *and*

$$T_n = T * \varrho^{(1/n)}$$

then $\|T - T_n\| \to 0$ *as* $n \to \infty$.[6] *Since* $T_n \in \mathscr{E}_\alpha$ *then this implies that* $\mathscr{E}_\alpha$ *is dense in* S^*_α.

Proof. Since

$$\tilde{T}_n(x) = \tilde{T}(x)\tilde{\varrho}^{(1/n)}(x)$$

then

$$\|T - T_n\|^2 = \int \frac{|\tilde{T}(x)|^2}{|x|^\alpha} |1 - \tilde{\varrho}^{(1/n)}(x)|^2 \, dx$$
$$= \int\limits_{|x|<R} + \int\limits_{|x|>R} \frac{|\tilde{T}(x)|^2}{|x|^\alpha} |1 - \tilde{\varrho}^{(1/n)}(x)|^2 dx.$$

But $|\tilde{\varrho}^{(1/n)}(x)| \leq 1$ and $\tilde{\varrho}^{(1/n)}(x) \to 1$ uniformly on any compact set. Therefore, first choosing a sufficiently large R, and then n, one can make $\|T - T_n\|$ arbitrarily small. ▯

The results obtained can be formulated in the form of the following theorem.

Theorem 6.1. *The Hilbert space* S^*_α *if distributions with finite energy is the completion of the space* $\mathscr{E}_\alpha$ *of complex measures with finite energy.*

The theory presented is easily extended to the case of kernels k, more general than the M. Riesz kernels $k_\alpha(x) = A(p, \alpha) |x|^{\alpha-p}$. It is sufficient to assume the following condition:

(A) The Fourier transform $\tilde{k}$ is a positive function in R^p, and $(\tilde{k})^{-1} \in S^*$.

We point out that by Theorem 0.10 $\tilde{k}(x)$ and $\tilde{k}^{-1}(x)$ are slowly increasing.

3. For any distribution T with compact support, the potential can be defined (cf. Lemma 0.6) as a distribution by the formula

$$U^T_\alpha = k_\alpha * T. \tag{6.1.4}$$

We will show how to define the potential of any distribution with finite energy.

---

[6] For the definition of $\varrho^{(r)}$, cf. Introduction, No.6.

For this we consider the function

$$\tilde{U}(x) = \tilde{k}_\alpha(x)\,\tilde{T}(x) = \frac{\tilde{T}(x)}{|x|^\alpha}\,. \tag{6.1.4.'}$$

First of all, it is locally integrable, since it can be represented as the product $\tilde{T}(x) \cdot |x|^{-\alpha/2} \cdot |x|^{-\alpha/2}$, in which each factor is square integrable on any compact set.

Furthermore, it is slowly increasing, i.e. it belongs to S^*. Indeed,

$$\left(\int \frac{|\tilde{U}(x)|}{(1+|x|^2)^m}\,dx\right)^2 \leq \int \frac{|\tilde{T}(x)|^2}{|x|^\alpha}\,dx \cdot \int \frac{dx}{|x|^\alpha(1+|x|^2)^{2m}}\,,$$

so that for $4m + \alpha > p$ these integrals will be finite.

Consequently, there exists a unique distribution $U_\alpha^T \in S^*$ such that

$$U_\alpha^T = \tilde{U}(x)\,.$$

This distribution U_α^T will be called the potential of $T \in S_\alpha^*$. By the Corollary to Theorem 0.12 for T with compact support, this definition coincides with (6.1.4).

Theorem 6.2. *The potential U_α^T of a distribution with finite energy is an absolutely continuous complex measure. Its density $U_\alpha^T(x)$ can be defined quasi-everywhere in R^p such that*

$$U_\alpha^{T_1+T_2}(x) = U_\alpha^{T_1}(x) + U_\alpha^{T_2}(x) \tag{6.1.5}$$

quasi-everywhere and

$$(T, \nu) = \int U_\alpha^T(x)\,\overline{d\nu(x)} \tag{6.1.6}$$

for any $\nu \in \mathscr{E}_\alpha$.

Proof. For any $T \in S_\alpha^*$, there is a sequence $\{\nu_n\}$ of complex measure in $\mathscr{E}_\alpha$ which strongly converges to T (Theorem 6.1). Without loss of generality, we may assume that the potentials $U_\alpha^{\nu_n}(x)$ are continuous (cf. Corollary to Lemma 1.2′ in Ch. I, § 4).

Using the method employed in the proof of Theorem 3.6 (the finite energy case), we see that from the sequence $\{U_\alpha^{\nu_n}(x)\}$ one can select a subsequence which converges quasi-everywhere in R^p.

Moreover, there is a decreasing sequence $\{G^{(k)}\}$ of open sets such that $C_\alpha(G^{(k)}) \to 0$, and convergence of this subsequence is uniform outside any $G^{(k)}$.

The limit function $V(x)$ is hence defined quasi-everywhere in R^p. It is not difficult to show that $V(x)$ does not depend on the choice of $\{\nu_n\}$ or the convergent subsequence of the potentials $\{U_\alpha^{\nu_n}(x)\}$.

Then, if for two distributions $T_1, T_2 \in S_\alpha^*$ we construct functions $V_1(x)$ and $V_2(x)$, it is easy to see that $V_1(x) + V_2(x)$ will correspond to $T_1 + T_2$.

Now we prove that for any complex measure $\nu \in \mathscr{E}_\alpha$

$$(T, \nu) = \int V(x)\, \overline{d\nu(x)}. \tag{6.1.7}$$

It is sufficient to establish (6.1.7) assuming that ν is a measure, with T, and hence, $V(x)$ real.

Suppose that μ is the restriction of the measure ν to the set $\{x\colon V(x) \geq 0\}$. Clearly, $\mu \in \mathscr{E}_\alpha^+$ and if we prove that

$$(T, \mu) = \int V(x)\, d\mu(x), \tag{6.1.7'}$$

then this will imply (6.1.7). First suppose that the measure is finite, i.e. $\mu(1) < \infty$. Denote by μ_k the restriction of μ to $G^{(k)}$; we will show that $||\mu_k|| \to 0$. Since the measure μ is C-absolutely continuous, and $C_\alpha(G^{(k)}) \to 0$ then $\mu_k \dot{\to} 0$. Since the measures μ_k are monotone decreasing, it is easy to establish that the $||\mu_k||$ are also monotone decreasing, and hence, have a limit. But then the sequence $\{\mu_k\}$ is fundamental, since

$$||\mu_{n+k} - \mu_k||^2 = ||\mu_{n+k}||^2 + ||\mu_k||^2 - 2(\mu_{n+k}, \mu_k)$$
$$\leq ||\mu_{k+n}||^2 + ||\mu_k||^2 - 2(\mu_{k+n}, \mu_{k+n}) = ||\mu_k||^2 - ||\mu_{k+n}||^2.$$

Lemma 1.5 (Ch. I, No. 13) shows that $||\mu_k|| \to 0$.

We now point out the following relation:

$$\left|\int U_\alpha^{\nu_n}(x)\, d\mu(x) - \int_{CG^{(k)}} U_a^{\nu_n}(x)\, d\mu(x)\right| = \left|\int U_a^{\nu_n}(x)\, d\mu_k(x)\right|$$
$$\leq ||\nu_n|| \cdot ||\mu_k||.$$

Since

$$\lim_{n\to\infty} \int U_a^{\nu_n}(x)\, d\mu(x) = \lim_{n\to\infty} (\nu_n, \mu) = (T, \mu),$$

and

$$\lim_{n\to\infty} \int_{CG^{(k)}} U_\alpha^{\nu_n}(x)\, d\mu(x) = \int_{CG^{(k)}} V(x)\, d\mu(x),$$

since on $CG^{(k)}$ convergence is uniform, and $\mu(1) < \infty$, then after passing to the limit $n \to \infty$ we obtain

$$\left|(T, \mu) - \int_{CG^{(k)}} V(x)\, d\mu(x)\right| \leq ||T|| \cdot ||\mu_k||,$$

from which

$$\lim_{k\to\infty} \int_{CG^{(k)}} V(x)\, d\mu(x) = (T, \mu).$$

But $V(x) \geq 0$ μ-almost everywhere, and hence,

$$\int V(x)\,d\mu(x) = (T, \mu).$$

If μ does not satisfy the finiteness condition, then μ can be represented as the strong limit of a sequence $\{\mu_i\}$ of finite measures (where, for example, $\mu_{(i)}$ is the restriction of μ to the ball $|x| \leq i$); then in the equation

$$\int V(x)\,d\mu_i(x) = (T, \mu_i)$$

one can pass to the limit $i \to \infty$.

Thus (6.1.7) is proved.

In order to conclude the proof of the theorem, we must also show that $V(x)$ is the density for U_α^T, i.e. that for any $\varphi \in D$,

$$U_\alpha^T(\varphi) = \int V(x)\,\overline{\varphi(x)}\,dx.$$

But this follows, since by (6.1.3′) and (6.1.4′)

$$U_\alpha^T(\varphi) = \tilde{U}_\alpha^T(\tilde{\varphi}) = \int \frac{\tilde{T}(x)}{|x|^\alpha}\,\overline{\tilde{\varphi}(x)}\,dx = (T, \varphi), \tag{6.1.8}$$

and the latter, by (6.1.7) is equal to

$$\int V(x)\,\overline{\varphi(x)}\,dx.$$

The theorem is proved. ☐

4. Let $K \subset R^p$ be a compact set. Denote by $S_{\alpha K}^* \subset S_\alpha^*$ the set of all distributions with finite energy whose support belongs to K.

Lemma 6.5. *$S_{\alpha K}^*$ is a subspace of the Hilbert space S_α^*.*

Proof. Since the linearity of $S_{\alpha K}^*$ is clear, we need only prove that $S_{\alpha K}^*$ is closed. For this, it is sufficient to establish that strong convergence of T_n to T implies weak convergence, i.e. that for any function $\varphi \in D$

$$T_n(\varphi) \to T(\varphi).$$

By the continuity of the Fourier operator and its inverse, it is sufficient to verify that

$$\tilde{T}_n(\varphi) \to \tilde{T}(\varphi), \quad \varphi \in D.$$

But

$$\tilde{T}_n(\varphi) - \tilde{T}(\varphi) = \int [\tilde{T}_n(x) - \tilde{T}(x)]\,\varphi(x)\,dx,$$

and therefore,

$$|\tilde{T}_n(\varphi) - \tilde{T}(\varphi)|^2 \leq \int \frac{|\tilde{T}_n(x) - \tilde{T}(x)|^2}{|x|^\alpha}\,dx \int |x|^\alpha |\varphi(x)|^2\,dx$$
$$= \|T_n - T\|^2 \int |x|^\alpha\,|\varphi(x)|^2\,dx,$$

which implies the required result. ☐

Denote by P_K the projection operator onto $S^*_{\alpha K}$. As we shall now see, the operator P_K is the analogue of the balayage operator on K (cf. Ch. IV, § 4).

Theorem 6.3. *Almost everywhere on the set $\mathring{K}$ of interior points of K*

$$U^T_\alpha(x) = U^{P_K T}_\alpha(x). \tag{6.1.9}$$

Proof. First of all, from the definition of projection, it follows that for any $T_1 \in S^*_{\alpha K}$

$$(P_K T, T_1) = (T, T_1) \tag{6.1.10}$$

Then, we note that any function $\varphi \in D$ belongs to S^*_α, since

$$\int \frac{|\tilde{\varphi}(x)|^2}{|x|^\alpha}\, dx < \infty.$$

Therefore, if the support of φ is contained in $\mathring{K}$, then $\varphi \in S^*_{\alpha K}$, and by (6.1.10)

$$(P_K T, \varphi) = (T, \varphi),$$

by (6.1.8) this can be rewritten in the form

$$U^{P_K T}_\alpha(\varphi) = U^T_\alpha(\varphi).$$

This proves the theorem. ☐

Now consider a distribution $T \in S^*_\alpha$, for which $U^T_\alpha \in D$, and $U^T_\alpha(x) \equiv 1$ on $\mathring{K}$. Such a T, of course, exists. In fact, if $\varphi(x) \in D$ and $\varphi(x) \equiv 1$ on $\mathring{K}$, then, putting $T = k_{-\alpha} * \varphi$, by (1.1.12), we will have $U^T_\alpha(x) \equiv \varphi(x)$. The T defined by this formula belongs to S^*_α since $\tilde{T}(x) = |x|^\alpha\, \tilde{U}^T_\alpha(x) \in L^2_{(\alpha)}$.

By Theorem 6.3 the distribution

$$\Gamma = P_K T \in S^*_{\alpha K}$$

has the property that $U^\Gamma_\alpha(x) = 1$ almost everywhere on $\mathring{K}$.

This distribution Γ does not depend on the choice of T, satisfying the above conditions.

In fact, Γ is the unique solution of the minimization problem for $||T_1 - T||^2$ in the class of distributions $T_1 \in S^*_{\alpha K}$. But

$$\begin{aligned} ||T_1 - T||^2 &= ||T_1||^2 - 2 \operatorname{Re} (T_1, T) + ||T||^2 \\ &= ||T_1||^2 - 2 \operatorname{Re} T_1(1) + ||T||^2, \end{aligned}$$

and therefore the minimization of $||T_1 - T||^2$ is equivalent to the minimization of

$$||T_1||^2 - 2 \operatorname{Re} T_1(1).^7$$

---

[7] Also equivalent to the minimization of $|| T_1||^2$ under the conditions $T_1 \in S^*_{\alpha K}$, $\operatorname{Re} T_1(1) = 1$ (cf. Theorem 2.3).

By (6.1.10)

$$(\Gamma, T_1) = (T, T_1) = \overline{T}_1(1), \quad T_1 \in S^*_{\alpha K}.$$

In particular, putting $T_1 = \Gamma$, we obtain that

$$\Gamma(1) = ||\Gamma||^2.$$

The distribution Γ is called the *equilibrium distribution* on the compact set K, and the quantity $\Gamma(1)$ is called the *spectral measure* of K.

Of course, in the Robin case, i.e. for $0 < \alpha \leq 2$, the equilibrium measure γ (cf. Ch, II, n⁰ 3) will coincide with Γ, and the spectral measure with the capacity.

5. Consider the case of a Newtonian potential, $\alpha = 2$, $p \geq 3$; we shall establish a connection between the distributions with finite energy and the Hilbert space BL_0 introduced in Ch. I, § 4. This connection is described in the following theorem.

Theorem 6.4. *The Newtonian potential* $U_2^T(x)$, $T \in S_2^*$ *is a function of the class* BL_0 *in the whole space* R^p, *and*

$$||T||^2 = \frac{1}{4\pi^2} \int |\text{grad } U_2^T(x)|^2 \, dx. \tag{6.1.11}$$

Conversely, any function of BL_0 *coincides almost everywhere with the potential of a distribution in* S_2^*.

Proof. From the construction of $U_2^T(x)$ (cf. proof of Theorem 6.2), it follows that this function will be continuous outside an open set with arbitrarily small capacity. Since capacity does not increase by projection (Theorem 2.9), we easily conclude from this, that $U_2^T(x)$ will be continuous on almost all lines parallel to the axes $0x_j (j = 1, 2, \ldots, p)$. Consider the distribution $\frac{\partial}{\partial x_j} U_2^T(x)$; we will show that it is an ordinary function belonging to $L^2(R^p)$. Indeed, letting $\mathfrak{F}$ denote the Fourier operator, we obtain

$$\mathfrak{F}\left[\frac{\partial}{\partial x_j} U_2^T(x)\right] = 2\pi i x_j \mathfrak{F}[U_2^T(x)] = 2\pi i x_j \frac{\tilde{T}(x)}{|x|^2}.$$

Since

$$\frac{\tilde{T}(x)}{|x|} \in L^2(R^p),$$

then

$$\mathfrak{F}\left[\frac{\partial}{\partial x_j} U_2^T(x)\right] \in L^2(R^p).$$

By the Plancherel theorem

$$\frac{\partial}{\partial x_j} U_2^T(x) \in L^2(R^p)$$

and

$$\int \left| \frac{\partial}{\partial x_j} U_2^T(x) \right|^2 dx = 4\pi^2 \int \frac{x_j^2}{|x|^2} \frac{|\tilde{T}(x)|^2}{|x|^2} dx.$$

Summing these equations over j, we obtain

$$\int \sum_j \left| \frac{\partial}{\partial x_j} U_2^T(x) \right|^2 dx = 4\pi^2 \int \frac{|\tilde{T}(x)|^2}{|x|^2} dx,$$

i.e. (6.1.11).

We will show that $U_2^T(x)$ is absolutely continuous on almost all lines parallel to the axis $0x_j$ ($j = 1, 2, \ldots, p$). Indeed, $\frac{\partial}{\partial x_j} U_2^T(x)$ is integrable on any compact set, and hence for almost all lines, parallel to the axis $0x_j$, it is locally integrable along x_j. Consider a line parallel to $0x_j$ with the property that on it, $U_2^T(x)$ is continuous.

For any $\varphi(x_j) \in D$

$$\int \frac{\partial}{\partial x_j} U_2^T(x)\, \varphi(x_j)\, dx_j = -\int U_2^T(x)\, \varphi'(x_j)\, dx_j.$$

Selecting a sequence of functions $\varphi(x_j)$ of the form

$$\varphi_n(x_j) = \chi(x_j) * \varrho^{(1/n)}(x_j),$$

where

$$\chi(x_j) = \begin{cases} 0 & \text{outside } (a, b), \\ 1 & \text{within } (a, b), \end{cases}$$

we obtain as $n \to \infty$

$$U_2^T(x)\,\big|_{x_j=b} - U_2^T(x)\,\big|_{x_j=a} = \int_a^b \frac{\partial}{\partial x_j} U_2^T(x)\, dx_j,$$

which proves the absolute continuity of $U_2^T(x)$ on the given line.

Thus, $U_2^T(x) \in BL$. According to the definition of BL_0, we must also check that

$$\lim_{r\to\infty} \int U_2^T(x)\, d\varepsilon^{(r)}(x) = 0.$$

But

$$\left| \int U_2^T(x)\, d\varepsilon^{(r)}(x) \right| = |(T, \varepsilon^{(r)})| \le \|T\| \cdot \|\varepsilon^{(r)}\|,$$

and it is easy to see that $\|\varepsilon^{(r)}\| \to 0$ as $r \to \infty$.

We shall now return to the proof of the second part of the theorem.

Suppose that $F(x) \in BL_0$. We form the functions $f_j(x) = \frac{\partial}{\partial x_j} F(x)$ and the distribution

$$T = -\frac{1}{4\pi^2} \Delta F = -\frac{1}{4\pi^2} \sum_{j=1}^{n} \frac{\partial}{\partial x_j} f_j(x).$$

Clearly, $T \in S^*$ and

$$\mathfrak{F}[T] = \frac{1}{2\pi i} \sum_{j=1}^{n} x_j \, \mathfrak{F}[f_j].$$

But

$$\mathfrak{F}[f_j] \in L^2(R^p),$$

from which it is clear that

$$\mathfrak{F}[T] \in L^2_{(2)},$$

i.e. it is square-integrable with weight $|x|^{-2}$. This means that $T \in S_2^*$, i.e. has finite energy. Then it is easy to check that

$$\Delta U_2^T(x) = -4\pi^2 T = \Delta F,$$

and therefore $U_2^T(x) - F(x)$ is a harmonic function. But its first derivatives belong to $L^2(R^p)$, and hence, it is a constant. Since this constant belongs to BL_0, it must be zero. Thus, $F(x) = U_2^T(x)$ and the theorem is proved. ▯

From (6.1.11) it follows that for $T_1, T_2 \in S_2^*$

$$(T_1, T_2) = \frac{1}{4\pi^2} \int \operatorname{grad} U_2^{T_1} \cdot \operatorname{grad} \overline{U_2^{T_2}} dx.$$

Thus, the mapping $T \to U_2^T(x)$ is a linear isometry of S_2^* onto BL_0.

Remark. From the proof of Theorem 6.4 it is clear that any distribution $T \in S_2^*$ can be represented in the form

$$T = -\sum_{j=1}^{n} \frac{\partial}{\partial x_j} A_j(x) = -\operatorname{div} \boldsymbol{I}(x), \tag{6.1.12}$$

where the vector $\boldsymbol{I}(x)$ with components $A_j(x)$ belongs to $L^2(R^p)$ (i.e. all its components $A_j(x) \in L^2(R^p)$). Here

$$\boldsymbol{I}(x) = \frac{1}{4\pi^2} \operatorname{grad} U_2^T(x) \tag{6.1.13}$$

and

$$||T||^2 = 4\pi^2 \int |\boldsymbol{I}(x)|^2 \, dx.$$

It is not difficult to show that even without condition (6.1.13) formula (6.1.12) with $\boldsymbol{I}(x) \in L^2(R^p)$ defines a distribution with finite energy. In this case

$$||T||^2 \leq 4\pi^2 \int |\boldsymbol{I}(x)|^2 \, dx,$$

where the equality sign holds only in case $\boldsymbol{I}(x)$ is a gradient.

The theory developed in this section extends to the case of Green potentials in a region $G \subset R^p$. The definition of a distribution with finite energy encounters certain difficulties, since one cannot use the Fourier transform. This can be circumvented by considering these functions as weak limits of fundamental sequences of complex measures μ in G with finite energy.

Further, one can define the potential $U_g^T(x)$ (also by considering sequences of complex measures μ_n, which strongly converge to T) and establish the connection with the space of functions BL in G. We shall not go into the details of these constructions, referring the reader to the original work of J. Deny [4].

§ 2. Kernels of More General Type

In this section we shall give a survey of some research devoted to the construction of a general potential theory, based on a kernel $K(x, y)$ which is taken more or less to be an arbitrary function in $R^p \times R^p$ (or in $X \times X$, where X is a locally compact topological space) or in certain cases, a measure.

The expression

$$U_K^\mu(x) = \int K(x, y)\, d\mu(y).$$

will be called the potential of the measure μ.

In contrast to the above, all references to the literature will be given in the text.

6. First of all, we list the properties of measures and their potentials studied above, which it would be desirable to preserve, even partially, in a more general theory.

(1) *Positiveness* of the kernel:

$$K(x, y) \geq 0 \quad \text{in} \quad X \times X.$$

(2) *Lower semicontinuity* of the kernel.

(3) The existence of a prescribed *singularity* for $x = y$.

(4) If X is an abelian group, then the kernel $K(x, y)$ is *invariant* under translation, i.e. $K(x, y) = k(x - y)$.

(5) The existence of an *inverse operator* L: $L^{(x)}[K(x, y)] = \varepsilon_y$.

(6) *Strict positive definiteness:* for any signed measure with compact support

$$\iint_{X \times X} K(x, y)\, d\nu(x)\, d\nu(y) \geq 0,$$

where equality holds only if $\nu \equiv 0$.

(7) *Regularity* of the kernel (continuity principle): the continuity of the potential of a measure with compact support $U_K^\mu(x)$ on the support $S(\mu)$ implies its continuity in X (cf. Theorem 1.7).

(8) *First maximum principle:* if

$$U_K^\mu(x) \leq M \text{ on } S(\mu)$$

then this inequality holds everywhere in X.

(9) *Generalized maximum principle:* if

$$U_K^\mu(x) \leq M \quad \text{on} \quad S(\mu),$$

then

$$U_K^\mu(x) \leq hM, \qquad x \in X,$$

where h is a certain positive constant.

(10) *Second maximum principle* (principle of domination): if

$$U_K^\mu(x) \leq U_K^\lambda(x)$$

on $S(\mu)$, then this inequality holds everywhere.

To formulate other principles we introduce the notion "K-approximately everywhere".

Suppose μ is any measure; put

$$V(\mu) = \sup_{S(\mu)} U_K^\mu(x).$$

Then, for any set $E \subset X$, put

$$V_*(E) = \inf V(\mu), \quad \mu(1) = 1, \quad S(\mu) \subset E.$$

If $E = \emptyset$, then by definition, we will take $V_*(E) = \infty$.

We shall say that a property P holds approximately everywhere (K-approximately everywhere), if $V_*(e) = \infty$ for the set e where P does not hold.

(11) *Equilibrium principle:* for any compact set $F \subset X$ there exists a constant $C_K(F)$ and a measure μ, $\mu(1) = 1$, $S(\mu) \subset F$, such that

$$U_K^\mu(x) = C_K(F)$$

approximately everywhere on F and

$$U_K^\mu(x) \leq C_K(F)$$

everywhere in X.

The constant $C_K(F)$ is called the *K-capacity* of F. Independent of the presence or absence of the equilibrium principle, the K-capacity of F can be defined as $\sup \mu(1)$ over all measures μ satisfying the conditions

$$V(\mu) \leq 1, \quad S(\mu) \subset F.$$

(12) The *weak principle of equilibrium* is formulated just as (11), but without the last requirement.

(13) *Balayage principle:* for any compact set $F \subset X$ and measure μ with finite potential, there exists a measure μ', $S(\mu') \subset F$ such that

$$U_K^{\mu'}(x) = U_K^{\mu}(x)$$

approximately everywhere on F and

$$U_K^{\mu'}(x) \leq U_K^{\mu}(x)$$

everywhere.

If we eliminate the last requirement, then we obtain the *weak balayage principle.*

(14) *Principle of balayage onto an open set:* for any set $G \subset X$ open and relatively compact, and any measure μ with finite potential, there is a measure μ', $S(\mu') \subset \overline{G}$, such that

$$U_K^{\mu'}(x) = U_K^{\mu}(x)$$

everywhere on G and

$$U_K^{\mu'}(x) \leq U_K^{\mu}(x)$$

everywhere on X.

Of course, the properties listed, some in the form of "principles", are not independent.

One of the most important problems in potential theory is to establish the connections between these principles.

Another natural problem is to find reasonably wide classes of kernels which satisfy some of these principles.

7. Suppose that the kernel, $K(x, y)$ is symmetric, positive, finite for $x \neq y$, and continuous in the extended sense (i.e. is a continuous map of $X \times X$ onto $[0, +\infty]$). Then the following diagram indicates some of the relations between the "principles":

$$(8) \underset{\not\leftarrow}{\rightarrow} (9) \underset{\not\leftarrow}{\rightarrow} (7).$$

In this connection we make several remarks. The assertion $(8) \rightarrow (9)$ is trivial. That the converse is false is shown by M. Riesz potentials for $\alpha > 2$.

It is interesting that (9) holds for a rather wide class of kernels depending on the distance $|x - y|$. Namely, T. Ugaheri [2] has proved that if

$$K(x, y) = \varphi(|x - y|);\ x, y \in R^p,$$

where $\varphi(t) \geq 0$, is continuous and for $t \rightarrow +0$ monotonically tends to $+\infty$, then (9) holds.

N. Ninomiya [7] strengthened this result, proving that (9) holds for kernels

$$K(x, y) = k(x - y), \quad x, y \in R^p,$$

continuous for $x \neq y$, if the set

$$E_s = \{x : k(x) \geq s)\}$$

for $s \geq s_0$ is a convex compact set containing 0 and for any two points $x_1, x_2 \in \partial E_s$

$$\frac{|x_1|}{|x_2|} < \text{const}.$$

Analogous results can be found in H. Cartan and J. Deny [1] and G. Choquet [2].

Also, the implication (9) → (7) is proved analogous to Theorem 1.7. The converse is false, as seen by the counter-example $K(x, y) = |x - y|$.

The details of the above diagram and many examples of kernels can be found in the memoir by M. Ohtsuka [7].

G. Choquet [2] gives a deeper analysis of the connections between (9) and (7). Denote by O_c $(c > 0)$ the set of points x, for any of whose neighborhoods $V(x)$ there is a measure μ, $S(\mu) \subset V(x)$ with potential U_K^μ satisfying the condition

$$\sup_{y \in V(x)} U_K^\mu(y) > c \sup_{y \in S(\mu)} U_K^\mu(y).$$

Then, we set

$$O_\infty = \bigcap_{c > 0} O_c.$$

All these sets are closed.

The condition

$$O'_\infty = \emptyset$$

is necessary for the regularity of a kernel. We point out that the condition $O_\infty = \emptyset$ is not necessary for this.

If $O_\infty = \emptyset$ then on any compact set F, (9) holds, and this already implies (7), i.e. the regularity of the kernel.

The importance of the regularity requirement on a kernel for potential theory was established by G. Choquet in [3] and [5]. He proved, in particular, that for regular kernels hold Theorem 3.6[8] and the theorem analogous to the convergence Theorem 3.8.[9] In connection with the latter theorem, cf. also the paper by M. Brelot and G. Choquet [2].

---

[8] Note that the second proof of this theorem presented in Ch. III, No. 4, and borrowed from G. Choquet [5], serves precisely to establish this more general theorem.

[9] To prove this theorem, G. Choquet also introduces the condition $K(x, x) = \infty$ at the points of O_∞.

The assumption of symmetry for the kernel was introduced only for convenience of exposition. If it is omitted, then in addition to $K(x, y)$ one must also consider the *transposed* kernel

$$\check{K}(x, y) = K(y, x)$$

and require that it, too, be regular.

If the kernel $K(x, y)$ is positive, symmetric and *continuous*, then N. Ninomiya [8] has established the following:

$$(8) \rightleftarrows (11); \qquad (10) \rightleftarrows (13).$$

Moreover, each of the principles (8), (10), (11), or (13) implies (6).[10]

N. Ninomiya obtains these results by studying the minimum of the functional

$$\Phi(\mu, \lambda) = \frac{||\mu||^2 \, ||\lambda||^2}{(\mu, \lambda)^2}$$

in the class of measures μ concentrated on F_1, and measures λ concentrated on F_2 where $F_1 \cap F_2 = \emptyset$.

If one allows a symmetric and continuous kernel $K(x, y)$ to have a singularity for $x = y$, namely, if $K(x, y) \leq +\infty$, then it can be shown that (6) and (12) imply (11) and that (6) and the weak balayage principle imply (13) (cf. N. Ninomiya [10]).

H. Cartan and J. Deny [1] study kernels $K(x, y)$ in $R^p \times R^p$ which satisfy the invariance condition (4); they also assume that the function $k(x)$ satisfies condition (A) of No. 2 (note that this implies (6)). They show that for such kernels $(10) \rightleftarrows (13)$. Moreover, these principles are equivalent to the following: for any two measures μ_1, μ_2 with finite energy

$$\inf (U_K^{\mu_1}, U_K^{\mu_2})$$

is the potential of a measure also with finite energy (cf. Theorem 1.31).

Furthermore, subjecting the kernel $k(x)$ to the conditions:

(a) $k(x) > 0$ and continuous for $x \neq 0$,

(b) the set $E_s = \{x: k(x) \geq s\}$ for $s \geq s_0$ is a convex compact set,

(c) for all x in a neighborhood of 0,

$$k(x) \leq 2^h k(2x), \quad h < p.$$

H. Cartan and J. Deny prove that (8) and (10) hold together if, and only if, there exists a family of measures $\{\sigma\}$ such that $\sigma \neq \varepsilon$, $\sigma(1) = 1$, $U_k^\sigma(x) \leq k(x)$ everywhere in R^p, and for any neighborhood V of the point 0 there is a measure $\sigma = \sigma_V$ for which $U_k^\sigma(x) = k(x)$ outside V.

---

[10] R. E. Edwards [1] shows how to obtan (6) for Green kernels by using (13).

Note that in the classival theory and for M. Riesz measures for $\alpha < 2$, the family $\{\sigma\}$ is obtained by balayage of ε onto CV.

8. B. Fuglede [1] constructs a potential theory by postulating for a symmetric kernel $K(x, y)$ the basic results of the Cartan theory presented in Ch. I, § 4.

He calls a kernel $K(x, y)$ *perfect* if it is positive definite, if the space $\mathscr{E}_K^+$ is complete and the strong topology in $\mathscr{E}_K^+$ is stronger than the vague topology of measures (cf. Lemma 1.3).

The last requirement easily implies (6), i.e. that the kernel is strict positive definite.

Next, the kernel $K(x, y)$ is called *consistent* if it is positive definite, and every fundamental sequence (or generalized sequence) in $\mathscr{E}_K^+$ vaguely convergent to a measure μ, is strongly convergent to μ (cf. Lemma 1.5).

The above two classes of kernels are related as follows: *a kernel is perfect if, and only if, it is consistent and strict positive definite.*

Green kernels, M. Riesz kernels for all α, $0 < \alpha < p$, Deny kernels (i.e. kernels satisfying condition (A), § 1) are all perfect.

If the space X is compact, and if (6) and (7) hold, then a kernel will be perfect (if (6) is replaced by requiring the kernel to be positive definite, then it will be consistent).

If a kernel $K(x, y)$ in a locally compact space X is continuous for $x \neq y$, then (8) will imply that it is consistent on any compact subspace of X (*K-consistency*).

Assume that the space X is normal and each of its open set is an F_σ (such spaces are called perfectly normal). Then all the results of Ch. II, § 2 can be obtained for any consistent kernel.

Suppose that X is a locally compact abelian group. Assume that the kernel $K(x,y)$ is invariant,

$$K(x, y) = k(x - y),$$

symmetric and positive. Then if $K(x, y)$ is positive definite, it is K-consistent.

Further, if $h(x) \geq 0$ is lower semicontinuous, then the kernel

$$K(x, y) = k(x - y), \quad k = h * \check{h},$$

will be consistent and the cone $\mathscr{E}_K^+$ will be complete.

This theorem easily implies that M. Riesz kernels are perfect and the same for *Aronszajn-Smith kernels* [1], which are defined in R^p by the function

$$k_\alpha(x) = (2\pi)^{-p/2}\, 2^{1-\alpha/2}\, \Gamma^{-1}(\alpha/2)\, \frac{K_{(p-\alpha)/2}(|x|)}{|x|^{(p-\alpha)/2}}$$

where $0 < \alpha < \infty$, and $K_\lambda(x)$ denotes the modified Bessel function of the third kind (i.e. $K_\lambda(x) = \frac{\pi i}{2} e^{\pi i \lambda/2} H_\lambda^{(1)}(ix)$, cf. Courant-Hilbert [1]).

9. We shall assume that X is a locally compact abelian group and we will take the coarse point of view. Then for kernels we can take not only point functions, but also, for instance, a measure $\varkappa$; defining the potential of the measure $\varkappa$ by the formula

$$U_\varkappa^\mu = \varkappa * \mu.$$

J. Deny, in the papers [6, 7, 11] studies the classes of kernels $\varkappa$, for which the balayage principle holds in the form (13)[11] or (14).

Since (14) → (13) on any perfectly normal group we mean (14) speaking about the balayage principle.

The usefulness of introducing kernels $\varkappa$ which are measures is apparent from the important fact that the set of all kernels for which balayage is possible, is vaguely closed.

This question is connected with the construction of the inverse operator of (5) and with the corresponding generalization of the notion of superharmonicity.

Assume that there is a fundamental system $\mathfrak{B}$ of compact neighborhoods of zero, where to each $V \in \mathfrak{B}$ there corresponds a measure σ_V such that

(i) $\varkappa * \sigma_V \leq \varkappa$ and $\varkappa * \sigma_V \not\equiv \varkappa$,

(ii) on CV $\varkappa = \varkappa * \sigma_V$,

(iii) as $n \to \infty$, $\varkappa * (\sigma_V)^n \dashrightarrow 0$.[12]

In this case the kernel $\varkappa$ will be called *associated*, or, briefly, an *a-kernel.*

It can be shown that for a given fundamental system $\mathfrak{B}$ the system of measures $\{\sigma_V\}$ if it exists at all, is uniquely determined (to within a positive factor).

The system of measures $\{\sigma_V\}$ gives rise to the notion of harmonicity and superharmonicity. A measure U for which $U * \sigma \leq U$ (respectively, $= U$) for any $\sigma \in \{\sigma_V\}$ will be called *superharmonic* (respectively, *harmonic*).

The kernel $\varkappa$ itself and any measure potential $U_\varkappa^\mu$ will be superharmonic.

These definitions are related, essentially, not to the system $\{\sigma_V\}$ but to the kernel $\varkappa$. In other words, if $\{\sigma_{V'}\}$ is any other system of measures,

---

[11] Of course, in the sense of the coarse theory, i.e. without mention of an exceptional set of null capacity.

[12] σ^n denotes the n-fold convolution $\sigma * \sigma * \cdots * \sigma$; $\sigma^0 = \varepsilon$.

satisfying conditions (i), (ii), and (iii), then the classes of superharmonic (harmonic) measures determined by the systems $\{\sigma_V\}$ and $\{\sigma_{V'}\}$ coincide.

An important type of a-kernel are the so-called *elementary kernels* (e-kernels) defined by a series of the form

$$\varkappa = \frac{1}{a} \sum_{n=0}^{\infty} \sigma^n ,$$

where $a > 0$, and the measure σ is such that the above series vaguely converges. In this case, the system $\{\sigma_V\}$ consists of the single measure σ, which can be verified directly. If $\sigma(1) < 1$, then the kernel $\varkappa$ satisfies condition (A) of § 1. Under this assumption the theory is substantially simpler, since one can immediately write down the inverse operator L:

$$L = a(\varepsilon - \sigma) .$$

The superharmonicity of the measure U can be written in the form $L * U \geq 0$. It is not difficult to show that a superharmonic function which is majorized by a potential of measure, is itself the potential of the measure $\mu = L * U$ (cf. Corollary 1 to Theorem 1.23). This implies that

$$\inf (U_\varkappa^{\mu_1}, U_\varkappa^{\mu_2}) = U_\varkappa^{\lambda}; \quad \lambda \geq 0$$

(cf. Theorem 1.31).

In [7] J. Deny shows that the assumption $\sigma(1) < 1$ is superfluous. One of his basic results concludes that every a-kernel is the vague limit of a sequence (generalized if necessary) of e-kernels.

Further, it turns out that the balayage principle (14) holds for any e-kernel (under the condition $\sigma(1) \leq 1$ this is proved in [6] and in the general case in [7]). This implies that the balayage principle also holds for any a-kernel. It is not known whether there exist kernels, not belonging to the vague closure of the class of e-measures, for which (14) holds.

In any case, if (14) holds for $\varkappa$, and $\varkappa$ has a bounded density outside any neighborhood of zero and satisfies condition (A), then $\varkappa$ belongs to the vague closure of e-kernels with $\sigma(1) < 1$.

The condition $\sigma(1) \leq 1$ (for any $\sigma \in \{\sigma_V\}$) is necessary and sufficient in order that the corresponding a-kernel satisfy the equilibrium principle for open sets, i.e. that for any open set ω with compact closure there exists a measure γ, $S(\gamma) \subset \overline{\omega}$ for which

$$\varkappa * \gamma = 1 \quad \text{in} \quad \omega, \quad \varkappa * \gamma \leq 1 \text{ everywhere.}$$

Here are some interesting examples of e-kernels and their vague limits. Let

$$\sigma = \mathfrak{F}[te^{-\lambda|x|^\alpha}], \ \lambda > 0$$

If $0 < \alpha \leq 2$, $t > 0$, then σ is a measure and

$$\sigma(1) = te^{-\lambda}$$

We shall assume that $t < 1$; then $\sigma(1) < 1$.

Suppose $\varkappa$ is the e-kernel constructed from this measure σ and the constant $a > 0$. Noting that

$$\mathfrak{F}[\varkappa] = a^{-1}(1 - te^{-\lambda|x|^\alpha})^{-1}$$

we let a tend to ∞, t to 1, λ to 0 in such a way that $a(1-t) \to A \geq 0$, $a\lambda \to B \geq 0$ (where A and B are not simultaneously equal to zero). Then

$$\mathfrak{F}[\varkappa] \to \frac{1}{A + B|x|^\alpha} \text{ in } S^*,$$

and hence

$$\varkappa \dot{\to} \mathfrak{F}^{-1}\left[\frac{1}{A + B\,|x|^\alpha}\right].$$

Thus,

$$\mathfrak{F}^{-1}\left[\frac{1}{A + B|x|^\alpha}\right]$$

is the vague limit of e-kernels, and consequently, is an a-kernel.

For $A = 0$ this construction yields M. Riesz kernels, and for $\alpha = 2$, $A > 0$, it gives the fundamental solution of the Helmholtz equation $\Delta v = 4\pi^2 \frac{A}{B} v$, satisfying the radiation condition, i.e. a kernel proportional to

$$|x|^{-(p-2)/2} K_{(p-2)/2}\left(2\pi\sqrt{\frac{A}{B}}\,|x|\right)$$

$\left(\text{for } p = 3 \text{ we obtain the kernel } \frac{1}{|x|}\, e^{-2\pi\sqrt{\frac{A}{B}}\,|x|}\right)$.

Analogously, starting with the measure

$$\sigma = \mathfrak{F}\left[t \exp\left(-\int_0^2 |x|^\alpha d\mu(\alpha)\right)\right],$$

where $t > 0$ and μ is an arbitrary measure on $[0,2]$ one can show that

$$\varkappa = \mathfrak{F}\left[\left(\int_0^2 |x|^\alpha d\mu(\alpha)\right)^{-1}\right]$$

is an a-kernel[13]. Furthermore J. Deny [11] (cf. also the note by G. Choquet and J. Deny [4]) establishes a close connection between a-kernels and so called Hunt kernels.

---

[13] One needs only that $\left(\int_0^2 |x|^\alpha\, d\mu(\alpha)\right)^{-1}$ is a locally integrable function (for $p \geq 3$ this always holds).

By a *Hunt kernel* we mean a measure $\varkappa$ which admits a representation

$$\varkappa = \int_0^\infty \alpha_t dt,$$

where the measures α_t form a *weakly continuous semigroup*, i.e.

(i) $\alpha_t * \alpha_s = \alpha_{t+s}$ $(s \geq 0, t \geq 0)$,

(ii) $\alpha_0 = \varepsilon$,

(iii) $\alpha_t \dot{\rightarrow} \alpha_s$ as $t \rightarrow s$.

The Deny result says that *every Hunt kernel is an a-kernel and, conversely, every a-kernel is a Hunt kernel.*

The proof of the first part of the assertion is based on the following. For any positive q, the kernel $\varkappa + \frac{1}{q}\varepsilon$ will be elementary. This follows from the formula

$$\varkappa + \frac{1}{q}\varepsilon = \frac{1}{q}\sum_{n=0}^{\infty}(qR_q)^n,$$

where

$$R_q = \int_0^\infty e^{-qt}\alpha_t dt$$

is the resolvent of the kernel $\varkappa$.

Hence for $\varkappa + \frac{1}{q}\varepsilon$ and therefore also for $\varkappa$ the principle of balayage (14) holds. Moreover, and this is highly non-trivial, a Hunt kernel satisfies the principle of balayage for *any* open set. After this, the associated family of measures $\{\sigma_V\}$ is constructed by the balayage of ε onto the open set CV.

The second part of the theorem requires the construction of the semigroup α_t for a given a-kernel. This is not difficult to do if the kernel is elementary:

$$\varkappa = \frac{1}{a}\sum_{n=0}^{\infty}\sigma^n.$$

If we put

$$\alpha_t = e^{-at}\sum_{n=0}^{\infty}\frac{a^n t^n \sigma^n}{n!} = e^{-at}e^{at\sigma},$$

then formally,

$$\varkappa = \int_0^\infty \alpha_t dt,$$

and this formal expression can be justified.

In the case of an arbitrary a-kernel, for every neighborhood $V \in \mathfrak{V}$, one constructs the e-kernel

$$\varkappa_V = \frac{1}{a_V} \sum_{n=0}^{\infty} (\sigma_V)^n,$$

where $a_V > 0$ is choosen so that the measure

$$\tau_V = a_V(\varepsilon - \sigma_V) * \varkappa$$

is normalized by the condition $\tau_V(1) = 1$. By property (iii) in the definition of $\{\sigma_V\}$,

$$\varkappa = \tau_V * \varkappa_V,$$

and by (ii) $\tau_V \overset{\mathfrak{V}}{\to} \varepsilon$, Hence, representing the elementary kernel $\varkappa_V$ in the form

$$\varkappa_V = \int_0^{\infty} \alpha_t^{(V)} dt,$$

one needs to pass to the limit over $\mathfrak{V}$. To prove that this is possible is rather complicated; we shall not do this here (cf. J. Deny [11]).

10. In classical potential theory the kernels $k_2(x - y)$ and $G(x, y)$ are, respectively, the fundamental solution and the Green function for the Laplace operator; indubitably, it would be interesting to extend this to the case of more general differential operators of elliptic type.

Consider in R^p ($p \geq 3$) any bounded region G.

Suppose given a *differential operator*

$$\mathfrak{A}u \equiv \sum_{i,k=1}^{p} a_{ik}(x) \frac{\partial^2 u}{\partial x_i \partial x_k} + \sum_{i=1}^{p} b_i(x) \frac{\partial u}{\partial x_i} + c(x)\, u,$$

where the coefficients $a_{ik}(x) = a_{ki}(x)$, $b_i(x)$, $c(x)$ are functions in G satisfying the regularity conditions which we specify below.

We shall assume that it is *elliptic* in G; without loss of generality, we assume that the form $\sum_{i,k} a_{ik}(x)\, \xi_i \xi_k$ is positively defined for all $x \in G$.

We introduce the *characteristic function*

$$H(x, y) = \frac{1}{\sqrt{A(y)}} \left[\sum_{i,k} A_{ik}(y)\, (x_i - y_i)\, (x_k - y_k)\right]^{\frac{2-p}{p}}$$

where $|A_{ik}||$ is the matrix inverse to $||a_{ik}||$ and A its determinant.

The function $F(x, y)$ will be called the *fundamental solution* of the equation $\mathfrak{A}u = 0$, if for $x \in G \setminus \{y\}$

$$\mathfrak{A}^{(x)} F(x, y) = 0,$$

and as $x \to y$, $F(x, y)$ has the same singularity as $H(x, y)$. This means that there is a $\lambda > 0$ such that as $x \to y$

$$F - H = O(|x - y|^{\lambda+2-p});$$

$$\frac{\partial}{\partial x_i}(F - H) = O(|x - y|^{\lambda+1-p});$$

$$\frac{\partial^2}{\partial x_i \partial x_k}(F - H) = O(|x - y|^{\lambda-p}).$$

Finally, the *Green function* $G(x; y)$ wll be the fundamental solution which also satisfies the boundary condition

$$G(x, y)|_{x \in \partial G} = 0.$$

If one assumes sufficiently smooth coefficients ($a_{ik} \in C^{(4)}$, $b_i \in C^{(3)}$, $c \in C^{(2)}$) and a sufficiently smooth ∂G, then one can assert the existence of a fundamental solution in G (cf. Yu. I. Lyubich [1]).

The conditions on the coefficients can be substantially weakened (it is sufficient to assume they satisfy the Hölder condition in $\overline{G}$), if on assumes that the Dirichlet problem

$$\mathfrak{A}u = 0, \quad u|_{\partial G} = \varphi,$$

has a unique solution (cf. C. Miranda [1]).

In this connection, let us point out two cases when this uniqueness holds:

(i) $c(x) \leq 0$ in G,

(ii) mes G is sufficiently small.

In case (i) the uniqueness theorem is implied by the following *Hopf principle* (cf. C. Miranda [1]).

If the coefficients of $\mathfrak{A}$ are continuous in $\overline{G}$ and $c(x) \leq 0$, then no solution of the equation $\mathfrak{A}u = 0$ distinct from a constant, can have within G a negative minimum or a positive maximum.

The following conditions are sufficient for the existence of a Green function in the region G:

(i) $\dfrac{\partial a_{ik}}{\partial x_j}$, b_i, c satisfy the Hölder condition in $\overline{G}$,

(ii) the surface ∂G satisfies the Lyapunov conditions,

(iii) the uniqueness theorem holds for the region G.

If in place of (iii) one imposes the stronger condition: $c(x) \leq 0$ in G then from the Hopf principle it easily follows that $G(x, y)$ is non-negative in G. Moreover, it will even be positive in G (cf. Brelot [32], p. 64).'

To this point we have assumed that G is a bounded region. If G coincides with the whole space R^p, then we restrict $G(x, y)$ to mean the fundamental solution of the equation $\mathfrak{A}u = 0$ which tends to zero at infinity. If such a $G(x, y)$ exists and $c(x) \leq 0$ everywhere in R^p, then $G(x, y) > 0$.

Sufficient conditions for the existence of such a "principal" fundamental solution are:

(i) the coefficients of $\mathfrak{A}$ are bounded in R^p and satisfy the Hölder condition (for a_{ik} uniformly in R^p),

(ii) $A(x) > \alpha > 0$,

(iii) $c(x) \leq 0$, and $c(x) < \beta < 0$ outside a certain compact set.

Under these conditions, for a "principal" fundamental solution we even have

$$G(x, y) = O(e^{-a|x-y|}), \quad \frac{\partial G}{\partial x_j} = O(e^{-a|x-y|}).$$

Note, finally, that the nature of the singularity $H(x, y)$ is such that, for any compact subset $K \subset G$ there exist two constants $m_1(K) > 0$, $m_2(K) > 0$ such that for $x, y \in K$

$$\frac{m_1}{|x-y|^{p-2}} \leq G(x, y) \leq \frac{m_2}{|x-y|^{p-2}} \tag{6.2.1}$$

11. Let us turn to potential theory with kernel $G(x, y)$. For any measure μ with compact support in G, the potential

$$U_G^\mu(x) = \int_{S(\mu)} G(x, y)\, d\mu(y)$$

is a non-negative function satisfying the equation

$$\mathfrak{A} U_G^\mu = 0$$

outside $S(\mu)$.

It is natural to call $U_G^\mu(x)$ an $\mathfrak{A}$-harmonic function outside $S(\mu)$.

Furthermore, we will call a function $f(x)$, $x \in G$, *$\mathfrak{A}$-superharmonic*, if

(a) $-\infty < f(x) \leq +\infty$,

(b) $f(x)$ is lower semicontinuous,

(c) for any region $\omega \subset \overline{\omega} \subset G$ and function $h(x)$ $\mathfrak{A}$-harmonic in $\overline{\omega}$, the inequality

$$f(x) \geq h(x), \quad x \in \partial\omega$$

implies that this inequality holds in ω. It is not difficult to see that the potential $U_G^\mu(x)$ of a measure with compact support is an $\mathfrak{A}$-superharmonic function.

Also, if $f(x) \in C^{(2)}(G)$, then $f(x)$ is $\mathfrak{A}$-superharmonic if, and only if, $\mathfrak{A}f \leq 0$. The potentials of measures are characterized in the class of

$\mathfrak{A}$-superharmonic functions by the property, that for them, zero is *the best* $\mathfrak{A}$-harmonic minorant (cf. Theorem 1.24′).

Introducing G-capacity using the kernel $G(x, y)$ (for example, after (Vallée-Poussin, cf. Ch. II, No. 6), one can see that by (6.2.1) G-capacity and Newtonian capacity have the same order on any compact set $K \subset G$. In particular, the class of sets with Newtonian capacity zero coincides with the class of sets of G-capacity zero.

This result coincides, in essence, with the theorem that the class of $\mathfrak{A}$-polar sets is identical to the class of Δ-polar sets (cf. R.-M. Hervé [2], Theorem 36.1).

Without special assumptions on the coefficient $c(x)$, one can still establish that the generalized maximum principle (9) holds on any compact set $K \subset G$: there exists a constant λ such that for any measure μ, $S(\mu) \subset K$, the inequality

$$U_G^\mu(x) \leq M \quad \text{on} \quad S(\mu)$$

implies that

$$U_G^\mu(x) = \lambda M \quad \text{on} \quad K.$$

This is sufficient for the regularity of the kernel $G(x, y)$ (cf. R.-M. Hervé [2]).

If we suppose that $c(x) \leq 0$, then, repeating exactly the proofs of Theorems 1.8, 1.9, and 1.10, we will be able to obtain the maximum principle (8) for the kernel $G(x, y)$.

In conclusion we note the following. We have assumed up to now that the boundary ∂G is sufficiently smooth. This assures the existence of the Green function and the solution of the Dirichlet problem in the classical sense. If the smoothness condition is omitted, then one can introduce the notion of a *generalized* solution of the Dirichlet problem and a generalized Green function, which will satisfy prescribed boundary conditions at all *regular* points of G. A basic fact is that *the class of regular points of the Dirichlet problem posed for the operator* $\mathfrak{A}$, *coincides with the class of regular points for the operator* Δ (cf. G. Tautz [1] and O. A. Olejnik [1]).

However, there is another way to construct the kernel $G(x, y)$, which is used by R.-M. Hervé [2]; it illustrates the axiomatic approach to the theory of superharmonic functions and potential theory.

After introducing the notion of $\mathfrak{A}$-harmonicity and $\mathfrak{A}$-superharmonicity, one can introduce the notion of potential, without having introduced a kernel. Namely, we will call an $\mathfrak{A}$-potential any $\mathfrak{A}$-superharmonic function for which zero is the best $\mathfrak{A}$-harmonic minorant. Then, the support of an $\mathfrak{A}$-superharmonic function (in particular, an $\mathfrak{A}$-potential) will be defined as the complement (in G) of the largest open set in which the given function is $\mathfrak{A}$-harmonic.

R.-M. Hervé [2] has proved that for any point $y \in G$ $\mathfrak{A}$-potentials with the point support $\{y\}$ are proportional. Normalizing this potential at the point y as for a Green function, we obtain a kernel analogous to $G(x, y)$.

It is characteristic of this approach that the kernel appears not at the beginning of the theory, but much later.

§ 3. Dirichlet Spaces

In Ch. I, § 4 for the classical case $\alpha = 2$ a connection was established between potentials with finite energy and the Hilbert space BL_0.

At that time the quadratic form

$$||u||^2 = \int_{R^p} |\operatorname{grad} u|^2 \, dx$$

which defines the Hilbert structure in BL_0 was related by the classical Green formulas with the Laplace operator, and hence, with the Newtonian kernel $|x|^{2-p}$.

For the more general case $0 < \alpha < 2$ we did not pursue the analogous connections.

Now we shall pose the question in an even more general form.

Is it possible to distinguish the class of Hilbert spaces of functions in R^p (more precisely, equivalence classes of functions), which have a certain "potential theory"?

The following exposition is based on the lectures of J. Deny [8, 9, 10]. The basic ideas, as J. Deny points out, are due to A. Beurling. Cf. also A. Beurling and J. Deny [1].

12. By a *contraction* we mean a mapping T of the complex plane z into itself which satisfies the following conditions

$$T(0) = 0,$$

$$|T(z_1) - T(z_2)| \leq |z_1 - z_2|.$$

As an example of a contraction take projection on any convex, closed set[14] containing the point 0.

Thus, the mappings $\operatorname{Re} z, (\operatorname{Re} z)^+$ will be contractions. Here are some other examples of contractions: $|z|$, $(\operatorname{Re} z - a)^+$ for $a > 0$.

Let H denote a Hilbert space whose elements are complex locally integrable functions in R^p (more precisely, classes of equivalent func-

---

[14] Recall that the projection of the point z on a closed convex set F is the point $Pz \in F$ for which $|Pz - z| \leq |\zeta - z|$, $\zeta \in F$.

tions). We will call H a *Dirichlet space*[15] if the following axioms are fulfilled:

(a) $||u_n|| \to 0$ implies that

$$\int_K |u_n(x)|\, dx \to 0$$

for any compact subset $K \subset R^p$,

(b) $\Phi \cap H$ is dense in H and in Φ,

(c) if $u \in H$, then for any contraction T we have

$$Tu \in H \quad \text{and} \quad ||Tu|| \leq ||u||$$

(d) the shift operators U_x defined by the equations

$$U_x u(y) = u(y - x)$$

are unitary operators in H, and as $x \to 0$

$$||U_x u - u|| \to 0,$$

i.e. U_x strongly converges to the unit operator I.

From (d) it follows that the vector function $U_x u$ of the argument x is continuous and bounded. This implies that for any finite measure

$$\int (U_x u)\, d\mu(x) = \int u(y - x)\, d\mu(x) \in H.$$

In particular, if $u \in H$, $f \in \Phi$, then $u * f \in H$.

The convolution of any function $u \in \Phi \cap H$ with any function of D yields a function belonging to $D \cap H$. Consequently, the set of functions $D \cap H$, in any case, is not empty.

In the classical case, the space H is obtained by completion of the space D of all infinitely differentiable functions with compact support in the norm

$$||u||^2 = \int_{R^p} |\operatorname{grad} u|^2\, dx.$$

In this case the potential $U_2^\nu(x)$ of a signed measurs ν with finite energy belongs to H and has the property that

$$(U_2^\nu, U_2^{\nu_1}) = \int \overline{U_2^{\nu_1}(x)}\, d\nu(x)$$

for any other potential $U_2^{\nu_1}$. Therefore, by Lemma 1.1 for any sufficiently smooth function φ with compact support

$$(U_2^\nu, \varphi) = \int \overline{\varphi(x)}\, d\nu(x)$$

and by continuity, this is preserved for any function $\varphi \in \Phi \cap H$.

---

[15] We differ slightly from Deny's terminology which in the presence of (d) calls H a special Dirichlet space, reserving the term "Dirichlet space" for a Hilbert space H which satisfies (a), (b), (c).

This gives rise in the general case to the following definition.

The element $u^\nu \in H$ will be called the *potential of the signed measure* ν, if for any function $\varphi \in \Phi \cap H$

$$(u^\nu, \varphi) = \int \overline{\varphi(x)}\, d\nu(x). \tag{6.3.1}$$

Note that these relations need only be verified for non-negative functions of $\Phi \cap H$. In fact, if $\varphi \in \Phi \cap H$, then $(\operatorname{Re}\varphi)^+$, $(\operatorname{Re}\varphi)^-$, $(\operatorname{Im}\varphi)^+$, $(\operatorname{Im}\varphi)^- \in \Phi \cap H$; this follows, since the operations $T_1\varphi = \operatorname{Re}\varphi$ and $T_2\varphi = (\operatorname{Re}\varphi)^+$ are contractions.

From (b) it follows that a signed measure ν is uniquely determined by its potential.

The set of all potentials is dense in H. In fact, if f is any bounded, measurable function, with compact support, then the expression

$$\int u(x)\, \overline{f(x)}\, dx$$

will, by axiom (a), be a continuous linear functional in H. Hence, there exists an element $u^f \in H$ such that

$$(u, u^f) = \int u(x)\, \overline{f(x)}\, dx, \quad u \in H.$$

This element, by (6.3.1) is the potential of an absolutely continuous signed measure with density $f(x)$. But the set of such elements is dense in H, since otherwise, there would exist an element $u \in H$, distinct from the null element, such that

$$(u, u^f) = \int u(x)\, \overline{f(x)}\, dx = 0$$

for any function $f(x)$ of the type described above. But this is impossible.

The following lemma *characterizes* potentials of measures.

Lemma 6.6. *Potentials of measures are positive functions. An element $u \in H$ is a potential of measure if, and only if, for any $v \in H$,* $\operatorname{Re} v \geq 0$,

$$\|u + v\| \geq \|u\|, \tag{6.3.2}$$

or, what is the same thing,

$$\operatorname{Re}(u, v) \geq 0. \tag{6.3.2'}$$

Proof. Suppose that u is a potential of measure and $v \in \Phi \cap H$, with $\operatorname{Re} v \geq 0$. By (6.3.1)

$$\operatorname{Re}(u, v) = \operatorname{Re} \int \overline{v(x)} d\mu(x) = \int \operatorname{Re} v(x)\, d\mu(x) \geq 0.$$

Since $\Phi \cap H$ is dense in H, this inequality holds for any $v \in H$, $\operatorname{Re} v \geq 0$. This easily implies (6.3.2).

We will show that (6.3.2) implies that $u(x) = |u(x)| \geq 0$. Consider the set

$$U = \{u + v : v \in H, \operatorname{Re} v \geq 0\} = \{w : w \in H, \operatorname{Re}(w - u) \geq 0\}.$$

This set is convex and closed, and from (6.3.2) it follows that the element $u \in U$ has least norm. It is known that there is only one element in U with least norm. On the other hand, $|u| \in U$, and from (c) it follows that $\||u|\| \leq \|u\|$. Hence, $u = |u|$.

It remains to be shown that (6.3.2) or (6.3.2′) imply that u is a potential of measure. Condition (6.3.2′) shows that the expression (v, u) with variable $v \in \Phi \cap H$ is a positive linear functional.[16] By (b) this functional extends by continuity to the whole of Φ and positiveness is preserved. But then (cf. Theorem 0.2′) there is a measure μ such that

$$(v, u) = \int v(x)\, d\mu(x), \quad v \in \Phi \cap H,$$

and this is equivalent to (6.3.1). ▯

We now prove a theorem which in a certain sense generalizes the equilibrium principle (11) of § 2.

Theorem 6.5. (*Condenser theorem*). *Suppose ω_0, ω_1 are open sets where ω_1 is relatively compact and $\bar{\omega}_0 \cap \bar{\omega}_1 = \emptyset$. Then there exists a potential $u^\nu(x)$ of a signed measure ν which satisfies the properties:*

(*i*) $0 \leq u^\nu(x) \leq 1$ *almost everywhere in* R^p;

(*ii*) $u^\nu(x) = 0$ *almost everywhere in* ω_0, $u^\nu(x) = 1$ *almost everywhere in* ω_1;

(*iii*) $S(\nu^+) \subset \bar{\omega}_1$, $S(\nu^-) \subset \bar{\omega}_0$.

Proof. Consider the convex set $A = \{f(x) \colon f \in H; \operatorname{Re} f \geq 1$ almost everywhere on ω_1, $\operatorname{Re} f \leq 0$ almost everywhere on $\omega_0\}$. It is not empty because of (b), and (a) easily implies that A is closed. Suppose that $u \in A$ is an element with least norm; as is known, this element is unique. Denote by T the operator in the complex plane which projects on the segment $0 \leq \operatorname{Re} z \leq 1$. As indicated at the beginning of this section, T is a contraction, and hence (cf. (c)) $\|Tu\| \leq \|u\|$. But $Tu \in A$ and by uniqueness $Tu = u$. This shows that almost everywhere $0 \leq u \leq 1$, and since $u \in A$, we have $u = 1$ almost everywhere on ω_1 and $u = 0$ almost everywhere on ω_0.

We will show that u is a potential and that (iii) holds.

Put

$$V = \{v : \operatorname{Im} v = 0, v \in \Phi \cap H, S(v) \subset C\bar{\omega}_0, v \geq 0 \text{ on } \omega_1\}$$

$$W = \{w : \operatorname{Im} w = 0, w \in \Phi \cap H, S(w) \subset C\bar{\omega}_1, w \geq \text{ on } \omega_0\}.$$

[16] The real valuedness of (u, v) for real u and v wil be proved below; cf. p. 381.

We show that these classes are nontrivial, i.e. contain functions distinct from zero. Consider, for example, V. Let ω_2 be a neighborhood of $\bar{\omega}_1$ not intersecting $\bar{\omega}_0$ and $v_0(x) \in \Phi$ such that $v_0(x) = 1$ on ω_1 and $S(v_0) \subset \omega_2$.

By (b) one can find a function $v_1(x) \in \Phi \cap H$, satisfying the condition $|v_0(x) - v_1(x)| < \varepsilon < 1/2$ everywhere. In particular, for $x \in \bar{\omega}_0$ $|v_1(x)| < \varepsilon$. Hence, the function $v(x) = (\text{Re}\, v_1 - \varepsilon)^+$, which belongs to $\Phi \cap H$, with $v(x) = 0$ for $x \in \bar{\omega}_0$, and $v(x) > 1 - 2\varepsilon > 0$ for $x \in \omega_1$, will belong to the class V. An analogous argument can be carried out for W.

For any constant $h \geq 0$ and any $v \in V$ $u + hv \in A$; therefore $||u + hv|| \geq ||u||$ and hence, Re $(u, v) \geq 0$. But since u, v are real, it follows that (u, v) is also real [17] so that $(u, v) \geq 0$. This shows that the functional $f(v) = (u, v)$ is determined only by the values of $v(x)$ on ω_1. In fact, if $v \in V$ and $v(x) = 0$ on ω_1, then $-v \in V$ and therefore $f(v) = 0$.

By (b) it can be extended by continuity to the space $C_{\bar{\omega}_1}$. Consequently, there is a measure ν^+, $S(\nu^+) \subset \bar{\omega}_1$ such that

$$(u, v) = \int v(x)\, d\nu^+(x) \quad v \in V.$$

Analogously,[18] one proves the existence of a measure ν^- such that

$$(u, w) = -\int w(x)\, d\nu^-(x), \quad w \in W.$$

Now suppose that $0 \leq \varphi(x) \in \Phi \cap H$. For any $\varepsilon > 0$ one can find functions v', v'' such that on ω_1

$$\varphi - \varepsilon \leq v' \leq \varphi \leq v'' \leq \varphi + \varepsilon.$$

We construct v'' for example. Suppose $\psi(x) \in \Phi$ satisfies the conditions $\psi(x) = \varphi(x) + \varepsilon$ on ω_1 and $S(\psi) \subset \omega_2$, where ω_2, as above, is a neighborhood of $\bar{\omega}_1$ not intersecting $\bar{\omega}_0$.

According to (b) there exists a function $v_0(x) \in \Phi \cap H$ which satisfies the inequality $|\psi(x) - v_0(x)| < \varepsilon/2$ everywhere. Then, it is not difficult to see that $v'' = (\text{Re}\, v_0 - \varepsilon/2)^+ \in V$ will satisfy the conditions imposed. Repeating this construction with $\psi(x) = \varphi(x)$ on ω_1, we obtain v'.

Analogously one can find functions $w', w'' \in W$ such that on ω_0 we have the inequalities

$$\varphi - \varepsilon \leq w' \leq \varphi \leq w'' \leq \varphi + \varepsilon.$$

---

[17] Indeed, since the operation Re is a contraction, by (c) we have $||u|| \leq ||u + iv||$. i.e.

$$0 \leq 2\, \text{Im}\, (u, v) + ||v||^2.$$

Since v can be replaced by λv, where λ is real, Im $(u, v) = 0$.

[18] One need only replace $C_{\bar{\omega}_1}$ by the space of functions with compact support which belong to $C_{\bar{\omega}_0}$.

Now note that for any constant $h \geq 0$

$$u + h(v'' + w' - \varphi) \in A, \qquad u + h(\varphi - v' - w'') \in A.$$

From this, as above, we conclude that

$$(u, v'' + w' - \varphi) \geq 0 \quad (u, \varphi - v' - w'') \geq 0,$$

i.e.

$$(u, v' + w'') \leq (u, \varphi) \leq (u, v'' + w')$$

or

$$\int v'(x)\, d\nu^+ - \int w''(x)\, d\nu^- \leq (u, \varphi) \leq \int v''(x)\, d\nu^+ - \int w'(x)\, d\nu^-.$$

If now we pass to the limit $\varepsilon \to 0$ then, in view of the fact that the supports of v', v'', w', w'' are all contained in a certain compact set, we obtain

$$(u, \varphi) = \int \varphi(x)\, d\nu^+ - \int \varphi(x)\, d\nu^-.$$

This proves the theorem. ▯

Using methods based on similar ideas, one can prove the following two theorems.

Theorem 6.6. *If $u_1, u_2, \ldots, u_n$ are potentials of measures, then*

$$u = \min(u_1, u_2, \ldots, u_n)$$

will also be a potential of measure.

Theorem 6.7. *If u^μ is the potential of a measure μ and ω is an arbitrary open set, then there exists a potential $u^{\mu'}$ of a measure μ' which satisfies the following conditions:*

(i) $u^{\mu'}(x) \leq u^\mu(x)$ almost everywhere,

(ii) $u^{\mu'}(x) = u^\mu(x)$ almost everywhere in ω,

(iii) $S(\mu') \subset \overline{\omega}$.

13. For the integral representation of norms in Dirichlet spaces one must introduce a class of functions which is closely related to the positive definite functions in R^p.

Recall, that a continuous function $s(x)$, $x \in R^p$, is called *positive definite*, if for any natural n, any $x_1, x_2, \ldots, x_n \in R^p$ and any complex numbers $\xi_1, \xi_2, \ldots, \xi_n$

$$\sum_{i,j=1}^{n} s(x_i - x_j)\, \xi_i \bar{\xi}_j \geq 0. \tag{6.3.3}$$

According to the well-known Bochner theorem (cf. S. Bochner [1]) $s(x)$ is positive definite if, and only if, it is representable in the form

$$s(x) = \int_{R^p} e^{2\pi i x y} d\mu(y),$$

where μ is a finite measure in R^p.

Condition (6.3.3) can also be written in the equivalent integral for m

$$\iint s(x_1 - x_2)\, \varphi(x_1)\, \overline{\varphi(x_2)}\, dx_1\, dx_2 \geq 0, \qquad (6.3.3')$$

where $\varphi(x)$ is an arbitrary function in Φ. This situation permits a natural extension of the notion of positive definite function. Specifically, a function $s(x)$ will be called *positive definite in the generalized sense,* if it is locally integrable and (6.3.3′) holds. Such functions are also (generalized) Fourier transforms of measures, not necessarily of finite measures (cf. L. Schwartz [1], vol. II, p. 132).

Now denote by (N) the class of functions $\lambda(x)$ continuous in R^p such that for any natural n, any $x_1, x_2, \ldots, x_n \in R^p$ and complex $\xi_1, \xi_2, \ldots, \xi_n$

$$\sum_{i,j=1}^{n} [\lambda(x_i) + \overline{\lambda(x_j)} - \lambda(x_i - x_j)]\, \xi_i \bar{\xi}_j \geq 0. \qquad (6.3.4)$$

For $p = 1$ the class (N) was examined by M. G. Krejn [1] who also obtained an integral representation for functions in this class (cf. below Lemma 6.9). For $p > 1$ J. Schoenberg [1] pointed out important special types of functions of the class (N).

(6.3.4) implies the following properties of functions $\lambda(x) \in (N)$:

(α) $\lambda(0) \geq 0$,

(β) $\operatorname{Re} \lambda(x) \geq \lambda(0)$,

(γ) $\lambda(-x) = \overline{\lambda(x)}$,

(δ) $\lambda(x) + C \in (N)$ for $C \geq -\lambda(0)$.

Indeed, (γ) follows since the quadratic form in (6.3.4) is real, (α) follows from (6.3.4) if we take $x_i = 0$ $(i = 1, 2, \ldots, n)$, and to obtain (β) set $n = 2$, $x_1 = x$, $x_2 = 0$. To prove (δ) it is sufficient to check that $\lambda(x) - \lambda(0) \in (N)$. This is so if in (6.3.4) one replaces n by $2n$, and then sets $x_{n+i} = 0$, $\xi_{n+i} = -\xi_i$ $(i = 1, 2, \ldots, n)$.

We also point out the assertion

(ε) if $C \geq 0$, and $s(x)$ is positive definite, then[19]

$$\lambda(x) = C + s(0) - s(x) \in (N)$$

---

[19] To prove (ε), write (6.3.3) for $n = 2m$, then put $x_{2k} = 0$ and $\xi_{2k-1} = -\xi_{2k}$.

and the inequality $|\lambda(2x)| \leq 4\,|\lambda(x)|$,[20] which implies that $\lambda(x) = O(|x|^2)$ as $x \to \infty$.

Furthermore, the class (N) is the convex cone in the space of all functions continuous in R^p, closed with respect to compact convergence.

Lemma 6.7. *$\lambda(x)$ belongs to the class (N) if, and only if, $\lambda(0) \geq 0$ and $e^{-t\lambda(x)}$ is positive definite for any $t \geq 0$.*

Proof. Necessity follows from the following algebraic theorem (for the proof, cf. G. Polya and G. Szegö, Problems and Theorems in Analysis, Part 2, sec. VII, problem 36).

If the Hermitian form $\Sigma a_{ij}\,\xi_i\bar{\xi}_j$ is positive definite, then the form $\Sigma\, e^{ta_{ij}}\xi_i\bar{\xi}_j$ will also be positive definite, where $t \geq 0$.

Indeed, if $\lambda(x) \in (N)$, then $\lambda(0) \geq 0$, and setting $\xi_i = \eta_i\, e^{t\lambda(x_i)}$ in the form $\Sigma\, e^{-t\lambda(x_i - x_j)}\,\xi_i\bar{\xi}_j$ we obtain, using (γ), that

$$\sum e^{-t\lambda(x_i - x_j)}\xi_i\bar{\xi}_j = \sum e^{t[\lambda(x_i) + \overline{\lambda(x_j)} - \lambda(x_i - x_j)]}\eta_i\bar{\eta}_j \geq 0.$$

We now prove the sufficiency of the conditions of the lemma. From (ε) it follows that $e^{-t\lambda(0)} - e^{-t\lambda(x)} \in (N)$ and hence

$$\frac{1}{t}\,[1 - e^{-t(\lambda(x) - \lambda(0))}] \in (N).$$

Since this holds for any $t > 0$, then, passing to the limit $t \to +0$ since (N) is closed, we obtain $\lambda(x) - \lambda(0) \in (N)$. By the condition, $\lambda(0) \geq 0$ and therefore $\lambda(x) \in (N)$. ▯

Lemma 6.8. *$\lambda(x)$ belongs to (N) if, and only if, it has the representation*

$$\lambda(x) = C + \lim_{n\to\infty}\big(s_n(0) - s_n(x)\big), \tag{6.3.5}$$

where $C \geq 0$, $s_n(x)$ are positive definite functions and convergence is uniform on compact sets.

Proof. Sufficiency follows from (ε), and necessity from Lemma 6.7 if we put

$$C = \lambda(0),\ \ s_n(x) = n \exp \frac{1}{n}\,[\lambda(0) - \lambda(x)]. \qquad ▯$$

Corollary 1. *If $\lambda(x) \in (N)$, then for $0 < \alpha \leq 1$*

$$\lambda^{\alpha}(x) \in (N).$$[21]

---

[20] To check this, in (6.3.4) set $n = 2$, $x_1 = x$, $x_2 = -x$.

[21] Consider the principal branch of this multivalued function.

Suppose first that $\lambda(x) = C + s(0) - s(x)$ where $C > 0$, and $s(x)$ is positive definite. Since $|s(x)| < C + s(0)$ then

$$\lambda^\alpha(x) = [C + s(0)^\alpha [1 - \varphi_1(x)]^\alpha$$
$$= [C + s(0)]^\alpha \left\{1 - \sum_{n=1}^{\infty} (-1)^{n-1} \binom{\alpha}{n} \varphi_1^n(x)\right\},$$

where $\varphi_1(x) = [C + s(0)]^{-1} s(x)$ is a positive definite function. Since all the coefficients in the last series are positive,

$$\varphi_2(x) = \sum_{n=1}^{\infty} (-1)^{n-1} \binom{\alpha}{n} \varphi_1^n(x)$$

is a positive definite function, where $\varphi_2(0) = 1 - [1 - \varphi_1(0)]^\alpha < 1$. By ($\varepsilon$) $\lambda^\alpha(x) \in (N)$. Since the class (N) is closed, the condition $C > 0$ can be replaced by $C \geq 0$. For the same reason, in view of Lemma 6.8, the result holds also for any $\lambda(x) \in (N)$.

Corollary 2. *If $\lambda(x) \in (N)$ and $\lambda^{-1}(x)$ is locally integrable, then $\lambda^{-1}(x)$ is a positive definite function (in the generalized sense).*

We first verify the assertion for the case $\lambda(x) = C + s(0) - s(x)$ where $C > 0$; this is done just as in the proof of Corollary 1.

Then Lemma 6.8 implies that it holds for any $\lambda(x) \in (N)$, if only $\lambda(0) > 0$.

In the general case we consider $\lambda(x) + \eta \in (N)$ $(\eta > 0)$ and from the positive definiteness of $(\lambda(x) + \eta)^{-1}$ under the condition of local integrability of $\lambda^{-1}(x)$, it easily follows that the latter function is positive definite (in the generalized sense).

Corollary 3. *For $0 < \alpha \leq 2$ the function $|x|^\alpha$ belongs to (N).*

Indeed since for any $t \geq 0$, $e^{-t|x|^2}$ is a positive definite function (this follows, since $\mathfrak{F}[e^{-t|x|^2}] > 0$), then by Lemma 6.7 $|x|^2 \in (N)$. Now it is sufficient to apply Corollary 1.

Finally, restricting ourselves for simplicity to the case of real functions, we present an integral representation for functions of the class (N).

Lemma 6.9. *The real function $\lambda(x)$ belongs to the class (N) if, and only if, it has a representation*

$$\lambda(x) = C + Q(x) + \int_{R^p} (1 - \cos 2\pi xy)\, d\mu(y) \tag{6.3.6}$$

where $C \geq 0$, $Q(x)$ is a non-negative quadratic form, and μ is a measure on $R^p \setminus \{0\}$, symmetric relative to the point 0 and such that

$$\int \frac{|y|^2}{1 + |y|^2} d\mu(y) < \infty.$$

Proof. If $\lambda(x) = C + s(0) - s(x)$, then the Bochner theorem implies

$$\lambda(x) = C + \int_{R^p} (1 - \cos 2\pi xy)\, d\mu(y).$$

Consequently, by Lemma 6.8 in the general case

$$\lambda(x) = C + \lim_{n\to\infty} \int_{R^p} (1 - \cos 2\pi xy)\, d\mu_n(y),$$

where μ_n are finite symmetric measures in R^p.

Form the measures

$$\sigma_n(e) = \int_e \frac{|y|^2}{1+|y|^2}\, d\mu_n(y);$$

we will show that they are uniformly bounded. Since

$$\lambda(x) = C + \lim_{n\to\infty} \int_{R^p} (1 - \cos 2\pi xy) \frac{1+|y|^2}{|y|^2}\, d\sigma_n(y) \tag{6.3.7}$$

uniformly for $|x| < 1$, for these x and sufficiently large n

$$\int_{R^p} (1 - \cos 2\pi xy) \frac{1+|y|^2}{|y|^2}\, d\sigma_n(y) < \lambda(x) - C + 1.$$

Averaging sides over the ball $|x| < 1$, we obtain the inequality

$$\int_{R^p} M(|y|)\, d\sigma_n(y) < Q_1,$$

and it is possible to show that $\inf M(|y|) = m > 0$. This yields

$$\int_{R^p} d\sigma_n(y) < \frac{Q_1}{m} \tag{6.3.7'}$$

as claimed.

Furthermore, we assert that for any $\varepsilon > 0$, one can find an $N(\varepsilon)$ such that

$$\int_{|y|>N} d\sigma_n(y) < \varepsilon \tag{6.3.7''}$$

for all n. Indeed, since $C = \lambda(0)$, and since $\lambda(x)$ is continuous and convergence is uniform in (6.3.5), we can find a $\delta = \delta(\varepsilon) < 1$ such that for $|x| < \delta$ and any n we will have the inequality

$$\int_{R^p} (1 - \cos 2\pi xy) \frac{1+|y|^2}{|y|^2}\, d\sigma_n(y) < \varepsilon$$

and moreover

$$\int_{|y|>1/\delta} (1 - \cos 2\pi xy) \frac{1+|y|^2}{|y|^2}\, d\sigma_n(y) < \varepsilon.$$

Average the left side over the ball $|x| < \delta$; since

$$m^{(\delta)}(1 - \cos 2\pi xy) = m^{(1)}(1 - \cos 2\pi x\delta y)$$
$$= M(\delta\,|y|^2)\,\frac{\delta^2\,|y|^2}{1 + \delta^2\,|y|^2} \geq \frac{m\delta^2\,|y|^2}{1 + \delta^2\,|y|^2}\,, \qquad m > 0,$$

then

$$\int_{|y|>1/\delta} m\delta^2\,\frac{1 + |y|^2}{1 + \delta^2|y|^2}\,d\sigma_n(y) < \varepsilon,$$

which implies

$$\int_{|y|>1/\delta} d\sigma_n(y) < \frac{2\varepsilon}{m}\,.$$

We now fix $\eta > 0$ and rewrite (6.3.7) in the form

$$\lambda(x) = C + \lim_{n\to\infty} \int_{|y|<\eta} (1 - \cos 2\pi xy)\,\frac{1 + |y|^2}{|y|^2}\,d\sigma_n(y)$$
$$+ \lim_{n\to\infty} \int_{|y|\geq\eta} (1 - \cos 2\pi xy)\,\frac{1 + |y|^2}{|y|^2}\,d\sigma_n(y)\,.$$

Here, by (6.3.7′) and Theorem 0.6 on vague compactness, we may assume that $\sigma_n \rightharpoonup \sigma$ and select η such that the set $\{y\colon |y| \geq \eta\}$ will be σ-almost clopen. By Theorem 0.5′ and (6.3.7″)

$$\lim_{n\to\infty} \int_{|y|\geq\eta} (1 - \cos 2\pi xy)\,\frac{1 + |y|^2}{|y|^2}\,d\sigma_n(y)$$
$$= \int_{|y|\geq\eta} (1 - \cos 2\pi xy)\,\frac{1 + |y|^2}{|y|^2}\,d\sigma(y).$$

Further,

$$\int_{|y|<\eta} (1 - \cos 2\pi xy)\,\frac{1 + |y|^2}{|y|^2}\,d\sigma_n(y)$$
$$= \int_{|y|<\eta} \frac{1 - \cos 2\pi xy}{(xy)^2}\,(1 + |y|^2)\,\frac{(xy)^2}{|y|^2}\,d\sigma_n(y)$$
$$= 2\pi^2 \int_{|y|<\eta} \frac{(xy)^2}{|y|^2}\,d\sigma_n(y) + O(\eta^2),$$

where by (6.3.7′) the constant in $O(\eta^2)$ does not depend on n. Thus

$$\int_{|y|<\eta} (1 - \cos 2\pi xy)\,\frac{1 + |y|^2}{|y|^2}\,d\sigma_n(y) = \sum_{i,j=1}^{p} b_{ij}(n, \eta)\,x_i x_j + O(\eta^2),$$

where

$$b_{ij}(n, \eta) = 2\pi^2 \int_{|y|<\eta} \frac{y_i y_j}{|y|^2} \, d\sigma_n(y).$$

For these coefficients we have by (6.3.7′) the bound

$$|b_{ij}(n, \eta)| \leq 2\pi^2 \frac{Q_1}{m}.$$

which is uniform in n and η. Hence, after choosing a subsequence of numbers n, we may assume that

$$\lim_{n\to\infty} b_{ij}(n, \eta) = b_{ij}(\eta)$$

and

$$\lim_{n\to\infty} \int_{|y|<\eta} (1 - \cos 2\pi xy) \frac{1 + |y|^2}{|y|^2} \, d\sigma_n(y) = \sum_{i,j=1}^{p} b_{ij}(\eta) x_i x_j + O(\eta^2).$$

Now in the equation

$$\lambda(x) = C + \sum_{i,j=1}^{p} b_{ij}(\eta) x_i x_j$$
$$+ \int_{|y|\geq\eta} (1 - \cos 2\pi xy) \frac{1 + |y|^2}{|y|^2} \, d\sigma(y) + O(\eta^2)$$

we can pass to the limit over a certain sequence $\eta \to 0$ and this gives the formula (6.3.6) with

$$d\mu(y) = \frac{1 + |y|^2}{|y|^2} \, d\sigma(y), \qquad y \neq 0.$$

Since the form $\sum_{i,j} b_{ij}(n, \eta) \, x_i x_j$ is non-negative, the form $Q(x)$ is non-negative too.

The lemma is proved. ☐

We point out that for an arbitrary function in (N), in a similar way one can obtain the representation

$$\lambda(x) = C + iL(x) + Q(x) + \int_{R^p} \left\{1 - e^{-2\pi ixy} - \frac{2\pi ixy}{1 + |y|^2}\right\} d\mu(y),$$

where $C \geq 0$ $L(x)$ is a real linear function, and μ is a measure concentrated on $R^p \setminus \{0\}$.

In what follows we shall also need the following lemma:

Lemma 6.10. *In the representation* (6.3.6) *if* $Q(x) \equiv 0$, *then as* $x \to \infty$ $\lambda(x) = o(|x|^2)$.

Proof. By the condition

$$\lambda(x) = C + \int_{R^p} (1 - \cos 2\pi xy) \frac{1 + |y|^2}{|y|^2} d\sigma(y),$$

where σ is a finite measure in $R^p \setminus \{0\}$. Fixing $\varepsilon > 0$, we take $\delta = \delta(\varepsilon) < 1$ such that

$$\int_{|y|<\delta} d\sigma(y) < \varepsilon.$$

Then

$$\lambda(x) \leq C + 2 \int_{|y|<\delta} \frac{1 - \cos 2\pi xy}{|y|^2} d\sigma(y) + M(\varepsilon)$$

$$< C + 8\pi^2 |x|^2 \varepsilon + M(\varepsilon),$$

whence

$$\frac{\lambda(x)}{|x|^2} < 8\pi^2\varepsilon + \frac{C + M(\varepsilon)}{|x|^2} < 8\pi^2\varepsilon + \varepsilon$$

for sufficiently large $|x|$. ▯

14. We now look for an integral representation for the norm in a Dirichlet space. Here we impose one more restriction on the space H.[22] Specifically, we shall assume that the following axiom holds:

(e) If $u(x) \in H$, then $u(2x) \in H$, where

$$||u(2x)|| \leq 2^{-\frac{1}{2}(p-2)} ||u(x)||.$$

As we are about to show, this will permit us to evaluate the norm of the potential of a *spherical condenser*.

We apply Theorem 6.5, taking for ω_1 the open ball $|x| < 2^{-n}$ and for ω_0 the exterior of the ball $|x| \leq 2^{-n+1}$. The corresponding potential will be denoted by $u_n(x)$ and its signed measure ν_n.

Lemma 6.11. *We have the inequalities*

$$||u_{n+1}|| \leq 2^{-\frac{n}{2}(p-2)} ||u_1||$$

and

$$\nu_{n+1}^+(1) \leq 2^{-n(p-2)} Q,$$

where Q depends only on p.

---

[22] In the papers [8] and [9] of J. Deny, whose content is presented in this section, this axiom is not introduced, even though from Theorem 6.9, it clearly must be satisfied in every Dirichlet space in which the norm admits a representation (6.3.8) on an everywhere dense set of elements. Unfortunately, we were not able to establish that part of Theorem 6.8 which proves that the function $\lambda(y)$ constructed below belongs to the class (N); hence we were forced to introduce axiom (e).

Proof. We introduce, as in the proof of Theorem 6.5, the class of functions

$$A_1 = \{f(x): f \in H;\ \mathrm{Re}\, f \geq 1 \text{ almost everywhere for } |x| < 1/2,$$
$$\mathrm{Re}\, f \leq 0 \text{ almost everywhere for } |x| > 1\};$$

let A_n denote the analogous class for a spherical condenser with radii 2^{-n} and 2^{-n+1}. Recall that

$$||u_1|| = \min_{f \in A_1} ||f||, \qquad ||u_{n+1}|| = \min_{f \in A_{n+1}} ||f||.$$

By (e) since $f(x) \in A_1$, it follows that $f(2^n x) \in A_{n+1}$. Hence

$$||u_{n+1}|| = \min_{f \in A_{n+1}} ||f(x)|| \leq \min_{f \in A_1} ||f(2^n x)||$$
$$\leq 2^{-\frac{n}{2}(p-2)} \min_{f \in A_1} ||f(x)|| = 2^{-\frac{n}{2}(p-2)} ||u_1||.$$

Further, we saw in the proof of Theorem 6.5, that for any real continuous function $v(x) \in H$, non-negative for $|x| < 2^{-n}$ and equal to zero for $|x| > 2^{-n+1}$,

$$(u_n, v) = \int v(x)\, d\nu_n^+(x).$$

Hence, if $v(x) \geq 1$ for $|x| < 2^{-n}$ (that such functions exist in H was also shown in the proof of the theorem), then

$$\nu_n^+(1) \leq (u_n, v) \leq ||u_n|| \cdot ||v||.$$

Suppose $v_1(x)$ is any fixed function of the indicated type, corresponding to the value $n = 1$. Then by (e), $v_1(2^n x) \in H$ and

$$\nu_{n+1}^+(1) \leq ||u_{n+1}|| \cdot ||v_1(2^n x)|| \leq 2^{-n(p-2)} ||u_1|| \cdot ||v_1||.$$

The lemma is proved. □

Theorem 6.8. *For any function $u \in \Phi \cap H$ we have*

$$||u||^2 = \int_{R^p} |\tilde{u}(y)|^2 \lambda(y)\, dy, \tag{6.3.8}$$

where $\tilde{u} = \mathfrak{F}[u]$ is the Fourier transform of the function u, and λ is a real function of the class (N) for which the inverse λ^{-1} is locally integrable.

Proof. Consider two spherical condensers with radii 2^{-n}, 2^{-n+1} and 2^{-m}, 2^{-m+1} respectively.

Suppose $f \in \Phi$; taking (d) into consideration, we obtain

$$(u_n * f, u_m) = \int (U_x u_n, u_m) f(x)\, dx = \int (u_n, U_{-x} u_m) f(x)\, dx$$
$$= \int \overline{(U_x u_m, u_n)} f(-x)\, dx = (u_n, u_m * \hat{f}),$$

where $\hat{f}(x) = \overline{f(-x)}$. Since $u_n * f \in \Phi$, $u_m * \hat{f} \in \Phi$ then the relation obtained can be rewritten in the following way:

$$\int (u_n * f)\, dv_m = \int (u_m * \check{f})\, dv_n \quad (\check{f}(x) = f(-x)),$$

or

$$\int \check{f}(y)\, dy \int u_n(x+y)\, dv_m(x) = \int \check{f}(y)\, dy \int u_m(x-y)\, dv_n(x). \quad (6.3.9)$$

But $\check{f}(y)$ is an arbitrary function in Φ, and hence

$$u_n * \check{v}_m = \check{u}_m * v_n. \quad (6.3.10)$$

Formula (6.3.9) will permit us to conclude that

$$v_n^-(1) \leq v_n^+(1),\ v_n^+(1) - v_n^-(1) = O(2^{-np}). \quad (6.3.11)$$

Indeed, in (6.3.9) take $\check{f}(y) \in \Phi^+$ such that the support $S(\check{f})$ belongs to the unit ball and $\int \check{f}(y)\, dy = 1$. Then for $2^{-m} > 2^{-n} + 1$ $(m < 0)$ we will have

$$\int \check{f}(y)\, dy \int u_n(x+y)\, dv_m(x) = \int \check{f}(y)\, dy \int u_n(x+y)\, dv_m^+(x) \geq 0$$

and

$$\int \check{f}(y)\, dy \int u_m(x-y)\, dv_n(x) = \int \check{f}(y)\, dy \int u_m(x-y)\, dv_n^+(x)$$
$$- \int \check{f}(y)\, dy \int u_m(x-y)\, dv_n^-(x)$$
$$= v_n^+(1) - \int \check{f}(y)\, dy \int u_m(x-y)\, dv_n^-(x),$$

so that

$$\int \check{f}(y)\, dy \int u_m(x-y)\, dv_n^-(x) \leq v_n^+(1).$$

But

$$\int \check{f}(y)\, dy \int u_m(x-y)\, dv_n^-(x) \geq v_n^-(\{x : |x| < 2^{-m} - 1\}),$$

and letting $m \to -\infty$, we obtain that $v_n^-(1) \leq v_n^+(1)$.

Then, from the equation

$$v_n^+(1) - \int \check{f}(y)\, dy \int u_m(x-y)\, dv_n^-(x) = \int \check{f}(y)\, dy \int u_n(x+y)\, dv_m^+(x)$$

it follows that

$$\begin{aligned} \nu_n^+(1) - \nu_n^-(1) &\le \int d\nu_m^+(x) \int u_n(x+y)\, \check{f}(y)\, dy \\ &\le \max \check{f}(y) \int d\nu_m^+(x) \int u_n(x+y)\, dy \\ &= \max \check{f}(y)\, \nu_m^+(1) \int u_n(y)\, dy, \end{aligned}$$

and since $m < 0$ can be assume fixed, we have for $n \to \infty$

$$\nu_n^+(1) - \nu_n^-(1) = O\,(2^{-np}).$$

Thus, both formulas (6.3.11) are proved.

Let us return to (6.3.10). Since ν_n^+ is a measure with compact support, then in view of (6.3.11) we obtain that $\nu_n \in S^*$, and using the corollary to Theorem 0.12, we obtain from (6.3.10) that

$$\tilde{u}_n(y)\, \tilde{\tilde{\nu}}_m(y) = \tilde{\tilde{u}}_m(y)\, \tilde{\nu}_n(y),$$

or

$$\tilde{u}_n(y)\, \overline{\tilde{\nu}_m(y)} = \overline{\tilde{u}_m(y)}\, \tilde{\nu}_n(y),$$

where all functions appearing in this relation are continuous. Rewriting it in the form

$$\frac{\tilde{\nu}_n(y)}{\tilde{u}_n(y)} = \frac{\overline{\tilde{\nu}_m(y)}}{\overline{\tilde{u}_m(y)}},$$

we see that the ratio

$$\lambda(y) = \frac{\tilde{\nu}_n(y)}{\tilde{u}_n(y)} \tag{6.3.12}$$

is real and does not depend on n. The function $\lambda(y)$ is defined for any $y \in R^p$ since n may be taken so large that $\tilde{u}_n(y) \neq 0$.

We will show that this function satisfies all the conditions of Theorem 6.8. The basic difficulty is in verifying that $\lambda(y) \in (N)$. First, note that

$$\lim_{n\to\infty} \frac{\tilde{u}_n(y)}{\tilde{u}_n(0)} = 1,$$

uniformly on any compact set, where $\tilde{u}_n(0) = \int u_n(x)\, dx$ has order 2^{-np}. Therefore,

$$\lambda(y) = \lim_{n\to\infty} \left[\frac{\tilde{\nu}_n^-(0) - \tilde{\nu}_n^-(y)}{\tilde{u}_n(0)} - \frac{\tilde{\nu}_n^+(0) - \tilde{\nu}_n^+(y)}{\tilde{u}_n(0)} + \frac{\tilde{\nu}_n^+(0) - \tilde{\nu}_n^-(0)}{\tilde{u}_n(0)} \right] \tag{6.3.13}$$

or, since $\lambda(y)$ is real,

$$\lambda(y) = \lim_{n\to\infty} \left[\frac{\tilde{\nu}_n^-(0) - \operatorname{Re}\tilde{\nu}_n^-(y)}{\tilde{u}_n(0)} - \frac{\tilde{\nu}_n^+(0) - \operatorname{Re}\tilde{\nu}_n^+(y)}{\tilde{u}_n(0)} + \frac{\tilde{\nu}_n^+(0) - \tilde{\nu}_n^-(0)}{\tilde{u}_n(0)} \right].$$

By (6.3.11) the latter term is non-negative and bounded. Hence, for a certain subsequence $n' \to \infty$, the limit

$$\lim_{n'\to\infty} \frac{\tilde{\nu}_{n'}^{+}(0) - \tilde{\nu}_{n'}^{-}(0)}{\tilde{u}_{n'}(0)} = C \geq 0$$

exists (below we shall simply write n instead of n').

Let us consider the function

$$\frac{\tilde{\nu}_n^{+}(0) - \operatorname{Re} \tilde{\nu}_n^{+}(y)}{\tilde{u}_n(0)} = \frac{1}{\tilde{u}_n(0)} \int_{|x|\leq 2^{-n}} (1 - \cos 2\pi xy)\, d\nu_n^{+}(x)$$

$$= \int_{|x|\leq 2^{-n}} (1 - \cos 2\pi xy)\, d\mu_n^{+}(x),$$

where $\mu_n^{+} = \nu_n^{+}/\tilde{u}_n(0)$. Since by Lemma 6.11 $\nu_n^{+}(1) = O(2^{-n(p-2)})$, then $\mu_n^{+}(1) = O(2^{2n})$ and

$$\int_{|x|\leq 2^{-n}} |x|^2\, d\mu_n^{+}(x) \leq 2^{-2n} \mu_n^{+}(1) - O(1).$$

This shows that the measures $d\sigma_n(x) = |x|^2\, d\mu_n^{+}(x)$ are uniformly bounded. With the same reasoning as in the proof of Lemma 6.9 and noting that the support of σ_n shrinks to the point 0, we obtain that for a certain subsequence of numbers (we will again call it $\{n\}$) the limit

$$\lim_{n\to\infty} \frac{\tilde{\nu}_n^{+}(0) - \operatorname{Re} \tilde{\nu}_n^{-}(y)}{\tilde{u}_n(0)} = Q(y),$$

exists, uniformly on any compact set, where $Q(y)$ is a non-negative quadratic form.

For this same subsequence, $\{n\}$, by (6.3.13) the limit

$$\lim_{n\to\infty} \frac{\tilde{\nu}_n^{-}(0) - \operatorname{Re} \tilde{\nu}_n^{-}(y)}{\tilde{u}_n(0)} = \lambda_1(y),$$

exists, and the convergence is again uniform on any compact set. Since $\tilde{\nu}_n^{-}(y)$ as the Fourier transform of a finite measure is a positive definite function, then $\operatorname{Re} \tilde{\nu}_n^{-}(y)$ is also positive definite; hence, by Lemma 6.8, $\lambda_1(y) \in (N)$. Consequently, in view of (6.3.13) and Lemma 6.9, we can write the equation

$$\lambda(y) = C_1 + Q_1(y) - Q(y) + \int_{R^p} (1 - \cos 2\pi yx)\, d\mu(x),$$

where $C_1 \geq 0$ and $Q_1(y)$ is a non-negative quadratic form.

Now to verify that $\lambda(y) \in (N)$, it is sufficient to establish that $Q_1(y) - Q(y) \geq 0$.

Suppose that $Q_1(y) - Q(y_0) < 0$. Since for $t \in R^1$

$$Q_1(ty_0) - Q(ty_0) = t^2[Q_1(y_0) - Q(y_0)],$$

while by Lemma 6.10

$$\int_{R^p} (1 - \cos 2\pi tyx)\, d\mu(x) = o(t^2) \quad (t \to \infty),$$

we conclude that under the assumption just made, there exists a point y_1 such that $\lambda(y_1) < 0$. We will show that this is impossible.

For this we shall establish formula (6.3.8) for any function $u \in D \cap H$ (in No. 12 we saw that this set is not empty). We normalize the potential $u_n(x)$, putting

$$v_n(x) = \frac{u_n(x)}{\tilde{u}_n(0)};$$

the signed measure corresponding to this potential $v_n(x)$ will be equal to

$$\mu_n = \frac{\nu_n}{\tilde{u}_n(0)}.$$

If $u(x)$ is an element of $D \cap H$, then using the Parseval equality, we obtain

$$||u * v_n||^2 = (u * v_n, u * v_n) = (v_n * u * \hat{u}, v_n) = \int (v_n * u * \hat{u})\,(x)\, d\mu_n\,(x)$$
$$= \int \tilde{v}_n(y)\, |\tilde{u}(y)|^2\, \overline{\tilde{\mu}_n(y)}\, dy = \int |\tilde{v}(y)|^2\, |\tilde{u}(y)|^2\, \lambda(y)\, dy.$$

Now suppose that $n \to \infty$. Since $v_n \dashrightarrow \varepsilon$, and the operator $U_x u(y) = u(y - x)$ is strongly continuous (cf. (d)), then $u * v_n \to u$ in H and $||u * v_n|| \to ||u||$. On the other hand, $\tilde{u}|(y)|\, \lambda(y)$ is a rapidly decreasing, and hence, integrable function in R^p. If one takes into account that $\tilde{v}_n(y) \to 1$ and $|\tilde{v}_n(y)| \leq 1$, then we immediately obtain equation (6.3.8).

Let us return to the assumption that $\lambda(y_1) < 0$. Since $\lambda(y)$ is continuous, there is a neighborhood T of the point y_1 in which $\lambda(y) < 0$. We will show that in $D \cap H$ there is a function $u_T(x)$ whose Fourier transform $\tilde{u}_T(y)$ has support contained in T. Setting $u(x) = u_T(x)$ in (6.3.8), we will obtain a contradiction; this will show that $\lambda(y) \in (N)$.

To construct $u_T(x)$ we begin with an arbitrary function $f(y)$ which has $p + 1$ continuous derivatives and such that $S(f) \subset T$. Then the function $g(x) = \mathfrak{F}^{-1}[f]$ is infinitely differentiable (it is even an entire function) and $\int_{R^p} |g(x)|\, dx < \infty$. Suppose $u(x) \in \Phi \cap H$; then it is easy to see that $u_T = u * g$ satisfies the given requirements.

Now that $\lambda(y) \geq 0$ is established, it is easy to prove formula (6.3.8) for any function $u \in \Phi \cap H$. For this it is sufficient to show that in the equation

$$||u * v_n||^2 = \int |\tilde{v}_n(y)|^2 |\tilde{u}(y)|^2 \lambda(y)\, dy$$

it is corred to pass to the limit $n \to \infty$ under the integral sign. But this follows from the Fatou Lemma.

It remains to be shown that $\lambda^{-1}(y)$ is locally integrable. Note that by formula (6.3.8) the Fourier operator $\mathfrak{F}$ establishes an isometric correspondence between $\Phi \cap H$ and a certain linear manifold in the space $L^2(\lambda)$ of functions quadratically integrable in R^p with weight $\lambda(y)$. Since $\Phi \cap H$ is dense in H, then by continuity $\mathfrak{F}$ can be extended to all of H. We shall denote this extension by $\mathfrak{F}_1$ and will show that $\mathfrak{F}_1[H] \equiv L^2(\lambda)$.

First of all, the completeness of H implies that $\mathfrak{F}_1[H]$ is closed in $L^2(\lambda)$.

Let $D^{p+1} \subset \Phi$ be the set of functions having $p + 1$ continuous derivatives. Since D^{p+1} is dense in $L^2(\lambda)$, from the inclusion $\mathfrak{F}_1[H] \supset D^{p+1}$ it will follow that $\mathfrak{F}_1[H] \equiv L^2(\lambda)$. If $g(y) \in D^{p+1}$, then $g(y) = \mathfrak{F}[h]$, where $\int_{R^p} |h(x)|\, dx < \infty$. Therefore, if $f(x) \in \Phi \cap H$, then $f * h \in H$ and $\tilde{f}(y)\, g(y)$

$\in \mathfrak{F}_1[H]$. Put

$$f(x) = u_\alpha(x) \left(\int u_\alpha(x)\, dx\right)^{-1}$$

and let α tend to zero. Then $\tilde{f}(y) \to 1$ uniformly on the support of $g(y)$, and hence $g(y) \in \mathfrak{F}_1[H]$.

Now we reason as follows. Suppose that $\nu(x)$ is an arbitrary, bounded, and measurable function with compact support. We have seen that there exists a potential u^ν of the signed measure ν. If $\tilde{u}^\nu(y) = \mathfrak{F}[u^\nu] \in L^2(\lambda)$ then for any function $\varphi(x) \in \Phi \cap H$ we will have

$$\int \tilde{u}^\nu(y)\, \overline{\tilde{\varphi}(y)}\, \lambda\, dy = (u^\nu, \varphi) = \int \nu(x)\, \overline{\varphi(x)}\, dx = \int \tilde{\nu}(y)\, \overline{\tilde{\varphi}(y)}\, dy.$$

Since $\Phi \cap H$ is dense in H, this equation holds for any function $\varphi(x) \in H$ and hence, for any $\tilde{\varphi}(y) \in L^2(\lambda)$. Thus,

$$\tilde{u}^\nu(y)\, \lambda(y) = \tilde{\nu}(y)$$

almost everywhere in R^p. Since $\tilde{\nu}(y)$ is an analytic function, $\lambda(y)$ can have only isolated zeros and almost everywhere

$$\tilde{u}^\nu(y) = \frac{\tilde{\nu}(y)}{\lambda(y)}. \tag{6.3.14}$$

But

$$||u^\nu||^2 = \int |\tilde{u}^\nu(y)|^2 \lambda(y)\, dy = \int \frac{|\tilde{\nu}(y)|^2}{\lambda(y)}\, dy,$$

and since $\tilde{\nu}(y)$ can be chosen to be bounded below on a given compact set $K \subset R^p$, we have

$$\int_K \frac{dy}{\lambda(y)} < \infty .$$

The theorem is proved. $\square$

15. Theorem 6.8 can be turned around.

Theorem 6.9. *Suppose that $\lambda(y)$ is an arbitrary real function of the class (N) such that $\lambda^{-1}(y)$ is locally integrable. Then $\mathfrak{F}[L^2(\lambda)]$ is a Dirichlet space satisfying axiom (e).*

Proof. First of all, any function $\tilde{u}(y) \in L^2(\lambda)$ is slowly increasing, which gives meaning to the notation $\mathfrak{F}[L^2(\lambda)]$. Indeed, since $\lambda^{-1}(y)$, being a positive definite function (in the generalized sense) belongs to S^* (cf. Introduction, § 3) and is also non-negative, then by Theorem 0.10 it is slowly increasing, i.e. for some n

$$\int \frac{dy}{\lambda(y)\,(1+|y|^2)^n} < \infty .$$

But in that case

$$\left| \int \frac{|\tilde{u}(y)|^2}{(1+|y|^2)^n} dy \right|^2 \leq \int |\tilde{u}(y)|^2 \lambda(y)\, dy \cdot \int \frac{dy}{\lambda(y)\,(1+|y|^2)^{2n}} < \infty .$$

If u_1, u_2 are two elements of $\mathfrak{F}[L^2(\lambda)]$ then, setting

$$(u_1, u_2) = \int \mathfrak{F}^{-1}[u_1]\, \overline{\mathfrak{F}^{-1}[u_2]}\, \lambda(y)\, dy$$

we easily see that $\mathfrak{F}[L^2(\lambda)]$ becomes a Hilbert space.

Since $\lambda(y)$ is also slowly increasing (specifically, $\lambda(y) = O(|y|^2)$), then $S \subset L^2(\lambda)$, and hence $S \subset \mathfrak{F}[L^2(\lambda)]$. Since convergence in S is stronger than convergence in $L^2(\lambda)$, we conclude that the sets of functions D and S are dense in $\mathfrak{F}[L^2(\lambda)]$.

Now suppose that $u(x) \in D$, and T is a contraction. We will show that $Tu \in \mathfrak{F}[L^2(\lambda)]$ and $||Tu|| \leq ||u||$.

Set

$$\lambda_k(y) = \lambda(0) + k\left[1 - e^{\frac{1}{k}(\lambda(0)-\lambda(y))}\right].$$

As $k \to \infty$ these bounded functions monotonically increasing tend to $\lambda(y)$. Then (δ), (ε) and Lemma 6.7 imply that $\lambda(y) \in (N)$. By Lemma 6.9

$$\lambda_k(y) = \lambda(0) + \int (1 - \cos 2\pi zy)\, d\mu_k(z),$$

and hence

$$\int |\tilde{u}(y)|^2 \lambda_k(y) dy = \lambda(0) \int |u(x)|^2\, dx + \frac{1}{2} \int d\mu_k(z) \int |u(x+z) - u(x)|^2\, dx .$$

This shows that

$$\int |\widetilde{Tu}(y)|^2 \lambda_k(y)\, dy \leq \int |\tilde{u}(y)|^2 \lambda_k(y)\, dy,$$

and after passing the limit $k \to \infty$

$$\int |\widetilde{Tu}(y)|^2 \lambda(y)\, dy \leq \int |\tilde{u}(y)|^2 \lambda(y)\, dy,$$

as asserted.

We now show that for any $u(x) \in D$ and any compact set K

$$\int_K |u(x)|\, dx \leq A(K)\, ||u||. \tag{6.3.15}$$

Take a function $\varphi(x)$, $0 \leq \varphi(x) \in D$, $\varphi(x) \geq 1$ on K. Then, putting $v(x) = |u(x)|$ and noting that v comes from u by a contraction, we will have

$$\left(\int_K |u(x)|\, dx\right)^2 \leq \left(\int v(x)\, \varphi(x)\, dx\right)^2 = \left(\int \tilde{v}(y)\, \overline{\tilde{\varphi}(y)}\, dy\right)^2$$
$$\leq \int |\tilde{v}(y)|^2 \lambda(y) dy \int \frac{|\tilde{\varphi}(y)|^2}{\lambda(y)}\, dy \leq ||u||^2 \int \frac{|\tilde{\varphi}(y)|^2}{\lambda(y)}\, dy,$$

which proves (6.3.15).

Since D is dense in $\mathfrak{F}[L^2(\lambda)]$, then (6.3.15) implies that all the elements of the space $\mathfrak{F}[L^2(\lambda)]$ are functions and satisfy axioms (a), (b), (c). Axiom (d) is also clearly satisfied and we have only to verify axiom (e).

Suppose that $u(x) \in \mathfrak{F}[L^2(\lambda)]$ and $\mathfrak{F}^{-1}(u] = \tilde{u}(y) \in L^2(y)$. Then

$$\mathfrak{F}^{-1}[u\,(2x)] = 2^{-p}\tilde{u}\left(\frac{y}{2}\right),$$

and we must show that

$$2^{-2p} \int_{R^p} \left|\tilde{u}\left(\frac{y}{2}\right)\right|^2 \lambda(y)\, dy \leq 2^{-p+2} \int_{R^p} |\tilde{u}(y)|^2 \lambda(y)\, dy.$$

But this follows from

$$\int_{R^p} \left|\tilde{u}\left(\frac{y}{2}\right)\right|^2 \lambda(y)\, dy = \int_{R^p} |\tilde{u}(z)|^2 \lambda(2z)\, 2^p dz \leq 2^{p+2} \int_{R^p} |\tilde{u}(z)|^2 \lambda(z)\, dz.$$

The theorem is proved. ▯

Corollary 1. *Any Dirichlet space H which satisfies axiom* (e), *consists of all functions which are Fourier transforms of functions of the space $L^2(\lambda)$, where $\lambda \in (N)$, i.e.*

$$H \equiv \mathfrak{F}[L^2(\lambda)].$$

Corollary 2. *If $u \in D$,*[23] *then*

$$||u||^2 = C\int |u(x)|^2\, dx + \sum_{i,j=1}^{p} a_{ij} \int \frac{\partial u}{\partial x_i} \frac{\partial u}{\partial x_j}\, dx$$
$$+ \frac{1}{2}\int d\mu(z) \int |u(x+z) - u(x)|^2\, dx, \qquad (6.3.16)$$

where $C \geq 0$, the form $Q(\xi) = \sum_{i,j}^{p} a_{ij}\xi_i\bar{\xi}_j$ is non-negative, and μ is a certain symmetric measure on $R^p \setminus \{0\}$ such that

$$\int \frac{|y|^2}{1+|y|^2}\, d\mu(y) < \infty.$$

To obtain (6.3.16) it is sufficient to substitute (6.3.6) into (6.3.8) and use the Parseval equality.

Remark. According to Corollary 2, No. 13, the function $\lambda^{-1}(y)$ will be positive definite. The Bochner-Schwartz theorem asserts that

$$\lambda^{-1}(x) = \mathfrak{F}[\varkappa]$$

where $\varkappa$ is a certain measure. In that case, (6.3.14) shows that

$$u^{\nu}(x) = \varkappa * \nu$$

and we see that the measure $\varkappa$ plays the role of the kernel in classical potential theory.

---

[23] From the preceding corollary it is clear that $D \subset H$, since $\lambda(x) = O(|x|^2)$ and hence $\mathfrak{F}[D] \subset S \subset L^2(\lambda)$.

Appendix

1. We shall compute the integral

$$I(y) = \int\limits_{|x| \leq r} \frac{dx}{(r^2 - |x|^2)^{\alpha/2} \, |x - y|^{p-\alpha}}$$

for $|y| \leq r$ and will show that it is equal to a constant.

Suppose that $|y| < r$. Consider the point transformation (imaginary inversion with center at y), in which any point $x \in R^p$ is mapped into the point x^* such that the points x, y, x^* lie on one line, y separates x and x^* and

$$|x - y| \cdot |x^* - y| = r^2 - |y|^2.$$

It is not difficult to see that

(a) Points of the sphere $x = r$ are mapped into points of the same sphere,

(b) the point y is mapped to ∞,

(c) $\dfrac{dx}{|x - y|^p} = \dfrac{dx^*}{|x^* - y|^p}$,

(d) $$\frac{|x - y|^2}{(r^2 - |x|^2)\,(r^2 - |y|^2)} = \frac{1}{|x^*|^2 - r^2}.$$

We shall verify (d). Consider the plane passing through $0, x, y, x^*$ and introduce complex coordinates there. The transformation above can be written in the form

$$w = \frac{a\bar{z} + b}{c\bar{z} + d},$$

and then (d) expresses the invariance of the cross-ratio of four points (namely, x, y and their symmetric counterparts relative to the circle $|x| = r$).

After the indicated inversion, for I we obtain the expression

$$I = \int\limits_{|x^*| \geq r} \frac{(r^2 - |y|^2)^{\alpha/2} \, dx^*}{(|x^*|^2 - r^2)^{\alpha/2} |x^* - y|^p}.$$

We introduce polar coordinates in R^p and direct the axis along the line $0y$. Then

$$I = (r^2 - |y|^2)^{\alpha/2} \cdot 2\pi \prod_{k=1}^{p-3} \int_0^\pi \sin^k\theta \, d\theta$$

$$\times \int_r^\infty \int_0^\pi \frac{\varrho^{p-1} \sin^{p-2}\theta \, d\varrho \, d\theta}{(\varrho^2 - r^2)^{\alpha/2} (\varrho^2 - 2\varrho|y| \cos\theta + |y|^2)^{p/2}}.$$

Setting

$$r = r_1|y|, \quad \varrho = \varrho_1 |y|,$$

we obtain

$$I = (r_1^2 - 1)^{\alpha/2} \cdot 2\pi \prod_{k=1}^{p-3} \int_0^\pi \sin^k\theta \, d\theta$$

$$\times \int_{r_1}^\infty \int_0^\pi \frac{\varrho_1^{p-1} \sin^{p-2}\theta \, d\varrho_1 \, d\theta}{(\varrho_1^2 - r_1^2)^{\alpha/2} (\varrho_1^2 - 2\varrho_1 \cos\theta + 1)^{p/2}}.$$

The integral

$$i_p = \int_0^\pi \frac{\sin^{p-2}\theta \, d\theta}{(\varrho_1^2 - 2\varrho_1 \cos\theta + 1)^{p/2}}$$

can be computed with the following change of variable:[1]

$$\frac{\sin\theta}{\sqrt{\varrho_1^2 - 2\varrho_1 \cos\theta + 1}} = \frac{\sin\alpha}{\varrho_1}.$$

From the latter we obtain

$$d\theta = \left(1 - \frac{\cos\alpha}{\sqrt{\varrho_1^2 - \sin^2\alpha}}\right) d\alpha,$$

and then, after elementary computations, it turns out that

$$i_p = \frac{1}{\varrho_1^{p-2} (\varrho_1^2 - 1)} \int_0^\pi \sin^{p-2}\alpha \, d\alpha.$$

Conseqeuently,

$$I = (r_1^2 - 1)^{\alpha/2} \, 2\pi \prod_{k=1}^{p-2} \int_0^\pi \sin^k\theta \, d\theta \int_{r_1}^\infty \frac{\varrho_1 \, d\varrho_1}{(\varrho_1^2 - r_1^2)^{\alpha/2} (\varrho_1^2 - 1)}$$

$$= \frac{\pi^{p/2}}{\Gamma(p/2)} (r_1^2 - 1)^{\alpha/2} \int_0^\infty \frac{dt}{t^{\alpha/2}(t + r_1^2 - 1)} = \frac{\pi^{p/2}}{\Gamma(p/2)} \int_0^\infty \frac{dt_1}{t_1^{\alpha/2} (t_1 + 1)}$$

$$= \frac{\pi^{p/2}}{\Gamma(p/2)} \mathrm{B}(\alpha/2, 1 - \alpha/2) = \frac{\pi^{p/2+1}}{\Gamma(p/2) \sin \pi\alpha/2}.$$

---

[1] This substitution was suggested by S. I. Greenberg. The angle α has a simple geometric interpretation.

Thus, for $|y| < r$

$$\int\limits_{|x|\le r} \frac{dx}{(r^2 - |x|^2)^{\alpha/2}\,|x-y|^{p-\alpha}} = \frac{\pi^{p/2+1}}{\Gamma(p/2)\sin \pi\alpha/2}, \tag{A.1}$$

and, by continuity of the left-hand side in the variable y, this equation holds for $|y| = r$.

Since for $|y| > r$ the integral in (A.1) is a superharmonic function of y (cf. Theorem 1.4) which tends to zero as $y \to \infty$, then

$$\int\limits_{|x|\le r} \frac{dx}{(r^2 - |x|^2)^{\alpha/2}\,|x-y|^{p-\alpha}} < \frac{\pi^{p/2+1}}{\Gamma(p/2)\sin \pi\alpha/2}, \quad |y| > r. \tag{A.2}$$

2. Let us apply the Kelvin transformation with center 0 (cf. Ch. IV, § 5) to the measure λ with density

$$\lambda(x) = \begin{cases} (r^2 - |x|^2)^{-\alpha/2}, & |x| < r \\ 0, & |x| \ge r, \end{cases}$$

Replacing r by r^{-1}, we obtain a measure λ^* concentrated in the region $|x^*| \ge r$ with density

$$\lambda^*(x^*) = \begin{cases} 0, & |x|^* \le r, \\ r^\alpha(|x^*|^2 - r^2)^{-\alpha/2}|x^*|^{-p}, & |x^*| > r \end{cases}$$

(recall that $|x|^{-p}\,dx = |x^*|^{-p}\,dx^*$).

By (A.1) and (4.5.4) its potential in the region $|x^*| \ge r$ is

$$\int\limits_{|y|>r} \frac{r^\alpha\,dy}{(|y|^2 - r^2)^{\alpha/2}|y|^p|x^* - y|^{p-\alpha}} = \frac{\pi^{p/2+1}}{\Gamma(p/2)\sin \pi\alpha/2}\,\frac{1}{|x^*|^{p-\alpha}}, \tag{A.3}$$

while for $|x^*| < r$ by (A.2) we have

$$\int\limits_{|y|>r} \frac{r^\alpha\,dy}{(|y|^2 - r^2)^{\alpha/2}|y|^p\,|x^* - y|^{p-\alpha}} < \frac{\pi^{p/2+1}}{\Gamma(p/2)\sin \pi\alpha/2}\,\frac{1}{|x^*|^{p-\alpha}}. \tag{A.4}$$

We now carry out the inversion (and the corresponding Kelvin transformation of the measure λ), putting the center of inversion x_0 at any point of R^p and choosing the radius so that the points of the sphere $|x| = r$ will be mapped into points of the same sphere.

Suppose $|x_0| < r$. Then we have imaginary inversion, as described in No. 1. Using (c) and (d) we find the measure

$$\lambda^*_{x_0}(x^*) = \begin{cases} 0, & |x^*| \le r, \\ (r^2 - |x_0|^2)^{\alpha/2}\,(|x^*|^2 - r^2)^{-\alpha/2}\,|x^* - x_0|^{-p}, & |x^*| > r. \end{cases}$$

It differs from the Poisson kernel $\hat{P}_r(x^*, x_0)$, introduced in Ch. I, § 6, (1.6.11') only by the absence of the normalization constant. By (4.5.4), for $|x| \geq r$ we have

$$\int_{|y|>r} \frac{(r^2 - |x_0|^2)^{\alpha/2}\, dy}{(|y|^2 - r^2)^{\alpha/2}\, |y - x_0|^p\, |x - y|^{p-\alpha}} = \frac{\pi^{p/2+1}}{\Gamma(p/2) \sin \pi\alpha/2} \frac{1}{|x - x_0|^{p-\alpha}}. \tag{A.5}$$

If $|x_0| > r$, then we invert with respect to x_0 and with a real radius $\sqrt{|x_0|^2 - r^2}$. Here the points of the ball $|x| \leq r$ are mapped again into points of the same ball and we obtain the measure

$$\lambda^*_{x_0}(x^*] = \begin{cases} 0, & |x^*| \geq r, \\ (|x_0|^2 - r^2)^{\alpha/2}\, (r^2 - |x^*|^2)^{-\alpha/2}\, |x^* - x_0|^{-p}, & |x^*| < r, \end{cases}$$

which differs only by a constant factor from the Poisson kernel $\check{P}_r(x^*, x_0)$ (cf. Ch. I, § 6, (1.6.11)).

For $|x| \leq r$, (4.5.4) and (A.1) imply

$$\int_{|y|<r} \frac{(|x_0|^2 - r^2)^{\alpha/2}\, dy}{(r^2 - |y|^2)^{\alpha/2}\, |y - x_0|^p\, |x - y|^{p-\alpha}} = \frac{\pi^{p/2+1}}{\Gamma(p/2) \sin \pi\alpha/2} \frac{1}{|x - x_0|^{p-\alpha}}. \tag{A.6}$$

Comments and Bibliographic References

Introduction

§ 1

This section contains well-known facts. Special attention is given to vague convergence of measures and signed measures and to the properties of convolution

1. The proof of Theorem 0.1 can be found in P. Halmos [1], pp. 121–122.

2. For more details on the various topologies in a measure space, see the article by H. Cartan [6] and J. Dieudonné [1].

Theorem 0.2, in greater generality (for any locally compact space) is proved in the book by P. Halmos [1], pp. 240 and 242.

3. Theorem 0.5 and 0.6 are multidimensional analogues of the well-known Helly theorems.

For integra'ion of measures, cf. N. Bourbaki [2].

4. Lemma 0.3 is due to J. Deny [7].

§ 2

A more extensive knowledge of distribution theory can be obtained from the books of L. Schwartz [1] and of I. M. Gel'fand and G. E. Shilov [1, 2].

Our exposition treats in greater detail than usual the questions surrounding convolution of distributions (No. 8 and 9). These details turn out to be essential in what follows (cf. Ch. I, § 6).

6–7. All the theorems in these sections are due to L. Schwartz [1].

§ 3

The results of this section in the main are also due to L. Schwartz [1].

13. Theorem 0.12 is a generalization of a theorem of I. M. Gel'fand and G. E. Shilov; cf. [2], p. 180.

Theorem 0.13 is, perhaps, new.

Chapter I

§ 1

The kernels $k_\alpha(x)$ and their corresponding potentials are connected with the name of M. Riesz since in the memoir [1], he obtained the most important results concerning potentials and superharmonic functions of order $\alpha (0 < \alpha < 2)$. It should be mentioned, however, that before M. Riesz these kernels were studied by O. Frostman [1].

1. The direct computation of the Fourier transform carried out in the text shows that the assumption $0 < \operatorname{Re}\alpha < \frac{p+1}{2}$ is necessary for the convergence of the integral $g(y)$. This eliminates *a* slight imprecision in I. M. Gel'fand and G. E. Shilov [1], p. 222.

The proof of the composition formula (1.1.3) using the Fourier transform is given by J. Deny [4]. However, it was not sufficiently rigorous, since instead of referring to Theorem 0.13, it contained a reference to an unformulated "theorem on convolution".

2. The method of analytic continuation used for the definition of $k_\alpha(x)$ is due to M. Riesz [3]; cf. also I. M. Gel'fand and G. E. Shilov [1], pp. 110–114. We have applied it here to justify the formula for composition.

The role of the composition formula in a more general potential theory was pointed out by H. Cartan [1].

3. Formula (1.1.15) is due to M. Riesz [1]. The succeeding formulas are rather simple consequences of it.

§ 2

In a more classical form, the theory of super-(more precisely, sub-harmonic) functions is presented in the books by I. I. Privalov [1] and T. Rado [1].

The corresponding "coarse" theory, in which measures, not point functions are considered, was basically expounded by E. Szpilrajn [1], and then developed by J. Deny [7].

The axiomatic theory of superharmonic functions is due to M. Brelot [32, 29]; cf. also R.-M. Hervé [2].

4–5. In the principle of the harmonic minorant it is not necessary that $h(x)$ be continuous on ∂G. If this is not assumed, then the boundary condition should be written in the form

$$\varliminf_{x \to s \in \partial G} [f(x) - h(x)] \geq 0.$$

The same remark applies also to Theorem 1.1.

§ 3

6. The distinction between "coarse" and "thin" potential theory was emphasized by J. Deny [1]. The terminology is due to him.

7. M. Riesz [1] began the study of potentials of signed measures with non-compact supports. He, however, did not use the finiteness condition (1.3.10).

The proof of Theorem 1.3 given in the text is in a sense "elementary". A more general approach to this theorem is taken, for example, by M. Brelot [31].

Lemma 1.0 is the "coarse" variant of a generalized theorem of G. C. Evans pointed out by M. Riesz [1]. The proof given here is apparently new.

8. The elementary proof of Theorem 1.6 (maximum principle for a logarithmic potential), which we use, is due to M. Tsuji [3]. The theorem itself was first proved by Maria [1]. Theorem 1.6′ is an easy corollary.

Theorem 1.7 is a theorem of F. Vasilesco [1]. Cf. also G. C. Evans [4], vol. I.

Theorems 1.8 and 1.9 which lead to a simple proof of the maximum principle, are due to G. Choquet [3, 5]. But the role of Lusin's theorem in potential theory was noted earlier by Y. Yosida [1].

9. Theorem 1.11 and inequality (1.3.13) are due to O. Frostman [1].

In connection with the uniqueness theorem 1.12 and its diverse generalizations cf. Vallée-Poussin [2, 4]. We point out one generalization of the uniqueness theorem due to M. Riesz [1]:

If $\alpha < p$ is not even, then if the equation $\overset{\nu}{U}_\alpha(x) \equiv 0$ holds in one connected component of the set $CS(\nu)$, then $\nu \equiv 0$.

Lemma 1.1 enables the simplification of proofs of several theorems, in particular Theorem 1.12.

Theorem 1.13 is apparently due to Vallée-Poussin [5].

§ 4

The importance of the positive definiteness of the kernel for potential theory was already understood by Vallée-Poussin [2, 3] and O. Frostman [1]. However, the fundamental role of the concept of energy and the corresponding pre-Hilbert space structure in the space of signed measures and measures was revealed in the work of H. Cartan [1, 5] and then J. Deny [1, 4, 5].

11. The proof of the positive definiteness of M. Riesz kernels (Theorem 1.15) based on the composition formula, is due to H. Cartan [1]. We have extended it to the case of a logarithmic potential (Theorem 1.16).

12. Lemma 1.2 is due to H. Cartan [5]; the proof given differs from the original one.

Theorem 1.17 in the classical case $\alpha = 2$ was proved by H. Cartan [5]. The general case is considered here for the first time. It is technically more complicated.

13. Lemma 1.4, central to the proof of completeness of $\mathscr{E}_\alpha^+$ (Theorem 1.18) was proved in the classical case $\alpha = 2$ by H. Cartan [5]. Its role in more general theories is clarified by B. Fuglede [1]. The proof, applicable for any α and not relying on 1.17 is in M. Ohtsuka [7], p. 192.

Theorem 1.18 is a fundamental result of H. Cartan [5]. Theorem 1.19 is also due to him.

14. Theorem 1.20 is due to G. C. Evans [4], vol. I. The proof presented is a modernization of the original.

For all the properties of Green lines formuated in the text, cf. the memoir of M. Brelot and G. Choquet [1], Ch. IV.

15. The space BL was studied by O. Nikodym [1]. Theorem 1.21 may be new.

§ 5

The basic results of this section are Theorem 1.22, 1.23, and 1.26. The first of these is a classical theorem of F. Riesz [1], II, the rest are due to H. Cartan [5].

16 The proof here of Theorem 1.22 is due essentially to L. Schwartz [1], vol. II, p. 76.

17. The proof of Theorem 1.23 is due to H. Cartan [5]. It was only been slightly sharpened (Lemma 1.8).

18. The content of this section is also due to H. Cartan [5].

§ 6

The content of this section is a development of the ideas of M. Riesz [1]. Essential here is the utilization of the ideas of distribution theory.

19. Averaging with the help of the function $\varepsilon^{(r)}_\alpha(x)$ was introduced by M. Riesz [1].

20. Superharmonic functions of fractional order (in the text, "α-superharmonic functions") were considered by M. Riesz [1] and by O. Frostman [4]. The notion of α-harmonicity is introduced here possibly for the first time. Theorem 1.28 is also new.

21. The content of this section is new.

22. The basic Theorem 1.30 is due to M. Riesz [1]. The proof of this theorem is only sketched by M. Riesz. We have used some of his considerations. However, at the basis of our proof is another idea. It is the development of an idea of L. Schwartz applied by him in the proof of a theorem of F. Riesz (Theorem 1.22).

23. M. Riesz introduced the generalized Poisson kernels $\check{P}_r(y, x)$, $\hat{P}_r(y, x)$ Lemma 1.13 is new.

24. Lemmas 1.14 and 1.15 can be found in M. Riesz [1] (without proof).

25. Theorem 1.31 in the case of a Newtonian potential is due to H. Cartan [5].

Chapter II

§ 1

1. The definition of the equilibrium measure of a compact set by means of minimizing the energy integral is due to O. Frostman [1], who used only vague convergence of measures. The introduction of a Hilbert metric and the "geometrization" of the problem is due to H. Cartan [5].

2. For brevity we have introduced the term "C-absolutely continuous signed measure".

3. The term "approximately everywhere" corresponds to the French "à peu près partout".

The properties of equilibrium measure were studied by O. Frostman [1]. Theorem 2.3 is also known, but it is difficult to point out its source.

4. Theorem 2.4 is due to O. Frostman [1].

5. The convexity property of capacity was discovered by G. Choquet [1].

§ 2

7. Theorem 2.6 was proved by Vallée-Poussin [5].

8. Theorem 2.7 is due to B. Fuglede [1].

10. The basic Theorem 2.8 on capacitability is due to G. Choquet [1]. He, however, used the maximum principle (cf. end of No. 9) and therefore his proof was applicable only in the case $\alpha \leq 2$. The first to eliminate this constraint were N. Aronszajn and K. T. Smith [1]. Our exposition basically follows the paper of B. Fuglede [1].

The existence of non-capacitable sets follows from the that R^p can be decomposed into two completely imperfect sets A and B, i.e. sets which do not contain any perfect subset (cf. F. Hausdorff [1]). Indeed, any compact set contained in A or in B is countable, and hence

$$\underline{C}_\alpha(A) = \underline{C}_\alpha(B) = 0.$$

Moreover, the outer capacity of one of the sets A, B is infinite.

§ 3

11. Theorem 2.9, after we establish the connection between α-capacity and the transfinite diameter (cf. No. 12), becomes trivial. Nevertheless we think that the direct proof is of some interest.

12—13. The generalized transfinite diameter is introduced by G. Polya and G. Szegö [1].

For more details on the set of points $\{\xi_i^{(n)}\}$, the so called extremal points of a compact set, cf. papers of F. Leja [1, 2], J. Górski [1–7], and J. Górski and J. Siciak [1].

14. The computations, results of which are presented in the table, can be found in the book of N. N. Lebedev, I. P. Skal'skaya, Ya. S. Uflyand [1]. It should be noted that due to our normalization of the kernel $k_2(x)$, the results of the table differ from those in the above book by a factor of $A(3,2) = \pi$.

§ 4

15. A more detailed analysis of the cases $C_2(K) > 0$, $C_2(K) < 0$, $C_2(K) = 0$ can be found in the book of Vallée-Poussin [5].

16. The theorem on capacitability for logarithmic capacity was obtained by G. Choquet [6].

17. The proofs of the results in the table are in the book of G. Polya and G. Szegö [3].

18. For the relation between Green capacity and the modulus of a doubly-connected region, cf. M. Tsuji [2].

The bound (2.4.6) is new.

The Green capacity introduced in this section does not really differ from the condenser capacity introduced in G. Polya and G. Szegö [3].

Chapter III

§ 1

1. Theorem 3.1 for the case of a Newtonian potential was proved by J. Deny [2]. The proof in the general case is based on the same idea, but involves more complicated techniques.

For the case of a compact set Theorem 3.1 was proved earlier by G. C. Evans [5]; the construction of the desired potential using atomic measures is due to him.

In the general case of an arbitrary G_δ set E of null capacity, G. Choquet [11] in the Robin case ($0 < \alpha \leq 2$) succeeded in constructing a measure λ concentrated on E. For $\alpha > 2$ the question of existence of such a measure remains open.

2. Theorem 3.2 is due to R. Nevanlinna [1], (the case of a logarithmic potential).

§ 2

3–4. The first proof we present of Theorem 3.6 is due to H. Cartan [5] (for the Newtonian case), the second is due to G. Choquet [5].

§ 3

5. Theorem 3.8 is due to M. Brelot [2], where the condition "quasi-everywhere" is replace by "approximately everywhere". The most general formulation is due to M. Brelot and G. Choquet [2].

6. Theorems 3.9 and 3.10 in the case $\alpha = 2$ are due to H. Cartan [5]. The case of Green potentials is considered in M. Brelot and G. Choquet [2].

§ 4

9. Theorem 3.13 in the case of a compact set was proved by O. Frostman [1]. His proof differs from the proof in the text.

Lemma 3.2, which is the development of an idea of L. Ahlfors [1], is established in a paper by the author (cf. N. S. Landkof [8]).

As shown by L. Carleson [1], Theorem 3.13 does not admit a converse. He constructed a linear compact set with positive logarithmic capacity, whose h-measure is equal to zero for any function $h(r)$ which satisfies the condition

$$\int_0 \frac{h(r)}{r}\, dr < \infty.$$

11. Bounds of this type, as in Theorem 3.15, arise in H. Cartan [1′].

The reasoning by H. Cartan relating to the case of non-atomic measures, is incomplete, as pointed out by L. Ahlfors [1]. For the case of the plane, L. Ahlfors has given a complete proof. In the general case cf. N. S. Landkof [8].

Chapter IV

§ 1

The concept of an associated measure and the approach to the solution of the balayage problem, indicated in the text, was set forth in a seminar at the Institute of Mathematics and Mechanics of Harkov State University in 1948 and then in lectures on potential theory given at Harkov State University in 1954–56. Cf. also N. S. Landkof [9].

§ 2

The approximation technique used here arises with N. Wiener [2].

All the results for the case of the plane can be found in the book by Vallée-Poussin [5].

9. Theorem 4.7 is due to G. Bouligand [2], p. 39.

10. Theorem 4.9 was first proved (in another way) by O. Frostman [3].

11. The construction of the Green measure for an irregular point was given by O. Frostman [5] and, independently, by the author (N. S. Landkof [1]). O. Frostman [5] also proved that in formula (4.2.15) the parameter m can in fact be any number in the segment [0, 1].

§ 3

12. Formula (4.3.1), which gives the solution of the generalized Dirichlet problem, is in Vallée-Poussin [2]. Due to him also is the criterion for irregularity in terms of the equilibrium potential.

13. The solvability conditions, including the existence of a countable "solvability" set $D_{\partial G}$, were first established by M. V. Keldysh [1, 2]. Cf. also N. S. Landkof [1, 4]. The notion of a quasi-isolated irregular point and the proof that every solvability set contains all the quasi-isolated points were given by the author (cf. N. S. Landkof [1] and especially [4]).

14. The formula of M. Riesz is obtained in [1].

§ 4

The results of this section are due to the author (cf. N. S. Landkof [9]). An exception is Lemma 4.1 earlier proved by J. Deny [3].

16. A theorem similar to Theorem 4.13 was proved by M. Brelot [33].

§ 5

19. The Kelvin transformation is systematically exploited in the theory of balayage by M. Riesz [1]. Cf. also the older paper of G. Bouligand [1] where inversion is used. Lemma 4.3 is possibly new.

20. The Green measure for the interior and exterior of a sphere and the generalized Harnack inequality were obtained by M. Riesz [1].

21. The results of this section are new.

22. The results are due to H. Cartan [5] who confined himself to the case of a Newtonian potential.

23. The utilization of a variational problem for the Gauss integral was begun by O. Frostman [1, 2] and continued by Vallée-Poussin [2]. For the latest developments of this idea cf. N. Ninomiya [9, 10] and M. Ohtsuka [7].

Theorem 4.16 (in a more general form, but for Newtonian potentials) is due to H. Cartan [6].

§ 6

The results of this section are due to H. Cartan (cf. H. Cartan [6]). However, we prove them in a somewhat different fashion.

Chapter V

§ 1

Our exposition is characterized by the systematic use of the connection between the irregularity of a point and the existence of an equilibrium measure on an unbounded set.

1. Theorem 5.1 is new.

2. Properties (1) – (15) are established from different considerations by H. Cartan [6].

3. Theorem 5.2 was proved by N. Wiener [2]. The example is also his. The integral form of the N. Wiener criterion is due to O. D. Kellog and F. Vasilesco [1].

4. Theorem 5.4 is due to M. V. Keldysh [2]. The proof has been somewhat improved. Theorem 5.5 was proved by the author (cf. N. S. Landkof [6]); the proof here has been sharpened.

6. The theory of balayage is presented in detail for a logarithmic potential in the book of Vallée-Poussin [5].

§ 2

7. The representation is based on a note by N. S. Landkof [3].

8. The results of this section are new.

§ 3

The notion of thinness and Theorem 5.10 is due to M. Brelot [9]. The fine topology was introduced by H. Cartan [6].

Later development of these ideas is due to M. Brelot [34]; cf. also M. Brelot [31, 32].

§ 4

This section is based on the work of the author (cf. N. S. Landkof [6]).

12. Example 2 is due to Vallée-Poussin [4].

13. Example 5 was communicated to the author by M. A. Lavrentiev. Lemma 5.6 is due to O. D. Kellogg and F. Vasilesco [1].

§ 5

The basic results of this section are due to M. V. Keldysh and M. A. Lavrentiev [1, 2, 3, 4] and M. V. Keldysh [1, 2].

16. Theorem 5.14 is due essentially to N. Wiener [1]. The proof has been simplified by using the properties of the balayage operator.

17—18. Theorems 5.15, 5.16, 5.17, and Lemma 5.10 are due to M. V. Keldysh [2]. The proofs given here differ from the original. In the paper [2] M. V. Keldysh constructed an example of a region G whose boundary ∂G is a simple, closed Jordan surface and $(CG)_I = \emptyset$ but $(C\overline{G})_I$ has positive harmonic and null two-dimensional measure.

There also is proved the following stability criterion: If ∂G consists of a finite number of Jordan surfaces, then every solvable Dirichlet problem which is stable in G will be stable in $\overline{G}$.

19. Theorem 5.18 was proved in the work of J. Deny [3]; cf. also the papers of N. S. Landkof [5, 7]; an earlier, essentially equivalent result (for the case of a continuum not separating the space) was obtained by M. V. Keldysh [1].

Theorems 5.19 and 5.20 were first proved (by another method) by M. V. Keldysh and M. A. Lavrentiev [4].

20. Theorem 5.21 and the proof presented are due to A. A. Grončar [1]. Addition: Recently A. A. Grončar sharpened his result, proving that $c(x_0, \varrho) = \varrho$ for almost all $x_0 \in K$. Cf. DAN SSSR **165** No. 3, 479—481 (1965).

21—22. The results of this section are taken from the author's paper (cf. N. S. Landkof [7]).

Bibliography*

Ahiezer, N. I.

1. Elementy teorii elliptičeskih funkcii. M., Gostehizdat, 1948. (Elements of the theory of elliptic functions.)
2. Klassičeskaya problema momentov. M., Fizmatgiz, 1961. (The classical moment problem.)
   The classical moment problem and some related questions in analysis, Edinburgh: Oliver & Boyd 1965.

Ahlfors, L.

1. Ein Satz von H. Cartan und seine Anwendung auf die Theorie der meromorphen Funktionen. Soc. sci. Fenn. Comment. Phys.-math. **5**, No. 16 (1931).

Aleksandrov, A. D.

1. Additive set functions in abstract spaces. Matem. sb. **13**, 169–238 (1943).

Anger, G.

1. Stetige Potentiale und deren Verwendung für einen Neuaufbau der Potentialtheorie. Thesis, Dresden, 1957, 62.
2. Sur la rôle des potentiels continus dans les fondaments de la théorie du potentiel. Sém. Théorie du potent., 1958, Paris, No. 3, 30.
3. Ein funktionalanalytischer Aufbau der Potentialtheorie. Wiss. Z. Techn. Hochschule Dresden 8, **4** (1958–1959).

Aronszajn, N., Smith, K. T.

1. Functional spaces and functional completion. Ann. Inst. Fourier **6**, 125–185 (1955–1956).

Arsove, M. G.

1. Functions representable as differences of subharmonic functions. Trans. Amer. Math. Soc. **75**, 327–365 (1953).
2. Functions of potential type, ibid., 526–551.

Bader, R.

1. La théorie du potentiel sur une surface de Riemann. C. R. Acad. Sci. Paris, **228**, 2001–2002 (1949).

Banach, S.

1. Théorie des opérations linéaires. Monogr. Mathem. No. 1, Warszawa 1932.

Bauer, H.

1. Un problème de Dirichlet pour la frontière de Šilov d'un espace compact. C. R. Acad. Sci. Paris, **247**, 843–846 (1958).
2. Frontière de Šilov et problème de Dirichlet. Sém. Théor. potent., 1958–1959, t. 3.
3. Šilovscher Rand und Dirichletscher Problem. Ann. Inst. Fourier **11**, 89–136 (1961).

---

* This bibliography also includes some references not cited in the text of the book.

Beurling, A., Deny, J.

1. Espaces de Dirichlet, I: Le cas élémentaire. Acta Math. **99**, 203—224 (1958).

Bochner, S.

1. Lectures on Fourier integrals, Princeton: Princeton University Press 1959.

Bouligand, G.

1. Sur les fonctions harmoniques et bornées dans un domaine infini, nulles sur la frontière. C. R. Acad. Sci. Paris, **169**, 763 (1919).
2. Fonctions harmoniques. Principes de Picard et de Dirichlet. Memor. Sci. Math. **11** (1926).

Bourbaki, N.

1. Intégration. Actual. Sci. Ind., No. 1175, Paris, 1952.
2. Intégration des mésures Actual. Sci. Ind., No. 1244, Paris, 1957.

Brelot, M.

1. Etudes des fonctions sousharmoniques au voisinage d'un point. Actual. Sci. Ind., **139** (1934), Paris.
2. Sur le potentiel et les suites de fonctions sousharmonique. C. R. Acad. Sci. Paris, **207**, 836 (1938).
3. Fonctions sousharmoniques et balayage. Bull. classe Sci., Bruxelles, 5 sér., **24**, 301—312; 421—436 (1938).
4. Sur le problème de Dirichlet et les fonctions sousharmoniques. C. R. Acad. Sci. Paris, **206**, 1161 (1938).
5. Sur un balayage d'ensembles fermes. C. R. Acad. Sci. Paris, **207**, 1157 (1938).
6. Problème de Dirichlet et majorantes harmoniques. Bull. Sci. Math. **63** (1939).
7. Critères de regularité et de stabilité. Bull. classe Sci., Bruxelles **25** (1939).
8. Familles de Perron et problème de Dirichlet. Acta scient. Szeged **9**, 133—153 (1939).
9. Points irréguliers et transformations continues en théorie du potentiel. Journ. math. pures et appl. **19**, 319—337 (1940).
10. Sur la théorie autonome des fonctions sousharmoniques. Bull. Sci. Math. **65**, 72—98 (1941).
11. Sur la rôle du point à l'infini dans la théorie des fonctions harmoniques. Ann. Ecole Norm. Sup. **61**, 301—322 (1944).
12. Sur les ensembles effillés. Bull. Sci. Math. **68**, 12—36 (1944).
13. Minorantes sousharmoniques, extrémales et capacités. J. math. pures et appl. **24**, 1—32 (1945).
14. Fonctions sousharmoniques, presques sousharmoniques ou sous-médianes. Ann. Univ. Grenoble **21**, 75—90 (1945).
15. Sur la mesure harmonique et le problème de Dirichlet. Bull. Sci. math. **69**, 153—156 (1945).
16. Sur le problème de Dirichlet. C. R. Acad. Sci. Paris **221**, 654—656 (1945).
17. Sur le problème de Dirichlet ramifié et la représentation conforme. C. R. Acad. Sci. Paris **222**, 851—852 (1946).
18. Étude générale des fonctions harmoniques ou sousharmoniques positive au voisinage d'un point frontière irrégulier. Ann. Univ. Grenoble **22**, 205—219 (1964).
19. Le problème de Dirichlet ramifié. ibid., 167—200.
20. Sur le principe des singularités positives et la topologie de R. S. Martin. ibid. **23**, 113—138 (1948).

21. Sur l'allure des fonctions harmoniques et sousharmoniques à la frontière. Math. Nachr. **4**, 298–307 (1950).
22. La théorie moderne du potentiel. Ann. Inst. Fourier **4**, 113–140 (1952).
23. Etude et extension du principe de Dirichlet. ibid., **5** (1953–1954).
24. A new proof of the fundamental theorem of Kellogg-Evans on the set of irregular points in the Dirichlet problem. Rend. Circ. Mat. Palermo **4**, 112–122 (1955).
25. Nouvelle démonstration du théorème fondamental sur la convergence des potentiels. Ann. Inst. Fourier **6**, 361–368 (1955–1956).
26. Le problème de Dirichlet. Axiomatique et frontière de Martin. J. math. pures et appl. **35**, 297–335 (1956).
27. Un axiomatique des fonctions sousharmoniques. C. R. Acad. Sci. Paris **245**, 1688–1690 (1957); **246**, 2334–2337 (1958).
27'. Un axiomatique générale du problème de Dirichlet dans les espaces localement compacts. Sém. Théor. potent., 1957, t. I, No. 6, 16.
28. Extension axiomatiques des fonctions sousharmoniques. C. R. Acad. Sci. Paris, **245**, 1688–1690 (1957); **246**, 2334–2337 (1958).
29. Axiomatique des fonctions harmoniques et sousharmoniques dans un espace localement compact. Sém. Théor. potent., 1958, t. 2, No. 1.
30. La convergence des fonctions surharmoniques et des potentiels généralisés. C. R. Acad. Sci. Paris, **246**, 2709–2712 (1958).
31. Eléments de la théorie classique du potentiel. Les cours de Sorbonne. Paris, 1959.
32. Lectures on potential theory. Tata Inst., Bombay, 1960.
33. Sur un théorème de prolongement fonctionel de M. Keldysh concernant le problème de Dirichlet. J. Analyse math. **8**, 273–283 (1960/61).
34. Introduction axiomatique de l'effilement. Annali mat. pura appl. **57**, 77–95 (1962).

Brelot, M., Choquet, G.
1. Espaces et lignes de Green. Ann. Inst. Fourier **3**, 199–263 (1952).
2. Le théorème de convergence en théorie du potentiel. J. Madras Univ. **B 27**, 277–286 (1957).

Carleson, L.
1. On the connection between Hausdorff measures and capacity. Ark. f. Mat. **3**, 403–406 (1958).

Cartan, H.
1. Sur les fondaments de la théorie du potentiel. Bull Soc. Math. France **69**, 71–96 (1941).
1'. Sur les systèmes des fonctions holomorphes à variétés linéaires et leurs applications. Ann. École Norm. Sup. **3**, 255–346 (1928).
2. Capacité extérieure et suites convergentes de potentiels. C. R. Acad. Sci. Paris **214**, 944–946 (1942).
3. Sur les suites de potentiels de masses ponctuelles. ibid., **214**, 994-996 (1942).
4. La théorie geénéral du potentiel dans les espace homogènes. Bull. Sci. Math. **66**, 126–132, 136–144 (1942).
5. Théorie du potentiel newtonien: énergie, capacité, suites de potentiels. Bull. Soc. Math. France **73**, 74–106 (1945).
6. Théorie générale du balayage en potentiel newtonien. Ann. Univ. Grenoble **22**, 221–280 (1946).

Cartan, H., Deny, J.
1. Le principe du maximum en théorie du potentiel et la notion de fonction surharmonique. Acta scient. Szeged **12A**, 81–100 (1950).

Choquet, G.

1. Theory of capacities. Ann. Inst. Fourier **5**, 131—295 (1955).
2. Les royaux réguliers en théorie du potentiel. C. R. Acad. Sci. Paris, **243**, 635—638 (1956).
3. Sur les fondaments de la theorie fine du potentiel. ibid., **244**, 1606—1609 (1957).
4. Potentiels sur un ensembles de capacité nulle. Suites de potentiels. ibid., 1707—1710.
5. Sur les fondaments de la théorie fine du potentiel. Sém. Théorie potent., 1957, Paris, No. 1.
6. Capacibilité en potentiel logarithmique. Bull. classe Sci., Bruxelles, **44**, 321—326 (1958).
7. L'integrale d'énergie en théorie du potentiel. Sém. Théorie potent., 1958—1959, Paris, No. 3.
8. Ensembles K-analytiques et K-sousliniens. Cas général et cas métrique. Ann. Inst. Fourier **9**, 75—83 (1959).
9. Forme abstraite du théorème de capacitabilité. ibid., 83—89.
10. Sur les points d'effilement d'un ensemble. Application à l'étude de la capacité. ibid., 91—102.
11. Sur les G_δ de capacité nulle. ibid., 103—110.

Choquet, G., Deny, J.

1. Aspects lineaires de la théorie du potential, I. Etudes des modeles finis. C. R. Acad. Sci. Paris, **242**, 222—225 (1956).
2. Aspects linéaires de la théorie du potentiel, II. Théorème de dualité et applications. ibid., **243**, 764—767 (1956).
3. Modéles finis en théorie du potentiel. J. Anal. Math. **5**, 77—135 (1956/1957).
4. Aspects linéaires de la théorie du potentiel. Noyaux de composition satisfaisant au principe du balayage sur tout ouvert. C. R. Acad. Sci. Paris, **250**, 4260—4262 (1960).

Courant, R.

1. Partial differential equations, New York-London 1962.

Courant, R., Hilbert, D.

1. Methods of mathematical physics, Vol. 1, New York-London: Interscience Publishers 1953.

Day, M. M.

1. Normed linear spaces, Berlin-Göttingen-Heidelberg: Springer 1958.

Deny, J.

1. Sur l'espace des distributions d'énergie finie et un théorème de H. Cartan. C. R. Acad. Sci. Paris **222**, 1374—1376 (1946).
2. Sur les infinis d'un potentiel. ibid., **224**, 524—525 (1947).
3. Systèmes totaux de fonctions harmoniques. Ann. Inst. Fourier, **1**, 103—112 (1949).
4. Les potentiels d'énergie finie. Acta Math. **82**, 107—183 (1950).
5. Sur la définition de l'énergie en théorie du potentiel. Ann. Inst. Fourier **2**, 83—99 (1950).
6. Le balayage. Meddel. Lunds. Univ. Mat. Sem., tome dedié à M. Riesz, 47—61 (1952).
7. Familles fondamentaux. Noyaux associés. Ann. Inst. Fourier **3**, 73—101 (1952).
8. Les deux aspects de la théorie du potentiel. Sém. Bourbaki (1957), No. 148, 18.

9. Sur les espaces de Dirichlet. Sém. Théorie potent. (1957), No. 5, 14.
10. Formes et espaces de Dirichlet. Sém. Bourbaki (1959—1960), No. 187, 11.
11. Noyaux de convolution de Hunt de nouveau associés à une famille fondamentale. Ann. Inst. Fourier **12**, 643—667 (1962).

Deny, J., Lions, J. L.
1. Les espaces du type de Beppo Levi. Ann. Inst. Fourier **5**, 305—370 (1953/1954).

Dieudonné, J.
1. Sur la convergence des suites de mesures de Radon. An. Acad. Brasil Ci. 23, 21—38, 277—282 (1961).

Edwards, R. E.
1. Cartan's balayage theory for hyperbolic Riemann surfaces. Ann. Inst. Fourier **8**, 263—272 (1958).

Erdös, Gillis
1. Note on the transfinite diameter. Journ. Lond. Math. Soc. **12**, 185—192 (1937).

Evans, G. C.
1. Fundamental points of potential theory. Rice Inst. Pamphl., VII, 252—329 (1920).
2. The logarithmic potential. Discontinuous Dirichlet and Neuman problems. N. J., 1927.
3. Applications of Poincaré's sweeping-out process. Proc. Nat. Acad. Sci. USA **19**, 457—461 (1933).
4. On potential of positive mass, I. Trans. Amer. Math. Soc. **37**, 226—253 (1935); II, **38**, 201—236 (1936).
5. Potentials and positively infinite singularities of harmonic functions. Monatsh. f. Math. u. Phys. **43** (1936).
6. Modern methods of analysis in potential theory. Bull. Amer. Math. Soc. **43**, 481—502 (1937).

Frostman, O.
1. Potentiel d'équilibre et capacité des ensembles avec quelques applications à la théorie des fonctions. Meddel. Lunds Univ. Mat. Sem. **3** (1935).
2. La méthode de variation de Gauss et les fonctions sousharmoniques. Acta Szeged. **8**, 149—159 (1937).
3. Sur le balayage des masse. Acta Szeged **9** (1938).
4. Sur les fonctions surharmoniques d'ordre fractionnaire. Ark. f. Mat., Astr. Fys., **26A**, No. 16 (1939).
5. Les points irréguliers dans la théorie du potential et le critère de Wiener. Meddel. Lunds. Univ. Mat. Sem. **4** (1939).
6. Distributions de masses normées par la métrique de L^p. Meddel. Lunds. Univ. Mat. Sem., tome dédié a M. Riesz, 90—100 (1952).

Fuglede, B.
1. On the theory of potentials in locally compact spaces. Acta Math. **103**, 139—215 (1960).
2. Capacitabilité des ensembles souslinies en théorie du potentiel. Ann. Inst. Fourier **12** (1962).

Gel'fand, I. M., Šilov, G. E.
1. Obobščennye funcii i deĭstviya nad nimi. M., Fizmatgiz., 1958. (Generalized functions and operations on them.)
2. Prostranstva osnovnnyh i obobščennyh funkciĭ. M., Fizmatgiz, 1958. (Spaces of fundamental and generalized functions.)

Goluzin, G. M.

1. Geometričeskaya teoriya funkciĭ kompleksnogo peremennogo. M., Gostehizdat, 1952. (The geometric theory of functions of a complex variable.)

Gončar, A. A.

1. O ravnomernom približenii nepreryvnyh funkciĭ garmoničeskimi. Izv. AN SSSR, ser. matem. **27**, 1239—1250 (1963). (On the uniform approximation of continuous functions by harmonic functions.)

Górski, J.

1. Méthode des points extrémaux de résolution du problème de Dirichlet dans l'espace. Ann. Pol. Math. **1**, 418—429 (1955).
2. Sur certaines propriétés des points extremaux liés à un domaine plane. ibid. **3**, 32—36 (1956).
3. Sur la représentation conforme d'un domaine multiplement connexe. ibid. **3**, 218—224 (1957).
4. Les suites de points extremaux liés aux ensemble dans l'espace à 3 dimensions. ibid. **4**, 14—20 (1957).
5. Distributions restreint des points extremaux liés aux ensemble dans l'espace. ibid. **4**, 325—339 (1958).
6. Une rémarque sur la méthode des points extremaux de F. Leja. ibid. **7**, 63—69 (1959).
7. The method of extremal points and Dirichlet's problem in the space of two complex variables. Arch. Rat. Mech. Anal. **4**, 412—427 (1960).

Górski, J., Siciak, J.

1. Certaines théorèmes concernant la répartition des points extrémaux dans les ensembles plans. Ann. Pol. Math. **4**, 21—29 (1957).

Goursat, E.

1. Cours d'analyse mathématique, t. III. 3e éd., Paris: Gauthier-Villars 1923.

Hällström, G.

1. Reguläre und irreguläre Randpunkte der Greenschen Funktion in der Ebene. Acta Soc. Sci. Fenn. (A), **3**, No. 5 (1944).

Halmos, P.

1. Measure theory, New York: Van Nostrand 1950.

Hausdorff, F.

1. Teoriya množestv. M., Gostehizdrat, 1937. (Set theory.)

Hervé, R.-M.

1. Développement sur une théorie axiomatique des fonctions surharmoniques. C. R. Acad. Sci. Paris **238**, 179—181 (1959).
2. Recherches axiomatiques sur la théorie des fonctions surharmoniques et du potentiel. Ann. Inst. Fourier **12**, 415—571 (1962).

Hervé, R.-M., Brelot, M.

1. Introduction de l'effillement dans une théorie axiomatique du potentiel. C. R. Acad. Sci. Paris **247**, 1956—1959 (1958).

Kametani, S.

1. Positive definite integral quadratic forms and generalized potentials. Proc. Imp. Acad. Tokyo **20**, 7—14 (1944).

Kantorovič, L. V., Akilov, G. P.

1. Funkcional'nyĭ analiz v normirovannyh prostranstvah. M., Fizmatgiz, 1959. (Functional Analysis and Normed Spaces.)

Keldysh, M. V.

1. O razrešimosti i ustoĭčivosti zadači Dirihle. DAN SSSR **18**, 315—318 (1938). (On the solvability and stability of the Dirichlet problem.)

2. O razrešimosti i ustoĭčivosti zadači Dirihle. UMN, st. ser. **8**, 171—292 (1941). (On the solvability and stability of the Dirichlet problem.)
3. O zadači Dirihle. DAN SSSR **32**, 308—309 (1941). (On the Dirichlet problem.)

Keldysh, M. V., Lavrent′ev, M. A.
1. Sur les suites des polynômes harmonique. C. R. Acad. Sci. Paris, **202**, 1149—1157 (1936).
2. Sur la problème de Dirichlet. ibid., **204**, 1788—1790 (1937).
3. Ob ustoĭčivosti rešeniya zadači Dirihle. Izv. AN SSSR, ser. matem., 551—593 (1937). (On the stability of the solution to the Dirichlet problem.)
4. Sur les suites convergentes de polynomes harmoniques. Trudy Matem. in-ta gruz. fil. AN SSSR **1**, 165—184 (1937).

Kellogg, O. D.
1. Foundations of potential theory, Berlin: Springer 1929.

Kellogg, O. D., Vasilesco, F.
1. A contribution to the theory of capacity. Amer. J. Math. **51** (1929).

Kishi, M.
1. On a theorem of Ugaheri. Proc. Jap. Acad. **32**, 314—319 (1956).
2. Interior limit of a sequence of potentials. ibid. **33**, 3 (1957).
3. Capacities of Borelian sets and the continuity of potentials. Nagoya Math. J. **12**, 195—219 (1957).
4. On the capacitability of analytic sets. Proc. Jap. Acad. **35**, 158—160 (1959).
5. On a topological property of a family of potentials. Jap. J. Math. **29**, 26—31 (1959).
6. A continuity theorem in the potential theory. Proc. Jap. Acad. **36**, 65—67 (1960).
7. Capacitability of analytic sets. Nagoya math. Journ. **16**, 91—109 (1960).
8. Unicity principles in the potential theory. Osaka Math. J. **13**, 41—74 (1961).

Krein, M. G.
1. O logarifme bezgranično razložimyh ermitovo položitel′nyh funkcii. DAN SSSR **25**, 99—102 (1940). (On the logarithm of infinitely decomposable Hermite positive functions.)

Kunugui, K.
1. Sur quelques points de la théorie du potentiel. (I) Proc. Jap. Acad. **21**, 234—239 (1945), (II) ibid. **23**, 95—100 (1947).
2. Étude sur la théorie du potentiel généralisé. Osaka Math. J. **2**, 63—103 (1950).

Landkof, N. S.
1. O strukture množestva irregulyarnyh toček v zadače Dirihle. DAN SSSR **28**, 291—293 (1940). (On the structure of the set of irregular points in the Dirichlet problem.)
2. O raspoloženii irregulyarnyh toček obobščennoĭ zadačĭ Dirihle. DAN SSSR **39**, 367—370 (1943). (On the distribution of irregular points in the generalized Dirichlet problem.)
3. O nekotoryh harakteristikah irregulyarnyh toček v zadače Dirihle. Matem. sb. **19**, 175—182 (1946). (On certain characteristics of irregular points in the Dirichlet problem.)
4. O razrešimosti obobščennoi zadači Dirihle. Izv. AN SSSR, ser. matem., **11**, 181—196 (1947). (On the solvability of the generalized Dirichlet problem.)

5. O plotnosti nekotoryh sistem garmoničeskih funkciĭ v prostranstve funkciĭ, nepreryvnyh na množestve. DAN SSSR **55**, 7—8 (1947). (On the density of certain systems of harmonic functions in the space of continuous functions on a set.)
6. Nekotorye novye svoĭstva množestva irregulyarnyh toček obobščennoi zadači Dirihle. Zapiski Har′k. in-ta matem. i HMO (4), **21**, 145—163 (1949). (Certain new properties of the set of irregular points in the generalized Dirichlet problem.)
7. Approksimaciya nepreryvnyh funkciĭ garmoničeskimi. Matem. sb. **25**, 95—106 (1949). (The approximation of continuous functions by harmonic functions.)
8. Emkosti i mery Hausdorfa. Ocenki potencialov. UMN **20**, 189—195 (1965.) (Capacities and Hausdorf measures. Evaluation of the potentials.)
9. Nekotorye primeneniya funkcional′nogo analiza v teorii garmoničeskih funkcii. Teoriya funkciĭ, funkcional′nyi analiz i ih priloženiya **1**, 182—193 (1965). (Certain applications of functional analysis in the theory of harmonic functions.)
10. Ob odnom tipe irregulyarnyh toček. Teoriya funkciĭ, funkcional′nyĭ analiz i ih priloženiya **2**, 138—141 (1966). (On irregular points of a certain type.)
11. O potencialah M. Rissa I, Zapiski meh.-matem. f-ta HGU i HMO **32**, 117—142 (1966). (On M. Riesz potentials I.)

Lebedev, N. N., Skal′skaya, I. P., Uflyand, Ya. S.

1. Sbornik zadač po matematičeskoi fizike. M., Gostehizdat, 1955. (A collection of problems in mathematical physics.)

Leja, F.

1. Distributions libres et restreintes des points extrémaux dans les ensembles plans. Ann. Pol. Math. **3**, 147—156 (1957).
2. Propriétés des points extrémaux des ensembles plans et leur application à la répresentation conforme, ibid., 319—342.
3. Sur certaines suites liés aux ensembles plans et leur application à la répresentation conforme, ibid. **4**, 8—13 (1957).

Lyubič, Yu. I.

1. O susčestvovanii „v bol′šom‘ fundamental′nyh rešenii lineinyh elliptičeskih uravnenii vtorogo poryadka. Matem. Sb. **57**, 45—58 (1962). (On the existence "in the large" of fundamental solutions to linear partial differential equations of second-order.)

Maria

1. The potential of a positive mass and the weight function of Wiener. Proc. Nat. Acad. USA **20**, 485—489 (1934).

Matsushita, S.

1. Laplacien local et la décomposition de F. Riesz. Proc. Jap. Acad. **32**, 436—440 (1956).
2. Generalized laplacian and balayage theory. J. Inst. Polytechn. Osaka City Univ. (A 8), **1**, 59—90 (1957).

Miranda, K.

1. Partial differential equations of elliptic type, 2nd Ed., Berlin-Heidelberg-New York: Springer 1970.

Monna, A. F.

1. Het problem van Dirichlet. Nieuw Arch. v. Wisk. **19** (1938).
2. On the Dirichlet problem and the method of sweeping-out. Proc. Kon. Nederl. Akad. Wet. **43**, 491—498 (1939).

3. Sur une généralisation du problème de Dirichlet pour ensembles bornés mesurable (B) quelconques. ibid., 745—752.
4. Sur la capacité des ensembles, ibid. **43**, 81—86 (1940).
5. Sur les notions de «points stable» et «points réguliers» dans le problème, de Dirichlet. ibid., 87—90.
6. Extension du problème de Dirichlet pour ensembles quelconques. ibid. 497—511.

Nevanlinna, R.
1. Eindeutige analytische Funktionen, Berlin: Springer 1936.

Nikodym, O.
1. Sur une classe des fonctions considerées dans l'étude du problème de Dirichlet. Fund. Math. **21**, 129—150 (1933).
2. Sur un théorème de S. Zaremba concernant les fonctions harmoniques. J. math. pures appl. **12**, 95—108 (1933).

Ninomiya, N.
1. Equilibrium potentials and energy integrals. Proc. Jap. Aad. **26**, 1—16 (1950).
2. Sur une suite convergente de distributions de masses et leurs potentiels correspondants. Math. Journ. Okayama Univ. **2**, 1—7 (1952).
3. Sur un ensemble de capacité nulle et l'infini d'un potentiel. ibid. **2**, 99—101 (1953).
4. Sur l'intégrale d'énergie dans la théorie du potentiel. J. Inst. Polytechn. Osaka City Univ. **5**, 97—100 (1954).
5. Une correction ... ibid. **6**, 79—82 (1955).
6. Sur le théorème du balayage et le théorème d'équilibre. ibid., 83—91.
7. Sur le principe de continuité dans la théorie du potentiel. ibid. **8**, 51—56 (1957).
8. Etude sur la théorie du potentiel pris par rapport au noyau symétrique. ibid., 147—179.
9. Méthode de variation du minimum dans la théorie du potentiel. Sém. Théor. potent. (1958—1959), No. 5, 9.
10. Sur le principe du maximum et le balayage. Japan J. Math. **29**, 68—77 (1959).

Ohtsuka, M.
1. Sur un espace complet de mesures positives dans la théorie du potentiel. Proc Jap. Acad. **32**, 311—313 (1956).
2. Sur un théorème de M. Kishi. ibid., 722—725.
3. Capacité d'ensemble produits. Nagoya Math. J. **12**, 95—130 (1957).
4. Les relations entre certaines principes en théorie du potentiel. Proc. Jap. Acad. **33**, 37—40 (1957).
5. On thin sets in potential theory. Sem. Anal. Func. Inst. Adv. Study, Princeton, 1957, **1**, 302—313.
6. Some examples in potential theory. Jap. J. Math. **29**, 101—110 (1959).
7. On potential in locally compact spaces. J. Sci. Hiroshima Univ., ser. **A 1**, 135—352 (1961).

Oleinik, O. A.
1. O zadače Dirihle dlya uravnenii elliptičeskogo tipa. Matem. sb. **24**, 3—14 (1949). (On the Dirichlet problem for elliptic equations.)

Pòlya, G., Szegö, G.
1. Über den transfiniten Durchmesser (Kapazitätskonstante) von ebenen und räumlichen Punktmengen. J. f. reine angew. Math. **165** (1931).

2. Inequalities for the capacity of a condenser. Amer. J. Math. **67**, 1–32 (1945).

3. Isoperimetric inequalities in mathematical physics, Princeton 1951.

Privalov, I. I.

1. Subgarmoničeskie funkcii. M., Gostehizdat, 1937. (Subharmonic functions.)

2. Graničnye zadači teorii garmoničeskih i subgarmoničeskih funkciĭ v prostranstve. Matem. sb. **3**, 3–25 (1938). (Boundary-value problems in the theory of harmonic and subharmonic functions in a space.)

3. Priloženiya ponyatiya garmoničeskoi mery k nekotorym zadačam teorii subgarmoničeskih funkcii. Matem. sb. **3**, 535–541 (1938). (Applications of the concept of a harmonic measure to certain problems in the theory of subharmonic functions.)

Radó, T.

1. Subharmonic functions. Ergebn. d. Math., Berlin 1937.

Riesz, F.

1. Sur les fonctions subharmoniques et leur rapport à la théorie du potentiel. I. Acta Math. **48**, 329–343 (1926), II. Acta Math. **54**, 321–360 (1930).

Riesz, F., Nagy, B.

1. Leçons d'analyse fontionelle, 2e éd., Budapest: Akadémiai Kiadó 1953.

Riesz, M.

1. Intégrales de Riemann–Liouville et potentiels. Acta Szeged **9**, 1–42 (1938).

2. Rectification an travail «Intégrales de ...» ibid. **9**, 116–118 (1939).

3. L'integrale de Riemann–Liouville et le problème de Cauchy. Acta Math. **81**, 1–233 (1949).

Schoenberg, J.

1. On certain metric spaces arising from Euclidean spaces by a change of metric and their imbedidng in Hilbert space. Ann. of Math. **38**, 787–793 (1937).

Schwartz, L.

1. Théorie des distributions, z. I, II, Paris, 1950–1951.

Smith, K. T.

1. Mean values and continuity of Riesz potentials. Comm. pure appl. math. **9**, 569–576 (1956).

Sonin, N. Ya.

1. Issledovaniya o cilindričeskih funkciyah i special'nyh polinomah, M., Gostehizdat, 1954. (Investigations of cylindrical functions and special polynomials.)

Szpilrajn, E.

1. Remarques sur les fonctions sousharmoniques. Ann. of Math. **34** (1933).

Szybiak, A.

1. On the density of the equilibrium distributions of plane sets. Ann. Pol. Math. **6**, 41–49 (1959).

Tautz, G.

1. Zur Theorie der ersten Randwertaufgabe. Math. Nachr. **2**, 279–303 (1949)

Thompson, J. M.

1. Distribution of mass for averages of Newtonian potential functions. Bull. Amer. Math. Soc. **41**, 744–752 (1935).

Toth, L. F.

1. Lagerung in der Ebene auf der Kugel und im Raum, 2. Aufl., Berlin-Heidelberg-New York: Springer 1971.

Tsuji, M.

1. A simple proof of Dirichlet principle. Journ. Math. Soc. Japan. **7**, 67–75 (1955).
2. On the modulus of a ring domain. Comment. Math. Univ. St. Pauli **4**, 1–3 (1955).
3. Potential theory in modern function theory, Tokyo 1959.

Ugaheri, T.

1. On the Newtonian capacity and the linear measure. Proc. Imp. Acad. Tokyo **18**, 602–605 (1942).
2. On the general potential and capacity. Jap. Journ. Math. **20**, 37–43 (1950).
3. On the general capacities and potentials. Bull. Tokyo Inst. Tech., ser. B, 149–179 (1953).

Ullmann, J. L.

1. Regularity criteria in potential theory. Lect. Funct. compl. var., Ann. Arbor Univ. Mich. Press 385–386 (1955).

Ursell, H. D.

1. Note on the transfinite diameter. J. Lond. Math. Soc. **13**, 34–37 (1938).

de la Vallée-Poussin, C.

1'. L'extension de la méthode du balayage. Annal. Inst. H. Poincaré **2**, 169–232 (1932).

1. Propriérés des fonctions harmoniques dans un domaine ouvert limité par des surfaces à courbure borné. Annal. R. Scuola Norm. Super. Pisa, ser. II, **2** (1933).
2. Les nouvelles méthodes de la théorie du potentiel et le problème généralisé de Dirichlet. Actual. Sci. Ind., No. 578, Paris, 1937.
3. Potentiel et problème généralisé de Dirichlet. Math. Gazette **22** (1938).
4. Points irréguliers. Détermination des masses par les potentiels. Bull. classe Sci., Bruxelles (5) **24**, 368–384, 672–689 (1938).
5. Le potentiel logarithmique, balayage et répresentation conforme, Paris: Louvian 1949.
6. Cours d'analyse infinitésimale, t. 1, Paris: Gauthier-Villars 1914.

Vasilesco, F.

1. Sur la continuité du potentiel à travers les masses. C. R. Acad. Sci. Paris **200**, 1173–1174 (1935).
2. La notion de point irregulier dans le problème de Dirichlet. Actual. Sci. Ind., No. 660, Paris, 1938.
3. Sur quelques formes plus générales des critères de régularité de Vallée-Poussin et Wiener. C. R. Acad. Sci. Paris **215**, 249–251 (1942).
4. Sur quelques critères nouveaux de stabilité. C. R. Acad. Sci. Paris, **215**, 296–297 (1942).
5. Sur quelques critères généraux de regularité et de stabilité. Ann. Sci. Ecole Norm. Sup. (3) **59**, 275–295 (1942).
6. Sur une notion nouvelle de capacité d'un ensemble. C. R. Acad. Sci. Paris **216**, 191–193 (1943).

Wiener, N.

1. Certain notions in potential theory. J. Math. Phys. Mass. Inst. Techn. **3**, 24–51 (1924).
2. The Dirichlet problem. ibid., 127–146.
3. Note on a paper of O. Perron. ibid. **4**, 21–32 (1925).

Yosida, Y.

1. Sur le principe du maximum dans la théorie du potentiel. Proc. Imp. Acad. Japan **17**, 476–478 (1941).

Zaremba, S.

1. Sur un problème mixte relatif à l'équation de Laplace. Bull. Acad. Sci. Cracovie (A), 313–344 (1910).

Index

DG 721/29/71

Die Grundlehren der mathematischen Wissenschaften in Einzeldarstellungen mit besonderer Berücksichtigung der Anwendungsgebiete

Eine Auswahl

---

67. Byrd/Friedman: Handbook of Elliptic Integrals for Engineers and Scientists. 2nd edition
68. Aumann: Reelle Funktionen. 2. Aufl.
74. Boerner: Darstellungen von Gruppen. 2. Aufl.
76. Tricomi: Vorlesungen über Orthogonalreihen. 2. Aufl.
77. Behnke/Sommer: Theorie der analytischen Funktionen einer komplexen Veränderlichen. Nachdruck der 3. Aufl.
78. Lorenzen: Einführung in die operative Logik und Mathematik. 2. Aufl.
86. Richter: Wahrscheinlichkeitstheorie. 2. Aufl.
87. van der Waerden: Mathematische Statistik. 3. Aufl.
94. Funk: Variationsrechnung und ihre Anwendung in Physik und Technik. 2. Aufl.
97. Greub: Linear Algebra. 3rd edition
99. Cassels: An Introduction to the Geometry of Numbers. 2nd printing
104. Chung: Markov Chains with Stationary Transition Probabilities. 2nd edition
107. Köthe: Topologische lineare Räume I. 2. Aufl.
114. MacLane: Homology. Reprint of 1st edition
116. Hörmander: Linear Partial Differential Operators. 3rd printing
117. O'Meara: Introduction to Quadratic Forms. 2nd printing
120. Collatz: Funktionalanalysis und numerische Mathematik. Nachdruck der 1. Aufl.

121./122. Dynkin: Markov Processes

123. Yosida: Functional Analysis. 3rd edition
124. Morgenstern: Einführung in die Wahrscheinlichkeitsrechnung und mathematische Statistik. 2. Aufl.
125. Itô/McKean jr.: Diffusion Processes and Their sample Paths
126. Lehto/Virtanen: Quasikonforme Abbildungen
127. Hermes: Enumerability, Decidability, Computability. 2nd edition
128. Braun/Koecher: Jordan-Algebren
129. Nikodým: The Mathematical Apparatus for Quantum Theories
130. Morrey jr.: Multiple Integrals in the Calculus of Variations
131. Hirzebruch: Topological Methods in Algebraic Geometry. 3rd edition
132. Kato: Perturbation Theory for Linear Operators
133. Haupt/Künneth: Geometrische Ordnungen
134. Huppert: Endliche Gruppen I
135. Handbook for Automatic Computation. Vol. 1/Part a: Rutishauser: Description of ALGOL 60
136. Greub: Multilinear Algebra
137. Handbook for Automatic Computation. Vol. 1/Part b: Grau/Hill/Langmaack: Translation of ALGOL 60
138. Hahn: Stability of Motion
139. Doetsch/Schäfke/Tietz: Mathematische Hilfsmittel des Ingenieurs. 1. Teil
140. Collatz/Nicolovius/Törnig: Mathematische Hilfsmittel des Ingenieurs. 2. Teil
141. Mathematische Hilfsmittel des Ingenieurs. 3. Teil
142. Mathematische Hilfsmittel des Ingenieurs. 4. Teil
143. Schur: Vorlesungen über Invariantentheorie
144. Weil: Basic Number Theory
145. Butzer/Berens: Semi-Groups of Operators and Approximation

146. Treves: Locally Convex Spaces and Linear Partial Differential Equations
147. Lamotke: Semisimpliziale algebraische Topologie
148. Chandrasekharan: Introduction to Analytic Number Theory
149. Sario/Oikawa: Capacity Functions
150. Iosifescu/Theodorescu: Random Processes and Learning
151. Mandl: Analytical Treatment of One-dimensional Markov Processes
152. Hewitt/Ross: Abstract Harmonic Analysis. Vol. 2: Structure and Analysis for Compact Groups, Analysis on Locally Compact Abelian Groups
153. Federer: Geometric Measure Theory
154. Singer: Bases in Banach Spaces I
155. Müller: Foundations of the Mathematical Theory of Electromagnetic Waves
156. van der Waerden: Mathematical Statistics
157. Prohorov/Rozanov: Probability Theory
159. Köthe: Topological Vector Spaces I
160. Agrest/Maksimov: Theory of Incomplete Cylindrical Functions and Their Applications
161. Bhatia/Szegö: Stability Theory of Dynamical Systems
162. Nevanlinna: Analytic Functions
163. Stoer/Witzgall: Convexity and Optimization in Finite Dimensions I
164. Sario/Nakai: Classification Theory of Riemann Surfaces
165. Mitrinović: Analytic Inequalities
166. Grothendieck/Dieudonné: Eléments de Géométrie Algebriqué I
167. Chandrasekharan: Arithmetical Functions
168. Palamodov: Linear Differential Operators with Constant Coefficients
170. Lions: Optimal Control of Systems Governed by Partial Differential Equations
171. Singer: Best Approximation in Normed Linear Spaces by Elements of Linear Subspaces
172. Bühlmann: Mathematical Methods in Risk Theory
173. F. Maeda/S. Maeda: Theory of Symmetric Lattices
174. Stiefel/Scheifele: Linear and Regular Celestial Mechanics. Perturbed Two-body Motion—Numerical Methods—Canonical Theory
175. Larsen: An Introduction of the Theory of Multipliers
176. Grauert/Remmert: Analytische Stellenalgebren
177. Flügge: Practical Quantum Mechanics I
178. Flügge: Practical Quantum Mechanics II
179. Giraud: Cohomologie non abélienne
180. Landkof: Foundations of Modern Potential Theory
181. Lions/Magenes: Non-Homogeneous Boundary Value Problems and Applications I
182. Lions/Magenes: Non-Homogeneous Boundary Value Problems and Applications II
183. Lions/Magenes: Non-Homogeneous Boundary Value Problems and Applications III
184. Rosenblatt: Markov Processes. Structure and Asymptotic Behavior
185. Rubinowicz: Sommerfeldsche Polynommethode
186. Wilkinson/Reinsch: Handbook for Automatic Computation II, Linear Algebra
187. Siegel/Moser: Lectures on Celestial Mechanics
188. Warner: Harmonic Analysis on Semi-Simple Lie Groups I
189. Warner: Harmonic Analysis on Semi-Simple Lie Groups II
190. Faith: Algebra: Rings, Modules, and Categories I
191. Faith: Algebra: Rings, Modules, and Categories. II
192. Mal'cev: Algebraic-Systems
193. Pólya/Szegö: Problems and Theorems in Analysis. Vol. 1
194. Igusa: Theta Functions